Texts in Applied Mathematics 32

Springer
New York
Berlin
Heidelberg
Barcelona
Budapest
Hong Kong
London
Milan
Paris
Singapore
Tokyo

Texts in Applied Mathematics

1. *Sirovich:* Introduction to Applied Mathematics.
2. *Wiggins:* Introduction to Applied Nonlinear Dynamical Systems and Chaos.
3. *Hale/Koçak:* Dynamics and Bifurcations.
4. *Chorin/Marsden:* A Mathematical Introduction to Fluid Mechanics, 3rd ed.
5. *Hubbard/West:* Differential Equations: A Dynamical Systems Approach: Ordinary Differential Equations.
6. *Sontag:* Mathematical Control Theory: Deterministic Finite Dimensional Systems, 2nd ed.
7. *Perko:* Differential Equations and Dynamical Systems, 2nd ed.
8. *Seaborn:* Hypergeometric Functions and Their Applications.
9. *Pipkin:* A Course on Integral Equations.
10. *Hoppensteadt/Peskin:* Mathematics in Medicine and the Life Sciences.
11. *Braun:* Differential Equations and Their Applications, 4th ed.
12. *Stoer/Bulirsch:* Introduction to Numerical Analysis, 2nd ed.
13. *Renardy/Rogers:* A First Graduate Course in Partial Differential Equations.
14. *Banks:* Growth and Diffusion Phenomena: Mathematical Frameworks and Applications.
15. *Brenner/Scott:* The Mathematical Theory of Finite Element Methods.
16. *Van de Velde:* Concurrent Scientific Computing.
17. *Marsden/Ratiu:* Introduction to Mechanics and Symmetry.
18. *Hubbard/West:* Differential Equations: A Dynamical Systems Approach: Higher-Dimensional Systems.
19. *Kaplan/Glass:* Understanding Nonlinear Dynamics.
20. *Holmes:* Introduction to Perturbation Methods.
21. *Curtain/Zwart:* An Introduction to Infinite-Dimensional Linear Systems Theory.
22. *Thomas:* Numerical Partial Differential Equations: Finite Difference Methods.
23. *Taylor:* Partial Differential Equations: Basic Theory.
24. *Merkin:* Introduction to the Theory of Stability of Motion.
25. *Naber:* Topology, Geometry, and Gauge Fields: Foundations.
26. *Polderman/Willems:* Introduction to Mathematical Systems Theory: A Behavioral Approach.
27. *Reddy:* Introductory Functional Analysis with Applications to Boundary-Value Problems and Finite Elements.
28. *Gustafson/Wilcox:* Analytical and Computational Methods of Advanced Engineering Mathematics.
29. *Tveito/Winther:* Introduction to Partial Differential Equations: A Computational Approach.
30. *Gasquet/Witomski:* Fourier Analysis and Applications: Filtering, Numerical Computation, Wavelets.
31. *Brémaud:* Markov Chains: Gibbs Fields, Monte Carlo Simulations, and Queues.
32. *Durran:* Numerical Methods for Wave Equations in Geophysical Fluid Dynamics.

Dale R. Durran

Numerical Methods for Wave Equations in Geophysical Fluid Dynamics

With 93 Illustrations

 Springer

Dale R. Durran
Atmospheric Sciences
University of Washington
Box 351640
Seattle, WA 98195-1640
USA

Series Editors

J.E. Marsden
Control and Dynamical Systems, 107–81
California Institute of Technology
Pasadena, CA 91125
USA

L. Sirovich
Division of Applied Mathematics
Brown University
Providence, RI 02912
USA

M. Golubitsky
Department of Mathematics
University of Houston
Houston, TX 77204-3476
USA

W. Jäger
Department of Applied Mathematics
Universität Heidelberg
Im Neuenheimer Feld 294
69120 Heidelberg
Germany

Mathematics Subject Classification (1991): 65Mxx, 86A10, 76-01, 35L05

Library of Congress Cataloging-in-Publication Data
Durran, Dale R.
    Numerical methods for wave equations in geophysical fluid dynamics
/ Dale R. Durran.
        p.   cm.—(Texts in applied mathematics ; 32)
    Includes bibliographical references and index.
    ISBN 0-387-98376-7 (alk. paper)
        1. Fluid dynamics—Methodology.   2. Geophysics—Methodology.
3. Wave equation.   4. Numerical analysis.   5. Differential equations,
Partial—Numerical solutions.   I. Title.   II. Series
QC809.F5D87   1998
550′.1′532059—dc21   98-24739

Printed on acid-free paper.

Production managed by A. Orrantia; manufacturing supervised by Joe Quatela.
Photocomposed copy prepared from the author's LaTeX files.
Printed and bound by R.R. Donnelley and Sons, Harrisonburg, VA.
Printed in the United States of America.

9 8 7 6 5 4 3 2 1

ISBN 0-387-98376-7 Springer-Verlag New York Berlin Heidelberg   SPIN 10656146

To every hand that's touched the Wall

Series Preface

Mathematics is playing an ever more important role in the physical and biological sciences, provoking a blurring of boundaries between scientific disciplines and a resurgence of interest in the modern as well as the classical techniques of applied mathematics. This renewal of interest, both in research and teaching, has led to the establishment of the series: *Texts in Applied Mathematics (TAM)*.

The development of new courses is a natural consequence of a high level of excitement on the research frontier as newer techniques, such as numerical and symbolic computer systems, dynamical systems, and chaos, mix with and reinforce the traditional methods of applied mathematics. Thus, the purpose of this textbook series is to meet the current and future needs of these advances and encourage the teaching of new courses.

TAM will publish textbooks suitable for use in advanced undergraduate and beginning graduate courses, and will complement the *Applied Mathematical Sciences (AMS)* series, which will focus on advanced textbooks and research level monographs.

Preface

This book is designed to serve as a textbook for graduate students or advanced undergraduates studying numerical methods for the solution of partial differential equations governing wave-like flows. Although the majority of the schemes presented in this text were introduced in either the applied-mathematics or atmospheric-science literature, the focus is not on the nuts-and-bolts details of various atmospheric models but on fundamental numerical methods that have applications in a wide range of scientific and engineering disciplines. The prototype problems considered include tracer transport, shallow-water flow and the evolution of internal waves in a continuously stratified fluid.

A significant fraction of the literature on numerical methods for these problems falls into one of two categories, those books and papers that emphasize theorems and proofs, and those that emphasize numerical experimentation. Given the uncertainty associated with the messy compromises actually required to construct numerical approximations to real-world fluid-dynamics problems, it is difficult to emphasize theorems and proofs without limiting the analysis to classical numerical schemes whose practical application may be rather limited. On the other hand, if one relies primarily on numerical experimentation it is much harder to arrive at conclusions that extend beyond a specific set of test cases. In an attempt to establish a clear link between theory and practice, I have tried to follow a middle course between the theorem-and-proof formalism and the reliance on numerical experimentation. There are no formal proofs in this book, but the mathematical properties of each method are derived in a style familiar to physical scientists. At the same time, numerical examples are included that illustrate these theoretically derived properties and facilitate the intercomparison of various methods.

A general course on numerical methods for geophysical fluid dynamics might draw on portions of the material presented in Chapters 2 through 6. Chapter 2 describes the largely classical theory of finite-difference approximations to the one-way wave equation (or alternatively the constant-wind-speed advection equation). The extension of these results to systems of equations, several space dimensions, dissipative flows and nonlinear problems is discussed in Chapter 3. Chapter 4 introduces series-expansion methods with emphasis on the Fourier and spherical-harmonic spectral methods and the finite-element method. Finite-volume methods are discussed in Chapter 5 with particular attention devoted to methods for simulating the transport of scalar fields containing poorly resolved spatial gradients. Semi-Lagrangian schemes are analyzed in Chapter 6. Both theoretical and applied problems are provided at the end of each chapter. Those problems that require numerical computation are marked by an asterisk.

In addition to the core material in Chapters 2 through 6, the introduction in Chapter 1 discusses the relation between the equations governing wave-like geophysical flows and other types of partial differential equations. Chapter 1 concludes with a short overview of the strategies for numerical approximation that are considered in detail throughout the remainder of the book. Chapter 7 examines schemes for the approximation of slow moving waves in fluids that support physically insignificant fast waves. The emphasis in Chapter 7 is on atmospheric applications in which the slow wave is either an internal gravity wave and the fast waves are sound waves, or the slow wave is a Rossby wave and the fast waves are both gravity waves and sound waves. Chapter 8 examines the formulation of wave-permeable boundary conditions for limited-area models with emphasis on the shallow-water equations in one and two dimensions and on internally stratified flow.

Many numerical methods for the simulation of internally stratified flow require the repeated solution of elliptic equations for pressure or some closely related variable. Due to the limitations of my own expertise and to the availability of other excellent references I have not discussed the solution of elliptic partial differential equations in any detail. A thumbnail sketch of some solution strategies is provided in Section 7.1.3; the reader is referred to Chapter 5 of Ferziger and Perić (1997) for an excellent overview of methods for the solution of elliptic equations arising in computational fluid dynamics.

I have attempted to provide sufficient references to allow the reader to further explore the theory and applications of many of the methods discussed in the text, but the reference list is far from encyclopedic and certainly does not include every worthy paper in the atmospheric science or applied mathematics literature. References to the relevant literature in other disciplines and in foreign language journals is rather less complete.[1]

---

[1] Those not familiar with the atmospheric science literature may be surprised by the number of references to *Monthly Weather Review*, which despite its title, has become the primary American journal for the publication of papers on numerical methods in atmospheric science.

This book would not have been written without the generous assistance of several colleagues. Christopher Bretherton, in particular, provided many perceptive answers to my endless questions. J. Ray Bates, Byron Boville, Michael Cullen, Marcus Grote, Robert Higdon, Randall LeVeque, Christoph Schär, William Skamarock, Piotr Smolarkiewicz, and David Williamson all provided very useful comments on individual chapters. Many students used earlier versions of this manuscript in my courses in the Atmospheric Sciences Department at the University of Washington, and their feedback helped improve the clarity of the manuscript. Two students to whom I am particularly indebted are Craig Epifanio and Donald Slinn. I am also grateful to James Holton for encouraging me to undertake this project.

It is my pleasure to acknowledge the many years of support for my numerical modeling efforts provided by the Mesoscale Dynamic Meteorology Program of the National Science Foundation. Additional support for my atmospheric simulation studies has been provided by the Coastal Meteorology ARI of the Office of Naval Research. Part of this book was completed while I was on sabbatical at the Laboratoire d'Aérologie of the Université Paul Sabatier in Toulouse, France, and I thank Daniel Guedalia and Evelyne Richard for helping make that year productive and scientifically stimulating.

As errors in the text are identified, they will be posted on the web at http://www. atmos.washington.edu/methods.for.waves, which can be accessed directly or via Springer's home page at http://www.springer-ny.com. I would be most grateful to be advised of any typographical or other errors by electronic mail at dale.durran@atmos.washington.edu.

Seattle, Washington   DALE R. DURRAN

Cover art: The three curves plot solutions to the linearized Rossby–adjustment problem. The governing equations and physical parameters for this problem are identical to those given in Problem 12 of Chapter 3, except that the spatial domain is -400 km $\leq x \leq 400$ km with open lateral boundaries, and the initial condition for the free-surface displacement is $h(x, t = 0) = \arctan(x/20 \text{ km})$. The curves shown are plots of $u(x, t = 943 \text{ s})$, $u(x, t = 1222 \text{ s})$, and $u(x, t = 1501 \text{ s})$ on an artisically cropped portion of the sub-domain $x > 0$.

Contents

Series Preface   vii

Preface   ix

1   Introduction   **1**
   1.1   Partial Differential Equations—Some Basics   2
       1.1.1   First-Order Hyperbolic Equations   4
       1.1.2   Linear Second-Order Equations
             in Two Independent Variables   7
   1.2   Wave Equations in Geophysical Fluid Dynamics   11
       1.2.1   Hyperbolic Equations   12
       1.2.2   Filtered Equations   20
   1.3   Strategies for Numerical Approximation   26
       1.3.1   Approximating Calculus with Algebra   26
       1.3.2   Marching Schemes   29
   Problems   33

2   Basic Finite-Difference Methods   **35**
   2.1   Accuracy and Consistency   36
   2.2   Stability and Convergence   39
       2.2.1   The Energy Method   41
       2.2.2   Von Neumann's Method   43
       2.2.3   The Courant–Fredrichs–Lewy Condition   45
   2.3   Time-Differencing   47
       2.3.1   The Oscillation Equation: Phase-Speed
             and Amplitude Error   48

2.3.2   Single-Stage Two-Level Schemes   50

2.3.3   Multistage Methods   53

2.3.4   Three-Level Schemes   56

2.3.5   Controlling the Leapfrog Computational Mode   60

2.3.6   Higher-Order Schemes   65

2.4   Space-Differencing   72

2.4.1   Differential–Difference Equations
and Wave Dispersion   73

2.4.2   Dissipation, Dispersion, and the Modified Equation . . .   80

2.4.3   Artificial Dissipation   82

2.4.4   Compact Differencing   86

2.5   Combined Time- and Space-Differencing   89

2.5.1   The Discrete-Dispersion Relation   91

2.5.2   The Modified Equation   94

2.5.3   The Lax–Wendroff Method   95

2.6   Summary Discussion of Elementary Methods   99

Problems   101

3   Beyond the One-Way Wave Equation   107

3.1   Systems of Equations   107

3.1.1   Stability   108

3.1.2   Staggered Meshes   113

3.2   Three or More Independent Variables   117

3.2.1   Scalar Advection in Two Dimensions   117

3.2.2   Systems of Equations in Several Dimensions   126

3.3   Splitting into Fractional Steps   129

3.3.1   Split Explicit Schemes   130

3.3.2   Split Implicit Schemes   132

3.3.3   Stability of Split Schemes   134

3.4   Diffusion, Sources, and Sinks   136

3.4.1   Pure Diffusion   136

3.4.2   Advection and Diffusion   138

3.4.3   Advection with Sources and Sinks   144

3.5   Linear Equations with Variable Coefficients   147

3.5.1   Aliasing Error   148

3.5.2   Conservation and Stability   154

3.6   Nonlinear Instability   159

3.6.1   Burgers's Equation   159

3.6.2   The Barotropic Vorticity Equation   163

Problems   167

4 Series-Expansion Methods   **173**
  4.1 Strategies for Minimizing the Residual 173
  4.2 The Spectral Method 176
      4.2.1 Comparison with Finite-Difference Methods 177
      4.2.2 Improving Efficiency Using the Transform Method . . . 184
      4.2.3 Conservation and the Galerkin Approximation 189
  4.3 The Pseudospectral Method 191
  4.4 Spherical Harmonics 195
      4.4.1 Truncating the Expansion 197
      4.4.2 Elimination of the Pole Problem 200
      4.4.3 Gaussian Quadrature and the Transform Method 202
      4.4.4 Nonlinear Shallow-Water Equations 207
  4.5 The Finite-Element Method 212
      4.5.1 Galerkin Approximation with Chapeau Functions 214
      4.5.2 Petrov–Galerkin and Taylor–Galerkin Methods 216
      4.5.3 Quadratic Expansion Functions 219
      4.5.4 Hermite-Cubic Expansion Functions 226
      4.5.5 Two-Dimensional Expansion Functions 231
  Problems 234

5 Finite Volume Methods   **241**
  5.1 Conservation Laws and Weak Solutions 243
      5.1.1 The Riemann Problem 244
      5.1.2 Entropy-Consistent Solutions 246
  5.2 Finite-Volume Methods and Convergence 249
      5.2.1 Monotone Schemes 251
      5.2.2 TVD Methods 252
  5.3 Discontinuities in Geophysical Fluid Dynamics 254
  5.4 Flux-Corrected Transport 257
      5.4.1 Flux Correction: The Original Proposal 259
      5.4.2 The Zalesak Corrector 260
      5.4.3 Iterative Flux Correction 263
  5.5 Flux-Limiter Methods 263
      5.5.1 Ensuring That the Scheme Is TVD 264
      5.5.2 Possible Flux Limiters 267
      5.5.3 Flow Velocities of Arbitrary Sign 271
  5.6 Approximation with Local Polynomials 272
      5.6.1 Godunov's Method 272
      5.6.2 Piecewise-Linear Functions 274
  5.7 Two Spatial Dimensions 277
      5.7.1 FCT in Two Dimensions 277
      5.7.2 Flux-Limiter Methods for Uniform 2-D Flow 279
      5.7.3 Nonuniform Nondivergent Flow 282
      5.7.4 A Numerical Example 284
      5.7.5 When Is a Flux Limiter Necessary? 291

5.8   Schemes for Positive Definite Advection   292
    5.8.1   An FCT Approach   293
    5.8.2   Antidiffusion via Upstream Differencing   294
5.9   Curvilinear Coordinates   296
Problems   297

6   Semi-Lagrangian Methods   **303**
6.1   The Scalar Advection Equation   305
    6.1.1   Constant Velocity   305
    6.1.2   Variable Velocity   310
6.2   Forcing in the Lagrangian Frame   313
6.3   Systems of Equations   318
    6.3.1   Comparison with the Method of Characteristics   318
    6.3.2   Semi-implicit Semi-Lagrangian Schemes   320
6.4   Alternative Trajectories   324
    6.4.1   A Noninterpolating Leapfrog Scheme   325
    6.4.2   Interpolation via Parametrized Advection   327
6.5   Eulerian or Semi-Lagrangian?   330
Problems   331

7   Physically Insignificant Fast Waves   **335**
7.1   The Projection Method   336
    7.1.1   Forward-in-Time Implementation   337
    7.1.2   Leapfrog Implementation   339
    7.1.3   Solving the Poisson Equation for Pressure   340
7.2   The Semi-implicit Method   342
    7.2.1   Large Time Steps and Poor Accuracy   343
    7.2.2   A Prototype Problem   345
    7.2.3   Semi-implicit Solution of the Shallow-Water Equations .   347
    7.2.4   Semi-implicit Solution of the Euler Equations   350
    7.2.5   Numerical Implementation   356
7.3   Fractional-Step Methods   359
    7.3.1   Complete Operator Splitting   359
    7.3.2   Partially Split Operators   365
7.4   Summary of Schemes for Nonhydrostatic Models   371
7.5   The Hydrostatic Approximation   372
7.6   Primitive Equation Models   374
    7.6.1   Pressure and σ Coordinates   375
    7.6.2   Spectral Representation of the Horizontal Structure . . .   379
    7.6.3   Vertical Differencing   381
    7.6.4   Energy Conservation   383
    7.6.5   Semi-implicit Time-Differencing   387
Problems   389

Contents xvii

8 **Nonreflecting Boundary Conditions**   **395**
  8.1   One-Dimensional Flow   397
        8.1.1   Well-Posed Initial–Boundary Value Problems.   397
        8.1.2   The Radiation Condition   400
        8.1.3   Time-Dependent Boundary Data   401
        8.1.4   Reflections at an Artificial Boundary—
                The Continuous Case   402
        8.1.5   Reflections at an Artificial Boundary—
                The Discretized Case   403
        8.1.6   Stability in the Presence of Boundaries   409
  8.2   Two-Dimensional Shallow-Water Flow   412
        8.2.1   One-Way Wave Equations   414
        8.2.2   Numerical Implementation   419
  8.3   Two-Dimensional Stratified Flow   419
        8.3.1   Lateral Boundary Conditions   420
        8.3.2   Upper Boundary Conditions   424
        8.3.3   Numerical Implementation of the Radiation Upper
                Boundary Condition   429
  8.4   Wave-Absorbing Layers   431
  8.5   Summary   436
  Problems   437

Appendix   Numerical Miscellany   **439**
  A.1   Finite-Difference Operator Notation   439
  A.2   Tridiagonal Solvers   440
        A.2.1   Code for a Tridiagonal Solver   440
        A.2.2   Code for a Periodic Tridiagonal Solver   441

Bibliography   **443**

Index   **457**

1
Introduction

The possibility of deterministic weather prediction was suggested by Vilhelm Bjerknes as early as 1904. Around the time of the First World War, Lewis Richardson actually attempted to produce such a forecast by manually integrating a finite-difference approximation to the equations governing atmospheric motion. Unfortunately, his calculations did not yield a reasonable forecast. Moreover, the human labor required to obtain this disappointing result was so great that subsequent attempts at deterministic weather prediction had to await the introduction of a high-speed computational aid. In 1950 a team of researchers, under the direction of Jule Charney and John von Neumann at the Institute for Advanced Study, at Princeton, journeyed to the Aberdeen Proving Ground, where they worked for approximately twenty-four hours to coax a one-day weather forecast from the first general-purpose electronic computer, the ENIAC.[1] The first computer-generated weather forecast was surprisingly good, and its success led to the rapid growth of a new meteorological subdiscipline, "numerical weather prediction." These early efforts in numerical weather prediction also began a long and fruitful collaboration between numerical analysts and atmospheric scientists.[2] The use of numerical models in atmospheric and oceanic science has subsequently expanded into almost all areas of current research. Numerical models are currently employed to study phenomena as diverse as global climate change, the interaction of ocean currents with bottom topography, and the development of rotation in tornadic thunderstorms.

---

[1] ENIAC is an acronym for Electronic Numerical Integrator and Calculator.

[2] Further details about these early weather prediction efforts may be found in Bjerknes (1904), Richardson (1922), Charney et al. (1950), Burks and Burks (1981), and Thompson (1983).

Many of the phenomena simulated with atmospheric and oceanic models can be classified as wave-like flows if the terminology "wave-like" is used in the general sense suggested by Whitham (1974), who defined a wave as "any recognizable signal that is transferred from one part of a medium to another with a recognizable velocity of propagation." The purpose of this book is to present the fundamental mathematical aspects of a wide variety of numerical methods for the simulation of wave-like flow. The methods to be considered are typically those that have seen some use in real-world atmospheric or ocean models, but the focus is on the essential properties of each method and not on the details of any specific model. The fundamental character of each scheme will be examined in standard fluid-dynamical problems like tracer transport, shallow-water waves, and waves in an internally stratified fluid. These are the same prototypical problems familiar to many applied mathematicians, fluid dynamicists, and practitioners in the larger discipline of computational fluid dynamics.

Most of the problems under investigation in the atmospheric and oceanic sciences involve fluid systems with low viscosity and weak dissipation. The equations governing these flows are often nonlinear, but their solutions almost never develop energetic shocks or discontinuities. Nevertheless, regions of scale collapse do frequently occur as the velocity field stretches and deforms an initially compact fluid parcel. The numerical methods that will be examined in this book may therefore be distinguished from the larger family of algorithms in computational fluid mechanics in that they are particularly appropriate for low-viscosity flows, but are not primarily concerned with the treatment of shocks.

It is assumed that the reader has already been exposed to the derivation of the equations describing fluid flow and tracer transport. These derivations are given in a general fluid-dynamical context in Batchelor (1967), Yih (1977), and Bird et al. (1960), and in the context of atmospheric and oceanic science in Gill (1982), Holton (1992), and Pedlosky (1987). The mathematical properties of these equations and commonly used simplifications, such as the Boussinesq approximation, will be briefly reviewed in this chapter. The chapter concludes with a brief overview of the numerical methods that will be considered in more detail throughout the remainder of the book.

1.1   Partial Differential Equations—Some Basics

Different types of partial differential equations require different solution strategies. It is therefore helpful to begin by reviewing some of the terminology used to describe various types of partial differential equations. The *order* of a partial differential equation is the order of the highest-order partial derivative that appears in the equation. Numerical methods for the solution of time-dependent problems are often designed to solve systems of partial differential equations in which the time derivatives are of first order. These numerical methods can be used to solve partial differential equations containing higher-order time derivatives by defining new unknown functions equal to the lower-order time derivatives of the original unknown function and expressing the result as system of partial differential equations in which all time derivatives are of order one. For example, the second-order partial differential equation

$$\frac{\partial^2 \psi}{\partial t^2} + \psi \frac{\partial \psi}{\partial x} = 0$$

can be expressed as the first-order system

$$\frac{\partial v}{\partial t} + \psi \frac{\partial \psi}{\partial x} = 0,$$

$$\frac{\partial \psi}{\partial t} - v = 0.$$

In geophysical applications it is seldom necessary to actually formulate first-order-in-time equations using this procedure, because suitable first-order-in-time systems can usually be derived from fundamental physical principles.

The accurate numerical solution of equations describing wave-like flow becomes more difficult if the solution develops significant perturbations on spatial scales close to the shortest scale that can be resolved by the numerical model. The possibility of waves developing small-scale perturbations from smooth initial data increases as the governing partial differential equation becomes more nonlinear. A partial differential equation is *linear* if it is linear in the unknown functions and their derivatives, in which case the coefficients multiplying each function or derivative depend only on the independent variables. As an example,

$$\frac{\partial u}{\partial t} + x^3 \frac{\partial u}{\partial x} = 0$$

is a linear first-order partial differential equation, whereas

$$\left(\frac{\partial u}{\partial t}\right)^2 + \sin\left(u \frac{\partial u}{\partial x}\right) = 0$$

is a nonlinear first-order partial differential equation.

Analysis techniques and solution procedures developed for linear partial differential equations can be generalized most easily to the subset of nonlinear partial differential equations that are quasi-linear. A partial differential equation of order p is *quasi-linear* if it is linear in the derivatives of order p; the coefficient multiplying each pth derivative can depend on the independent variables and all derivatives of the unknown function through order $p - 1$. Two examples of quasi-linear partial differential equations are

$$\frac{\partial u}{\partial t} + u^3 \frac{\partial u}{\partial x} = 0$$

and the vorticity equation for two-dimensional nondivergent flow

$$\frac{\partial \nabla^2 \psi}{\partial t} + \frac{\partial \psi}{\partial x} \frac{\partial \nabla^2 \psi}{\partial y} - \frac{\partial \psi}{\partial y} \frac{\partial \nabla^2 \psi}{\partial x} = 0,$$

where $\psi(x, y, t)$ is the stream function for the nondivergent velocity field and

$$\nabla^2 = \frac{\partial^2}{\partial x^2} + \frac{\partial^2}{\partial y^2}.$$

1.1.1   First-Order Hyperbolic Equations

Many waves can be mathematically described as solutions to hyperbolic partial differential equations. One simple example of a hyperbolic partial differential equation is the general first-order quasi-linear equation

$$A(x, t, u)\frac{\partial u}{\partial t} + B(x, t, u)\frac{\partial u}{\partial x} = C(x, t, u), \tag{1.1}$$

where A, B, and C are real-valued functions with continuous first derivatives. This equation is hyperbolic because there exists a family of real-valued curves in the x-t plane along which the solution can be locally determined by integrating ordinary differential equations. These curves, called *characteristics*, may be defined with respect to the parameter s by the relations

$$\frac{dt}{ds} = A, \qquad \frac{dx}{ds} = B. \tag{1.2}$$

The identity

$$\frac{du}{ds} = \frac{\partial u}{\partial t}\frac{dt}{ds} + \frac{\partial u}{\partial x}\frac{dx}{ds}$$

can then be used to express (1.1) as the ordinary differential equation

$$\frac{du}{ds} = C. \tag{1.3}$$

Given the value of u at some arbitrary point (x_0, t_0), the coordinates of the characteristic curve passing through (x_0, t_0) can be determined by integrating the ordinary differential equations (1.2). The solution along this characteristic can be obtained by integrating the ordinary differential equation (1.3). A unique solution to (1.1) can be determined throughout some local region of the x-t plane by specifying data for u along any noncharacteristic line.

In physical applications where the independent variable t represents time, the particular solution of (1.1) is generally determined by specifying initial data for u along the line $t = 0$. In such applications A is nonzero, and any perturbation in the distribution of u at the point (x_0, t_0) translates through a neighborhood of x_0 at the speed

$$\frac{dx}{dt} = \frac{B}{A}.$$

The solutions to (1.1) are wave-like in the general sense that the perturbations in u travel at well-defined velocities even though they may distort as they propagate.

The evolution of the solution is particularly simple when $C = 0$ and B/A is some constant value c, in which case (1.1) reduces to

$$\frac{\partial u}{\partial t} + c\frac{\partial u}{\partial x} = 0. \tag{1.4}$$

If $u(x, 0) = f(x)$, the solution to the preceding is $f(x - ct)$, implying that the initial perturbations in u translate without distortion at a uniform velocity c. Equation (1.4), which is often referred to as the *one-way wave equation* or the *constant-wind-speed advection equation*, is the simplest mathematical model for wave propagation. Although it is quite simple, (1.4) is a very useful prototype problem for testing numerical methods because solutions to more complex linear hyperbolic systems can often be expressed as the superposition of individual waves governed by one-way wave equations.

A system of partial differential equations in two independent variables is hyperbolic if it has a complete set of characteristic curves that can in principle be used to locally determine the solution from appropriately prescribed initial data. As a first example, consider a constant-coefficient linear system of the form

$$\frac{\partial u_r}{\partial t} + \sum_{s=1}^{n} a_{rs}\frac{\partial u_s}{\partial x} = 0, \qquad r = 1, 2, \ldots, n. \tag{1.5}$$

This system may be alternatively written as

$$\frac{\partial \mathbf{u}}{\partial t} + \mathbf{A}\frac{\partial \mathbf{u}}{\partial x} = \mathbf{0},$$

where uppercase boldface letters represent matrices and lowercase boldface letters denote vectors. The system is hyperbolic if there exist bounded matrices $\mathbf{T}$ and $\mathbf{T}^{-1}$ such that $\mathbf{T}^{-1}\mathbf{A}\mathbf{T} = \mathbf{D}$, where $\mathbf{D}$ is a diagonal matrix with real eigenvalues d_{jj}. When the system is hyperbolic, it can be transformed to

$$\frac{\partial \mathbf{v}}{\partial t} + \mathbf{D}\frac{\partial \mathbf{v}}{\partial x} = \mathbf{0} \tag{1.6}$$

by defining $\mathbf{v} = \mathbf{T}^{-1}\mathbf{u}$. Since $\mathbf{D}$ is a diagonal matrix, each element v_j of the vector of unknown functions may be determined by solving a simpler scalar equation of the form (1.4). Each diagonal element of $\mathbf{D}$ is associated with a family of characteristic curves along which the perturbations in v_i propagate at speed $dx/dt = d_{jj}$. The wave-like character of the solution can be demonstrated by Fourier transforming (1.6) with respect to x to obtain

$$\frac{\partial \hat{\mathbf{v}}}{\partial t} + ik\mathbf{D}\hat{\mathbf{v}} = \mathbf{0}, \tag{1.7}$$

where $\hat{\mathbf{v}}$ is the Fourier transform of $\mathbf{v}$ and k is the wave number, or dual variable. In order to satisfy (1.7), the jth component of $\mathbf{v}$ must be a wave of the form

$\exp ik(x - d_{jj}t)$. Every solution to the original system (1.5) is a linear superposition of these waves.

Now consider the general first-order linear system

$$\frac{\partial \mathbf{u}}{\partial t} + \mathbf{A}\frac{\partial \mathbf{u}}{\partial x} + \mathbf{Bu} + \mathbf{c} = \mathbf{0},$$

where the coefficient matrices are smooth functions of x and t. This system is hyperbolic throughout some region R of the x-t plane if for all x and t in R there exist bounded matrices $\mathbf{T}^{-1}$ and $\mathbf{T}$ such that $\mathbf{D}(x, t) = \mathbf{T}^{-1}(x, t)\mathbf{A}(x, t)\mathbf{T}(x, t)$ is a diagonal matrix with real eigenvalues. Again, let $\mathbf{u} = \mathbf{Tv}$. Then

$$\frac{\partial \mathbf{v}}{\partial t} + \mathbf{D}\frac{\partial \mathbf{v}}{\partial x} + \tilde{\mathbf{B}}\mathbf{v} + \mathbf{T}^{-1}\mathbf{c} = \mathbf{0}, \tag{1.8}$$

where

$$\tilde{\mathbf{B}} = \mathbf{T}^{-1}\left(\frac{\partial \mathbf{T}}{\partial t} + \mathbf{A}\frac{\partial \mathbf{T}}{\partial x} + \mathbf{BT}\right).$$

The solution to (1.8) may be obtained via the iteration

$$\frac{\partial \mathbf{v}^{n+1}}{\partial t} + \mathbf{D}\frac{\partial \mathbf{v}^{n+1}}{\partial x} + \tilde{\mathbf{B}}\mathbf{v}^n + \mathbf{T}^{-1}\mathbf{c} = \mathbf{0} \tag{1.9}$$

(Courant and Hilbert 1953, p. 476). Since $\mathbf{D}$ is diagonal, the preceding is a set of decoupled scalar relations for the components v_i^{n+1}, each of which is a simple first-order hyperbolic partial differential equation.

To generalize the preceding definition of a hyperbolic system to problems with three or more independent variables, consider the system of partial differential equations

$$\frac{\partial \mathbf{u}}{\partial t} + \left(\mathbf{A}_1\frac{\partial}{\partial x_1} + \mathbf{A}_2\frac{\partial}{\partial x_2} + \cdots + \mathbf{A}_m\frac{\partial}{\partial x_m}\right)\mathbf{u} = \mathbf{0} \tag{1.10}$$

and suppose that the coefficient matrices are constant. Unlike the two-independent-variable case, it is not usually possible to find a transformation that simultaneously diagonalizes all the coefficient matrices in (1.10) and thereby generates a set of decoupled scalar equations. Instead, take the Fourier transform of (1.10) with respect to each spatial coordinate to obtain

$$\frac{\partial \hat{\mathbf{u}}}{\partial t} + i\mathbf{P}(\mathbf{k})\hat{\mathbf{u}} = \mathbf{0}, \qquad \text{where} \qquad \mathbf{P}(\mathbf{k}) = \sum_{q=1}^{m} \mathbf{A}_q k_q$$

and $\mathbf{k} = (k_1, k_2, \ldots, k_m)$ is a real-valued vector of the wave number (or dual variable) with respect to each spatial coordinate.

The system (1.10) will be hyperbolic if all its solutions are the linear superposition of waves of the form $\exp i(\mathbf{k} \cdot \mathbf{x} - \omega t)$, where $\omega(\mathbf{k})$ is a real-valued frequency. This will be the case if $\mathbf{P}(\mathbf{k})$ has a complete set of real eigenvalues for any nonzero wave number vector $\mathbf{k}$, or equivalently, if for every $\mathbf{k}$ such that $\|\mathbf{k}\| = 1$ there exist bounded matrices $\mathbf{T}^{-1}(\mathbf{k})$ and $\mathbf{T}(\mathbf{k})$ such that $\mathbf{D}(\mathbf{k}) = \mathbf{T}^{-1}(\mathbf{k})\mathbf{P}(\mathbf{k})\mathbf{T}(\mathbf{k})$ is a diagonal matrix with real eigenvalues.

The definition of a hyperbolic system in several space dimensions is extended to the case where the coefficient matrices in (1.10) are smooth functions of $\mathbf{x}$ and t by requiring that at every point $(\mathbf{x}, t)$ throughout some domain R there exist bounded matrices $\mathbf{T}^{-1}(\mathbf{k}, \mathbf{x}, t)$ and $\mathbf{T}(\mathbf{k}, \mathbf{x}, t)$ such that for all real vectors $\mathbf{k}$ of unit length, $\mathbf{T}^{-1}(\mathbf{k}, \mathbf{x}, t)\mathbf{P}(\mathbf{k}, \mathbf{x}, t)\mathbf{T}(\mathbf{k}, \mathbf{x}, t)$ is a diagonal matrix with real eigenvalues (Gustafsson et al. 1995, p. 221). Since all symmetric matrices may be transformed to real-valued diagonal matrices, the matrix $\mathbf{P}$ will be symmetric and the original system (1.10) will be hyperbolic if all the coefficient matrices $\mathbf{A}_q$ are symmetric. The easiest way to show that many multidimensional systems are hyperbolic is to transform them to equivalent systems in which all the coefficient matrices are symmetric.

1.1.2 Linear Second-Order Equations in Two Independent Variables

Not all waves are solutions to hyperbolic equations. Hyperbolic equations can be compared with two other fundamental types of partial differential equations, *parabolic* and *elliptic* equations, by considering the general family of linear second-order partial differential equations in two independent variables

$$a(x, y)u_{xx} + 2b(x, y)u_{xy} + c(x, y)u_{yy} + L(x, y, u, u_x, u_y) = 0. \quad (1.11)$$

In the preceding, the subscripts denote partial derivatives, and L is a linear function of u, u_x, and u_y whose coefficients may depend on x and y. New independent variables η and ξ can be defined that transform (1.11) into one of three canonical forms. The particular form that can be achieved depends on the number of families of characteristic curves associated with (1.11). In those regions of the x-y plane where $b^2 - ac > 0$ there are two independent families of characteristic curves; the equation is hyperbolic, and it can be transformed to the canonical form

$$u_{\xi\eta} + \tilde{L}(\xi, \eta, u, u_\xi, u_\eta) = 0. \quad (1.12)$$

There is one family of characteristic curves and the equation is parabolic in those regions where $b^2 - ac = 0$, in which case (1.11) can be transformed to

$$u_{\eta\eta} + \tilde{L}(\xi, \eta, u, u_\xi, u_\eta) = 0. \quad (1.13)$$

In those regions where $b^2 - ac < 0$, there are no real-valued characteristic curves; the equation is elliptic, and it transforms to

$$u_{\xi\xi} + u_{\eta\eta} + \tilde{L}(\xi, \eta, u, u_\xi, u_\eta) = 0. \quad (1.14)$$

In each of the preceding, $\tilde{L}(\xi, \eta, u, u_\xi, u_\eta)$ is a linear function of u, u_ξ, and u_η with coefficients that may depend on ξ and η.

In order to carry out the transformation, the various partial derivatives of u with respect to x and y in (1.11) must be replaced by derivatives with respect to ξ and η. Differentiating $u[\xi(x, y), \eta(x, y)]$ yields

$$u_x = u_\xi \xi_x + u_\eta \eta_x,$$
$$u_{xx} = u_{\xi\xi} \xi_x^2 + 2u_{\xi\eta} \xi_x \eta_x + u_{\eta\eta} \eta_x^2 + u_\xi \xi_{xx} + u_\eta \eta_{xx},$$
$$u_{xy} = u_{\xi\xi} \xi_x \xi_y + u_{\xi\eta} \left(\xi_x \eta_y + \xi_y \eta_x \right) + u_{\eta\eta} \eta_x \eta_y + u_\xi \xi_{xy} + u_\eta \eta_{xy},$$

along with similar expressions for u_y and u_{yy} that may be substituted into (1.11) to obtain

$$A(\xi, \eta)u_{\xi\xi} + 2B(\xi, \eta)u_{\xi\eta} + C(\xi, \eta)u_{\eta\eta} + \tilde{L}(\xi, \eta, u, u_\xi, u_\eta) = 0, \qquad (1.15)$$

where

$$A(\xi, \eta) = a\xi_x^2 + 2b\xi_x\xi_y + c\xi_y^2,$$
$$B(\xi, \eta) = a\xi_x \eta_x + b \left(\xi_x \eta_y + \xi_y \eta_x \right) + c\xi_y \eta_y,$$
$$C(\xi, \eta) = a\eta_x^2 + 2b\eta_x \eta_y + c\eta_y^2.$$

The new coordinates must be chosen such that the Jacobian

$$\xi_x \eta_y - \xi_y \eta_x$$

is nonzero throughout the domain to guarantee that the transformation between (x, y) and (ξ, η) is unique and has a unique inverse. This coordinate transformation does not change the classification of the partial differential equation as hyperbolic, parabolic, or elliptic because, as can be shown by direct substitution,

$$B^2 - AC = (b^2 - ac)(\xi_x \eta_y - \xi_y \eta_x)^2, \qquad (1.16)$$

implying that for nonsingular transforms the sign of $b^2 - ac$ is inherited by $B^2 - AC$.

Now consider the hyperbolic case, for which the canonical form (1.12) is obtained by choosing ξ and η to make $A(\xi, \eta) = C(\xi, \eta) = 0$. $A(\xi, \eta)$ will be zero when

$$a\xi_x^2 + 2b\xi_x\xi_y + c\xi_y^2 = 0,$$

or if $a \neq 0$,

$$\frac{\xi_x^2}{\xi_y^2} + 2\frac{b}{a}\frac{\xi_x}{\xi_y} + \frac{c}{a} = 0. \qquad (1.17)$$

Assuming again that $a \neq 0$, the condition $C(\xi, \eta) = 0$ requires that η_x/η_y be a root of the same quadratic equation, i.e.,

$$\frac{\eta_x^2}{\eta_y^2} + 2\frac{b}{a}\frac{\eta_x}{\eta_y} + \frac{c}{a} = 0. \qquad (1.18)$$

Since $b^2 - ac > 0$, these roots are real and distinct. Denoting the roots by $-v_1$ and $-v_2$, one may choose transformed coordinates such that $dy/dx = v_1$ along lines of constant ξ and $dy/dx = v_2$ along lines of constant η. This choice of coordinates satisfies (1.17) and (1.18) because

$$\frac{dy}{dx}\bigg|_{\xi} = -\frac{\xi_x}{\xi_y} = v_1 \quad \text{and} \quad \frac{dy}{dx}\bigg|_{\eta} = -\frac{\eta_x}{\eta_y} = v_2.$$

Those curves along which either ξ or η is constant are the characteristic curves for the hyperbolic equation (1.11).

After zeroing A and C, the canonical form (1.12) is obtained by dividing (1.15) by $B(\xi, \eta)$, which must be nonzero by (1.16) because

$$\xi_x \eta_y - \xi_y \eta_x = \xi_y \eta_y (v_2 - v_1) \neq 0.$$

In the case $a = 0$, a similar expression for the transformed coordinates can be obtained by dividing the relations $A(\xi, \eta) = 0$ and $C(\xi, \eta) = 0$ by c instead of a. If both a and c are zero, the partial differential equation is placed in canonical form simply by dividing by b (which is nonzero because $b^2 - ac > 0$).

If $\tilde{L}$ is zero, the canonical hyperbolic equation (1.12) has solutions of the form $g(\xi)$ and $h(\eta)$. One circumstance in which $\tilde{L}$ is zero occurs when a, b, and c are constant and $L = 0$ in (1.11). Then the characteristics are the straight lines

$$\xi = y - v_1 x \quad \text{and} \quad \eta = y - v_2 x,$$

and there exist solutions of the form $g(y - v_1 x)$ and $h(y - v_2 x)$. When (1.11) serves as a mathematical model for wave-propagation problems, it usually includes a second-order derivative with respect to time. Suppose, therefore, that $a \neq 0$ and that x represents the time coordinate. Then the speed of signal propagation along the characteristics is given by their slope in the y-x plane, which is v_1 for the constant-ξ characteristics and v_2 for the constant-η characteristics.

In the parabolic case with $a \neq 0$, the quadratic equation (1.17) has the double root $-b/a$, and there is a single characteristic defined such that

$$\frac{dy}{dx}\bigg|_{\xi} = -\frac{\xi_x}{\xi_y} = \frac{b}{a}. \tag{1.19}$$

Let $\eta(x, y)$ be any simple function such that

$$\xi_x \eta_y - \xi_y \eta_x \neq 0.$$

These choices for ξ and η imply that $A = 0$ and $B^2 - AC = 0$, which in turn implies that $B = 0$. The canonical parabolic form (1.13) is obtained by dividing (1.15) by C, which must be nonzero, or else neither (1.11) nor (1.15) will be a second-order differential equation. If, on the other hand, $a = 0$ in (1.11), a similar transformation can be performed after dividing through by c, which must be nonzero if a second-order partial derivative is present in (1.11) because $b^2 = b^2 - ac = 0$.

When parabolic partial differential equations describe time-dependent physical systems, such as the diffusion of heat along a rod, the second-order partial derivative is usually computed with respect to a spatial coordinate. Letting x represent the spatial coordinate and y the time coordinate, the one-dimensional heat equation becomes

$$\frac{\partial^2 \psi}{\partial x^2} - \frac{\partial \psi}{\partial y} = 0,$$

which is in the general form (1.11) with $b = c = 0$. According to (1.19), the characteristic curves for the heat equation have slope $dy/dx = 0$, i.e., they are lines parallel to the spatial coordinate (which in contrast to the hyperbolic example is now x).

If the partial differential equation is elliptic, then $b^2 - ac < 0$, and there are no real-valued functions that satisfy (1.17) and (1.18). Provided that a, b, and c are analytic,[3] a transformation can always be found that zeros B and sets $A = C = 1$, thereby obtaining the canonical form (1.14). (See Carrier and Pearson 1988 or Kevorkian 1990 for further details.) If a, b, and c are constant, the transformation to canonical form may be accomplished by choosing

$$\xi = \frac{bx - ay}{(ac - b^2)^{1/2}}, \qquad \eta = x,$$

and dividing the resulting equation by a.

Since elliptic partial differential equations do not have real-valued characteristics, their solutions do not generally include wave-like perturbations that propagate through the domain at well-defined velocities. Nevertheless, elliptic equations describing the spatial distribution of a physical parameter such as pressure can be coupled with other time-dependent equations to yield a problem with wave-like solutions. As noted by Whitham (1974), linearized surface gravity waves in a flat-bottomed basin of infinite horizontal extent and depth H are governed by the elliptic partial differential equation

$$\frac{\partial^2 p}{\partial x^2} + \frac{\partial^2 p}{\partial y^2} + \frac{\partial^2 p}{\partial z^2} = 0, \tag{1.20}$$

subject to the upper and lower boundary conditions

$$\frac{\partial^2 p}{\partial t^2} + g \frac{\partial p}{\partial z} = 0 \qquad \text{at} \quad z = 0,$$

---

[3]Let $z = x + iy$ be a complex variable in which x and y are real-valued. The function $f(z)$ is *analytic* if its derivative

$$\frac{df}{dz} = \lim_{\Delta z \to 0} \frac{f(z + \Delta z) - f(z)}{\Delta z}$$

exists and is uniquely defined as Δz goes to zero along any arbitrary path in the complex plane. If $f = u + iv$ where u and v are real-valued, a necessary condition for f to be analytic is that u and v satisfy the Cauchy–Riemann conditions

$$u_x = v_y, \qquad u_y = -v_x.$$

$$\frac{\partial p}{\partial z} = 0 \qquad \text{at} \quad z = -H.$$

The wave-like character of the solution is produced by the time-dependent upper boundary condition.

The elliptic nature of (1.20) does not follow from the the preceding classification scheme, which requires the evaluation of $b^2 - ac$ and is directly applicable only to linear second-order partial differential equations in two independent variables. In order to generalize this classification scheme to equations with n independent variables, consider the family of linear second-order partial differential equations of the form

$$\sum_{i=1}^{n} \sum_{j=1}^{n} a_{ij} \frac{\partial^2 u}{\partial x_i \partial x_j} + \sum_{i=1}^{n} b_i \frac{\partial u}{\partial x_i} + cu + d = 0. \tag{1.21}$$

If a_{ij}, b_i, c, and d are constants, there exists a one-to-one transformation to a new set of independent variables ξ_i such that the second-order terms in the preceding equation become

$$\sum_{i=1}^{n} A_{ii} \frac{\partial^2 u}{\partial \xi_i{}^2}.$$

If all the A_{ii} are nonzero and have the same sign, (1.21) is elliptic. If all the A_{ii} are nonzero and all but one have the same sign, (1.21) is hyperbolic. If at least one of the A_{ii} is zero, (1.21) is parabolic.

1.2   Wave Equations in Geophysical Fluid Dynamics

The wave-like motions of primary interest in geophysical fluid dynamics are the physical transport of scalar variables by the motion of fluid parcels, oscillatory motions associated with buoyancy perturbations (gravity waves), and oscillatory motions associated with potential vorticity perturbations (Rossby waves). Acoustic waves (sound waves) also propagate through all geophysical fluids, but in many applications these are small-amplitude perturbations whose detailed structure is of no interest. Both inviscid tracer transport and the propagation of sound waves are mathematically described by hyperbolic partial differential equations. Gravity waves and Rossby waves are also solutions to hyperbolic systems of partial differential equations, but some of the fluid properties essential for the support of these waves are represented in the governing equations by terms involving the zero-order derivatives of the unknown variables. These zero-order terms play no role in the classification of the governing equations as hyperbolic, and simpler nonhyperbolic systems of partial differential equations, such as the Boussinesq equations, can be derived whose solutions closely approximate the gravity-wave and Rossby-wave solutions to the original hyperbolic system. These simpler systems will be referred to as *filtered* equations.

1.2.1 Hyperbolic Equations

The concentration of a nonreactive chemical constituent is approximately governed by the first-order linear hyperbolic equation

$$\frac{\partial \psi}{\partial t} + u\frac{\partial \psi}{\partial x} + v\frac{\partial \psi}{\partial y} + w\frac{\partial \psi}{\partial z} = S, \tag{1.22}$$

where $\psi(x, y, z, t)$ is the mixing ratio of the chemical (in nondimensional units such as grams per kilogram or parts per billion) and $S(x, y, z, t)$ is the sum of all sources and sinks. This equation is an approximation because the molecular diffusivity of air is assumed to be negligible, in which case the transport of ψ is produced entirely by the velocity field. The characteristic curves associated with (1.22) are identical to the fluid parcel trajectories determined by the ordinary differential equations

$$\frac{dx}{dt} = u, \qquad \frac{dy}{dt} = v, \qquad \frac{dz}{dt} = w. \tag{1.23}$$

In geophysics the transport of a quantity by the velocity field is commonly referred to as *advection*;[4] both (1.22) and the one-way-wave equation (1.4) are "advection equations."

Equations describing the inviscid transport and chemical reactions among a family of chemical constituents can be written as the system

$$\frac{\partial \mathbf{c}}{\partial t} + u\frac{\partial \mathbf{c}}{\partial x} + v\frac{\partial \mathbf{c}}{\partial y} + w\frac{\partial \mathbf{c}}{\partial z} = \mathbf{s},$$

where $\mathbf{c}$ is a vector whose components are the concentration of each individual chemical species and $\mathbf{s}$ is a vector whose components are the net sources and sinks of each species. In general the sources and sinks depend on $\mathbf{c}$ but not on the derivatives of $\mathbf{c}$, so the preceding is a first-order linear hyperbolic system whose solution could be obtained by integrating a coupled system of ordinary differential equations along the family of characteristic curves defined by (1.23).

When diffusion is included, the mathematical model for nonreactive chemical transport becomes

$$\frac{\partial \psi}{\partial t} + u\frac{\partial \psi}{\partial x} + v\frac{\partial \psi}{\partial y} + w\frac{\partial \psi}{\partial z} - S$$
$$= \frac{\partial}{\partial x}\left(\kappa\frac{\partial \psi}{\partial x}\right) + \frac{\partial}{\partial y}\left(\kappa\frac{\partial \psi}{\partial y}\right) + \frac{\partial}{\partial z}\left(\kappa\frac{\partial \psi}{\partial z}\right), \tag{1.24}$$

which is a linear second-order parabolic partial differential equation. If this equation is derived strictly from first principles, κ represents a molecular diffusivity. The molecular diffusivities of air and water are so small that the contribution from the terms involving the second derivatives are important only when the

---

[4]In many disciplines the terms "convection" and "advection" are essentially interchangeable. In geophysics, however, the term "convection" is generally reserved for the description of thermally forced circulations.

fluctuations in ψ occur on much smaller scales than those of primary interest in most geophysical problems. Thus in most geophysical applications the solution to (1.24) is essentially identical to that for the inviscid problem, and the numerical techniques suitable for the approximation of (1.24) are almost identical to those for the purely hyperbolic problem (1.22).

When computing numerical solutions to either (1.22) or (1.24), there will be limits on the spatial and temporal scales at which the velocity field can be represented in any finite data set. The influence of the unresolved velocity perturbations on the distribution of the tracer is not directly computable, but is often parametrized by replacing κ by an *eddy diffusivity*, κ_e. The eddy diffusivity is supposed to represent the tendency of random unresolved velocity fluctuations to spread the distribution of ψ away from the centerline of the smooth air-parcel trajectories computed from the resolved-scale velocity field. Eddy diffusivities are much larger than the molecular diffusivity, but even when κ is replaced by a typical eddy diffusivity, the terms on the right side of (1.22) remain relatively small, and the basic character of the solution is still wave-like. Nevertheless, some eddy-diffusivity parametrizations do generate large values for κ_e in limited regions of the flow. High values of κ_e might, for example, be found in the planetary boundary layer where strong subgrid-scale motions are driven by thermal and mechanical turbulence. Large κ_e might also be parametrized to develop in regions where vigorous subgrid-scale motions are generated through Kelvin–Helmholtz instability. In these limited areas of high eddy diffusivity, the solutions to the parametrized problem may no longer be wave-like.

Now consider the nonlinear shallow-water equations

$$\frac{\partial u}{\partial t} + u\frac{\partial u}{\partial x} + v\frac{\partial u}{\partial y} + g\frac{\partial h}{\partial x} - fv = 0, \tag{1.25}$$

$$\frac{\partial v}{\partial t} + u\frac{\partial v}{\partial x} + v\frac{\partial v}{\partial y} + g\frac{\partial h}{\partial y} + fu = 0, \tag{1.26}$$

$$\frac{\partial h}{\partial t} + u\frac{\partial h}{\partial x} + v\frac{\partial h}{\partial y} + h\left(\frac{\partial u}{\partial x} + \frac{\partial v}{\partial y}\right) = 0, \tag{1.27}$$

where u and v are the horizontal velocity components, h is the fluid depth, and f is the Coriolis parameter. This is a system of quasi-linear first-order differential equations. If one is concerned only with smooth solutions, the fundamental properties of the shallow-water system may be determined from the linearized versions of (1.25)–(1.27). Consider, therefore, a geostrophically balanced basic-state flow such that

$$fV = g\frac{\partial H}{\partial x} \quad \text{and} \quad fU = -g\frac{\partial H}{\partial y},$$

where U and V are constant and H is linear in x and y. The first-order perturbations satisfy

$$\frac{\partial \mathbf{u}}{\partial t} + \mathbf{A}_1\frac{\partial \mathbf{u}}{\partial x} + \mathbf{A}_2\frac{\partial \mathbf{u}}{\partial y} + \mathbf{B}\mathbf{u} = \mathbf{0}, \tag{1.28}$$

where

$$
\mathbf{u} = \begin{pmatrix} u' \\ v' \\ h' \end{pmatrix}, \qquad\qquad \mathbf{B} = \begin{pmatrix} 0 & -f & 0 \\ f & 0 & 0 \\ fV/g & -fU/g & 0 \end{pmatrix},
$$

$$
\mathbf{A}_1 = \begin{pmatrix} U & 0 & g \\ 0 & U & 0 \\ H & 0 & U \end{pmatrix}, \qquad \mathbf{A}_2 = \begin{pmatrix} V & 0 & 0 \\ 0 & V & g \\ 0 & H & V \end{pmatrix}.
$$

As discussed in connection with (1.10), the preceding system will be hyperbolic if any linear combination of the coefficient matrices, $k_1\mathbf{A}_1 + k_2\mathbf{A}_2$, can be transformed to a real diagonal matrix through multiplication by bounded transformation matrices. Such transformation matrices always exist when the coefficient matrices are symmetric. Thus an easy way to demonstrate that the preceding system is hyperbolic is to exhibit a change of variables that renders $\mathbf{A}_1$ and $\mathbf{A}_2$ symmetric. A suitable transformation is obtained by letting $\mathbf{v} = \mathbf{S}^{-1}\mathbf{u}$, where

$$
\mathbf{S}^{-1} = \begin{pmatrix} c & 0 & 0 \\ 0 & c & 0 \\ 0 & 0 & g \end{pmatrix},
$$

and $c(x, y) = \sqrt{gH}$. Then (1.28) becomes

$$
\frac{\partial \mathbf{v}}{\partial t} + \tilde{\mathbf{A}}_1 \frac{\partial \mathbf{v}}{\partial x} + \tilde{\mathbf{A}}_2 \frac{\partial \mathbf{v}}{\partial y} + \tilde{\mathbf{B}}\mathbf{v} = 0,
$$

where

$$
\tilde{\mathbf{A}}_1 = \mathbf{S}^{-1}\mathbf{A}_1\mathbf{S} = \begin{pmatrix} U & 0 & c \\ 0 & U & 0 \\ c & 0 & U \end{pmatrix}, \qquad \tilde{\mathbf{A}}_2 = \mathbf{S}^{-1}\mathbf{A}_2\mathbf{S} = \begin{pmatrix} V & 0 & 0 \\ 0 & V & c \\ 0 & c & V \end{pmatrix},
$$

$$
\tilde{\mathbf{B}} = \mathbf{S}^{-1}\left[\mathbf{A}_1\frac{\partial \mathbf{S}}{\partial x} + \mathbf{A}_2\frac{\partial \mathbf{S}}{\partial y} + \mathbf{B}\mathbf{S}\right] = \begin{pmatrix} 0 & -f & 0 \\ f & 0 & 0 \\ \frac{1}{2}fV/c & -\frac{1}{2}fU/c & 0 \end{pmatrix}.
$$

The symmetry of $\tilde{\mathbf{A}}_1$ and $\tilde{\mathbf{A}}_2$ imply that the linearized shallow-water equations are a hyperbolic system.

The wave solutions to this hyperbolic system do not, however, propagate exactly along the characteristic curves unless f is zero. The relationship between the paths followed by propagating waves and the characteristics is most easily investigated by considering plane waves propagating parallel to the x-axis in a basic state with no mean flow. Let the Coriolis parameter have the constant value f_0 and define a vector of new unknown functions

$$
\mathbf{v} - \begin{pmatrix} u - gh/c \\ v \\ u + gh/c \end{pmatrix},
$$

which transforms (1.28) to

$$\frac{\partial \mathbf{v}}{\partial t} + \begin{pmatrix} -c & 0 & 0 \\ 0 & 0 & 0 \\ 0 & 0 & c \end{pmatrix} \frac{\partial \mathbf{v}}{\partial x} + \begin{pmatrix} 0 & -f_0 & 0 \\ f_0/2 & 0 & f_0/2 \\ 0 & -f_0 & 0 \end{pmatrix} \mathbf{v} = \mathbf{0}.$$

The characteristics for this system are the curves satisfying $dx/dt = \pm c$ and $dx/dt = 0$.

Wave solutions to (1.28) have the form

$$(u', v', h') = \Re \left\{ (u_0, v_0, h_0)e^{i(kx - \omega t)} \right\}, \tag{1.29}$$

provided that the frequency ω and wave number k satisfy the dispersion relation

$$\omega^2 = c^2 k^2 + f_0^2, \tag{1.30}$$

as may be demonstrated by substituting (1.29) into (1.28). Lines of constant phase, such as the locations of the troughs and crests, propagate at the *phase speed* ω/k, which from (1.30) is

$$\frac{\omega}{k} = \pm c \left(1 + \frac{f_0^2}{c^2 k^2} \right)^{1/2}.$$

A compact group of waves travels at the group velocity $\partial \omega / \partial k$, which can also be computed from (1.30):

$$\frac{\partial \omega}{\partial k} = \pm c \left(1 + \frac{f_0^2}{c^2 k^2} \right)^{-1/2}.$$

In the limit $|k| \gg f_0/c$, the phase speed and group velocity both approach the slope of a characteristic along which $|dx/dt| = c$. Nevertheless, for any finite value of k,

$$\left| \frac{\partial \omega}{\partial k} \right| < c < \left| \frac{\omega}{k} \right|,$$

and neither the lines of constant phase nor the wave groups follow trajectories that coincide with the characteristic curves. Note that the magnitude of the group velocity, which is the rate at which energy propagates in a wave, is bounded by c. The maximum rate of energy propagation can therefore be determined without considering the zero-order coefficient matrix $\mathbf{B}$ in (1.28).

The loose connection between wave propagation and the characteristics in the preceding example can disappear altogether if the Coriolis parameter is a function of the spatial coordinate. Then a second type of wave, the Rossby wave, may appear as an additional solution. If f increases linearly in proportion to y, Rossby-wave solutions may exist with phase speeds in the negative-x direction (Holton 1992; Pedlosky 1987). Neither the phase speeds nor the group velocities of these waves have any relation to the characteristic curves. It is not surprising that Rossby waves do not propagate along the characteristics, because the terms involving the undifferentiated functions of u and v play no role in the determination of the characteristics of (1.28), yet those same terms are essential for the maintenance of the Rossby waves.

The Euler equations governing inviscid isentropic motion in a density stratified fluid provide another example of a hyperbolic system that supports a type of wave whose propagation is completely unrelated to the characteristics. The Euler equations for the inviscid isentropic motion of a perfect gas can be expressed in the form

$$\frac{d\mathbf{v}}{dt} + \frac{1}{\rho}\nabla p = -g\mathbf{k}, \tag{1.31}$$

$$\frac{\partial \rho}{\partial t} + \nabla \cdot (\rho\mathbf{v}) = 0, \tag{1.32}$$

$$\frac{d\theta}{dt} = 0, \tag{1.33}$$

where Coriolis forces have been neglected,

$$\frac{d(\ )}{dt} = \frac{\partial(\ )}{\partial t} + \mathbf{v} \cdot \nabla(\ ),$$

$\mathbf{v}$ is the three-dimensional velocity vector, ρ is density, p is pressure, g is the gravitational acceleration, $\mathbf{k}$ is a unit vector directed opposite to the gravitational restoring force, and θ is the potential temperature, which is related to the entropy, S, such that

$$S = c_p \ln\theta + \text{constant}.$$

Conservation of momentum is required by (1.31), conservation of mass by (1.32), and conservation of entropy by (1.33).

As written above, the Euler equations constitute a system of five equations involving six unknowns. In atmospheric applications, the system may be closed using the equation of state for a perfect gas

$$p = \rho R T \tag{1.34}$$

and the definition of the potential temperature

$$\theta = T (p/p_0)^{-R/c_p}$$

to arrive at the diagnostic equation

$$p = p_0 \left(\frac{R}{p_0}\rho\theta\right)^{c_p/c_v}. \tag{1.35}$$

In the preceding, T is the temperature, p_0 is a constant reference pressure, R is the gas constant for dry air, c_p is the specific heat at constant pressure, and c_v is the specific heat at constant volume.

The Euler equations are a quasi-linear system of first-order partial differential equations. The fundamental character of the smooth solutions to this system can be determined by linearizing these equations about a horizontally uniform isothermally stratified basic state. Simpler basic states can be obtained by neglecting gravitational forces and the density stratification (Gustafsson et al. 1995, p. 136; see also Problem 3), but the isothermal basic state is of more geophysical relevance. As a preliminary step, p and ρ can be eliminated from (1.31)–(1.33) by introducing the nondimensional *Exner function* pressure defined as

$$\pi = (p/p_0)^{R/c_p}. \tag{1.36}$$

It follows that

$$\frac{1}{\rho}\nabla p = c_p \theta \nabla \pi,$$

so the momentum equation may written

$$\frac{d\mathbf{v}}{dt} + c_p \theta \nabla \pi = -g\mathbf{k}. \tag{1.37}$$

It also follows from (1.35) and (1.36) that

$$\pi = \left(\frac{R}{p_0}\rho\theta\right)^{R/c_v};$$

thus

$$\frac{d}{dt}\ln(\pi) = \frac{R}{c_v}\left[\frac{d}{dt}\ln(\rho) + \frac{d}{dt}\ln(\theta)\right],$$

or, using (1.32) and (1.33),

$$\frac{d\pi}{dt} + \frac{R\pi}{c_v}\nabla \cdot \mathbf{v} = 0. \tag{1.38}$$

Equations (1.33), (1.37), and (1.38) constitute a closed system of five equations in the five unknown variables, θ, π, and the three components of $\mathbf{v}$.

The essential properties of this system can be more simply examined in a two-dimensional context. Let x and z be the horizontal and vertical coordinates, and decompose the thermodynamic fields into a vertically varying basic state and a perturbation such that

$$\begin{aligned}
\pi(x, z, t) &= \overline{\pi}(z) + \pi'(x, z, t), \\
\theta(x, z, t) &= \overline{\theta}(z) + \theta'(x, z, t), \\
c_p\overline{\theta}\,\frac{d\overline{\pi}}{dz} &= -g.
\end{aligned} \tag{1.39}$$

The velocity components are decomposed as

$$u(x, z, t) = U + u'(x, z, t), \qquad w(x, z, t) = w'(x, z, t). \tag{1.40}$$

The basic-state vertical velocity is zero to ensure that the basic state is a steady solution to the nonlinear equations. Substituting these expressions for u, w, π, and θ into the two-dimensional versions of (1.33), (1.37), and (1.38), and neglecting second-order terms in the perturbation variables under the assumption that the perturbations are small-amplitude, one obtains the linear system

$$\left(\frac{\partial}{\partial t} + U\frac{\partial}{\partial x}\right)u' + c_p\overline{\theta}\frac{\partial\pi'}{\partial x} = 0, \tag{1.41}$$

$$\left(\frac{\partial}{\partial t} + U\frac{\partial}{\partial x}\right)w' + c_p\overline{\theta}\frac{\partial\pi'}{\partial z} = g\frac{\theta'}{\overline{\theta}}, \tag{1.42}$$

$$\left(\frac{\partial}{\partial t} + U\frac{\partial}{\partial x}\right)\theta' + \frac{\overline{\theta}}{g}N^2 w' = 0, \tag{1.43}$$

$$\left(\frac{\partial}{\partial t} + U\frac{\partial}{\partial x}\right)\pi' + w'\frac{\partial\overline{\pi}}{\partial z} + \frac{R\overline{\pi}}{c_v}\left(\frac{\partial u'}{\partial x} + \frac{\partial w'}{\partial z}\right) = 0, \tag{1.44}$$

where

$$N^2 = \frac{g}{\overline{\theta}}\frac{d\overline{\theta}}{dz}$$

is the square of the Brunt–Väisälä frequency.

Suppose that the reference state is isothermal. Then N^2 and the speed of sound $c_s = (c_p R T / c_v)^{1/2}$ are constant, and the preceding system can be simplified by removing the influence of the decrease in the mean density with height via the transformation

$$\tilde{u} = \left(\frac{\overline{\rho}}{\rho_0}\right)^{1/2}u', \qquad \tilde{\pi} = \left(\frac{\overline{\rho}}{\rho_0}\right)^{1/2}\frac{c_p\overline{\theta}}{c_s}\pi', \tag{1.45}$$

$$\tilde{w} = \left(\frac{\overline{\rho}}{\rho_0}\right)^{1/2}w', \qquad \tilde{\theta} = \left(\frac{\overline{\rho}}{\rho_0}\right)^{1/2}\frac{g}{N\overline{\theta}}\theta'. \tag{1.46}$$

Note that $\tilde{\theta}$ represents a scaled buoyancy and $\tilde{\pi}$ a scaled pressure. Let

$$\mathbf{v} = \begin{pmatrix} \tilde{u} & \tilde{w} & \tilde{\theta} & \tilde{\pi} \end{pmatrix}^T;$$

then the transformed equations have the form

$$\frac{\partial\mathbf{v}}{\partial t} + \mathbf{A}_1\frac{\partial\mathbf{v}}{\partial x} + \mathbf{A}_2\frac{\partial\mathbf{v}}{\partial z} + \mathbf{B}\mathbf{v} = \mathbf{0}, \tag{1.47}$$

in which

$$A_1 = \begin{pmatrix} U & 0 & 0 & c_s \\ 0 & U & 0 & 0 \\ 0 & 0 & U & 0 \\ c_s & 0 & 0 & U \end{pmatrix}, \qquad A_2 = \begin{pmatrix} 0 & 0 & 0 & 0 \\ 0 & 0 & 0 & c_s \\ 0 & 0 & 0 & 0 \\ 0 & c_s & 0 & 0 \end{pmatrix},$$

$$B = \begin{pmatrix} 0 & 0 & 0 & 0 \\ 0 & 0 & -N & -S \\ 0 & N & 0 & 0 \\ 0 & S & 0 & 0 \end{pmatrix}, \qquad S = c_s\left[\frac{1}{2\overline{\rho}}\frac{\partial\overline{\rho}}{\partial z} + \frac{1}{\overline{\theta}}\frac{\partial\overline{\theta}}{\partial z}\right].$$

Since the coefficient matrices for the first-order derivatives in (1.47) are symmetric, the linearized Euler equations are a hyperbolic system. The eigenvalues of $\mathbf{A}_1$ are U, U, $U + c_s$, and $U - c_s$; those of $\mathbf{A}_2$ are 0, 0, c_s, and $-c_s$. The eigenvalues involving c_s give the speed at which sound waves in an unstratified fluid propagate parallel to the x and z coordinate axes. As will be demonstrated below, sound waves in an isothermally stratified atmosphere actually propagate at slightly different speeds due to the influence of the zero-order term in (1.47). The remaining eigenvalues relate to the speed at which fluid parcels are advected horizontally and vertically by the mean flow. These eigenvalues have no relation to the propagation of gravity (or buoyancy) waves, which are the second type of fundamental wave motion supported by (1.47).

When the basic state is isothermal, S is constant, and wave solutions to (1.47) exist in the form

$$(\tilde{u}, \tilde{w}, \tilde{\theta}, \tilde{\pi}) = \Re\left\{(u_0, w_0, \theta_0, \pi_0)e^{i(kx+\ell z-\omega t)}\right\},\qquad(1.48)$$

provided that ω, k, and ℓ satisfy the dispersion relation

$$(\omega - Uk)^2 = \frac{c_s^2}{2}\left(k^2 + \ell^2 + \frac{N^2 + S^2}{c_s^2}\right)$$

$$\pm \frac{c_s^2}{2}\left[\left(k^2 + \ell^2 + \frac{N^2 + S^2}{c_s^2}\right)^2 - \frac{4N^2 k^2}{c_s^2}\right]^{1/2},\qquad(1.49)$$

which is obtained by substituting (1.48) into (1.47). As will be discussed in Section 7.2.4, the second term inside the square root is much smaller than the first term in most applications, so (1.49) can be separated into a pair of approximate dispersion relations for the sound waves and the gravity waves. The dispersion relation for the sound waves,

$$(\omega - Uk)^2 = c_s^2\left(k^2 + \ell^2\right) + S^2 + N^2,$$

is obtained by taking the positive root in (1.49). In a manner analogous to the effect of the Coriolis force on gravity waves in the shallow-water system, the terms involving the product of N or S with the zero-order derivatives of the unknown variables introduce a slight discrepancy between the phase speeds and group velocities of the actual sound waves and those that might be suggested by the eigenvalues of $\mathbf{A}_1$ and $\mathbf{A}_2$.

The dispersion relation for the gravity waves is obtained by taking the negative root in (1.49), which to a good approximation yields

$$(\omega - Uk)^2 = \frac{N^2 k^2}{k^2 + \ell^2 + (S^2 + N^2)/c_s^2}.\qquad(1.50)$$

Neither the phase speeds nor the group velocities of these waves have any relation to the eigenvalues of $\mathbf{A}_1$ and $\mathbf{A}_2$. Unlike sound waves, gravity waves do not even approximately propagate along the characteristics. There is no relation between the characteristics and the paths of the gravity-waves because some of the physical processes essential for gravity-wave propagation are mathematically represented by undifferentiated functions of the unknown variables, and as such exert no influence on the shape of the characteristics.

1.2.2   Filtered Equations

The Euler equations support sound waves, but sound waves have no direct influence on many types of atmospheric and oceanic motion. Analytic simplicity can often be achieved by approximating the Euler equations with alternative sets of *filtered* governing equations that do not support sound waves. As will be discussed in Chapter 7, eliminating the sound waves may also allow the resulting system of equations to be numerically integrated using a much larger time step than that which would be required for a similar numerical integration of the original Euler equations. These sets of filtered equations are not hyperbolic systems, but they support gravity waves that closely approximate the gravity-wave solutions to the full Euler equations. If the latitudinal variation of the Coriolis parameter is included, the filtered equations also support Rossby waves. The Coriolis parameter will, however, be neglected in the following discussion in order to present the essential ideas in the simplest context.

According to (1.35), the pressure perturbations in a perfect gas arise from variations in density and entropy. Variations in entropy play no fundamental role in the physics of sound wave propagation. Indeed, for the general class of fluids described by an equation of state of the form

$$\rho \equiv \rho(p, S),$$

the speed of sound is given by the square root of $(\partial p/\partial \rho)_S$ (Batchelor 1967, p. 166). In order to filter sound waves from the governing equations, it is therefore necessary to sever the link between density perturbations and pressure perturbations. This can be accomplished through any one of a family of related approximations that neglect terms involving the time variation of the density in the mass continuity equation (1.32).

One approximation that will filter sound waves is obtained by assuming that the flow is incompressible, in which case

$$\nabla \cdot \mathbf{v} = 0, \tag{1.51}$$

and mass conservation is replaced by volume conservation. The approximation of (1.32) by (1.51) is widely referred to as the *Boussinesq approximation*. Unfortunately, the term "Boussinesq approximation" has been used in two different senses. In some disciplines, the Boussinesq approximation refers only to the approximation of mass conservation by volume conservation. In the atmospheric and oceanic sciences, the Boussinesq approximation is generally understood to include both the preceding and additional approximations in the momentum equations that will be discussed in connection with (1.60). The latter definition, encompassing approximations to both the mass continuity and momentum equations, appears to be consistent with the actual approximations employed by Boussinesq (1903, pp. 157 and 174), and will be the form of the Boussinesq approximation referred to throughout this book.

A second approximation to the full compressible continuity equation is *anelastic* compressibility (Ogura and Phillips 1962; Lipps and Hemler 1982)

$$\nabla \cdot (\overline{\rho}\mathbf{v}) = 0, \tag{1.52}$$

in which the density involved in the mass budget is a steady reference-state density $\overline{\rho}(z)$ that varies only along the coordinate axis parallel to the gravitational restoring force. A third approximation is *pseudo-incompressibility* (Durran 1989)

$$\frac{\partial \hat{\rho}}{\partial t} + \nabla \cdot (\hat{\rho}\mathbf{v}) = 0, \tag{1.53}$$

in which $\hat{\rho}$ is determined by the time-varying potential temperature and the pressure in a steady reference state $\tilde{p}(x, y, z)$ via the equation of state

$$\tilde{p} = p_0 \left(\frac{R}{p_0} \hat{\rho}\theta \right)^{c_p/c_v}.$$

The pseudo-incompressible approximation neglects the influence of perturbation pressure on perturbation density in the mass budget. According to the preceding definition of $\hat{\rho}$, the term $\partial\hat{\rho}/\partial t$ in (1.53) is entirely determined by $\partial\theta/\partial t$. The pseudo-incompressible continuity equation may be written in the obviously diagnostic form

$$\nabla \cdot (\tilde{\rho}\tilde{\theta}\mathbf{v}) = 0 \tag{1.54}$$

by using the thermodynamic equation (1.33) to eliminate $\partial\theta/\partial t$ from (1.53) and defining steady reference fields of density $\tilde{\rho}(x, y, z)$ and potential temperature $\tilde{\theta}(x, y, z)$ such that the reference fields satisfy the equation of state,

$$\tilde{p} = p_0 \left(\frac{R}{p_0} \tilde{\rho}\tilde{\theta} \right)^{c_p/c_v}.$$

Note that if F_θ represents any thermal forcing or viscous terms that might appear on the right side of the thermodynamic equation in more general applications, (1.53) is unchanged but (1.54) becomes

$$\nabla \cdot (\tilde{\rho}\tilde{\theta}\mathbf{v}) = \hat{\rho}F_\theta.$$

The pseudo-incompressible system can be rigorously derived through scale analysis by assuming that the Mach number (U/c_s) and the perturbation of the total pressure about the reference pressure, $\tilde{p}$, are both small (Durran 1989).

In order for an approximate set of governing equations to provide a physically acceptable approximation to the dynamics of the unapproximated system, the approximate equations should conserve energy in the sense that the domain integral of the total energy should be equal to the divergence of an energy flux through the boundaries of the domain. The energy equation for the full compressible system is

$$\frac{\partial E}{\partial t} + \nabla \cdot [(E + p)\mathbf{v}] = 0, \tag{1.55}$$

where

$$E = \rho \left(\frac{\mathbf{v} \cdot \mathbf{v}}{2} + gz + c_v T \right)$$

is the total energy (kinetic plus potential plus internal) per unit volume in a compressible fluid. Similar energy equations can be obtained using the incompressible or pseudo-incompressible continuity equations without introducing additional approximations in the momentum equations.

If the flow is incompressible, the mass continuity equation breaks into the two separate relations

$$\frac{d\rho}{dt} = 0 \tag{1.56}$$

and (1.51); the thermodynamic equation is no longer required to close the system, and the governing equations are simply (1.31), (1.51), and (1.56). The energy equation for this system has the same form as that for the compressible system (1.55) except that the energy,

$$E_i = \rho \left(\frac{\mathbf{v} \cdot \mathbf{v}}{2} + gz \right),$$

does not include the term representing internal energy. The pseudo-incompressible system, which consists of (1.33), (1.37), and (1.54), conserves

$$E_{pi} = \hat{\rho} \left(\frac{\mathbf{v} \cdot \mathbf{v}}{2} + gz \right) + c_v \tilde{\rho} \tilde{T},$$

according to the energy equation

$$\frac{\partial E_{pi}}{\partial t} + \nabla \cdot \left[\left(E_{pi} + \hat{p} \right) \mathbf{v} \right] = 0,$$

where $\hat{p} = \tilde{p} + c_p \tilde{\rho} \tilde{\theta} \pi' \approx \tilde{p} + p' = p$ and $\pi' = (p/p_0)^{R/c_p} - (\tilde{p}/p_0)^{R/c_p}$. The energy flux in the pseudo-incompressible system differs from the energy flux in the full compressible system a factor of $\hat{\rho}/\rho$, because

$$c_v \tilde{\rho} \tilde{T} + \hat{p} = (c_v + R)\tilde{\rho}\tilde{\theta}\tilde{\pi} + c_p \tilde{\rho}\tilde{\theta}\pi' = c_p \tilde{\rho}\tilde{\theta}\pi = c_p \hat{\rho}\theta\pi$$

$$= c_p \hat{\rho} T = \frac{\hat{\rho}}{\rho} (c_v \rho T + p).$$

In contrast to the situation for the incompressible and pseudo-incompressible approximations, the pressure gradient terms in the momentum equations must be linearized and modified to obtain an energy-conservative system of anelastic equations. As a first step toward developing such a system, the thermodynamic variables are decomposed into a vertically varying reference state and a perturbation. This decomposition is also quite useful outside the context of the anelastic equations because in many geophysical fluids the gravitational acceleration and the vertical pressure gradient are nearly in balance. Both numerical accuracy and physical insight can be then be enhanced by splitting the pressure and density fields into steady hydrostatically balanced vertical profiles and finite-amplitude perturbations about those reference profiles such that

$$p(x, y, z, t) = \overline{p}(z) + p'(x, y, z, t),$$
$$\rho(x, y, z, t) = \overline{\rho}(z) + \rho'(x, y, z, t),$$
$$\frac{d\overline{p}}{dz} = -\overline{\rho}g.$$

After removing the hydrostatically balanced component of the pressure, the momentum equation (1.31) may be written without approximation as

$$\frac{d\mathbf{v}}{dt} + \frac{1}{\rho}\nabla p' = -g\frac{\rho'}{\rho}\mathbf{k}. \tag{1.57}$$

If the pressure gradients in the momentum equation are expressed in terms of π and θ, the hydrostatic reference state is removed by defining

$$\pi(x, y, z, t) = \overline{\pi}(z) + \pi'(x, y, z, t),$$
$$\theta(x, y, z, t) = \overline{\theta}(z) + \theta'(x, y, z, t),$$
$$c_p\overline{\theta}\frac{d\overline{\pi}}{dz} = -g, \tag{1.58}$$

in which case (1.37) becomes

$$\frac{d\mathbf{v}}{dt} + c_p\theta\nabla\pi' = g\frac{\theta'}{\overline{\theta}}\mathbf{k}. \tag{1.59}$$

The term on the right side of either (1.57) or (1.59) represents a buoyancy force. Note that since no approximations have been introduced in these equations, the pressure gradient terms in (1.57) and (1.59) remain nonlinear.

In addition to the previously discussed modifications to the mass continuity equation, the Boussinesq and anelastic approximations include additional simplifications to the momentum equations that linearize the pressure gradient terms in (1.57) and (1.59). The form of the *Boussinesq approximation* that is most common in geophysical fluid dynamics *neglects the effects of density variations on the mass balance in the continuity equation and on inertia in the momentum equations, but includes the effect of density variations on buoyancy forces* (Gill 1982, p. 130). Letting ρ_0 be a constant reference density, the Boussinesq form of the momentum equations may be written

$$\frac{d\mathbf{v}}{dt} + \frac{1}{\rho_0}\nabla p' = -g\frac{\rho'}{\rho_0}\mathbf{k}, \tag{1.60}$$

where the perturbation density continues to be defined as $\rho - \overline{\rho}(z)$ (rather than $\rho - \rho_0$). The resulting Boussinesq system, consisting of (1.51), (1.56), and (1.60), can be concisely expressed in terms of the Boussinesq pressure, buoyancy, and Brunt–Väisälä frequency,

$$P = \frac{p}{\rho_0}, \qquad b = -g\frac{\rho - \overline{\rho}}{\rho_0}, \qquad N_b^2 = -\frac{g}{\rho_0}\frac{d\overline{\rho}}{dz},$$

respectively, as

$$\frac{d\mathbf{v}}{dt} + \nabla P = b\mathbf{k}, \tag{1.61}$$

$$\frac{db}{dt} + N_b^2 w = 0, \tag{1.62}$$

$$\nabla \cdot \mathbf{v} = 0. \tag{1.63}$$

The Boussinesq system is governed by an energy equation of the form (1.55), except that the total "Boussinesq" energy is

$$E_b = \rho_0 \frac{\mathbf{v} \cdot \mathbf{v}}{2} + \rho g z.$$

Although the Boussinesq approximation provides a qualitatively correct mathematical model for the study of buoyancy effects in fluids, it is not quantitatively accurate in situations where there is a significant change in mean density over the depth of the fluid, as would be the case in any atmospheric layer that is more than a couple of kilometers deep. Somewhat better quantitative agreement between the Boussinesq equations and atmospheric flows can be obtained using the same Boussinesq system (1.61)–(1.63) with the pressure, buoyancy, and Brunt–Väisälä frequency defined as

$$P = c_p\theta_0\pi', \qquad b = g\frac{\theta - \overline{\theta}}{\theta_0}, \qquad N_b^2 = \frac{g}{\theta_0}\frac{d\overline{\theta}}{dz},$$

respectively, where θ_0 is a constant reference temperature. Using these definitions for P and b, the full momentum equation (1.59) will be well approximated by (1.61) whenever the full and basic-state potential temperatures are close to θ_0. In atmospheric applications, it is often easier to satisfy this constraint than to demand that ρ_0 be a good approximation to ρ in (1.57). Even if the reference state is nearly isentropic, some quantitative error in the Boussinesq solution will still be introduced by the incompressible continuity equation. The quantitative errors associated with Boussinesq approximations to deep atmospheric flows can be greatly diminished using either the anelastic or pseudo-incompressible approximations.indexanelastic approximation

An energy-conservative form of the anelastic equations was derived by Lipps and Hemler (1982) by writing the momentum equations in the form

$$\frac{d\mathbf{v}}{dt} + c_p\nabla(\overline{\theta}\pi') = g\frac{\theta'}{\overline{\theta}}\mathbf{k}, \tag{1.64}$$

where the hydrostatically balanced components of the Exner function pressure and the potential temperature have been removed using (1.58). The anelastic system can be derived from the pseudo-incompressible system by choosing a horizontally uniform hydrostatically balanced reference state, approximating the total pressure gradient as $c_p \overline{\theta} \nabla \pi'$, and neglecting $d\overline{\theta}/dz$ in both (1.54) and the momentum equations. The same approximation can be obtained by a rigorous, if somewhat delicate, scaling argument (Lipps 1990). The anelastic system consisting of equations (1.33), (1.52), and (1.64) provides a good approximation to the full compressible equations. The Lipps and Hemler anelastic system satisfies the energy equation

$$\frac{\partial E_a}{\partial t} + \nabla \cdot \left[(E_a + \check{p})\mathbf{v} \right] = 0,$$

where

$$E_a = \overline{\rho} \left(\frac{\mathbf{v} \cdot \mathbf{v}}{2} + gz + c_p \overline{\pi} \theta' \right) + c_v \overline{\rho T}$$

and $\check{p} = \overline{p} + c_p \overline{\rho \theta} \pi' \approx \overline{p} + p' = p$.

Simple wave solutions to the preceding filtered systems can be obtained by linearizing the two-dimensional form of each system about an appropriate basic-state flow with a constant horizontal wind speed U. Solutions to the two-dimensional Boussinesq system exist in the form

$$(u, w, P, b) = \Re \left\{ (u_0, w_0, P_0, b_0) e^{i(kx + \ell z - \omega t)} \right\},$$

provided that N_b^2 is constant and

$$(\omega - Uk^2) = \frac{N_b^2 k^2}{k^2 + \ell^2}.$$

These solutions are gravity waves, as may be seen by comparing the preceding dispersion relation with (1.50) in the limit $c_s \to 0$. There are no sound-wave solutions to the Boussinesq equations.

If the basic state is isothermally stratified, the prognostic variables in the two-dimensional anelastic and pseudo-incompressible systems can be transformed as per (1.45)–(1.46) to yield constant-coefficient linear systems of partial differential equations with wave solutions of the form (1.48). In the case of the anelastic equations, these waves satisfy the dispersion relation (1.50), which is an excellent approximation to the dispersion relation for gravity waves in the full compressible system. In the case of the pseudo-incompressible equations, the waves satisfy the dispersion relation

$$(\omega - Uk^2) = \frac{N^2 k^2}{k^2 + \ell^2 + S^2/c_s^2}.$$

Since in most applications $k^2 + \ell^2$ is much larger than the remaining terms in the denominator of (1.50), the preceding is also a very good approximation to the gravity-wave dispersion relation for the full compressible equations. The relative accuracy of the anelastic and pseudo-incompressible approximations cannot be judged solely on the basis of their dispersion relations. Nance and Durran (1994) and Nance (1997) compared the accuracy of several different systems of filtered equations and found that the pseudo-incompressible system and the anelastic system suggested by Lipps and Hemler are the most accurate, and that the anelastic system performs slightly better in the hydrostatic limit, whereas the pseudo-incompressible system gives slightly better accuracy when the flow is not hydrostatic.

1.3   Strategies for Numerical Approximation

A wide variety of different methods have been employed to obtain numerical solutions to the systems of partial differential equations discussed in the preceding sections of this chapter. Before delving into the details of these methods, we conclude this introductory chapter by comparing some of the most general properties of the various methods, including the manner in which each method approximates the value of the unknown function and estimates its derivatives. We will also consider some of the fundamental differences between the numerical algorithms used to solve elliptic and hyperbolic partial differential equations.

1.3.1   Approximating Calculus with Algebra

Digital computers are not designed to solve differential equations directly. Although the digital computer can perform algebraic operations such as addition and multiplication, it does not have any intrinsic ability to differentiate and integrate functions. As a consequence, every numerical method is designed to convert the original differential equation into a set of solvable algebraic equations. As part of this task the continuous functions associated with the original problem must be represented by a finite set of numbers that can be stored in a computer's memory or on disk. There are therefore two basic problems that must be addressed by every numerical scheme: how to represent the solution by a finite data set and how to compute derivatives. There are also two basic solution strategies: grid-point methods and series expansion methods.

In grid-point methods, each function is described by its value at a set of discrete grid points. Figure 1.1 shows how $f(x)$, a periodic function on the interval $[0, 2\pi]$, might be represented by its exact value at five different points along the x-axis. The spacing of the grid points can be chosen arbitrarily, although any variations in the grid spacing will affect the accuracy of the approximation. If a priori knowledge of the function's periodicity is available, a natural choice for the five pieces of information would be $(f(2\pi/5), f(4\pi/5), \ldots, f(2\pi))$. No assumption is made about the value of the approximate solution between the points on the numerical mesh. These methods are usually called *finite-difference* methods because

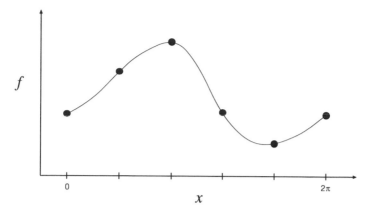

FIGURE 1.1. Grid-point approximation of a periodic function on the interval $[0, 2\pi]$. Individual points show the function values at intervals of $2\pi/5$.

derivatives are approximated using formulae such as

$$\frac{df}{dx}(x_0) \approx \frac{f(x_0 + \Delta x) - f(x_0 - \Delta x)}{2\Delta x},$$

which is a centered finite difference computable from data on a uniform mesh with grid interval Δx. Finite-difference methods will be discussed in Chapters 2 and 3.

Finite-volume methods are an important variation of the basic grid-point approach in which some assumption is made about the structure of the approximate solution between the grid points. In a finite-volume method the grid-point value f_j represents the average of the function $f(x)$ over the interval (or grid cell) $[(j-\frac{1}{2})\Delta x, (j+\frac{1}{2})\Delta x]$. Finite-volume methods are very useful for approximating solutions that contain discontinuities. If the solution being approximated is smooth, finite-difference and finite-volume methods yield essentially the same numerical schemes. It is sometimes mistakenly supposed that all grid-point methods necessarily generate approximations to the grid-cell average; however, only finite-volume methods have this property.

In order to completely define the numerical algorithm arising from a conventional finite-difference approximation, it is necessary to specify particular formulae for the finite differences (e.g., centered differencing, one-sided differencing, or one of the other options described in Chapter 2). In finite-volume methods, on the other hand, the derivatives are determined by the assumed structure of the approximate solution within each cell. In practice, finite-volume methods often require the computation of the fluxes through the edges of each grid cell rather than the evaluation of derivatives, but in order to compute these fluxes it is once again necessary to make some assumption about the structure of the solution within each grid cell. The approximate solution cannot simply be the piecewise linear function that interpolates the grid-point values, because then the value at an individual grid point will not equal the average of the piecewise linear approximation over

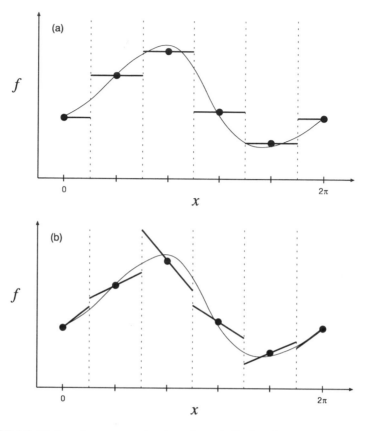

FIGURE 1.2. Finite-volume approximation of a periodic function on the interval $[0, 2\pi]$ using: (a) piecewise constant functions, and (b) piecewise linear functions.

the surrounding grid cell. Two possible finite-volume approximations to $f(x)$ are shown in Fig. 1.2. Accounting for periodicity, five pieces of information are again used to construct these approximations. Piecewise constant functions are used in the approximation shown in Fig. 1.2a; Fig. 1.2b shows the approximation obtained using piecewise linear functions defined such that

$$f(x) \approx f_j + \sigma_j(x - j\Delta x) \quad \text{for all} \quad x \in \left((j-\tfrac{1}{2})\Delta x, (j+\tfrac{1}{2})\Delta x\right),$$

where f_j is the average of the approximate solution over the grid cell centered at $j\Delta x$, and $\sigma_j = (f_{j+1} - f_j)/\Delta x$. The accuracy of the numerical approximations shown in Figs. 1.1 and 1.2 is poor because only five data points are used to resolve $f(x)$. Between 12 and 20 data points would be required to obtain a minimally acceptable approximation in most practical applications. Finite-volume methods will be discussed in Chapter 5.

   In series-expansion methods, the unknown function is approximated by a linear combination of a finite set of continuous expansion functions, and the data set describing the approximated function is the finite set of expansion coefficients.

Derivatives are computed analytically by differentiating the expansion functions. When the expansion functions form an orthogonal set, the series expansion approach is a *spectral method*. If the preceding periodic function were to be approximated by a spectral method using five pieces of data, a natural choice would be the truncated Fourier series

$$a_1 + a_2 \cos x + a_3 \sin x + a_4 \cos 2x + a_5 \sin 2x. \tag{1.65}$$

The five Fourier coefficients $(a_1, a_2, \ldots, a_5)$ need not be chosen such that the value of the Fourier series exactly matches the value of $f(x)$ at any specific point in the interval $0 \le x \le 2\pi$. Nevertheless, one possible way to choose the coefficients would be to require that (1.65) be identical to $f(x)$ at each of the five points used by the grid-point methods discussed previously. Another useful strategy is to choose the coefficients to minimize the x-integral of the square of the difference between the approximation expansion (1.65) and $f(x)$.

If the expansion functions are nonzero in only a small part of the total domain, the series expansion technique is a *finite-element method*. In the finite-element approach the function $f(x)$ is again approximated by a finite series of functions of the form $b_0 s_0(x) + b_1 s_1(x) + \cdots + b_5 s_5(x)$, but the functions s_n differ from the trigonometric functions in the spectral method because each individual function is zero throughout most of the domain. The simplest finite-element expansion functions are piecewise linear functions defined with respect to some grid. Each function is unity at one grid point, or node, and zero at all the other nodes. The values of the expansion function between the nodes are determined by linear interpolation using the values at the two nearest nodes. Six linear finite-element expansion functions suitable for approximating $f(x)$ might appear as shown in Fig. 1.3. Accounting for periodicity, the five pieces of information describing $f(x)$ would be the coefficients $(b_1, b_2, \ldots, b_5)$. When finite elements are constructed with piecewise-linear functions, the resulting numerical expressions are often similar to those obtained using grid-point methods. If finite elements are constructed from piecewise quadratic or cubic functions, however, the resulting formulae are quite different from those that arise naturally through finite differencing. Series expansion methods will be studied in Chapter 4.

The numerical solution is defined throughout the entire spatial domain at every time step, but in time-dependent problems the approximate solution is typically available at only a few time levels at any given step of the numerical simulation. As a consequence, the use of series expansions is generally restricted to the representation of functional variations along spatial coordinates. Time derivatives are almost always approximated by finite differences.

1.3.2 Marching Schemes

Suppose that numerical solutions are sought to the first-order linear system

$$\frac{\partial u}{\partial y} + \frac{\partial v}{\partial x} = 0, \qquad \frac{\partial v}{\partial y} + \gamma \frac{\partial u}{\partial x} = 0, \tag{1.66}$$

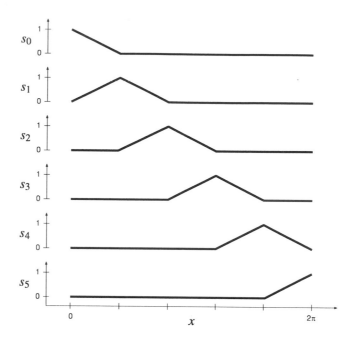

FIGURE 1.3. Six finite-element expansion functions, $s_0(x), s_1(x), \ldots, s_5(x)$.

throughout the domain $0 \le x \le 2\pi, 0 \le y \le Y$. Let the domain be periodic in x and suppose that boundary conditions are specified for $u(x, 0)$ and $v(x, 0)$. One possible finite-difference approximation to the preceding system is

$$\frac{u_j^{n+1} - u_j^n}{\Delta y} + \frac{v_{j+1}^n - v_{j-1}^n}{2\Delta x} = 0, \tag{1.67}$$

$$\frac{v_j^{n+1} - v_j^n}{\Delta y} + \gamma \frac{u_{j+1}^{n+1} - u_{j-1}^{n+1}}{2\Delta x} = 0, \tag{1.68}$$

where u_j^n and v_j^n denote the numerical approximations to $u(j\Delta x, n\Delta y)$ and $v(j\Delta x, n\Delta y)$. The boundary conditions on u and v at $y = 0$ can be used to specify u_j^0 and v_j^0. The numerical solution along the line $y = \Delta y$ can then be calculated by solving (1.67) for u_j^1 at every j, and using these values of u_j^1 to compute v_j^1 from (1.68). In principle, this procedure can be repeated to compute approximations to the solution at $y = 2\Delta y, 3\Delta y, \ldots$ and thereby sequentially evaluate the numerical solution throughout the entire domain.

Under what circumstances will this procedure yield an accurate approximation to the true solution? This question can be answered without any detailed knowledge of numerical analysis when $\gamma < 0$. The linear second-order partial differential equation

$$\frac{\partial^2 u}{\partial y^2} - \gamma \frac{\partial^2 u}{\partial x^2} = 0 \tag{1.69}$$

can be obtained by eliminating v from (1.66). As per the discussion of (1.11), this equation is hyperbolic if $\gamma > 0$, and it is elliptic if $\gamma < 0$. Suppose that the boundary conditions on u and v are

$$u(x, 0) = \frac{1}{N^2} \sin(Nx), \qquad v(x, 0) = 0, \tag{1.70}$$

where N is a positive integer. In the limit $N \to \infty$, the preceding boundary conditions become

$$u(x, 0) = 0, \qquad v(x, 0) = 0, \tag{1.71}$$

for which the exact solution to (1.66) is simply $u(x, y) = v(x, y) = 0$.

When (1.67) and (1.68) form an elliptic system (i.e., when $\gamma < 0$), the exact solution subject to the boundary conditions (1.70) is

$$u(x, y) = \frac{1}{N^2} \sin(Nx) \cosh(\beta N y), \qquad v(x, y) = \frac{\beta}{N^2} \cos(Nx) \sinh(\beta N y),$$

where $\beta = \sqrt{-\gamma}$ is a real constant. As $N \to \infty$, the difference between the boundary conditions (1.70) and (1.71) disappears, but the difference between the solutions generated by each boundary condition increases without bound along any line $y = y_0 > 0$. Arbitrarily small changes in the amplitude of the imposed boundary values can produce arbitrarily large changes in the amplitude of the interior solution. Under such circumstances there is no hope of accurately approximating the true solution by the finite-difference method (1.67)–(1.68) because the round-off errors incurred as (1.70) is evaluated to obtain numerical values for the grid points along $y = 0$ may generate arbitrarily large perturbations in the interior solution.

The mathematical problem of solving (1.66) subject to boundary conditions specified for $u(x, 0)$ and $v(x, 0)$ is not well-posed whenever $\gamma < 0$. A *well-posed problem* is one in which a unique solution to a given partial differential equation exists and depends continuously on the initial- and boundary-value data. When $\gamma < 0$, the preceding problem is not well-posed because the solution does not depend continuously on the boundary data. On the other hand, when $\gamma > 0$ the problem is hyperbolic, and the solution subject to (1.70) is

$$u(x, y) = \frac{1}{N^2} \sin(Nx) \cos\left(\sqrt{\gamma} N y\right), \quad v(x, y) = -\frac{\sqrt{\gamma}}{N^2} \cos(Nx) \sin\left(\sqrt{\gamma} N y\right).$$

In this case both $u(x, y) \to 0$ and $v(x, y) \to 0$ as $N \to \infty$. The interior solutions associated with the boundary conditions (1.70) and (1.71) approach each other as the difference between the two boundary conditions goes to zero, and small changes in the amplitude of the boundary data produce only small changes in the amplitude of the interior solution. As demonstrated in Gustafsson et al. (1995), the hyperbolic problem is well posed. When $\gamma > 0$, it is possible to obtain good approximations to the correct solution using (1.67) and (1.68), although as will be discussed in Chapter 2, the quality of the result depends on the parameter $\sqrt{\gamma}\Delta y / \Delta x$.

Physicists seldom worry about well-posedness, since properly formulated mathematical models of the physical world are almost always well-posed. The preceding example may be recognized as an initial value problem in which y represents time and x is the spatial coordinate. In contrast to their hyperbolic cousins, elliptic partial differential equations describe steady-state physical systems and do not naturally arise as initial value problems. When a real-world system is governed by an elliptic equation, physical considerations usually provide data for the dependent variables or their normal derivatives along each boundary, and the additional boundary-value data leads to a well-posed problem. The fact that elliptic partial differential equations are not well-posed as initial value problems may therefore be irrelevant to the physicist—but it is not irrelevant to the numerical analyst. Given a well-posed elliptic problem, such as (1.69) with $\gamma < 0$ and u specified at $y = 0$ and $y = Y$, could one expect to compute an accurate approximate solution on some numerical grid by starting with the known values along one boundary and stepping across the grid, one point at a time? The answer is no, an approach of this type is numerically unstable—indeed it mimics the not-well-posed formulation of an elliptic partial differential equation as an initial value problem. Practical methods for the numerical solution of elliptic partial differential equations are therefore not "marching" schemes. Instead of computing the solution at one point and then proceeding to the next, all the grid-point values must be simultaneously adjusted (perhaps through some iterative process) in order to adequately satisfy the governing differential equation and the boundary conditions. In contrast, hyperbolic partial differential equations do lend themselves to numerical solution via marching techniques.[5]

Another major difference in the numerical treatment of elliptic and hyperbolic equations arises in the specification of boundary conditions. As suggested by the preceding example, boundary conditions are usually imposed at every boundary as part of the natural formulation of an elliptic problem. Moreover, the incorporation of these boundary data into a numerical algorithm is generally straightforward. On the other hand, if one attempts to compute the solution to a hyperbolic problem in a limited spatial domain, the numerical algorithm may require boundary conditions in regions where none should actually be specified (i.e., at a boundary where all the characteristic curves are directed out of the domain). Improper boundary conditions may lead to instabilities or to nonuniqueness in the numerical solution of a hyperbolic system. Further discussion of boundary conditions will be presented in Chapter 8.

---

[5]L.F. Richardson, who explored the numerical solution of a variety of partial differential equations prior to his celebrated attempt at numerical weather prediction, coined the terms "jury" and "marching" methods to describe the basic difference between the numerical techniques suitable for the solution of elliptic equations and hyperbolic equations. The adjective "jury" alluded to the idea that one needed to adjust all the values in the numerical solution until the whole was "judged" to constitute a satisfactory approximation.

Problems

1. Suppose that
$$\frac{\partial^2 u}{\partial t^2} + a\frac{\partial^2 u}{\partial x \partial t} + b\frac{\partial^2 u}{\partial x^2} = 0$$
is a hyperbolic partial differential equation and that a and b are constants. Show that this equation can be transformed to a decoupled pair of first-order wave equations. What are the propagation speeds of the solutions to these first-order wave equations?

2. Show that when (1.11) is hyperbolic, it can be transformed to the alternative canonical form
$$u_{\xi\xi} - u_{\eta\eta} + \tilde{L}(\xi, \eta, u, u_\xi, u_\eta) = 0,$$
where $\tilde{L}(\xi, \eta, u, u_\xi, u_\eta)$ is once again a linear function of u, u_ξ, and u_η with coefficients that may depend on ξ and η. (*Hint:* start with (1.12) and define new independent variables equal to $\xi + \eta$ and $\xi - \eta$.)

3. If gravity and density stratification are neglected, the two-dimensional Euler equations for inviscid isentropic flow reduce to a system of four equations in the unknowns (u, w, ρ, p). Linearize this system about a basic state with constant (u_0, w_0, ρ_0, p_0) and show that the linearized system is hyperbolic. (*Hint:* transform the perturbation thermodynamic variables to $p'/(c\rho_0)$ and $\rho' - p'/c^2$, where $c^2 = \partial p/\partial \rho$ is the square of the speed of sound in the basic state.)

4. The pressure and density changes in compressible isentropic flow satisfy the relation
$$\frac{d\rho}{dt} = \frac{1}{c_s^2}\frac{dp}{dt}.$$
   (a) Derive the preceding relationship.
   (b) Show that the preceding relationship is approximated as
$$\frac{d\rho}{dt} = 0$$
in the incompressible system, as
$$\frac{d\rho}{dt} = \frac{\rho}{\bar{\rho}}\frac{d\bar{\rho}}{dt}$$
in the anelastic system, and as
$$\frac{d\rho}{dt} = \frac{\rho}{\tilde{\rho}}\frac{1}{\tilde{c}_s^2}\frac{d\tilde{p}}{dt}$$
in the pseudo-incompressible system (where the tilde denotes the steady reference field).

5. Show that the backward heat equation,

$$\frac{\partial \psi}{\partial t} = -\frac{\partial^2 \psi}{\partial x^2},$$

and the initial condition $\psi(x, 0) = f(x)$ do not constitute a well-posed problem on the domain $-\infty < x < \infty, t > 0$.

2
Basic Finite-Difference Methods

As discussed in the preceding chapter, there are two conceptually different ways to represent continuous functions on digital computers: as a finite set of grid-point values or as a finite set of series-expansion functions. The grid-point approach is used in conjunction with finite-difference methods, which were widely implemented on digital computers somewhat earlier than the series-expansion techniques. In addition, the theory for these methods is somewhat simpler than that for series-expansion methods. We will parallel this historical development by studying finite-difference methods in this chapter and deferring the treatment of series expansion methods to Chapter 4. Moreover, it is useful to understand finite-difference methods before investigating series-expansion techniques because even when series expansions are used to represent the spatial dependence of some atmospheric quantity, the time dependence is almost always discretized and treated with finite differences.

The derivative of a function $f(x)$ at the point x_0 could be defined in any of the following three ways:

$$\frac{df}{dx}(x_0) = \lim_{\Delta x \to 0} \frac{f(x_0 + \Delta x) - f(x_0)}{\Delta x}, \tag{2.1}$$

$$\frac{df}{dx}(x_0) = \lim_{\Delta x \to 0} \frac{f(x_0) - f(x_0 - \Delta x)}{\Delta x}, \tag{2.2}$$

$$\frac{df}{dx}(x_0) = \lim_{\Delta x \to 0} \frac{f(x_0 + \Delta x) - f(x_0 - \Delta x)}{2\Delta x}. \tag{2.3}$$

If the derivative of $f(x)$ is continuous at x_0, all three expressions produce the same unique answer. Suppose, however, that f is an approximation to some differentiable function that is defined on a discrete grid; then the preceding expressions must be evaluated using a finite value of Δx. The approximations to the true derivative obtained by evaluating the algebraic expressions on the right side of (2.1)–(2.3) using finite Δx are known as *finite differences*. The basic idea behind finite-difference methods is to convert the differential equation into a system of algebraic equations by replacing each derivative with a finite difference.

When Δx is finite, the finite-difference approximations (2.1)–(2.3) are not equivalent; they differ in their accuracy, and when they are substituted for derivatives in differential equations they generate different algebraic equations. The differences in the structures of these algebraic equations can have a great influence on the stability of the numerical solution. In this chapter, we will examine the stability and accuracy of basic finite-difference methods.

2.1   Accuracy and Consistency

The exact derivative can be calculated to within an arbitrarily small error using any one of (2.1)–(2.3) by insuring that Δx is sufficiently small. However, since computer capacities always place a practical limit on the numerical resolution, it is necessary to consider the case when Δx is small but finite and to inquire whether one of the finite-difference formulas (2.1)–(2.3) is likely to be more accurate than the others. If $f(x)$ is sufficiently smooth, this question can be answered by expanding the terms like $f(x_0 \pm \Delta x)$ in Taylor series about x_0 and substituting these expansions into the finite-difference formula. For example, when

$$f(x_0 + \Delta x) = f(x_0) + \Delta x \frac{df}{dx}(x_0) + \frac{(\Delta x)^2}{2}\frac{d^2 f}{dx^2}(x_0) + \frac{(\Delta x)^3}{6}\frac{d^3 f}{dx^3}(x_0) + \cdots$$

is substituted into (2.1), one finds that

$$\frac{f(x_0 + \Delta x) - f(x_0)}{\Delta x} - \frac{df}{dx}(x_0) = \frac{\Delta x}{2}\frac{d^2 f}{dx^2}(x_0) + \frac{(\Delta x)^2}{6}\frac{d^3 f}{dx^3}(x_0) + \cdots . \quad (2.4)$$

The right side of (2.4) is known as the *truncation error*. The lowest power of Δx in the truncation error determines the *order of accuracy* of the finite difference. Inspection of its truncation error shows that the one-sided difference (2.1) is first-order accurate. In contrast, the truncation error associated with the centered difference (2.3) is

$$\frac{(\Delta x)^2}{6}\frac{d^3 f}{dx^3}(x_0) + \frac{(\Delta x)^4}{120}\frac{d^5 f}{dx^5}(x_0) + \cdots ,$$

and the centered difference is therefore second-order accurate. If the higher-order derivatives of f are bounded in some interval about x_0 (i.e., f is "smooth") and the grid spacing is reduced, the error in the second-order difference (2.3) will approach zero more rapidly than the error in the first-order difference (2.1). The fact that the truncation error of the centered difference is of higher order does not, however, guarantee that it will always generate a more accurate estimate of the derivative. If the function is sufficiently rough and the grid spacing sufficiently coarse, neither formula is likely to produce a good approximation, and the superiority of one over the other will be largely a matter of chance.

Higher-order finite-difference approximations can be constructed by including additional grid points in the finite-difference formula. Suppose, for example, that one wishes to obtain a fourth-order approximation to df/dx by determining the five coefficients $a, b, \ldots, e$ that satisfy

$$\frac{df}{dx}(x_0) = af(x_0 + 2\Delta x) + bf(x_0 + \Delta x) + cf(x_0)$$

$$+ d\, f(x_0 - \Delta x) + ef(x_0 - 2\Delta x) + O\left[(\Delta x)^4\right]. \quad (2.5)$$

Expanding $f(x_0 \pm \Delta x)$ and $f(x_0 \pm 2\Delta x)$ in Taylor series, substituting those expansions into (2.5), and equating the coefficients of like powers of Δx yields five equations for the unknown coefficients:

$$u + b + c + d + e = 0,$$
$$2a + b - d - 2e = 1/\Delta x,$$
$$4a + b + d + 4e = 0,$$
$$8a + b - d - 8e = 0,$$
$$16a + b + d + 16e = 0.$$

The unique solution to this system requires $c = 0$ and yields an approximation to the derivative of the form

$$\frac{df}{dx}(x_0) = \frac{4}{3}\left(\frac{f(x_0 + \Delta x) - f(x_0 - \Delta x)}{2\Delta x}\right)$$

$$- \frac{1}{3}\left(\frac{f(x_0 + 2\Delta x) - f(x_0 - 2\Delta x)}{4\Delta x}\right) + O\left[(\Delta x)^4\right]. \quad (2.6)$$

Similar procedures can be used to generate even higher-order formulae, off-centered formulae, and formulae for irregular grid intervals.

As an alternative to the brute force manipulation of Taylor series, the derivation of higher-order finite-difference formulae can be facilitated by the systematic use of operator notation and simple lower-order formulae. A simpler derivation of (2.6) may be obtained by defining a finite-difference operator δ_{nx} such that

$$\delta_{nx} f(x) = \frac{f(x + n\Delta x/2) - f(x - n\Delta x/2)}{n\Delta x}. \quad (2.7)$$

Using this notation, the second-order centered difference satisfies

$$\delta_{2x} f = \frac{df}{dx} + \frac{(\Delta x)^2}{6}\frac{d^3 f}{dx^3} + O\left[(\Delta x)^4\right]. \quad (2.8)$$

From the definition of δ_{nx},

$$\delta_x^2 f = \delta_x(\delta_x f) = \frac{f(x + \Delta x) - 2f(x) + f(x - \Delta x)}{(\Delta x)^2},$$

and a conventional Taylor series analysis of the truncation error shows that

$$\delta_x^2 f = \frac{d^2 f}{dx^2} + O\left[(\Delta x)^2\right].$$

It follows that $\delta_{2x}\delta_x^2 f$ is a second-order approximation to the third derivative of f, since

$$\delta_{2x}\delta_x^2 f = \delta_{2x}\left(\frac{d^2 f}{dx^2} + O\left[(\Delta x)^2\right]\right) = \frac{d^3 f}{dx^3} + O\left[(\Delta x)^2\right].$$

Substitution of the preceding into (2.8) yields

$$\left(1 - \frac{(\Delta x)^2}{6}\delta_x^2\right)\delta_{2x} f = \frac{df}{dx} + O\left[(\Delta x)^4\right]. \tag{2.9}$$

Expansion of this formula via the operator definition (2.7) yields the centered fourth-order difference (2.6). Although it allows finite-difference equations to be expressed in a very compact form, operator notation will not be used for all finite-difference equations throughout the remainder of this book, but will be reserved for complicated formulae that become unwieldy when written in expanded form. Most of the finite-difference schemes considered in the remainder of this chapter are sufficiently simple that they will be expressed without using operator notation.

We now turn from the consideration of individual finite differences to examine the accuracy of an entire finite-difference scheme. Suppose that an approximation to the advection equation

$$\frac{\partial \psi}{\partial t} + c\frac{\partial \psi}{\partial x} = 0 \tag{2.10}$$

is to be obtained at the grid points $(n\Delta t, j\Delta x)$, where n and j are integers. It is convenient to represent the numerical approximation to $\psi((n\Delta t, j\Delta x)$ in the shorthand notation ϕ_j^n. One possible finite-difference formula for the numerical approximation of (2.10) is

$$\frac{\phi_j^{n+1} - \phi_j^n}{\Delta t} + c\frac{\phi_j^n - \phi_{j-1}^n}{\Delta x} = 0; \tag{2.11}$$

when $c > 0$, this is known as the "upstream" or "donor-cell" scheme. The accuracy of a finite-difference scheme is characterized by the residual error with which the solution of the continuous equation fails to satisfy the finite-difference formulation. Under the assumption that ψ is sufficiently smooth, its value at adjacent grid points can be obtained from a Taylor series expansion about $(n\Delta t, j\Delta x)$ and substituted into (2.11) to yield

$$\frac{\psi_j^{n+1} - \psi_j^n}{\Delta t} + c\frac{\psi_j^n - \psi_{j-1}^n}{\Delta x} = \frac{\Delta t}{2}\frac{\partial^2 \psi}{\partial t^2} - c\frac{\Delta x}{2}\frac{\partial^2 \psi}{\partial x^2} + \cdots. \qquad (2.12)$$

The right side of (2.12) is the truncation error of the finite-difference scheme. The order of accuracy of the scheme is determined by the lowest powers of Δt and Δx appearing in the truncation error. According to (2.12), the upstream scheme is first-order accurate in space and time. If the truncation error of the finite-difference scheme approaches zero as $\Delta t \to 0$ and $\Delta x \to 0$, the scheme is *consistent*. Inspection of (2.12) clearly shows that the upstream scheme is consistent. Although it is not difficult to design consistent difference schemes, this property should not be taken for granted. One sometimes encounters methods that require additional relations between Δt and Δx, such as $\Delta t/\Delta x \to 0$, in order to achieve consistency.

2.2   Stability and Convergence

The preceding measures of accuracy do not describe the difference between the numerical solution ϕ_j^n and the true solution $\psi(n\Delta t, j\Delta x)$, which, of course, is the most direct measure of the quality of the numerical solution. Before discussing this error one needs a way to measure its size, i.e., one needs to define a *norm*. The general mathematical notation for a norm is a pair of vertical bars $\| \ \|$. In the following we will be concerned with the maximum norm and the Euclidean, or ℓ_2, norm. The maximum norm, defined as

$$\|\phi\|_\infty = \max_{1 \le j \le N} |\phi_j|, \qquad (2.13)$$

is simply the extremum of the grid-point values. The Euclidean, or ℓ_2, norm is defined as

$$\|\phi\|_2 \equiv \left(\sum_{j=1}^{N} |\phi_j|^2 \Delta x \right)^{1/2}. \qquad (2.14)$$

If the constant scaling factor Δx is ignored, (2.14) is just the length of an N-dimensional vector (hence the name Euclidean norm). The inclusion of the Δx factor makes (2.14) a numerical approximation to the square root of the spatial integral of the function times its complex conjugate, $\phi\phi^*$, whence the name ℓ_2

norm.[1] It might appear that the maximum norm is the most natural one to compute when working with grid-point values; however, the ℓ_2 norm is also useful, because it is more closely related to conserved physical quantities, such as the total energy.

A finite-difference scheme is said to be *convergent* of order (p, q) if in the limit $\Delta x, \Delta t \to 0$,

$$\|\psi(n\Delta t, j\Delta x) - \phi_j^n\| = O\left[(\Delta t)^p\right] + O\left[(\Delta x)^q\right].$$

The relationship between convergence and consistency is described by the *Lax equivalence theorem*, which states that *if a finite-difference scheme is linear, stable, and accurate of order (p, q), then it is convergent of order (p, q)* (Lax and Richtmyer 1956). Lax's theorem shows that mere consistency is not enough to assure the convergence of a numerical method. The method must also be *stable*. There are a great number of consistent finite-difference methods that are utterly useless because they are unstable. It is common practice to describe a finite-difference scheme as "unstable" if it generates a numerical solution that grows much more rapidly than the true solution. When "stable" and "unstable" are used in this sense, some reference must be made to the properties of the true solution, and since the true solution can exhibit a wide range of different behaviors, one can arrive at several different criteria for "stability."

The fundamental definition of stability makes no reference to the properties of the true solution and only identifies the least-restrictive additional constraint that must be satisfied in order to ensure the convergence of solutions generated by a consistent finite-difference scheme. A consistent linear finite-difference scheme will be convergent, and the Lax equivalence theorem will be satisfied, provided that for any time T there exists a constant C_T such that

$$\|\phi^n\| \leq C_T \|\phi^0\| \quad \text{for all} \quad n\Delta t \leq T \tag{2.15}$$

and all sufficiently small values of Δt and Δx. In the preceding, C_T may depend on the time T, but not on Δt, Δx, or the number of time steps n. This definition leaves the numerical solution tremendous latitude for growth with time, but it rules out solutions that grow as a function of the number of time steps. If a difference scheme is unstable in the sense that it fails to satisfy (2.15), repeated reductions in Δt and Δx may generate an unbounded amplification in the numerical approximation to the true solution at time T. In such a situation, the numerical

---

[1] To better appreciate the notation used to represent the maximum and ℓ_2 norms, note that $\|\phi\|_\infty$ is essentially the integral

$$\left(\int |\phi|^\infty \, dx\right)^{1/\infty}$$

and $\|\phi\|_2$ is

$$\left(\int |\phi|^2 \, dx\right)^{1/2}.$$

solution could hardly be expected to converge to the true solution in the limit Δx, $\Delta t \to 0$.

The practical shortcoming of the preceding definition of stability is that it says nothing about the quality of the solution that might be obtained when using finite values of Δt and Δx; it only ensures that an accurate solution will be obtained in the limit Δx, $\Delta t \to 0$. Schemes that are stable according to the criteria (2.15) may, nevertheless, generate solutions that "blow up" in practical applications (see Section 3.4.3 for an example). In order to ensure that the numerical solution is qualitatively similar to the true solution when Δx and Δt are finite, it is often useful to impose stability constraints that are more stringent than (2.15). In many wave propagation problems, the norm of the true solution is constant with time, and in such instances it is appropriate to require that the numerical scheme satisfy

$$\|\phi^n\| \leq \|\phi^0\| \quad \text{for all} \quad n. \tag{2.16}$$

In contrast to (2.15), this condition is not necessary for convergence, and it cannot be sensibly imposed without specific knowledge about the boundedness of the solutions to the associated partial differential equation. Nevertheless, (2.16) is a perfectly reasonable constraint to impose in applications where the true solution is not growing with time, and unlike (2.15), it guarantees that the solution will not blow up.

It is relatively easy to formulate consistent difference schemes and to determine their truncation error and order of accuracy. The analysis of stability can, however, be far more difficult, particularly when the finite-difference scheme and the associated partial differential equation are nonlinear. Thus, our initial discussion of stability will be focused on the simplest case—linear finite-difference schemes for the approximation of linear partial differential equations with constant coefficients. Nonlinear equations and linear equations with variable coefficients will be considered in Chapter 3.

2.2.1   The Energy Method

In practice, the energy method is used much less frequently than the Von Neumann method, which will be discussed in the next section. Nevertheless, the energy method is important, because unlike the Von Neumann method, it can be applied to nonlinear equations and to problems without periodic boundaries. The basic idea behind the energy method is to find a positive definite quantity like $\sum_j (\phi_j^n)^2$ and show that this quantity is bounded for all n. If $\sum_j (\phi_j^n)^2$ is bounded, the solution is stable with respect to the ℓ_2-norm.

As an example, let us investigate the stability of the upstream finite-difference scheme (2.11). Defining $\mu = c\Delta t / \Delta x$, the scheme may be written as

$$\phi_j^{n+1} = (1 - \mu)\phi_j^n + \mu\phi_{j-1}^n. \tag{2.17}$$

Squaring both sides and summing over all j gives

$$\sum_j (\phi_j^{n+1})^2 = \sum_j \left[(1-\mu)^2 (\phi_j^n)^2 + 2\mu(1-\mu)\phi_j^n \phi_{j-1}^n + \mu^2 (\phi_{j-1}^n)^2 \right].$$

(2.18)

Assuming cyclic boundary conditions,[2]

$$\sum_j (\phi_{j-1}^n)^2 = \sum_j (\phi_j^n)^2,$$

(2.19)

and using the Schwarz inequality (which states that for two vectors $\mathbf{u}$ and $\mathbf{v}$, $|\mathbf{u} \cdot \mathbf{v}| \leq \|\mathbf{u}\| \|\mathbf{v}\|$),

$$\sum_j \phi_j^n \phi_{j-1}^n \leq \left[\sum_j (\phi_j^n)^2 \right]^{1/2} \left[\sum_j (\phi_{j-1}^n)^2 \right]^{1/2} = \sum_j (\phi_j^n)^2.$$

(2.20)

If $\mu(1-\mu) \geq 0$, all three coefficients in (2.18) are positive, and (2.19) and (2.20) may be used to construct the inequality

$$\sum_j (\phi_j^{n+1})^2 \leq \left[(1-\mu)^2 + 2\mu(1-\mu) + \mu^2 \right] \sum_j (\phi_j^n)^2 = \sum_j (\phi_j^n)^2, \quad (2.21)$$

which requires $\|\phi^{n+1}\|_2 \leq \|\phi^0\|_2$ and implies that the scheme is stable. The condition used to obtain (2.21),

$$\mu(1-\mu) \geq 0,$$

(2.22)

is therefore a sufficient condition for stability. Under the assumption that $\mu > 0$, division of (2.22) by μ leads to the relation $\mu \leq 1$, and the total constraint on μ is therefore $0 < \mu \leq 1$. A similar treatment of the case $\mu \leq 0$ leads to the contradictory requirement that $\mu \geq 1$ and provides no additional solutions. Thus, recalling the definition of μ and noting that $\mu = 0$ satisfies (2.22), the stability condition may be written

$$0 \leq \frac{c\Delta t}{\Delta x} \leq 1.$$

As is typical with most conditionally stable difference schemes, there is a maximum limit on the time step beyond which the scheme is unstable, and the stability limit becomes more severe as the spatial resolution is increased.

---

[2]If more general boundary conditions are imposed at the edges of the spatial domain, a rigorous stability analysis becomes much more difficult. The determination of stability in the presence of nonperiodic boundaries is discussed in Section 8.1.6.

2.2.2   Von Neumann's Method

One drawback of the energy method is that each new problem requires fresh insight in order to define an appropriate energy and to show that the finite-difference scheme preserves a bound on that energy. Von Neumann's method has the advantage that it can be applied by following a prescribed procedure; however, it is applicable only to linear finite-difference equations with constant coefficients.[3] The basic idea of the Von Neumann method is to represent the discretized solution at some particular time step by a finite Fourier series of the form

$$\phi_j^n = \sum_{k=-N}^{N} a_k^n e^{ikj\Delta x},$$

and to examine the stability of the individual Fourier components. The total solution will be stable if and only if every Fourier component is stable. The use of finite Fourier series is strictly appropriate only if the spatial domain is periodic. When problems are posed with more general boundary conditions, a rigorous stability analysis is more difficult, but the Von Neumann method still provides a useful way of weeding out obviously unsuitable schemes.

A key property of Fourier series is that individual Fourier modes are eigenfunctions of the derivative operator, i.e.,

$$\frac{d}{dx} e^{ikx} = ik e^{ikx}.$$

Finite Fourier series have an analogous property in that individual modes $e^{ikj\Delta x}$ are eigenfunctions of linear finite-difference operators. Thus, if the initial conditions for some linear, constant-coefficient finite-difference scheme are $\phi_j^n = e^{ikj\Delta x}$, after one iteration the solution will have the form

$$\phi_j^{n+1} = A_k e^{ikj\Delta x},$$

where A_k is a complex constant, known as the *amplification factor*, that is determined by the form of the finite-difference formulae. Since the analysis is restricted to linear constant-coefficient schemes, the amplification factor will not vary from time step to time step, and if a_k^n denotes the amplitude of the kth finite Fourier component at the nth time step, then

$$a_k^n = A_k a_k^{n-1} = (A_k)^n a_k^0.$$

---

[3]In order to apply Von Neumann's method to more general problems, the governing finite-difference equations must be linearized and any variable coefficients must be frozen at some constant value. The Von Neumann stability of the family of linearized, frozen-coefficient systems may then be examined. See Section 3.5.

It follows that the stability of each Fourier component is determined by the modulus of its amplification factor.

The *Von Neumann stability condition*, which is necessary and sufficient for the stability of a linear constant-coefficient finite-difference equation,[4] requires the amplification factor of every Fourier component resolvable on the grid to be bounded such that

$$|A_k| \leq 1 + \gamma \Delta t, \tag{2.23}$$

where γ is a constant independent of k, Δt, and Δx. This condition ensures that a consistent finite-difference scheme satisfies the minimum stability criteria for convergence in the limit Δx, $\Delta t \rightarrow 0$, (2.15). In applications where the true solution is bounded by the norm of the initial data, it is usually advantageous to enforce the more stringent requirement that

$$|A_k| \leq 1, \tag{2.24}$$

which will guarantee satisfaction of the stability condition (2.16). When the Von Neumann condition is satisfied, every finite Fourier component is stable, and the full solution, being a linear combination of the individual Fourier components, must also be stable.

As an illustration of the Von Neumann method, consider once again the finite-difference equation (2.17). The solutions to the associated partial differential equation (2.10) do not grow with time, so we will require $|A_k| \leq 1$. Substitution of an arbitrary Fourier component, of the form $e^{ikj\Delta x}$, into (2.17) yields

$$A_k e^{ikj\Delta x} = (1 - \mu)e^{ikj\Delta x} + \mu e^{ik(j-1)\Delta x}.$$

Dividing out the common factor $e^{ikj\Delta x}$ gives

$$A_k = 1 - \mu + \mu e^{-ik\Delta x}. \tag{2.25}$$

The magnitude of A_k is obtained by multiplying by its complex conjugate and taking the square root. Thus,

$$\begin{aligned}
|A_k|^2 &= (1 - \mu + \mu e^{-ik\Delta x})(1 - \mu + \mu e^{ik\Delta x}) \\
&= 1 - 2\mu(1 - \mu)(1 - \cos k\Delta x).
\end{aligned} \tag{2.26}$$

The Von Neumann condition (2.24) will therefore be satisfied if

$$1 - 2\mu(1 - \mu)(1 - \cos k\Delta x) \leq 1.$$

---

[4]The sufficiency of the Von Neumann condition holds only for single equations in one unknown. The stability of systems of finite-difference equations in several unknown variables is discussed in Section 3.1.

Since $1 - \cos k\Delta x > 0$ for all wave numbers except the trivial case $k = 0$, the preceding inequality reduces to

$$\mu(1 - \mu) \geq 0,$$

which is identical to the condition (2.22) obtained using the energy method. As discussed previously in connection with (2.22), this stability condition may be expressed as

$$0 \leq \frac{c\Delta t}{\Delta x} \leq 1.$$

Inspection of (2.26) shows that the $2\Delta x$ wave grows most rapidly in any integration performed with an unstable value of μ. Thus, as it "blows up," an unstable solution becomes dominated by large-amplitude $2\Delta x$ waves. Most other finite-difference approximations to the advection equation exhibit the same tendency: When solutions become unstable, they usually become contaminated by large-amplitude short waves. The upstream scheme is, nevertheless, unusual in that all waves become unstable for the same critical value of μ. In many other schemes, such as the leapfrog-time centered-space formulation (2.91), there exist values of μ for which only a few of the shorter wavelengths are unstable. One might suppose that such nominally unstable values of μ could still be used in numerical integrations if the initial data were filtered to remove all amplitude from the unstable finite Fourier components; however, even if the initial data have zero amplitude in the unstable modes, round-off error in the numerical computations will excite the unstable modes and trigger the instability.

2.2.3 The Courant–Fredrichs–Lewy Condition

The basic idea of the Courant–Fredrichs–Lewy (CFL) condition is that the solution of a finite-difference equation must not be independent of the data that determines the solution to the associated partial differential equation. The CFL condition can be made more precise by defining the *domain of influence* of a point (x_0, t_0) as that region of the x-t plane where the solution to some particular partial differential equation is influenced by the solution at (x_0, t_0). A related concept, the *domain of dependence* of a point (x_0, t_0), is defined as the set of points containing (x_0, t_0) within their domains of influence. The domain of dependence of (x_0, t_0) will therefore consist of all points (x, t) at which the solution has some influence on the solution at (x_0, t_0). A similar concept applicable to the discretized problem is the *numerical domain of dependence* of a grid point $(n_0\Delta t, j_0\Delta x)$, which consists of the set of all nodes on the space–time grid $(n\Delta t, j\Delta x)$ at which the value of the numerical solution influences the numerical solution at $(n_0\Delta t, j_0\Delta x)$. *The CFL condition requires that the numerical domain of dependence of a finite-difference scheme include the domain of dependence of the associated partial differential equation.* Satisfaction of the CFL condition is a necessary condition for stability, but is not sufficient to guarantee stability.

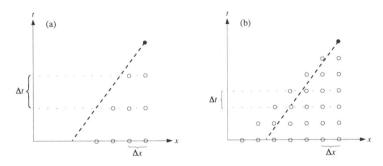

FIGURE 2.1. The influence of the time step on the relationship between the numerical domain of dependence of the upstream scheme (open circles) and the true domain of dependence of the advection equation (heavy dashed line): (a) unstable Δt, (b) stable Δt.

The nature of the CFL condition can be illustrated by considering the advection equation (2.10), which has general solutions of the form $\psi(x - ct)$. Thus the true domain of influence of a point (x_0, t_0) is the straight line

$$t = t_0 + \frac{1}{c}(x - x_0), \quad t \geq t_0.$$

The same "characteristic line" also defines the true domain of dependence of (x_0, t_0), except that one looks backward in time by requiring $t \leq t_0$. The true domain of dependence is plotted as a dashed line in Fig. 2.1, together with those grid points composing the numerical domain of dependence of the upstream finite-difference scheme (2.17). The two panels in this figure show the influence of two different time steps on the shape of the numerical domain of dependence. In Fig. 2.1a, the initial value of ψ along the x-axis, which determines the solution to the partial differential equation at $(n\Delta t, j\Delta x)$, plays no role in the determination of the finite-difference solution ϕ_j^n. The numerical solution can be in error by any arbitrary amount, and will not converge to the true solution as $\Delta t, \Delta x \to 0$ unless there is a change in the ratio $\Delta t/\Delta x$. Hence, the finite-difference method, which is consistent with the original partial differential equation, must be unstable (or else the Lax equivalence theorem would be violated).

The situation shown in Fig. 2.1b is obtained by halving the time step. Then the numerical domain of dependence contains the domain of dependence of the true solution, and it is possible for the numerical solution to be stable. In this example the CFL condition requires the slope of the characteristic curve to be greater than the slope of the left edge—and less than the slope of the right edge—of the domain of dependence. As evident from Fig. 2.1, the slope condition at the right edge of the domain is $1/c \leq \infty$, which is always satisfied. The slope condition at the left edge of the domain may be expressed as $\Delta t/\Delta x \leq 1/c$. If $c > 0$, this requires

$$c\Delta t/\Delta x \leq 1, \tag{2.27}$$

and the nonnegativity of Δt and Δx implies

$$c\Delta t/\Delta x \geq 0. \tag{2.28}$$

Simultaneous satisfaction of (2.27) and (2.28) is obtained when

$$0 \leq c\frac{\Delta t}{\Delta x} \leq 1.$$

In the case $c < 0$, similar reasoning leads to contradictory requirements, and the solution is unstable.

The preceding stability condition is identical to those already obtained using the energy and Von Neumann methods, but such agreement is actually rather unusual. The CFL condition is only a necessary condition for stability, and in many cases the sufficient conditions for stability are more restrictive than those required by the CFL condition. As an example, consider the following approximation to the advection equation,

$$\delta_{2t}\phi + c\left(\frac{4}{3}\delta_{2x}\phi - \frac{1}{3}\delta_{4x}\phi\right) = 0,$$

which uses the fourth-order accurate approximation to the spatial derivative (2.6). Since the spatial difference utilizes a five-grid-point-wide stencil, the CFL condition is satisfied when

$$\left|c\frac{\Delta t}{\Delta x}\right| \leq 2.$$

Yet the actual sufficient condition for stability is the much more restrictive condition

$$\left|c\frac{\Delta t}{\Delta x}\right| \leq 0.728,$$

which may be derived via a Von Neumann stability analysis.

2.3   Time-Differencing

A given partial differential equation can be approximated by an almost unlimited variety of different finite-difference formulae. In order to systematically examine the properties of various finite-difference schemes, let us begin by discussing possible approximations to the time derivative without explicitly considering the spatial derivatives. The primary reason for discussing time and space differencing separately is that they present the numerical analyst with rather different sets of practical problems. After the nth step of the integration, the numerical solution ϕ will be known at every point on the spatial mesh, and several grid-point values may be easily included in any finite-difference approximation to the spatial derivatives. It is easy, for example, to construct high-order centered approximations to spatial derivatives. In contrast, storage limitations dictate that ϕ be retained at as few time levels as possible, and the only time levels available are those from previous iterations. Thus, higher-order finite-difference approximations to the time derivative are inherently one-sided.

The following discussion of the effects of time-differencing on the numerical solution is transferable, after minor modification, to situations where the spatial derivatives are approximated by centered differences. In addition, this discussion provides an exact analysis of the influence of time-differencing on schemes, such as the spectral method, where finite differences are not used to evaluate the spatial derivatives. Nevertheless, one must be careful not to assume that space and time differences are completely independent. Indeed, techniques such as the Lax–Wendroff method cannot be properly analyzed without understanding the interaction between space truncation error and time truncation error. The combined effects of space- and time-differencing will be discussed, together with schemes like the Lax–Wendroff method, in Section 2.5.

Time-differencing formulae used in the numerical solution of partial differential equations are related, naturally enough, to the numerical methods used to integrate ordinary differential equations. In comparison with typical ordinary differential equation solvers, the methods used to integrate partial differential equations are of very low order. Low-order schemes are used for two basic reasons. First, the approximation of the time derivative is not the only source of finite-differencing error in the solution of partial differential equations; other errors arise through the approximation of the spatial derivatives. In many circumstances the largest errors in the solution are introduced through the numerical evaluation of the spatial derivatives, so it is pointless to devote additional computational resources to higher-order time-differencing. The second reason for using low-order methods is that practical limitations on computational resources often leave no other choice.

2.3.1   The Oscillation Equation: Phase-Speed and Amplitude Error

Hundreds of papers have been written investigating various techniques for the finite-difference solution of the advection equation (2.10), many of which are listed in the extensive review by Rood (1987). The vastness of this body of literature is a testament to the subtle tradeoffs involved in the selection of the "best" numerical method for even very simple equations. It might be supposed that the relative accuracy of different methods could be easily determined by comparing their respective truncation errors. The analysis of truncation error is, however, most effective at predicting the behavior of well-resolved waves, and the most serious errors are often found in the poorly resolved waves. The accuracy of both well resolved and poorly resolved waves can be better examined by extending the standard Von Neumann analysis to study the phase-speed and amplitude error in each Fourier component of the numerical solution.

If the spatial structure of the solution to the advection equation

$$\frac{\partial \psi}{\partial t} + c\frac{\partial \psi}{\partial x} = 0$$

is represented by a Fourier series (or a Fourier Integral), the amplitude of an individual Fourier mode $b_k(t)$ must satisfy

$$\frac{db_k}{dt} = -ikcb_k. \tag{2.29}$$

The preceding is an ordinary differential equation that can be used to examine the numerical error introduced by time-differencing in wave-propagation problems. Equation (2.29) is a specific example of the *oscillation equation*

$$\frac{d\psi}{dt} = i\kappa\psi, \tag{2.30}$$

where κ is a real constant representing a frequency. Integrating the oscillation equation over a time Δt yields

$$\psi(t_0 + \Delta t) = e^{i\kappa\Delta t}\psi(t_0) \equiv A_e\psi(t_0). \tag{2.31}$$

Here the last relation defines an "exact amplification factor" A_e, which is a complex number of modulus one. According to (2.31), ψ moves $\kappa\Delta t$ radians around a circle of radius $|\psi(t_0)|$ in the complex plane over the time interval Δt.

Suppose that a finite-difference scheme is used to compute an approximate solution ϕ to the oscillation equation. A numerical amplification factor may be defined such that $\phi^{n+1} = A\phi^n$, where ϕ^n is the numerical approximation to $\psi(n\Delta t)$. The standard Von Neumann stability analysis seeks a yes–no answer to the question, Is $|A| \le |A_e| = 1$? Additional information about the amplitude and phase-speed error in the numerical solution can, however, be obtained by writing the numerical amplification factor in the form $|A|e^{i\theta}$, where

$$|A| = (\Re\{A\}^2 + \Im\{A\}^2)^{1/2} \quad \text{and} \quad \theta = \arctan\left(\frac{\Im\{A\}}{\Re\{A\}}\right).$$

Phase-speed errors arise from the difference between the argument of the finite-difference amplification factor θ and the correct value of $\kappa\Delta t$. A useful way to characterize phase-speed error is through the relative phase change $R = \theta/(\kappa\Delta t)$, which is the ratio of the phase advance produced by one time step of the finite-difference scheme divided by the change in phase experienced by the true solution over the same time interval. If $R > 1$, the finite-difference scheme is *accelerating*; if $R < 1$, the scheme is *decelerating*.

Amplitude errors arise from the difference between the modulus of the finite-difference amplification factor $|A|$ and the correct value of unity. When $|A| = 1$, the scheme is *neutral*. If $|A| < 1$, the scheme is *damping*; and if $|A| > 1$, it is *amplifying*. Amplifying schemes are certainly unstable in the sense that they generate approximate solutions to the oscillation equation that blow up, whereas the correct solution remains bounded by $|\phi^0|$. A more subtle characterization of the stability of amplifying schemes, which will be considered in Sections 2.3.2 and 2.5, concerns the extent to which they can produce approximate solutions to ordinary and partial differential equations that converge in the limit $\Delta t \to 0$ and $\Delta x \to 0$.

2.3.2   Single-Stage Two-Level Schemes

The simplest techniques for the solution of the differential equation

$$\frac{d\psi}{dt} = F(\psi) \tag{2.32}$$

are members of the general family of single-stage two-time-level schemes, which may be written in the form

$$\frac{\phi^{n+1} - \phi^n}{\Delta t} = \alpha F(\phi^n) + \beta F(\phi^{n+1}). \tag{2.33}$$

Here ϕ^n is the numerical approximation to $\psi(n\Delta t)$, and $\alpha + \beta = 1$ for consistency with the original equation. The preceding general form includes several well-known elementary methods. When $\alpha = 1$, $\beta = 0$, the scheme is known as forward differencing or Euler's method. Backward differencing corresponds to the case $\alpha = 0$, $\beta = 1$, and the trapezoidal method is obtained when $\alpha = \beta = \frac{1}{2}$.

The finite-difference scheme (2.33) may be alternatively expressed in the form

$$\phi^{n+1} = \phi^n + \Delta t \left(\alpha F(\phi^n) + \beta F(\phi^{n+1}) \right), \tag{2.34}$$

which is analogous to the integrated form of the original differential equation

$$\psi[(n+1)\Delta t] = \psi(n\Delta t) + \int_{n\Delta t}^{(n+1)\Delta t} F(\psi(t))\, dt.$$

Although (2.33) and (2.34) are clearly equivalent, confusion sometimes arises in determining the accuracy of numerical methods expressed in the integrated form (2.34). If the numerical method is convergent of order r and if ψ is the solution to the continuous problem, expansion of ψ in a Taylor series and substitution of that series into the discrete derivative form (2.33) will yield a residual error of $O[(\Delta t)^r]$, whereas substitution into the discrete integral form (2.34) will yield an error of $O[(\Delta t)^{r+1}]$. Of course, if the solutions to (2.33) and (2.34) converge to the true solution as $\Delta t \to 0$, they must converge at the same rate, because the two schemes are algebraically equivalent. This rate of convergence is equal to the order of the *global truncation error*, which is the same order as the truncation error associated with (2.33). The $O[(\Delta t)^{r+1}]$ truncation error associated with the integral form (2.34) is the *local truncation error*, or *one-step error*, and represents the error introduced in each step of the integration, i.e., the difference between ψ^{n+1} as computed by one step of the finite-difference method and the exact solution to the differential equation at $t = (n+1)\Delta t$ subject to the initial condition

$\psi(n\Delta t) = \psi^n$. The rate at which the solution of a *stable* finite-difference scheme converges to the true solution as $\Delta t \to 0$ is one power of Δt lower than the order of the local truncation error because, roughly speaking, the global error generated by a stable scheme during an integration over a time interval T is the cumulative sum of $T/\Delta t$ local errors. When the term "truncation error" is used without qualification in this book, it will refer to the global truncation error. The truncation error of all members of the family of schemes (2.33) is $O(\Delta t)$, except for the trapezoidal method, which is $O\left[(\Delta t)^2\right]$.

Application of (2.34) to the oscillation equation (2.30) yields

$$(1 - i\beta\kappa\Delta t)\phi^{n+1} = (1 + i\alpha\kappa\Delta t)\phi^n.$$

The amplification factor for this scheme is

$$A \equiv \frac{\phi^{n+1}}{\phi^n} = \frac{1 + i\alpha\kappa\Delta t}{1 - i\beta\kappa\Delta t}.$$

Multiplying A by its complex conjugate gives

$$|A|^2 = \frac{1 + \alpha^2\kappa^2\Delta t^2}{1 + \beta^2\kappa^2\Delta t^2}$$

$$= 1 + (\alpha^2 - \beta^2)\frac{\kappa^2\Delta t^2}{1 + \beta^2\kappa^2\Delta t^2}. \tag{2.35}$$

Inspection of (2.35) shows that the scheme is neutral when $\alpha = \beta$, damping when $\alpha < \beta$, and amplifying when $\alpha > \beta$. The amplification produced when $\alpha > \beta$ is clearly unstable in the sense that approximate solutions computed with finite Δt can generate floating-point overflows on digital computers, whereas the magnitude of the correct solution is bounded by $|\phi^0|$. Note, however, that the amplification factor for forward differencing satisfies the general Von Neumann stability condition (2.23), since for $\Delta t \leq 1$,

$$|A|_{\text{forward}} = 1 + (\kappa\Delta t)^2 \leq |A|^2_{\text{forward}} \leq 1 + \kappa^2\Delta t.$$

to yield

$$|A|_{\text{forward}} \leq |A|^2_{\text{forward}} = 1 + (\kappa\Delta t)^2 \leq 1 + \kappa^2\Delta t$$

As a consequence, the amplifying solutions obtained using forward differencing do converge to the correct solution of the oscillation equation as $\Delta t \to 0$. Forward differencing also generates convergent approximations to most other ordinary differential equations, but convergence is not guaranteed, and (2.23) is not satisfied, when forward time differencing is used in conjunction with centered-difference approximations to the spatial derivative in the advection equation. This point is discussed further in Section 2.5.

The amplitude and phase errors in the approximate solution are functions of the *numerical resolution*. The solution to the governing differential equation (2.30) oscillates with a period $T = 2\pi/\kappa$. An appropriate measure of numerical resolution is the number of time steps per oscillation period, $T/\Delta t$. The numerical resolution is improved by decreasing the step size. In the limit of very good numerical resolution, $T/\Delta t \to \infty$ and $\kappa\Delta t \to 0$. Assuming good numerical resolution,

Taylor series expansions, such as

$$(1 + x)^{1/2} = 1 + \frac{x}{2} - \frac{x^2}{8} + \cdots \quad \text{for} \quad |x| < 1,$$

may be used to reduce (2.35) to

$$|A| \approx 1 + \tfrac{1}{2}(\alpha^2 - \beta^2)(\kappa \Delta t)^2.$$

It follows that

$$|A|_{\text{forward}} \approx 1 + \tfrac{1}{2}(\kappa \Delta t)^2 \quad \text{and} \quad |A|_{\text{backward}} \approx 1 - \tfrac{1}{2}(\kappa \Delta t)^2, \qquad (2.36)$$

indicating that the spurious amplitude changes introduced by both forward differencing and backward differencing are $O[(\kappa \Delta t)^2]$.

The relative phase change in the family of single-stage two-level schemes is

$$R = \frac{1}{\kappa \Delta t} \arctan \left(\frac{(\alpha + \beta)\kappa \Delta t}{1 - \alpha\beta(\kappa \Delta t)^2} \right).$$

Thus,

$$R_{\text{forward}} = R_{\text{backward}} = \frac{\arctan \kappa \Delta t}{\kappa \Delta t}, \qquad (2.37)$$

which ranges between 0 and 1, implying that both forward differencing and backward differencing are decelerating. Assuming, once again, that the numerical solution is well-resolved, the preceding expression for the phase-speed error may be approximated using the Taylor series expansions, such as

$$\arctan x = x - \frac{x^3}{3} + \frac{x^5}{5} - \cdots \quad \text{for} \quad |x| < 1,$$

to obtain

$$R_{\text{forward}} = R_{\text{backward}} \approx 1 - \frac{(\kappa \Delta t)^2}{3}.$$

The phase-speed error, like the amplitude error, is $O[(\Delta t)^2]$.

The trapezoidal scheme gives the best results; it generates no amplitude error, and its relative phase change is

$$R_{\text{trapezoidal}} = \frac{1}{\kappa \Delta t} \arctan \left(\frac{\kappa \Delta t}{1 - \kappa^2 \Delta t^2 / 4} \right).$$

For small values of $\kappa \Delta t$, this may be approximated using Taylor series expansions as

$$R_{\text{trapezoidal}} \approx \frac{1}{\kappa \Delta t} \arctan \left(\kappa \Delta t \left(1 + \frac{\kappa^2 \Delta t^2}{4} \right) \right) \approx 1 - \frac{\kappa^2 \Delta t^2}{12}.$$

As with forward differencing and backward differencing, the trapezoidal scheme retards the phase change of well-resolved oscillations. However, the deceleration is only $\frac{1}{4}$ as great as that produced by the other schemes.

Although the trapezoidal scheme is accurate and unconditionally stable, it suffers from one serious disadvantage: It requires the evaluation of $F(\phi^{n+1})$ during the computation of ϕ^{n+1}. A scheme such as the trapezoidal method, in which the calculation of ϕ^{n+1} depends on $F(\phi^{n+1})$, is known as an *implicit* method. If the calculation of ϕ^{n+1} does not depend on $F(\phi^{n+1})$, the scheme is *explicit*. In the case of the oscillation equation, implicitness is a trivial complication. However, if F is a nonlinear function, any implicit finite-difference scheme will convert the differential equation into a nonlinear algebraic equation for ϕ^{n+1}. In the general case, the solution to this nonlinear equation must be obtained by some iterative technique. Thus, implicit finite-difference schemes generally require much more computation per individual time step than do similar explicit methods. Some of that extra computation may be offset if the implicit method is unconditionally stable, in which case the step size is determined solely by accuracy considerations, and the implicit time step can sometimes be much larger than the maximum stable time step of comparable explicit schemes.

2.3.3 Multistage Methods

All stable schemes of the form (2.33) have the disadvantage that they are implicit. Moreover, all except the trapezoidal scheme are only of first order. Is there a stable, accurate scheme that is not implicit? One may attempt to construct such a scheme by evaluating the function F in (2.32) at additional points in the interval $(n\Delta t, (n+1)\Delta t)$ and using this extra information to improve the accuracy of the calculation. These schemes are often referred to as multistage methods, because each integration step may require the estimation of ψ at several intermediate times, or "stages," before a final approximation to $\psi((n+1)\Delta t)$ is obtained. Each stage involves an additional evaluation of F, the right side of the differential equation. The family of consistent two-stage schemes may be written in the general form

$$\tilde{\phi}^{n+\alpha} = \phi^n + \alpha \Delta t F(\phi^n),$$
$$\phi^{n+1} = \phi^n + \beta \Delta t F(\tilde{\phi}^{n+\alpha}) + (1-\beta)\Delta t F(\phi^n).$$

Here $\tilde{\phi}^{n+\alpha}$ is an "intermediate" approximation to $\psi[(n+\alpha)\Delta t]$. One might attempt to choose α and β to maximize the *order* of the local truncation error. This criterion does not produce a unique solution, but rather leads to the requirement $\alpha\beta = \frac{1}{2}$. The family of schemes satisfying $\alpha\beta = \frac{1}{2}$ compose the set of second-order *Runge–Kutta* methods. One particular member of the Runge–Kutta family is the *Heun* method, obtained by setting $\alpha = 1$, $\beta = \frac{1}{2}$. The Heun method creates a trapezoidal-like approximation to the integral of F, but differs from the true trapezoidal method because $F(\phi^{n+1})$ is replaced by the estimate $F(\tilde{\phi}^{n+1})$. Another Runge–Kutta scheme is the *midpoint* method, in which $\alpha = \frac{1}{2}$ and $\beta = 1$. An example of a non-Runge–Kutta scheme is the *Matsuno*, or *forward–backward*, method, for which $\alpha = 1$ and $\beta = 1$ (Matsuno 1966b).

If the preceding multistage formula is applied to the oscillation equation, the result is

$$\phi^{n+1} = \phi^n + \beta i \kappa \Delta t (\phi^n + \alpha i \kappa \Delta t \phi^n) + (1 - \beta) i \kappa \Delta t \phi^n. \tag{2.38}$$

The amplification factor is

$$A = 1 + i \kappa \Delta t - \alpha \beta (\kappa \Delta t)^2,$$

and

$$|A|^2 = 1 + (1 - 2\alpha\beta)(\kappa \Delta t)^2 + \alpha^2 \beta^2 (\kappa \Delta t)^4, \tag{2.39}$$

which shows that the set of second-order Runge–Kutta schemes (i.e., those schemes for which $\alpha\beta = \frac{1}{2}$) have $O[(\Delta t)^4]$ amplitude error, whereas the amplitude error in the two-stage non-Runge–Kutta schemes is $O[(\Delta t)^2]$. Unfortunately, all the Runge–Kutta schemes are unstable, since in the limit of good numerical resolution,

$$|A|_{\mathrm{R-K2}} \approx 1 + \tfrac{1}{8}(\kappa \Delta t)^4.$$

Although the Runge–Kutta schemes are unstable, the growth is $O[(\Delta t)^4]$. At a given step size, the erroneous amplification produced by the second-order Runge–Kutta methods will be much weaker than the $O[(\Delta t)^2]$ growth produced by forward time-differencing (see [2.36]). The slow growth generated by the Runge–Kutta methods can sometimes be tolerated if $\kappa \Delta t$ is sufficiently small and the total length of the integration is sufficiently short.[5]

Many physical systems contain several different modes, each oscillating at a different frequency. When simulating these systems, the highest-frequency components of the numerical solution are likely to be most seriously in error because of their poor numerical resolution. It is precisely these poorly resolved features that amplify most rapidly in the Runge–Kutta solutions. The amplification of the highest-frequency components can be prevented by choosing other values for α and β, although such a choice also increases the truncation error. According to (2.39), very low frequency oscillations ($\kappa \Delta t \ll 1$) will be stable whenever $\alpha\beta$ is greater than $\frac{1}{2}$. Matsuno (1966b) suggested setting $\alpha = 1$, $\beta = 1$, in which case (2.39) becomes

$$|A|^2_{\mathrm{Matsuno}} = 1 - (\kappa \Delta t)^2 + (\kappa \Delta t)^4. \tag{2.40}$$

The Matsuno scheme damps the solution whenever $0 < \kappa \Delta t < 1$. Differentiation of (2.40) with respect to $\kappa \Delta t$ shows that the maximum damping occurs when

---

[5]As explored in Problem 15, the auxiliary relation $O(\Delta t) \leq O[(\Delta x)^{4/3}]$ may be required to ensure the convergence of finite-difference approximations to the advection equation when the time difference is evaluated by a second-order Runge–Kutta scheme.

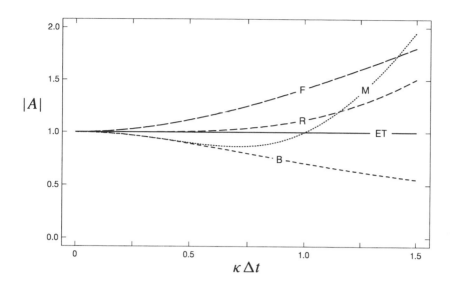

FIGURE 2.2. The modulus of the amplification factor $|A|$ as a function of temporal resolution $\kappa \Delta t$ for the true solution and five two-level schemes: exact solution and trapezoidal method (ET), forward differencing (F), backward differencing (B), second-order Runge–Kutta (R), and Matsuno (M).

$\kappa \Delta t = 1/\sqrt{2}$. Thus, if the time step is chosen such that $0 \le \kappa \Delta t \le 1/\sqrt{2}$ for all frequencies κ in the physical system, Matsuno time-differencing will preferentially damp the highest-frequency waves. The damping properties of the Matsuno scheme have been exploited to eliminate high-frequency gravity waves generated during the initialization of weather prediction models. The standard Matsuno scheme produces too much damping, however, for most nonspecialized applications. The fourth-order Runge–Kutta scheme (see Section 2.3.6) may also be used to preferentially damp high-frequency modes, and in most instances it would be a better choice than the Matsuno scheme because it is more efficient and far more accurate.

The amplitude errors generated by the preceding two-level schemes are compared in Fig. 2.2. The strong damping associated with the backward and Matsuno schemes is evident, along with the rapid amplification produced by forward differencing. These relatively large errors may be contrasted with the significantly weaker amplification produced by the second-order Runge–Kutta method and the neutral amplification of the trapezoidal method.

The relative phase change associated with the general two-stage scheme (2.38) is

$$R = \frac{1}{\kappa \Delta t} \arctan \left(\frac{\kappa \Delta t}{1 - \alpha \beta (\kappa \Delta t)^2} \right).$$

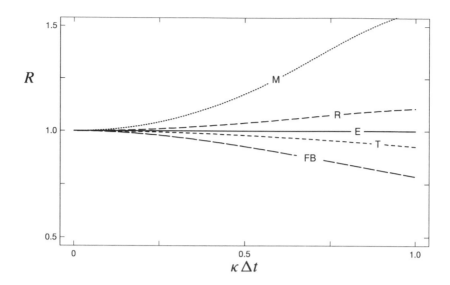

FIGURE 2.3. Relative phase change R as a function of temporal resolution $\kappa \Delta t$ for the true solution and five two-level schemes: exact solution (E), trapezoidal method (T), forward and backward differencing (FB), second-order Runge–Kutta (R), and Matsuno (M).

In the limit of good numerical resolution, the relative phase changes produced by second-order Runge–Kutta schemes and the Matsuno scheme are

$$R_{\text{R–K2}} \approx 1 + \tfrac{1}{6}(\kappa \Delta t)^2, \qquad R_{\text{Matsuno}} \approx 1 + \tfrac{2}{3}(\kappa \Delta t)^2.$$

The relative phase change for several two-level schemes is plotted as a function of temporal resolution in Fig. 2.3. The Matsuno and second-order Runge–Kutta schemes are accelerating, whereas the forward, backward, and trapezoidal schemes are decelerating.

2.3.4   Three-Level Schemes

As an alternative to multistage methods, information from several earlier time levels can be incorporated into the integration formula. This increases the storage requirements of the scheme, but it avoids the necessity of performing more than one evaluation of the right side per time step. According to the terminology developed for ordinary differential equations, these are *multistep* methods. A typical multistep ordinary differential equation solver might use data from a half dozen preceding time levels. The large storage requirements of many atmospheric and ocean models have, however, discouraged researchers from using data from more than two earlier time levels in their time-differencing schemes. Let us therefore consider the family of three-time-level schemes.

The general form for an explicit three-time-level method is

$$\phi^{n+1} = \alpha_1 \phi^n + \alpha_2 \phi^{n-1} + \beta_1 \Delta t F(\phi^n) + \beta_2 \Delta t F(\phi^{n-1}). \tag{2.41}$$

When formulating a three-level scheme, one seeks to improve upon the two-time-level methods, so it is reasonable to require that the global truncation error be of at least second order. The three-level scheme will be of at least second order if

$$\alpha_1 = 1 - \alpha_2, \qquad \beta_1 = \tfrac{1}{2}(\alpha_2 + 3), \qquad \beta_2 = \tfrac{1}{2}(\alpha_2 - 1), \tag{2.42}$$

where the coefficient α_2 remains a free parameter. One could choose α_2 to further reduce the truncation error, but the result is not a useful scheme (see problem 9). The most important explicit three-level schemes are obtained by choosing α_2 to minimize the amount of data that must be stored and carried over from the $n-1$ time level, i.e., by setting $\alpha_2 = 1$, in which case $\beta_2 = 0$, or by setting $\alpha_2 = 0$. If α_2 is set to one, (2.41) becomes the *leapfrog* scheme. The choice $\alpha_2 = 0$ gives the second-order *Adams–Bashforth* method. The remainder of this section will be devoted to an examination of these two schemes.

If the leapfrog scheme is applied to the oscillation equation, the result is

$$\phi^{n+1} = \phi^{n-1} + 2i\kappa \Delta t \phi^n. \tag{2.43}$$

Since the preceding is a linear finite-difference equation with constant coefficients, the amplification factor is constant from time step to time step and satisfies the quadratic equation

$$A^2 - 2i\kappa \Delta t A - 1 = 0.$$

The two roots are

$$A_\pm = i\kappa \Delta t \pm \left(1 - \kappa^2 \Delta t^2\right)^{1/2}. \tag{2.44}$$

In the limit of good numerical resolution, $\kappa \Delta t \to 0$ and $A_+ \to 1$; $A_- \to -1$. Evidently, the numerical solution is capable of behaving in two very different ways, or *modes*. The mode associated with A_+ is known as the *physical mode* because it approximates the solution to the original differential equation. The mode associated with A_- is referred to as the *computational mode*, since it arises solely as an artifact of the numerical computation. If $|\kappa \Delta t| \le 1$, the second term in (2.44) is real and $|A_+| = |A_-| = 1$; i.e., both the physical and the computational modes are stable and neutral. In the case $\kappa \Delta t > 1$,

$$|A_+| = |i\kappa \Delta t + i\left(\kappa^2 \Delta t^2 - 1\right)|^{1/2} > |i\kappa \Delta t| > 1,$$

and the scheme is unstable. When $\kappa \Delta t < -1$, a similar argument shows that $|A_-| > 1$. Note that when $\kappa \Delta t > 1$, A_+ lies on the positive imaginary axis in the complex plane, and thus each integration step produces a 90° shift in the phase of the oscillation. As a consequence, unstable leapfrog solutions grow with a period of $4\Delta t$.

The complete leapfrog solution can typically be written as a linear combination of the physical and computational modes. An exception occurs if $\kappa \Delta t = \pm 1$, in which case $A_+ = A_-$, and the physical and computational modes are not linearly independent. In such circumstances, the general solution to (2.43) has the form

$$\phi^n = C_1(i\kappa\Delta t)^n + C_2 n(i\kappa\Delta t)^n.$$

Since the magnitude of the preceding solution grows as function of time step, the leapfrog scheme is *not* stable when $|\kappa\Delta t| = 1$. Nevertheless, the $O(n)$ growth of the solution that occurs when $\kappa\Delta t = \pm 1$ is far slower than the $O(A^n)$ amplification that is produced when $|\kappa\Delta t| > 1$.

The source of the computational mode is particularly easy to analyze in the trivial case of $\kappa = 0$; then the analytic solution to the oscillation equation (2.30) is $\psi(t) = C$, where C is a constant determined by the initial condition at $t = t_0$. Under these circumstances, the leapfrog scheme reduces to

$$\phi^{n+1} = \phi^{n-1}, \tag{2.45}$$

and the amplification factor has the roots $A_+ = 1$, $A_- = -1$. The initial condition requires $\phi^0 = C$, which, according to the difference scheme (2.45), also guarantees that $\phi^2 = \phi^4 = \phi^6 = \cdots = C$. The odd time levels are determined by a second, computational, initial condition imposed on ϕ^1. In practice, ϕ^1 is often obtained from ϕ^0 by taking a single time step with a two-level method, and the resulting approximation to $\psi(t_0 + \Delta t)$ will contain some error E. It is obvious that in our present example, the correct choice for ϕ^1 is C, but in order to mimic the situation in a more general problem, suppose that $\phi^1 = C + E$. Then the numerical solution at any subsequent time will be the sum of two modes,

$$\phi^n = (C + E/2) - (-1)^n E/2.$$

to yield $\phi^n = (C + E/2) - (-1)^n E/2$.

Here, the first term represents the physical mode, and the second term represents the computational mode. The computational mode oscillates with a period of $2\Delta t$, and does not decay with time. In this example, the amplitude of the computational mode is completely determined by the error in the specification of the computational initial condition ϕ^1. Since there is no coupling between the physical and computational modes in solutions to linear problems, the errors in the initial conditions also govern the amplitude of the computational mode in leapfrog solutions to most linear equations. If the governing equations are nonlinear, however, the nonlinear terms introduce a coupling between ϕ_+ and ϕ_- that often amplifies the computational mode until it eventually dominates the solution. This spurious growth of the computational mode can be avoided by periodically discarding the solution at ϕ^{n-1} and taking a single time step with a two-level scheme, or by filtering the high-frequency components of the numerical solution. Various techniques for controlling the leapfrog scheme's computational mode will be discussed in Section 2.3.5.

The relative phase changes in the two leapfrog modes are

$$R_{\pm\text{leapfrog}} = \frac{1}{\kappa\Delta t} \arctan\left(\frac{\pm\kappa\Delta t}{(1 - \kappa^2\Delta t^2)^{1/2}}\right).$$

The computational mode and the physical mode oscillate in opposite directions. In the limit of good time resolution,

$$R_{+\text{leapfrog}} \approx 1 + \frac{(\kappa \Delta t)^2}{6},$$

showing that leapfrog time differencing is accelerating.

Now consider the second-order Adams–Bashforth method, which has the form

$$\phi^{n+1} = \phi^n + \Delta t \left(\tfrac{3}{2} F(\phi^n) - \tfrac{1}{2} F(\phi^{n-1}) \right). \tag{2.46}$$

The second-order Adams–Bashforth formula may be interpreted as numerical integration via the midpoint method, except that the value of the integrand at the midpoint, $F(\phi^{n+1/2})$, is obtained by linear extrapolation. Application of the Adams–Bashforth method to the oscillation equation yields

$$\phi^{n+1} = \phi^n + i\kappa \Delta t \left(\tfrac{3}{2} \phi^n - \tfrac{1}{2} \phi^{n-1} \right).$$

The amplification factor associated with this scheme is given by the quadratic

$$A^2 - \left(1 + \frac{3i\kappa \Delta t}{2} \right) A + \frac{i\kappa \Delta t}{2} = 0,$$

in which case

$$A_{\pm} = \frac{1}{2} \left(1 + \frac{3i\kappa \Delta t}{2} \pm \left(1 - \frac{9(\kappa \Delta t)^2}{4} + i\kappa \Delta t \right)^{1/2} \right). \tag{2.47}$$

As the numerical resolution increases, $A_+ \to 1$ and $A_- \to 0$. Thus, the Adams–Bashforth method damps the computational mode. The highly desirable damping of the computational mode is somewhat offset by a weak instability in the physical mode. This instability is revealed if (2.47) is approximated under the assumption that $\kappa \Delta t$ is small; then

$$A_+ = \left(1 - \frac{(\kappa \Delta t)^2}{2} - \frac{(\kappa \Delta t)^4}{8} - \cdots \right) + i \left(\kappa \Delta t + \frac{(\kappa \Delta t)^3}{4} + \cdots \right),$$

$$A_- = \left(\frac{(\kappa \Delta t)^2}{2} + \frac{(\kappa \Delta t)^4}{8} + \cdots \right) + i \left(\frac{\kappa \Delta t}{2} - \frac{(\kappa \Delta t)^3}{4} - \cdots \right),$$

and

$$|A_+|_{\text{A-B2}} \approx 1 + \tfrac{1}{4} (\kappa \Delta t)^4, \qquad |A_-|_{\text{A-B2}} \approx \tfrac{1}{2} \kappa \Delta t.$$

The modulus of the amplification factor of the physical mode exceeds unity by an $O[(\kappa \Delta t)^4]$ term. As was the case for the second-order Runge–Kutta methods, this weak instability can sometimes be tolerated if the length of the integration is limited and the time step is sufficiently small. The dependence of $|A_+|$ and $|A_-|$

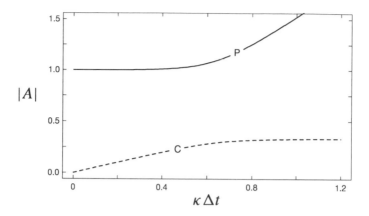

FIGURE 2.4. Modulus of the amplification factors for the second-order Adams–Bashforth scheme as a function of temporal resolution $\kappa \Delta t$. The solid and dashed lines represent the physical and the computational modes, respectively.

upon temporal resolution is plotted in Fig. 2.4. The relative phase change in the physical mode in the Adams–Bashforth method is

$$R_{\text{A-B2}} \approx 1 + \frac{5}{12}(\kappa \Delta t)^2,$$

where as before, it is assumed that $\kappa \Delta t \ll 1$.

Three-level schemes require information from two previous time levels, yet initial conditions for well-posed physical problems give information about the solution at only one time. It is therefore necessary to initialize the leapfrog and second-order Adams–Bashforth methods by taking a single time step using a two-level method. In most instances, a simple forward step is adequate. Although forward differencing is unstable, the amplification produced by a single step will generally not be large (see Problem 20). Moreover, even though the truncation error of a forward difference is $O(\Delta t)$, the execution of a single forward time step does not reduce the $O\left[(\Delta t)^2\right]$ global accuracy of leapfrog and Adams–Bashforth integrations. The basic reason that $O\left[(\Delta t)^2\right]$ accuracy is preserved is that forward differencing is used only over a Δt-long portion of the total integration. The contribution to the total error produced by the accumulation of $O\left[(\Delta t)^2\right]$ errors over a finite time interval is of the same order as the error arising from the accumulation of $O(\Delta t)$ errors over a time Δt.

2.3.5 Controlling the Leapfrog Computational Mode

The trapezoidal method is unconditionally stable, and it has the lowest truncation error of all the schemes presented in Sections 2.3.2–2.3.4. The weakness of the trapezoidal method is that it is implicit, which is often a serious disadvantage in computing numerical solutions to wave propagation problems. The best explicit scheme presented in the preceding sections might appear to be the leapfrog scheme. The leapfrog scheme is stable (unlike the second-order Runge–Kutta and Adams–Bashforth methods), it is of second order (unlike the Matsuno method), and it requires only one function evaluation per time step (unlike the Matsuno and Runge–Kutta schemes). The weakness of the leapfrog scheme is its undamped computational mode, which slowly amplifies during simulations of nonlinear wave propagation problems and generates an instability often known as "time splitting."

One way to control the growth of the computational mode is to periodically discard the data from the $n - 1$ time level (or alternatively, to average the n and $n - 1$ time-level solutions) and to restart the integration using a two-time-level method. Forward differencing is often used to reinitialize leapfrog integrations. Forward differencing is easy to implement, but since it is a first-order scheme, it degrades the second-order accuracy of the unadulterated leapfrog method. In addition, forward differencing is unstable and tends to amplify the high-frequency components of the solution. Moreover, it is difficult to quantify these adverse effects, since they vary according to the number of leapfrog steps between each forward step. Restarting with a second-order Runge–Kutta scheme is a far better choice, since this preserves second-order accuracy and produces less unstable amplification. The midpoint method is one second-order Runge–Kutta formulation that can be used to restart leapfrog integrations in complex numerical models without greatly complicating the model code. A midpoint–method restart may be implemented by taking a forward step of length $\Delta t/2$ followed by a single leapfrog step of length $\Delta t/2$.

In atmospheric science, it is common, though questionable, practice to control the computational mode through the use of a second-order time filter. Consider, therefore, the centered second-order time filter

$$\overline{\phi^n} = \phi^n + \gamma \left(\phi^{n+1} - 2\phi^n + \phi^{n-1} \right), \tag{2.48}$$

where ϕ^n denotes the solution at time $n\Delta t$ prior to time filtering, $\overline{\phi^n}$ is the solution after filtering, and γ is a positive real constant that determines the strength of the filter. The last term in (2.48) is the usual finite-difference approximation to the second derivative and preferentially damps the highest frequencies. Suppose that the unfiltered values are sampled from the exact solution to the oscillation equation; then

$$\overline{\phi^n} = \left(1 + \gamma \left(e^{i\kappa\Delta t} - 2 + e^{-i\kappa\Delta t} \right) \right) \phi^n.$$

Defining a *filter factor* $X = \overline{\phi^n}/\phi^n$, one obtains

$$X_{\text{centered}} = 1 - 2\gamma(1 - \cos\kappa\Delta t). \tag{2.49}$$

Since X_{centered} is real, it does not produce any change in the phase of the solution. In the limit $\kappa \Delta t \to 0$,

$$X_{\text{centered}} \approx 1 - \gamma (\kappa \Delta t)^2,$$

showing that well-resolved oscillations undergo an $O[(\Delta t)^2]$ damping. The centered filter has the greatest impact on the most poorly resolved component of the solution, the $2\Delta t$ oscillation. According to (2.49), each filter application reduces the amplitude of the $2\Delta t$ wave by a factor of $1 - 4\gamma$. If γ is specified to be $\frac{1}{4}$, each filtering operation will completely eliminate the $2\Delta t$ oscillation.

Robert (1966) and Asselin (1972) suggested a scheme to control the leapfrog computational mode by incorporating an approximate second-derivative time filter into the time integration cycle. They proposed following each leapfrog step

$$\phi^{n+1} = \overline{\phi^{n-1}} + 2\Delta t F(\phi^n)$$

by the filtering operation

$$\overline{\phi^n} = \phi^n + \gamma \left(\overline{\phi^{n-1}} - 2\phi^n + \phi^{n+1} \right). \tag{2.50}$$

A filter parameter of $\gamma = 0.06$ is typically used in global atmospheric models. Values of $\gamma = 0.2$ are common in convective cloud models; indeed, Schlesinger et al. (1983) recommend choosing γ in the range 0.25–0.3 for certain advection–diffusion problems.

If the Asselin-filtered leapfrog scheme is applied to the oscillation equation, the amplification factor is determined by the simultaneous equations

$$A^2 \phi^{n-1} = \overline{\phi^{n-1}} + 2i\kappa \Delta t A \phi^{n-1}, \tag{2.51}$$

$$\overline{A\phi^{n-1}} = A\phi^{n-1} + \gamma \left(\overline{\phi^{n-1}} - 2A\phi^{n-1} + A^2 \phi^{n-1} \right). \tag{2.52}$$

Under the assumption that $\overline{(A\phi^n)} = A(\overline{\phi^n})$, whose validity will be discussed shortly, (2.52) may be written

$$(A - \gamma)\overline{\phi^{n-1}} = A \left((1 - 2\gamma) + A\gamma \right) \phi^{n-1}. \tag{2.53}$$

Eliminating $\overline{\phi^{n-1}}$ between (2.51) and (2.53) yields

$$A_{\pm} = \gamma + i\kappa \Delta t \pm \left((1 - \gamma)^2 - \kappa^2 \Delta t^2 \right)^{1/2}, \tag{2.54}$$

which reduces to the result for the standard leapfrog scheme when $\gamma = 0$. In the limit of small $\kappa \Delta t$, the amplification factor for the Asselin-filtered physical mode becomes

$$A_{\text{Asselin-LF}} = 1 + i\kappa \Delta t - \frac{(\kappa \Delta t)^2}{2(1 - \gamma)} + O[(\kappa \Delta t)^4].$$

A comparison of this expression with the asymptotic behavior of the exact amplification factor

$$A_e = e^{i\kappa\Delta t} = 1 + i\kappa\Delta t - \frac{(\kappa\Delta t)^2}{2} - i\frac{(\kappa\Delta t)^3}{6} + O[(\kappa\Delta t)^4]$$

shows that the *local* truncation error of the Asselin-filtered leapfrog scheme is $O[(\kappa\Delta t)^2]$. In contrast, the local truncation error of the unfiltered leapfrog scheme ($\gamma = 0$) is $O[(\kappa\Delta t)^3]$. Thus, Asselin filtering degrades the *global* truncation error of the leapfrog scheme from second order to first order.

The preceding derivation was based on the assumption that $\overline{(A\phi^n)} = A(\overline{\phi^n})$. Is this justified? In practice, the initial condition is not time filtered; one simply defines $\overline{\phi^0} \equiv \phi^0$. Thus,

$$\overline{(A\phi^0)} - A(\overline{\phi^0}) = \overline{\phi^1} - \phi^1 = \gamma\left(\phi^0 - 2\phi^1 + \phi^2\right) \neq 0.$$

Nevertheless, an application of the Asselin time filter to ϕ^{n+1} gives

$$\begin{aligned}
\overline{(A\phi^n)} &= A\phi^n + \gamma\left(\overline{(A\phi^{n-1})} - 2A\phi^n + A^2\phi^n\right)\\
&= A\left(\phi^n + \gamma\left(\overline{\phi^{n-1}} - 2\phi^n + A\phi^n\right)\right) + \gamma\left(\overline{(A\phi^{n-1})} - A\overline{\phi^{n-1}}\right)\\
&= A(\overline{\phi^n}) + \gamma\left(\overline{(A\phi^{n-1})} - A\overline{\phi^{n-1}}\right),
\end{aligned}$$

from which it follows that

$$\overline{(A\phi^n)} - A(\overline{\phi^n}) = \gamma^n[\overline{\phi^1} - \phi^1]. \tag{2.55}$$

In all cases of practical interest, $n \gg 1$ and $\gamma \ll 1$; therefore, (2.55) implies that A may be factored out of the filtering operation with negligible error, and that (2.53) is indeed equivalent to (2.52).

In the limit of $\kappa\Delta t \ll 1$, the modulus of the amplification factor for the Asselin-filtered leapfrog scheme may be approximated as

$$|A_+|_{\text{Asselin-LF}} \approx 1 - \frac{\gamma}{2(1-\gamma)}(\kappa\Delta t)^2,$$

$$|A_-|_{\text{Asselin-LF}} \approx (1-2\gamma) + \frac{\gamma}{2 - 6\gamma + 4\gamma^2}(\kappa\Delta t)^2.$$

Like other first-order schemes, such as forward differencing and the Matsuno method, the physical mode in the Asselin-filtered leapfrog scheme has an $O[(\Delta t)^2]$ amplitude error. The behavior of the computational mode is also notable in that $|A_-|$ does not approach zero as $|\kappa\Delta t| \to 0$.

The asymptotic behavior of the relative phase change in the physical mode is

$$R_{+\text{Asselin-LF}} \approx 1 + \frac{1+2\gamma}{6(1-\gamma)}(\kappa\Delta t)^2.$$

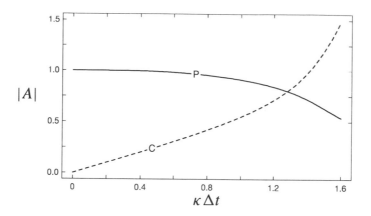

FIGURE 2.5. Modulus of the amplification factor for the leapfrog–trapezoidal method as a function of temporal resolution $\kappa\,\Delta t$. The solid and dashed lines represent the physical and the computational modes, respectively.

Asselin–Robert filtering increases the phase error; doubling it as γ increases from 0 to $\frac{1}{4}$.

The main problem with the Asselin-filtered leapfrog scheme is its first-order accuracy. There are two alternative techniques that control the leapfrog computational mode without sacrificing second-order accuracy—the leapfrog–trapezoidal method and the Magazenkov method. The leapfrog–trapezoidal method (Kurihara 1965; Zalesak 1979) is an iterative scheme in which a leapfrog predictor is followed by a trapezoidal correction step, i.e.,

$$\phi^* = \phi^{n-1} + 2\Delta t\,F(\phi^n),$$

$$\phi^{n+1} = \phi^n + \frac{\Delta t}{2}\left(F(\phi^n) + F(\phi^*)\right).$$

If this scheme is applied to the oscillation equation, the amplitude and relative phase changes in the physical mode are

$$|A|_{\text{LF-trap}} \approx 1 - \frac{(\kappa\,\Delta t)^4}{4}, \qquad R_{\text{LF-trap}} \approx 1 - \frac{(\kappa\,\Delta t)^2}{12},$$

where as usual, these approximations hold for small $\kappa\,\Delta t$. Leapfrog–trapezoidal integrations of the oscillation equation will be stable provided that $\kappa\,\Delta t \le \sqrt{2}$. The amplitude error associated with the leapfrog–trapezoidal scheme is plotted as a function of temporal resolution in Fig. 2.5.

Magazenkov (1980) suggested that the computational mode could be controlled by alternating each leapfrog step with a second-order Adams–Bashforth step. Since the Magazenkov method uses different schemes on the odd and even time steps, the amplification factor differs between the odd and even steps. In order to analyze the behavior of the Magazenkov method, it is therefore, best to consider the averaged effect of a combined leapfrog–Adams–Bashforth cycle.

Thus, for analysis purposes, the scheme will be written as a system of equations that maps (ϕ^{n-2}, ϕ^{n-1}) into (ϕ^n, ϕ^{n+1}),

$$\phi^n = \phi^{n-2} + 2\Delta t F(\phi^{n-1}), \qquad (2.56)$$

$$\phi^{n+1} = \left(\phi^{n-2} + 2\Delta t F(\phi^{n-1})\right)$$
$$+ \frac{\Delta t}{2}\left[3F\left(\phi^{n-2} + 2\Delta t F(\phi^{n-1})\right) - F(\phi^{n-1})\right]. \qquad (2.57)$$

When actually implementing the Magazenkov method, however, (2.57) would be replaced by the equivalent expression (2.46). Application of (2.56) and (2.57) to the oscillation equation yields a system of two equations in two unknowns,

$$\begin{pmatrix} 1 & 2i\kappa\Delta t \\ 1 + \frac{3}{2}i\kappa\Delta t & \frac{3}{2}i\kappa\Delta t - 3(\kappa\Delta t)^2 \end{pmatrix}\begin{pmatrix} \phi^{n-2} \\ \phi^{n-1} \end{pmatrix} = \begin{pmatrix} \phi^n \\ \phi^{n+1} \end{pmatrix}.$$

The coefficient matrix in the preceding equation determines the combined amplification and phase-shift generated by each pair of leapfrog and Adams–Bashforth time steps. The eigenvalues of the coefficient matrix are determined by the characteristic equation

$$\lambda^2 - \left(\frac{3i\kappa\Delta t}{2} - 3(\kappa\Delta t)^2 + 1\right)\lambda - \frac{i\kappa\Delta t}{2} = 0.$$

The eigenvalues are distinct and have magnitudes less than one when $|\kappa\Delta t| < \frac{2}{3}$, implying that the method is conditionally stable. For well-resolved physical-mode oscillations, the average amplitude and relative phase change per single time step are

$$|A|_{\mathrm{Mag}} = (|\lambda|)^{1/2} \approx 1 - \frac{(\kappa\Delta t)^4}{4}, \qquad R_{\mathrm{Mag}} = 1 + \frac{(\kappa\Delta t)^2}{6}.$$

The average amplitude error per single time step is plotted as a function of temporal resolution in Fig. 2.6.

2.3.6 Higher-Order Schemes

Relatively little attention has been devoted to the incorporation of third- or fourth-order time differencing into schemes for the numerical solution of partial differential equations. A major reason for the lack of interest in higher-order time differencing is that in many applications the errors in the numerical representation of the spatial derivatives dominate the time-discretization error, and as a consequence it might appear unlikely that the accuracy of the solution could be improved through the use of higher-order time differences. Several higher-order schemes do, nevertheless, have attractive stability characteristics that merit further discussion. Schemes of particular interest are the third-order Adams–Bashforth method and the third- and fourth-order Runge–Kutta methods. Whereas the second-order

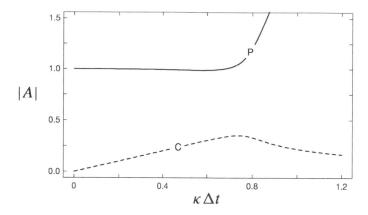

FIGURE 2.6. Modulus of the average amplification factor per single time step of the Magazenkov method plotted as a function of temporal resolution $\kappa \Delta t$. The solid and dashed lines represent the physical and the computational modes, respectively.

Runge–Kutta and Adams–Bashforth schemes produce amplifying solutions to the oscillation equation, their third- and fourth-order formulations are stable with strongly damped computational modes. As such, they offer additional possibilities for achieving better than second-order accuracy with a stable explicit time-differencing scheme. Moreover, they are better suited than the leapfrog–trapezoidal and Magazenkov schemes for the solution of a generalized oscillation equation in which κ has a positive imaginary part and the amplitude of the oscillation decays with time.

The distinctive advantage of the third-order Runge–Kutta scheme is the possibility of selecting a low-storage variant. As discussed earlier, there is no unique formula for the second-order Runge–Kutta method. Instead, the requirement of second-order accuracy leads to a family of schemes whose coefficients depend on the value of a free parameter. The coefficients of higher-order Runge–Kutta schemes are similarly nonunique. For example, the coefficients of the third- and fourth-order Runge–Kutta methods are determined by two independent parameters (Gear 1971, pp. 34–35). Williamson (1980) examined the subset of all possible third-order Runge–Kutta schemes that may be evaluated with minimal computer storage and recommended the following scheme:

$$
\begin{aligned}
q_1 &= \Delta t F(\phi^n), & \phi_1 &= \phi^n + q_1/3, \\
q_2 &= \Delta t F(\phi_1), -5q_1/9, & \phi_2 &= \phi_1 + 15q_2/16, \\
q_3 &= \Delta t F(\phi_2) - 153q_2/128, & \phi^{n+1} &= \phi_2 + 8q_3/15.
\end{aligned}
$$

In practical applications involving time-dependent partial differential equations, ϕ^n may be an extremely long vector of unknown variables (e.g., the velocity, temperature, and pressure at every node on a large three-dimensional mesh). It may therefore be difficult to store several copies of ϕ and $F(\phi)$ in the random access memory (RAM) of a digital computer. If m is the number of unknowns in ϕ, the Williamson–Runge–Kutta scheme economizes on storage by allowing the integration to proceed using only $2m$ storage locations, divided between the arrays q and ϕ, which are overwritten three times during each integration step. The storage requirement of the Williamson–Runge–Kutta scheme is identical to that of forward time-differencing and the standard leapfrog scheme, and is less than that required for the Asselin-filtered leapfrog scheme.

Finite-difference formulae for several time-differencing schemes are summarized in Table 2.1. Although it is not apparent from their most common names, most of the schemes shown in Table 2.1 are either Adams–Bashforth, Adams–Moulton, or Runge–Kutta schemes. Adams–Bashforth schemes are explicit multilevel methods whose first-order variant is the forward difference. Adams–Moulton methods are implicit multilevel methods that include backward and trapezoidal differencing as their first- and second-order representatives. One way to approximate the solution of the implicit algebraic equations generated by an Adams–Moulton method is to estimate ϕ^{n+1} using an Adams–Bashforth scheme and then substitute this estimate into the Adams–Moulton formula. The third-order Adams–Bashforth–Moulton corrector is listed in Table 2.1. The particular second-order Runge–Kutta scheme appearing in Table 2.1 is also the second-order Adams–Bashforth–Moulton predictor corrector.

Several important properties of the schemes listed in Table 2.1 are given in Table 2.2. The column labeled "storage factor" indicates the number of full arrays that must be allocated for each unknown variable in order to implement each scheme. Storage factors are not provided for the implicit methods listed in Table 2.2 because the storage factor for implicit methods can vary from problem to problem, depending on the numerical algorithm used to solve the implicit system. Inspection of Table 2.2 clearly reveals the low-storage advantage of the third-order Runge–Kutta scheme. This advantage may, however, be slightly exaggerated, since the storage factors listed in Table 2.2 are upper limits that allow each method to be programmed in a completely straightforward manner. In many instances, it is possible to utilize less memory than that suggested by the storage factor if newly computed quantities are initially placed in a small, temporary storage array. As an example, when integrating a partial differential equation with forward time differencing, it is not generally possible to write the newly computed ϕ_j^{n+1} directly into the storage occupied by ϕ_j^n, because ϕ_j^n may be required for the computation of ϕ_{j+1}^{n+1}. However, at some point in the integration cycle, ϕ_j^n will no longer be needed, and at that stage it may be overwritten by ϕ_j^{n+1}. During the interim between the calculation of ϕ_j^{n+1} and the last use of ϕ_j^n, ϕ_j^{n+1} may be held in a temporary storage array. In many applications, the temporary storage array can be much smaller than the full array required to hold a complete set of ϕ^n, and use of such a temporary array will reduce the storage factor by almost one unit.

In applications where storage is not a problem, the third-order Adams–Bashforth scheme

Method	Order	Formula
Forward	1	$\phi^{n+1} = \phi^n + hF(\phi^n)$
Backward	1	$\phi^{n+1} = \phi^n + hF(\phi^{n+1})$
Asselin Leapfrog	1	$\phi^{n+1} = \overline{\phi^{n-1}} + 2hF(\phi^n)$ $\overline{\phi^n} = \phi^n + \gamma(\overline{\phi^{n-1}} - 2\phi^n + \phi^{n+1})$
Leapfrog	2	$\phi^{n+1} = \phi^{n-1} + 2hF(\phi^n)$
Adams–Bashforth	2	$\phi^{n+1} = \phi^n + \dfrac{h}{2}\left[3F(\phi^n) - F(\phi^{n-1})\right]$
Trapezoidal	2	$\phi^{n+1} = \phi^n + \dfrac{h}{2}\left[F(\phi^{n+1}) + F(\phi^n)\right]$
Runge–Kutta	2	$q_1 = hF(\phi^n), \qquad \phi_1 = \phi^n + q_1$ $q_2 = hF(\phi_1) - q_1, \quad \phi^{n+1} = \phi_1 + q_2/2$
Magazenkov	2	$\phi^n = \phi^{n-2} + 2hF(\phi^{n-1})$ $\phi^{n+1} = \phi^n + \dfrac{h}{2}\left[3F(\phi^n) - F(\phi^{n-1})\right]$
Leapfrog–Trapezoidal	2	$\phi_1 = \phi^{n-1} + 2hF(\phi^n)$ $\phi^{n+1} = \phi^n + \dfrac{h}{2}\left[F(\phi_1) + F(\phi^n)\right]$
Adams–Bashforth	3	$\phi^{n+1} = \phi^n + \dfrac{h}{12}\left[23F(\phi^n) - 16F(\phi^{n-1}) + 5F(\phi^{n-2})\right]$
Adams–Moulton	3	$\phi^{n+1} = \phi^n + \dfrac{h}{12}\left[5F(\phi^{n+1}) + 8F(\phi^n) - F(\phi^{n-1})\right]$
ABM Predictor–Corrector	3	$\phi_1 = \phi^n + \dfrac{h}{2}\left[3F(\phi^n) - F(\phi^{n-1})\right]$ $\phi^{n+1} = \phi^n + \dfrac{h}{12}\left[5F(\phi_1) + 8F(\phi^n) - F(\phi^{n-1})\right]$
Runge–Kutta	3	$q_1 = hF(\phi^n), \qquad\qquad \phi_1 = \phi^n + q_1/3$ $q_2 = hF(\phi_1) - 5q_1/9, \qquad \phi_2 = \phi_1 + 15q_2/16$ $q_3 = hF(\phi_2) - 153q_2/128, \quad \phi^{n+1} = \phi_2 + 8q_3/15$
Runge–Kutta	4	$q_1 = hF(\phi^n), \qquad\qquad q_2 = hF(\phi^n + q_1/2)$ $q_3 = hF(\phi^n + q_2/2), \qquad q_4 = hF(\phi^n + q_3)$ $\phi^{n+1} = \phi^n + (q_1 + 2q_2 + 2q_3 + q_4)/6$

TABLE 2.1. Summary of methods for the solution of ordinary differential equations. The second- and third-order Runge–Kutta methods are low-storage variants; $h = \Delta t$.

Method	Storage Factor	Efficiency Factor	Amplification Factor	Phase Error	Max s
Forward	2	0	$1+\dfrac{s^2}{2}$	$1-\dfrac{s^2}{3}$	0
Backward	*	∞	$1-\dfrac{s^2}{2}$	$1-\dfrac{s^2}{3}$	∞
Asselin Leapfrog	3	< 1	$1-\dfrac{\gamma s^2}{2(1-\gamma)}$	$1+\dfrac{(1+2\gamma)s^2}{6(1-\gamma)}$	< 1
Leapfrog	2	1	1	$1+\dfrac{s^2}{6}$	1
Adams–Bashforth–2	3	0	$1+\dfrac{s^4}{4}$	$1+\dfrac{5}{12}s^2$	0
Trapezoidal	*	∞	1	$1-\dfrac{s^2}{12}$	∞
Runge–Kutta–2	2	0	$1+\dfrac{s^4}{8}$	$1+\dfrac{s^2}{6}$	0
Magazenkov	3	0.67	$1-\dfrac{s^4}{4}$	$1+\dfrac{s^2}{6}$	0.67
Leapfrog–Trapezoidal	3	0.71	$1-\dfrac{s^4}{4}$	$1-\dfrac{s^2}{12}$	1.41
Adams–Bashforth–3	4	0.72	$1-\dfrac{3}{8}s^4$	$1+\dfrac{289}{720}s^4$	0.72
Adams–Moulton–3	*	0	$1+\dfrac{s^4}{24}$	$1-\dfrac{11}{720}s^4$	0
ABM Predictor–Corrector–3	4	0.60	$1-\dfrac{19}{144}s^4$	$1+\dfrac{1243}{8640}s^4$	1.20
Runge–Kutta–3	2	0.58	$1-\dfrac{s^4}{24}$	$1+\dfrac{s^4}{30}$	1.73
Runge–Kutta–4	4[†]	0.70	$1-\dfrac{s^6}{144}$	$1-\dfrac{s^4}{120}$	2.82

[†] A storage factor of 3 may be achieved following the algorithm of Blum (1962).

TABLE 2.2. Characteristics of the schemes listed in Table 2.1. The amplification factor and relative phase change are for well-resolved solutions to the oscillation equation, and $s = \kappa\,\Delta t$. "Max s" is the maximum value of $\kappa\,\Delta t$ for which the solution is nonamplifying. The storage and efficiency factors are defined in the text. No storage factor is given for implicit schemes.

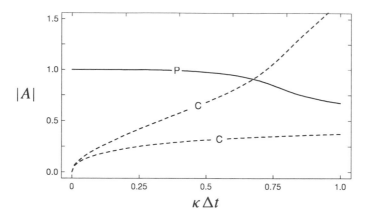

FIGURE 2.7. Modulus of the amplification factor for the third-order Adams–Bashforth method plotted as a function of temporal resolution $\kappa \Delta t$. The solid line represents the physical mode and the dashed lines the two computational modes.

$$\phi^{n+1} = \phi^n + \frac{\Delta t}{12} \left[23F(\phi^n) - 16F(\phi^{n-1}) + 5F(\phi^{n-2}) \right] \qquad (2.58)$$

can be an attractive alternative. The primary advantage of the third-order Adams–Bashforth scheme is its relative efficiency. In most practical applications involving partial differential equations, the bulk of the computational effort is associated with the evaluation of F, the function that determines the time derivative. Thus, a rough measure of the comparative efficiency of each method may be obtained by defining an efficiency factor as the maximum stable time step with which the oscillation equation can be integrated, divided by the number of evaluations of $F(\phi)$ that each scheme requires to perform a single integration step. Inspection of Table 2.2 shows that with the exception of the leapfrog scheme and its time-filtered variant, the third-order Adams–Bashforth scheme has the highest efficiency factor. The amplitude error in the third-order Adams–Bashforth solution to the oscillation equation is plotted in Fig. 2.7. Unlike the second-order Adams–Bashforth method, instability is not associated with unstable growth of the physical mode; instead, it is one of the computational modes that becomes unstable for $\kappa \Delta t > 0.724$.

Other schemes with efficiency factors almost as large as the third-order Adams–Bashforth method are the leapfrog–trapezoidal method and the fourth-order Runge–Kutta method. The leapfrog–trapezoidal scheme, being a lower-order scheme, is not a particularly attractive alternative. On the other hand, the fourth-order Runge–Kutta scheme is of higher order and potentially attractive, but its high efficiency factor is somewhat misleading. Figure 2.8 shows the amplification factor plotted as a function of temporal resolution for both the third- and fourth-order Runge–Kutta schemes. As shown in Fig. 2.8, once the time step ex-

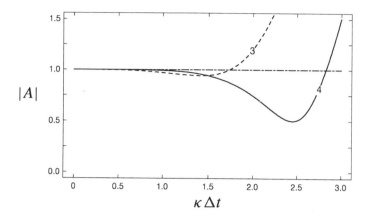

FIGURE 2.8. Modulus of the amplification factor plotted as a function of temporal resolution $\kappa \Delta t$ for higher-order Runge–Kutta solutions to the oscillation equation. Dashed line: third-order scheme; solid line: fourth-order method.

ceeds the maximum stable time step for the third-order scheme, the *fourth-order* method becomes highly damping. In some circumstances it may be desirable to selectively damp the highest-frequency modes, and in such cases the fourth-order Runge–Kutta method would appear to be much preferable to the first-order Matsuno method. On the other hand, if one wishes to avoid excessive damping of the the high-frequency components, it will not be possible to use the full stable time step of the fourth-order Runge–Kutta scheme, and as a result, the practical efficiency factor of the scheme will drop.

Tables 2.1 and 2.2 also list the order of accuracy of each scheme and give expressions for the relative phase change and amplitude error generated when each scheme is applied to the oscillation equation (2.30). The relationship between a scheme's order of accuracy and the orders of the amplitude and phase error is not entirely intuitive. As discussed in Section 2.3.1, the exact amplification factor for solutions to the oscillation equation is

$$A_e = e^{i\kappa k \Delta t} = 1 + i\kappa \Delta t - \frac{(\kappa \Delta t)^2}{2} - i \frac{(\kappa \Delta t)^3}{6} + \frac{(\kappa \Delta t)^4}{24} + \cdots.$$

The amplification factor A of an nth-order time-differencing scheme will match all terms in the preceding expression through order $(\kappa \Delta t)^n$. The amplitude error and the phase error characterize the errors in the modulus and the argument of A, respectively, and as such, their order of accuracy may differ from the general order of accuracy of the scheme. In particular, since amplitude and phase errors are special aspects of the total error, it is possible for either of these quantities to be smaller than the total error. The general relationship between the truncation error and the amplitude and phase errors may be stated as follows (Durran 1991):

If the oscillation equation (2.30) is integrated using a linear finite-difference scheme and if the truncation error of the resulting finite-difference approximation to the oscillation equation is of order r, then as $\kappa \Delta t \to 0$ the amplitude change in each step of the numerical solution is no worse than

$$1 + O[(\kappa \Delta t)^n], \quad \text{where} \quad \begin{cases} n = r + 1, & \text{if } r \text{ is odd;} \\ n \geq r + 2, & \text{if } r \text{ is even;} \end{cases}$$

and the relative phase change is no worse than

$$1 + O[(\kappa \Delta t)^m], \quad \text{where} \quad \begin{cases} m \geq r + 1, & \text{if } r \text{ is odd;} \\ m = r, & \text{if } r \text{ is even.} \end{cases}$$

Switching from an even- to an odd-order scheme increases the order of accuracy of the relative phase change without improving the order of accuracy of the amplitude error. Switching from odd to even order reduces the asymptotic amplitude error without altering the order of the error in the relative phase change.

2.4   Space-Differencing

Having examined the errors associated with time-differencing in Section 2.3, let us now consider the errors introduced when spatial derivatives are replaced with finite differences. In order to isolate the influence of the spatial differencing, the time dependence will not be discretized. Once again, our investigation will focus on the constant-wind-speed advection equation

$$\frac{\partial \psi}{\partial t} + c \frac{\partial \psi}{\partial x} = 0. \tag{2.59}$$

If the x-domain is periodic or unbounded, the spatial structure of the solution may be represented by a Fourier series or a Fourier integral, and a solution for each individual mode may be sought in the form of a traveling wave

$$\psi(x, t) = e^{i(kx - \omega t)}.$$

Here k is the *wave number*, and ω is the *frequency*. Substitution of this assumed solution into (2.59) shows that the traveling wave will satisfy the governing equation only if its frequency satisfies the *dispersion relation*

$$\omega = ck.$$

The wave travels with constant amplitude at a *phase speed* $\omega/k = c$. These waves are *nondispersive*, meaning that their phase speed is independent of the wave number. The energy associated with an isolated "packet" of waves propagates at the group velocity $\partial \omega / \partial k = c$, which is also independent of wave number. Readers unfamiliar with the concept of group velocity may wish to consult Gill (1982) or Whitham (1974).

2.4.1 *Differential–Difference Equations and Wave Dispersion*

Suppose that the spatial derivative in the advection equation is replaced with a second-order centered difference. Then (2.59) becomes the *differential–difference* equation:[6]

$$\frac{d\phi_j}{dt} + c\left(\frac{\phi_{j+1} - \phi_{j-1}}{2\Delta x}\right) = 0. \tag{2.60}$$

Individual wave-like solutions to this equation may be obtained in the form

$$\phi_j(t) = e^{i(kj\Delta x - \omega_{2c}t)}, \tag{2.61}$$

where ω_{2c} denotes the frequency associated with centered second-order spatial differencing. Substitution of (2.61) into the differential–difference equation yields

$$-i\omega_{2c}\phi_j = -c\left(\frac{e^{ik\Delta x} - e^{-ik\Delta x}}{2\Delta x}\right)\phi_j,$$

from which one obtains the dispersion relation

$$\omega_{2c} = c\frac{\sin k\Delta x}{\Delta x}. \tag{2.62}$$

Because ω_{2c} is real, there is no change in wave amplitude with time, and therefore no amplitude error. However, the phase speed,

$$c_{2c} \equiv \frac{\omega_{2c}}{k} = c\frac{\sin k\Delta x}{k\Delta x}, \tag{2.63}$$

is a function of k, so unlike the solutions to the original advection equation, these waves are dispersive. If the numerical resolution is good, $k\Delta x \ll 1$, and the Taylor series expansion $\sin x \approx x - x^3/6$ may be used to obtain

$$c_{2c} \approx c\left[1 - \tfrac{1}{6}(k\Delta x)^2\right],$$

showing that the phase-speed error is second-order in $k\Delta x$. Although the error for a well-resolved wave is small, the phase-speed error does become significant as the spatial resolution decreases. The least well-resolved wave on a numerical grid has wavelength $2\Delta x$ and wave number $k = \pi/\Delta x$. According to (2.63), *the phase speed of the $2\Delta x$ wave is zero.* Needless to say, this is a considerable error. The situation with the group velocity

$$\frac{\partial \omega_{2c}}{\partial k} = c\cos k\Delta x \tag{2.64}$$

is, however, even worse. The group velocity of well-resolved waves is approximately correct, but the group velocity of the poorly resolved waves is severely

---

[6]The set of differential–difference equations (2.60) for ϕ_j at every grid point constitute a large system of ordinary differential equations that could, in principle, be evaluated numerically using standard packages. This procedure, known as the *method of lines*, is usually not the most efficient approach.

retarded. The group velocity of the $2\Delta x$ wave is $-c$; its energy propagates backwards!

If the spatial derivative in the advection equation is replaced with a fourth-order centered difference, the resulting differential–difference equation

$$\frac{d\phi_j}{dt} + c\left[\frac{4}{3}\left(\frac{\phi_{j+1} - \phi_{j-1}}{2\Delta x}\right) - \frac{1}{3}\left(\frac{\phi_{j+2} - \phi_{j-2}}{4\Delta x}\right)\right] = 0 \qquad (2.65)$$

has wave solutions of the form (2.61), provided that the frequency ω_{4c} satisfies the dispersion relation

$$\omega_{4c} = \frac{c}{\Delta x}\left(\frac{4}{3}\sin k\Delta x - \frac{1}{6}\sin 2k\Delta x\right).$$

As is the case for centered second-order differences, there is no amplitude error, only phase-speed error. Once again, the waves are dispersive, and the phase speed of the $2\Delta x$ wave is zero. The phase-speed error of a well-resolved wave is, however, reduced to $O\left[(k\Delta x)^4\right]$, since for $k\Delta x$ small,

$$c_{4c} = \frac{\omega_{4c}}{k} \approx c\left(1 - \frac{(k\Delta x)^4}{30}\right).$$

The group velocity

$$\frac{\partial \omega_{4c}}{\partial k} = c\left(\frac{4}{3}\cos k\Delta x - \frac{1}{3}\cos 2k\Delta x\right) \qquad (2.66)$$

is also fourth-order accurate for well-resolved waves, but the group velocity of the $2\Delta x$ wave is $-5c/3$, an even greater error than that obtained using centered second-order differences.

The influence of spatial differencing on the frequency is illustrated in Fig. 2.9. As suggested by the preceding analysis, ω_{4c} approaches the true frequency more rapidly than ω_{2c} as $k\Delta x \to 0$, but both finite-difference schemes completely fail to capture the oscillation of $2\Delta x$ waves. The greatest advantages of the fourth-order difference over the second-order formulation are evident at "intermediate" wavelengths on the order of three to eight Δx. The improvements in the frequencies of these intermediate waves also generates a considerable improvement in their phase speeds and group velocities. The variation in the phase speed of a Fourier mode as a function of wave number is shown in Fig. 2.10. The improvement in the phase speed associated with an increase from second- to fourth-order accurate spatial differences is apparent even in the $3\Delta x$ wave. The fourth-order difference does not, however, improve the phase speed of the $2\Delta x$ wave. In fact, almost all finite-difference schemes fail to propagate the $2\Delta x$ wave. The basic problem is that there are only two possible configurations, differing by a phase angle of 180°, in which $2\Delta x$ waves can appear on a finite mesh. Thus, as shown

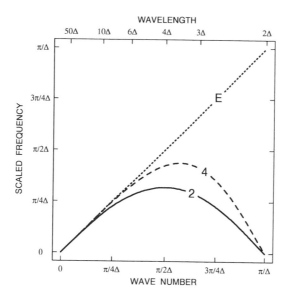

FIGURE 2.9. Scaled frequency (ω/c) as a function of wave number for the analytic solution of the advection equation (dotted line) and for corresponding differential–difference approximations using second- (solid line) and fourth-order (dashed line) centered differences.

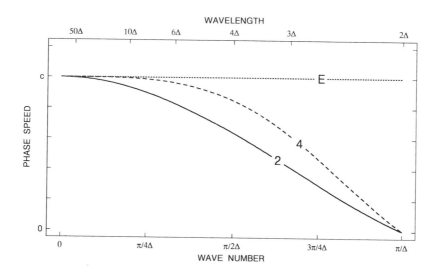

FIGURE 2.10. Phase speed as a function of numerical resolution for the analytic solution of the advection equation (dotted line) and for corresponding differential–difference approximations using second- (solid line) and fourth-order (dashed line) centered differences.

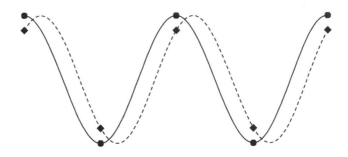

FIGURE 2.11. Misrepresentation of a $2\Delta x$ wave translating to the right as a decaying standing wave when the wave is sampled at fixed grid points on a numerical mesh. The grid-point values are indicated by dots at the earlier time and diamonds at the later time.

in Fig. 2.11, the grid-point representation of a translating $2\Delta x$ wave will be misinterpreted as a decaying standing wave.

The group velocities for the true solution and for the solutions of the second- and fourth-order differential–difference equations are plotted as a function of wave number in Fig. 2.12. The fourth-order scheme allows a better approximation of the group velocity for all but the shortest wavelengths. As discussed previously, the group velocity of the $2\Delta x$ wave produced by the fourth-order finite difference

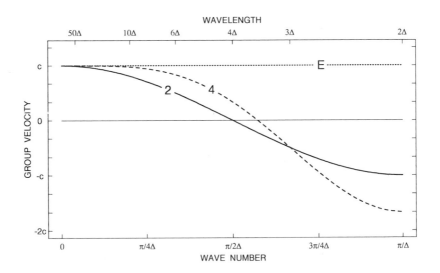

FIGURE 2.12. Group velocity as a function of numerical resolution for the analytic solution of the advection equation (dotted line) and corresponding differential–difference approximations using second- (solid line) and fourth-order (dashed line) centered differences.

is actually worse than that obtained with the second-order method. The degradation of the $2\Delta x$ group velocity in the higher-order scheme—or, equivalently, the increase in $\partial \omega / \partial k$ at $k = \pi/\Delta x$ in Fig. 2.9—is an unfortunate by-product of the inability of all finite-difference schemes to propagate the $2\Delta x$ wave and the otherwise desirable tendency of higher-order schemes to better approximate ω for wavelengths slightly longer than $2\Delta x$. In the absence of dissipation, the large negative group velocities associated with the $2\Delta x$ wave rapidly spread short-wavelength noise away from regions where $2\Delta x$ waves are forced.

One might attempt to improve the representation of extremely short waves by avoiding centered differences. If the spatial derivative in the advection equation is replaced with a first-order one-sided difference, (2.59) becomes

$$\frac{d\phi_j}{dt} + c\left(\frac{\phi_j - \phi_{j-1}}{\Delta x}\right) = 0. \tag{2.67}$$

Substitution of a wave solution of the form (2.61) into (2.67) yields the dispersion relation for the frequency associated with one-sided spatial differencing,

$$\omega_{1s} = \frac{c}{i\,\Delta x}\left(1 - e^{-ik\Delta x}\right) = \frac{c}{\Delta x}\left(\sin k\Delta x + i(\cos k\Delta x - 1)\right). \tag{2.68}$$

The real part of ω_{1s} is identical to the real part of ω_{2c}, and hence one-sided spatial differencing introduces the same dispersion error as centered second-order spatial differencing. Unlike centered differencing, however, the one-sided difference also generates amplitude error through the imaginary part of ω_{1s}. The amplitude of the differential–difference solution will grow or decay at the rate

$$\exp\left(-\frac{c}{\Delta x}(1 - \cos k\Delta x)t\right).$$

Thus, poorly resolved waves change amplitude most rapidly. If $c > 0$, the solution damps; the solution amplifies when $c < 0$. Note that if $c < 0$, the numerical domain of dependence does not include the domain of dependence of the original partial differential equation, so instability could also be predicted from the CFL condition.

A comparison of the performance of first-order, second-order, and fourth-order spatial differencing is provided in Fig. 2.13, which shows analytic solutions to the advection equation and numerical solutions to the corresponding differential–difference problem. The differential–difference equations are solved numerically on a periodic spatial domain using a fourth-order Runge–Kutta scheme to integrate (2.60), (2.65), and (2.67) with a very small time step.

Fig. 2.13a shows the distribution of ϕ that develops when the initial condition is a narrow spike, such that $\phi_j(t = 0)$ is zero everywhere except at the midpoint of the domain. Although the numerical domain is periodic, large-amplitude perturbations have not reached the lateral boundaries at the time shown in Fig. 2.13a. The narrow initial spike is formed by the superposition of many waves of different wavelengths; however, the Fourier components with largest amplitude are all of very short wavelength. The large diffusive error generated by one-sided differencing rapidly damps these short wavelengths and reduces the spike to a highly

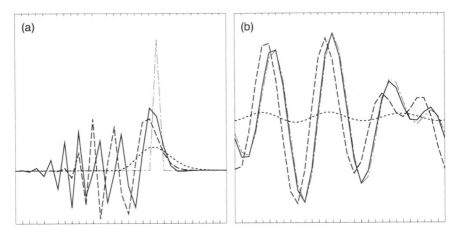

FIGURE 2.13. Exact solution and differential–difference solutions for (a) advection of a spike over a distance of five grid points, and (b) advection of the sum of equal-amplitude $7.5\Delta x$ and $10\Delta x$ sine waves over a distance of twelve grid points. Exact solution (dot-dashed), one-sided first-order (short-dashed), centered second-order (long-dashed), and centered fourth-order (solid). The distribution is translating to the right. Grid-point locations are indicated by the tick marks at the top and bottom of the plot.

smoothed low-amplitude disturbance. The second- and fourth-order centered differences also produce a dramatic distortion in the amplitude of the solution. Although the centered schemes preserve the amplitude of each individual Fourier component, the various components propagate at different speeds, and thus the superposition of these components ceases to properly represent the true solution. Consistent with the values of the group velocity given by (2.64) and (2.66), the energy in the shortest waves propagates back upstream from the initial location of the spike. As predicted by theory, the upstream propagation of the $2\Delta x$ wave is most rapid for the fourth-order method. Switching to a higher-order scheme does not improve the performance of finite-difference methods when they are used to model poorly resolved features like the spike in Fig. 2.13a; in fact, in many respects the fourth-order solution is worse than the second-order result.

The spike test is an extreme example of a common problem for which many numerical schemes are poorly suited, namely, the task of properly representing solutions with near discontinuities. As such the spike test provides a reference point that characterizes a scheme's ability to properly model poorly resolved waves. A second important reference point is provided by the test in Fig. 2.13b, which examines each scheme's ability to approximate features at an intermediate numerical resolution. The solution in Fig.2.13b is the sum of equal-amplitude $7.5\Delta x$ and $10\Delta x$ waves; in all other respects the problem is identical to that in Fig.2.13a. Unlike the situation with the spike test, the higher-order schemes are clearly superior in their treatment of the waves in Fig. 2.13b. Whereas the first-order difference generates substantial amplitude error and is distinctly inferior to the other two schemes, the second-order difference produces a reasonable approximation to the correct solution. Second-order centered differencing does, however, generate a noticeable lag in the phase speed of the disturbance (as in (2.63)). Moreover, since the phase lag of the $7.5\Delta x$ wave differs from that of the $10\Delta x$ wave in the second-order solution, the relative phase of the two waves changes during the simulation, and a significant error develops in the amplitude of the two rightmost wave crests. This example serves to emphasize that although centered differences do not produce amplitude errors in individual Fourier components, they still generate amplitude errors in the total solution. Finally, in contrast to the first- and second-order schemes, the errors introduced by fourth-order differencing are barely detectable at this time in the simulation.

The damping associated with the first-order upstream scheme (2.67) can be significantly reduced by using a higher-order one-sided difference. The differential–difference equation

$$\frac{d\phi_j}{dt} + \frac{c}{6}\left(\frac{2\phi_{j+1} + 3\phi_j - 6\phi_{j-1} + \phi_{j-2}}{\Delta x}\right) = 0 \qquad (2.69)$$

may be obtained by replacing the spatial derivative in the advection equation with a third-order difference. The dispersion relation associated with this differential–difference equation is

$$\omega_{3s} = \frac{c}{\Delta x}\left[\left(\frac{4}{3}\sin k\Delta x - \frac{1}{6}\sin 2k\Delta x\right) - \frac{i}{3}(1 - \cos k\Delta x)^2\right]. \qquad (2.70)$$

The real part of ω_{3s} is identical to that of ω_{4c}, and the phase-speed errors associated with the third- and fourth-order schemes are therefore identical. As was the case with first-order one-sided differencing, the sign of the imaginary part of ω_{3s} is determined by the sign of c such that solutions amplify for $c < 0$ and damp for $c > 0$. The damping associated with the third-order scheme is considerably less than that of the first-order scheme. According to (2.68) and (2.70),

$$\frac{\Im(\omega_{3s})}{\Im(\omega_{1s})} = \frac{1}{3}(1 - \cos k\Delta x).$$

As might be expected with a higher-order scheme, the well-resolved waves are damped much more slowly by the third-order approximation. Even the short waves show substantial improvement.

Some idea of the relative performance of the first-, second-, and third-order differences is provided in Fig. 2.14, which is identical to Fig. 2.13, except that the solid curve now represents the third-order solution. As indicated in Fig. 2.14, the damping produced by the third-order scheme is much weaker than that generated by first-order upstream differencing. Moreover, the third-order solution to the spike test is actually better than the second- and fourth-order results (compare Figs. 2.13a and 2.14a). In problems with extremely poor resolution, such as the spike test, the tendency of the third-order scheme to damp short wavelengths can

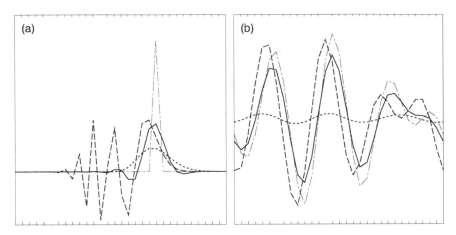

FIGURE 2.14. Exact solution and differential–difference solutions for (a) advection of a spike over a distance of five grid points, and (b) advection of the sum of equal-amplitude $7.5\Delta x$ and $10\Delta x$ sine waves over a distance of twelve grid points. Exact solution (dot-dashed), one-sided first-order (short-dashed), centered second-order (long-dashed), and one-sided third-order (solid).

be beneficial, because it largely eliminates the dispersive trail of waves found in the centered difference solutions. On the other hand, the damping of intermediate wavelengths is sufficiently weak that the third-order solution retains almost the same amplitude in the region of the spike as the "nondamping" second- and fourth-order schemes. The situation in Fig. 2.14b is somewhat different, and it is not entirely obvious whether the third-order results should be preferred over the second-order scheme. The third-order scheme clearly exhibits less phase-speed error, but it also shows more amplitude error than the centered second-order method.

2.4.2   Dissipation, Dispersion, and the Modified Equation

One way to estimate phase-speed and amplitude error is to derive the differential-difference dispersion relation, as described in the preceding section. Another way to characterize the relative magnitude of the these errors is to examine the lowest-order terms in the truncation error of the finite-difference formula. The truncation errors for each of the finite-difference approximations considered in the preceding section are as follows. One-sided, first-order:

$$\frac{\psi_j - \psi_{j-1}}{\Delta x} = \frac{\partial \psi}{\partial x} - \frac{\Delta x}{2} \frac{\partial^2 \psi}{\partial x^2} + \frac{\Delta x^2}{6} \frac{\partial^3 \psi}{\partial x^3} + O\left[(\Delta x)^3\right]. \qquad (2.71)$$

Centered, second-order:

$$\frac{\psi_{j+1} - \psi_{j-1}}{2\Delta x} = \frac{\partial \psi}{\partial x} + \frac{\Delta x^2}{6} \frac{\partial^3 \psi}{\partial x^3} + O\left[(\Delta x)^4\right].$$

One-sided, third-order:

$$\frac{2\psi_{j+1} + 3\psi_j - 6\psi_{j-1} + \psi_{j\,2}}{6\Delta x} = \frac{\partial \psi}{\partial x} + \frac{\Delta x^3}{12} \frac{\partial^4 \psi}{\partial x^4} - \frac{\Delta x^4}{30} \frac{\partial^5 \psi}{\partial x^5} + O\left[(\Delta x)^5\right].$$

Centered, fourth-order:

$$\frac{4}{3}\left(\frac{\psi_{j+1} - \psi_{j-1}}{2\Delta x}\right) - \frac{1}{3}\left(\frac{\psi_{j+2} - \psi_{j-2}}{4\Delta x}\right) = \frac{\partial \psi}{\partial x} - \frac{\Delta x^4}{30} \frac{\partial^5 \psi}{\partial x^5} + O\left[(\Delta x)^6\right].$$

If one of these formulae is used to determine the truncation error in a differential–difference approximation to the advection equation and the resulting scheme is $O\left[(\Delta x)^m\right]$ accurate, the same differential–difference scheme will approximate the *modified equation*

$$\frac{\partial \psi}{\partial t} + c\frac{\partial \psi}{\partial x} = a(\Delta x)^m \frac{\partial^{m+1} \psi}{\partial x^{m+1}} + b(\Delta x)^{m+1} \frac{\partial^{m+2} \psi}{\partial x^{m+2}} \qquad (2.72)$$

to $O\left[(\Delta x)^{m+2}\right]$, where a and b are rational numbers determined by the particular finite-difference formula. Thus, as $\Delta x \to 0$, the numerical solution to the differential–difference equation will approach the solution to the modified equation more rapidly than it approaches the solution to the advection equation. A qualitative description of the effects of the leading-order errors in the differential–difference equation may therefore be obtained by examining the prototypical response generated by each of the forcing terms on the right side of the modified equation (2.72).

The term with the even-order derivative in (2.72) introduces a forcing identical to that in the prototypical equation

$$\frac{\partial \xi}{\partial t} = (-1)^{m+1} \frac{\partial^{2m} \xi}{\partial x^{2m}},$$

whose solutions

$$\xi(x, t) = Ce^{ikx}e^{-k^{2m}t}$$

become smoother with time because the shorter-wavelength modes decay more rapidly than the longer modes. Thus, the term with the lowest-order even derivative produces amplitude error, or *numerical dissipation*, in the approximate solution of the advection equation. The odd-order derivative on the right side of (2.72) introduces a forcing identical to that in the prototypical equation

$$\frac{\partial \xi}{\partial t} = -\frac{\partial^{2m+1} \xi}{\partial x^{2m+1}},$$

whose solutions are waves of the form

$$\xi(x, t) = C e^{i(kx - \omega t)}, \quad \text{where} \quad \omega = (-1)^m k^{2m+1}.$$

For $m > 0$, these waves are dispersive, because their phase speed ω / k depends on the wave number k. As a consequence, the lowest-order odd derivative on the right side of (2.72) produces a wave-number-dependent phase speed error known as *numerical dispersion*.

Centered spatial differences do not produce numerical dissipation because there are no even derivatives in the truncation error of a centered difference scheme. Numerical dissipation is, however, produced by the leading-order term in the truncation error of the one-sided differences. There is a pronounced qualitative difference between the solutions generated by schemes with leading-order dissipative and leading-order dispersive errors. The modified equations associated with the preceding first- and second-order spatial differences both include identical terms in $\partial^3 \psi / \partial x^3$. As a consequence, both schemes produce essentially the same dispersive error. The dispersion errors in the third- and fourth-order schemes are also very similar because the truncation error associated with each of these schemes includes identical terms in $\partial^5 \psi / \partial x^5$. Yet, as was illustrated in Figs. 2.13 and 2.14, the impact of dispersion on even- and odd-order schemes is very different. Numerical dispersion is the only error in the centered even-order differences, so when short-wavelength modes are present, the dispersion is quite evident. In contrast, the numerical dispersion generated by the one-sided odd-order schemes is largely obscured by the lower-order dissipative errors that dominate the total error in these schemes.

2.4.3 Artificial Dissipation

As suggested by the test problems shown in Figs. 2.13 and 2.14, the lack of dissipation in centered-spatial differences can sometimes be a disadvantage. In particular, the error produced by the dispersion of poorly resolved Fourier components is free to propagate throughout the solution without loss of amplitude. It is therefore often useful to add scale-selective dissipation to otherwise nondissipative schemes in order to damp the shortest resolvable wavelengths. Moreover, in nonlinear problems it is often necessary to remove energy from the shortest spatial scales to prevent the development of numerical instabilities that can arise through the nonlinear interaction of short-wavelength modes (see Section 3.6).

The centered finite-difference approximations to even spatial derivatives of order two or higher provide potential formulae for scale-selective smoothers. Consider the isolated effect of a second-derivative smoother in an equation of the form

$$\frac{d\phi_j}{dt} = \gamma_2 \left(\phi_{j+1} - 2\phi_j + \phi_{j-1} \right), \tag{2.73}$$

where γ_2 is a parameter that determines the strength of the smoother. Substitution of solutions of the form

$$\phi_j = A(t) e^{ikj\Delta x} \tag{2.74}$$

into (2.73) yields

$$\frac{dA}{dt} = -2\gamma_2(1 - \cos k\Delta x)A,$$

implying that $2\Delta x$ waves are damped most rapidly, and that well-resolved waves undergo an $O[(k\Delta x)^2]$ dissipation. Indeed, if the second-derivative smoother is combined with the standard second-order centered difference,[7] the total truncation error in the smoothed difference becomes

$$\frac{\psi_{j+1} - \psi_{j-1}}{2\Delta x} - \gamma_2 \left(\psi_{j+1} - 2\psi_j + \psi_{j-1}\right)$$

$$= \frac{\partial\psi}{\partial x} - (\Delta x)^2 \left(\gamma_2 \frac{\partial^2\psi}{\partial x^2} - \frac{1}{6}\frac{\partial^3\psi}{\partial x^3}\right) + O\left[(\Delta x)^4\right].$$

Thus, the smoothed difference remains of second order, but the leading-order truncation error becomes both dissipative and dispersive. Note that as $\Delta x \to 0$, the preceding scheme will generate less dissipation than one-sided differencing, because as indicated by (2.71), one-sided differencing produces $O(\Delta x)$ dissipation. Furthermore, the addition of a separate smoother allows the dissipation rate to be explicitly controlled through the specification of γ_2.

Greater scale selectivity can be obtained using a fourth-derivative filter of the form

$$\frac{d\phi_j}{dt} = \gamma_4 \left(-\phi_{j+2} + 4\phi_{j+1} - 6\phi_j + 4\phi_{j-1} - \phi_{j-2}\right), \qquad (2.75)$$

or the sixth-derivative filter

$$\frac{d\phi_j}{dt} = \gamma_6 \left(\phi_{j+3} - 6\phi_{j+2} + 15\phi_{j+1} - 20\phi_j + 15\phi_{j-1} - 6\phi_{j-2} + \phi_{j-3}\right). \qquad (2.76)$$

Substituting a single wave of the form (2.74) into any of the preceding smoothers (2.73), (2.75), or (2.76) yields

$$\frac{dA}{dt} = -\gamma_n \left[2(1 - \cos k\Delta x)\right]^{n/2} A, \qquad (2.77)$$

where $n = 2$, 4, or 6 is the order of the derivative in each of the respective smoothers. In all cases, the $2\Delta x$ wave is damped most rapidly, and long waves are relatively unaffected. The actual scale selectivity of these filters is determined by the factor $(1 - \cos k\Delta x)^{n/2}$, which for well-resolved waves is $O\left[(k\Delta x)^n\right]$. This scale selectivity is illustrated in Fig. 2.15, in which the exponential decay rate as-

---

[7]If a dissipative filter is used in conjunction with leapfrog time-differencing, the terms involved in the filtering calculation must be evaluated at the $t - \Delta t$ time level to preserve stability. Time-differencing schemes appropriate for the simulation of diffusive processes are examined in Section 3.4.

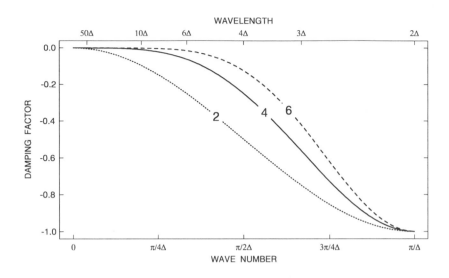

FIGURE 2.15. Normalized damping rate as a function of horizontal wave number for second- (dotted line), fourth- (solid line), and sixth-order (dashed-line) diffusive filters.

sociated with each smoother is plotted as a function of wave number. In order to facilitate the comparison of these filters, the decay rate of the $2\Delta x$ wave has been normalized to unity by choosing $\gamma_n = 2^{-n}$.

The test problems shown in Figs. 2.13 and 2.14 were repeated using fourth-order centered differencing in combination with fourth-order and sixth-order spatial smoothers and the results plotted in Fig. 2.16. The filtering coefficients were set such that $\gamma_4 = 0.2$, and $\gamma_6 = \gamma_4/4$; this choice for γ_6 insures that both filters will damp a $2\Delta x$ wave at the same rate. As evident in a comparison of Figs. 2.13a and 2.16a, both the fourth- and the sixth-order filters remove much of the dispersive train of short waves that were previously present behind the isolated spike in the unfiltered solution. Those waves that remain behind the spike in the smoothed solutions have wavelengths near $4\Delta x$. Since γ_4 and γ_6 have been chosen to damp $2\Delta x$ waves at the same rate, the $4\Delta x$ waves in the dispersive train are not damped as rapidly by the sixth-order smoother, and as is evident in Fig. 2.16a, the sixth-order smoother leaves more amplitude in the wave train behind the spike. Although the scale selectivity of the sixth-order smoother interferes with the damping of the dispersive wave train behind the spike, it significantly improves the simulation of the moderately resolved waves shown in Fig. 2.16b. The solution obtained using the sixth-order filter is almost perfect, whereas the fourth-order filter generates significant damping. In fact, the general character of the solution obtained with the fourth-order filter is reminiscent of that obtained with the third-order one-sided finite-difference approximation. This similarity is not coincidental; the phase-speed errors produced by the third- and fourth-order finite differences are identical, and the leading-order numerical dissipation in the third-

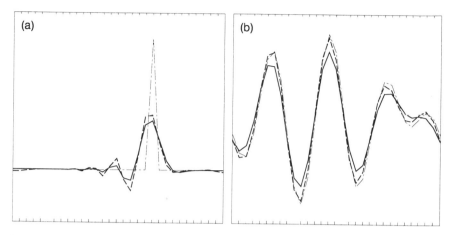

FIGURE 2.16. Exact solution and differential–difference solutions for (a) advection of a spike over a distance of five grid points, and (b) advection of the sum of equal-amplitude $7.5\Delta x$ and $10\Delta x$ sine waves over a distance of twelve grid points. Exact solution (dot-dashed), and fourth-order centered difference solutions in combination with a fourth-derivative filter (solid) or a sixth-derivative filter (dashed).

order difference, which is proportional to the fourth derivative, has the same scale selectivity as the fourth-order smoother.

Indeed, the proper choice of γ_4 will produce an exact equivalence between the solution obtained with the third-order scheme and the result produced by the combination of a fourth-order centered difference and a fourth-order smoother. The third-order differential–difference equation (2.69) can be expressed in a form that remains upstream independent of the sign of c as

$$\frac{d\phi_j}{dt} + \frac{c}{12\Delta x}\left(-\phi_{j+2} + 8(\phi_{j+1} - \phi_{j-1}) + \phi_{j-2}\right)$$

$$= -\frac{|c|}{12\Delta x}\left(\phi_{j+2} - 4\phi_{j+1} + 6\phi_j - 4\phi_{j-1} + \phi_{j-2}\right), \quad (2.78)$$

which is the combination of a fourth-order centered spatial difference and a fourth-order filter with a filter coefficient $\gamma_4 = |c|/(12\Delta x)$. Note that the value of the fourth-derivative filter in the preceding is an inverse function of Δx. The implicit Δx-dependence of the filtering coefficient in (2.78) makes the scheme $O\left[(\Delta x)^3\right]$, whereas the dissipation introduced by the explicit fourth-order filter (2.75) is $O\left[(\Delta x)^4\right]$.

In practical applications, the time derivatives in (2.73) and (2.75) are replaced by finite differences, and the maximum values for γ_2 and γ_4 will be determined by stability considerations. If the differencing is forward in time, the maximum useful smoothing coefficients are determined by the relations $\gamma_2\Delta t \leq 0.25$ and

$\gamma_4 \Delta t \leq 0.0625$. When $\gamma_2 \Delta t = 0.25$ (or $\gamma_4 \Delta t = 0.0625$) any $2\Delta x$ wave will be completely removed by a single application of the second-order (or fourth-order) filter.

2.4.4 Compact Differencing

Further improvements in the filtered solutions shown in Fig. 2.16 can be obtained by using more accurate finite-difference schemes. Simply switching to a higher-order explicit scheme, such as the centered sixth-order difference

$$\frac{df}{dx} = \frac{3}{2}\delta_{2x} f - \frac{3}{5}\delta_{4x} f + \frac{1}{10}\delta_{6x} f + O\left[(\Delta x)^6\right] \tag{2.79}$$

(where the operator δ_{nx} is defined by (2.7)), provides only marginal improvement. More significant improvements can be obtained using *compact differencing*, in which the desired derivative is given implicitly by a matrix equation. Our attention will be restricted to compact schemes in which this implicit coupling leads to tridiagonal matrices, since tridiagonal systems can be evaluated with modest computational effort (see Appendix).

The simplest compact scheme is obtained by rewriting the expression for the truncation error in the centered second-order difference (2.8) in the form

$$\delta_{2x} f = \left(1 + \frac{(\Delta x)^2}{6}\delta_x^2\right) \frac{df}{dx} + O\left[(\Delta x)^4\right]. \tag{2.80}$$

Expanding the finite-difference operators in the preceding expression yields the following $O\left[(\Delta x)^4\right]$ accurate expression for the derivative:

$$\frac{f_{j+1} - f_{j-1}}{2\Delta x} = \frac{1}{6}\left[\left(\frac{df}{dx}\right)_{j+1} + 4\left(\frac{df}{dx}\right)_j + \left(\frac{df}{dx}\right)_{j-1}\right]. \tag{2.81}$$

This scheme allows fourth-order-accurate derivatives to be calculated on a three-point stencil. At intermediate numerical resolution, the fourth-order compact scheme is typically more accurate than the sixth-order explicit difference (2.79).

If one is going to the trouble to solve a tridiagonal matrix, it can be advantageous to do a little extra work and use the sixth-order tridiagonal scheme. The formula for the sixth-order tridiagonal compact scheme may be derived by first noting that the truncation error in the fourth-order explicit scheme (2.9) is

$$\left(1 - \frac{(\Delta x)^2}{6}\delta_x^2\right)\delta_{2x} f = \frac{df}{dx} - \frac{(\Delta x)^4}{30}\frac{d^5 f}{dx^5} + O\left[(\Delta x)^6\right],$$

and the truncation error in the fourth-order compact scheme (2.80) is

$$\delta_{2x} f = \left(1 + \frac{(\Delta x)^2}{6}\delta_x^2\right)\frac{df}{dx} - \frac{(\Delta x)^4}{180}\frac{d^5 f}{dx^5} + O\left[(\Delta x)^6\right].$$

Eliminating the $O\left[(\Delta x)^4\right]$ term between these two expressions, one obtains

$$\left(1 + \frac{(\Delta x)^2}{30}\delta_x^2\right)\delta_{2x}f = \left(1 + \frac{(\Delta x)^2}{5}\delta_x^2\right)\frac{df}{dx} + O\left[(\Delta x)^6\right].$$

Expanding the operators in the preceding yields the following $O\left[(\Delta x)^6\right]$-accurate tridiagonal system for df/dx:

$$\frac{1}{15}\left(14\delta_{2x}f_j + \delta_{4x}f_j\right) = \frac{1}{5}\left[\left(\frac{df}{dx}\right)_{j+1} + 3\left(\frac{df}{dx}\right)_j + \left(\frac{df}{dx}\right)_{j-1}\right]. \quad (2.82)$$

When compact schemes are used to approximate partial derivatives in complex equations in which one must compute several different spatial derivatives, such as the multidimensional advection equation, it is simplest to solve either (2.81) or (2.82) as a separate tridiagonal system for each derivative. However, in very simple problems, such as the one-dimensional advection equation (2.59), the spatial derivatives in the compact formulae may be replaced directly by $-(1/c)\partial\psi/\partial t$. Thus, in order to analyze the phase-speed error associated with compact spatial differencing, the fourth-order compact approximation to the advection equation may be written

$$\frac{\phi_{j+1} - \phi_{j-1}}{2\Delta x} = \frac{-1}{6c}\left[\left(\frac{\partial\phi}{\partial t}\right)_{j+1} + 4\left(\frac{\partial\phi}{\partial t}\right)_j + \left(\frac{\partial\phi}{\partial t}\right)_{j-1}\right]. \quad (2.83)$$

Substitution of a wave solution of the form (2.61) into the preceding yields the following expression for the phase speed of the differential–difference solution:

$$c_{4c} = \frac{\omega_{4p}}{k} = \frac{3c}{2 + \cos k\Delta x}\left(\frac{\sin k\Delta x}{k\Delta x}\right). \quad (2.84)$$

The phase speeds for the sixth-order compact scheme,

$$c_{6c} = \frac{c}{3(3 + 2\cos k\Delta x)}\left(14\frac{\sin k\Delta x}{k\Delta x} + \frac{\sin 2k\Delta x}{2k\Delta x}\right),$$

may be obtained through a similar derivation. These phase speeds are plotted as a function of $k\Delta x$, together with the curves for second-, fourth-, and sixth-order explicit centered differences, in Fig. 2.17. It is apparent that the compact schemes are superior to the explicit schemes. In particular, the phase speeds associated with the sixth-order compact differencing are almost perfect for wavelengths as short as $4\Delta x$. Note that although the order of accuracy of a scheme determines the rate at which the phase-speed curves in Fig. 2.17 asymptotically approach the correct value as $k\Delta x \to 0$, the order of accuracy does not reliably predict a scheme's ability to represent the poorly resolved waves. Lele (1992) observed that a better treatment of the shorter waves can be obtained by perturbing the coefficients in

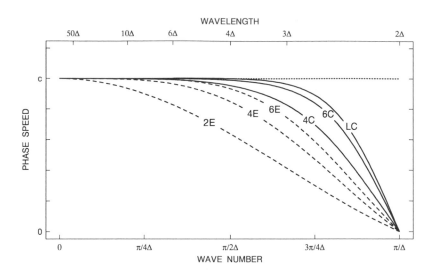

FIGURE 2.17. Phase speed as a function of numerical resolution for the analytic solution of the advection equation (dotted line) and for corresponding differential–difference approximations using second-, fourth-, and sixth-order explicit differences (dashed lines), fourth- and sixth-order compact differences (solid lines), and the low-phase-speed error fourth-order compact scheme of Lele (solid line labeled "LC").

the sixth-order compact scheme to create the fourth-order method

$$\frac{1}{12}\left(11\delta_{2x}f_j + \delta_{4x}f_j\right) = \frac{1}{24}\left[5\left(\frac{df}{dx}\right)_{j+1} + 14\left(\frac{df}{dx}\right)_{j} + 5\left(\frac{df}{dx}\right)_{j-1}\right].$$
(2.85)

The phase speeds associated with this differencing scheme are plotted as the solid curve labeled "LC"in Fig. 2.17. Observe that Lele's compact scheme produces phase-speed errors in a $3\Delta x$ wave that are comparable to the errors introduced in a $6\Delta x$ wave by explicit fourth-order differences.

The performance of Lele's compact scheme on the test problems considered previously in connection with Figs. 2.13, 2.14, and 2.16 is illustrated in Fig. 2.18. Since they accurately capture the frequency of very short waves while still failing to detect any oscillations at $2\Delta x$, compact schemes propagate the energy in the $2\Delta x$ wave backwards at very large group velocities (i.e., $-\partial\omega/\partial k$ is large near $k = 2\Delta x$). The preceding compact schemes are also nondamping because they are centered in space. It is therefore necessary to use a spatial filter in conjunction with these schemes when modeling problems with significant short-wavelength features. In these tests, a sixth-order filter (2.76) was used in combination with both the compact scheme (2.85) and the fourth-order explicit method. In all cases $\gamma_6 = 0.05$, which is the same value used in the computations shown in Fig. 2.16.

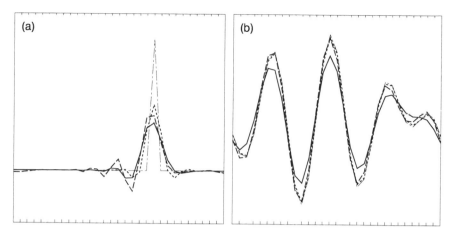

FIGURE 2.18. Exact solution and differential–difference solutions for (a) advection of a spike over a distance of five grid points, and (b) advection of the sum of equal-amplitude $7.5\Delta x$ and $10\Delta x$ sine waves over a distance of twelve grid points. Exact solution (dot-dashed), third-order one-sided solution (solid), fourth-order centered explicit solution (long dashed), and the solution obtained using Lele's low phase-speed-error compact scheme (short dashed). A sixth-order smoother, with $\gamma_6 = .05$ was used in combination with the fourth- and sixth-order differences.

In fact, the fourth-order solutions shown in these tests are identical to those shown previously in Fig. 2.16. Also plotted in Fig. 2.18 are the exact solution and the third-order one-sided solution (previously plotted in Fig. 2.14). As evident in Fig. 2.18a, the smoothed sixth-order compact scheme exhibits less of a $4\Delta x$ dispersive trail than either the third- or fourth-order scheme. Since the dissipation applied to the compact solution is identical to that used with the fourth-order scheme (and less than that inherent in the third-order method), the relative absence of dispersive ripples in compact solution indicates a relative lack of dispersive error at the $4\Delta x$ wavelength. This, of course, is completely consistent with the theoretical phase speed analysis shown in Fig. 2.17. The compact scheme also performs best on the two-wave test, Fig. 2.18b. Although the filtered compact scheme is the best-performing method considered in this section, it is also the most computationally burdensome. Other approaches to the problem of creating methods that can adequately represent short-wavelength features without sacrificing accuracy in smoother parts of the flow will be discussed in connection with the concept of flux-corrected transport in Chapter 5.

2.5   Combined Time- and Space-Differencing

The error introduced by time-differencing in ordinary differential equations was examined in Section 2.3. In Section 2.4, the error generated by spatial differencing was isolated and investigated through the use of differential–difference equations.

We now consider finite-difference approximations to the complete partial differential equation and analyze the total error that arises from the combined effects of both temporal and spatial differencing.

In some instances, the fundamental behavior of a scheme can be deduced from the characteristics of its constituent spatial and temporal differences. For example, suppose that the advection equation

$$\frac{\partial \psi}{\partial t} + c \frac{\partial \psi}{\partial x} = 0 \qquad (2.86)$$

is approximated using forward time-differencing in combination with centered spatial differencing. The result should be amplifying because forward time-differencing is amplifying and centered spatial differencing is neutral. Their combined effect will therefore produce amplification. On the other hand, it might be possible to combine forward time-differencing with one-sided space-differencing because the one-sided spatial difference is damping—provided that it is computed using "upstream" data. If this damping dominates the amplification generated by the forward time difference, it will stabilize the scheme. Further analysis would be required to determine the actual stability condition and the phase-speed error.

As another example, consider the use of leapfrog time-differencing and centered spatial differencing to approximate the advection equation. Since both differences are neutral, it seems likely that such a scheme would be conditionally stable. Once again, further analysis is required to determine the exact stability condition and the phase-speed error. In the absence of such analysis, the sign of the phase-speed error is in doubt, since the leapfrog scheme is accelerating, whereas centered spatial differencing is decelerating. Finally, suppose that leapfrog differencing is combined with one-sided spatial differences. The result should be unstable because the leapfrog solution consists of two modes (the physical and computational modes) each propagating in the opposite direction. If the one-sided difference is "upstream" with respect to one mode, it will be "downstream" with respect to the second mode, thereby amplifying the second mode.

Although as just noted, the forward-time and centered-space scheme

$$\frac{\phi_j^{n+1} - \phi_j^n}{\Delta t} + c \frac{\phi_{j+1}^n - \phi_{j-1}^n}{2\Delta x} = 0 \qquad (2.87)$$

will produce a nonphysical amplification of the approximate solution to the advection problem, one might wonder whether this amplification is sufficiently weak that the scheme nevertheless satisfies the more general Von Neumann stability condition

$$|A_k| \leq 1 + \gamma \Delta t, \qquad (2.88)$$

where γ is a constant independent of k, Δt, and Δx. If so, then (2.87) will still generate convergent approximations to the correct solution in the limit $\Delta x \to 0$, $\Delta t \to 0$, because it is a consistent approximation to the advection equation. Recall that as discussed in Section 2.3.2, forward differencing produces amplifying solutions that nevertheless converge to the correct solution of the oscillation equation as $\Delta t \to 0$. The amplification factor arising from a Von Neumann stability analysis of (2.87) satisfies

$$|A_k|^2 = 1 + \left(\frac{c\sin(k\Delta x)}{\Delta x}\right)^2 (\Delta t)^2.$$

Here, in contrast to the results obtained if ordinary differential equations are approximated with a forward difference, the coefficient of Δt includes a factor of $(\Delta x)^{-2}$ that cannot be bounded by a constant independent of Δx as $\Delta x \to 0$. As a consequence, the forward-time centered-space scheme does not satisfy the Von Neumann condition (2.88) and is both unstable in the sense that it generates growing solutions to a problem where the true solution is bounded, and unstable in the more general sense that it does not produce convergent solutions as $\Delta x \to 0$ and $\Delta t \to 0$. Note in particular that after N time steps the amplitude of a $4\Delta x$ wave increases by a factor of $(1 + \mu^2)^{N/2}$ (where $\mu = c\Delta t / \Delta x$). Thus, if a series of integrations are performed in which the space–time grid is refined while holding μ constant, the cumulative amplification of the $4\Delta x$ wave occurring over a fixed interval of physical time increases as $\Delta x \to 0$ and $\Delta t \to 0$.

2.5.1   The Discrete-Dispersion Relation

Although the preceding discussion suggests that useful deductions can be made by examining temporal and spatial differences independently, that discussion also reveals the need to rigorously analyze the combined effects of all finite differences in a specific formula in order to determine the complete behavior of the numerical solution. A useful tool in the analysis of errors in wave propagation problems is the *discrete-dispersion relation*, which is just the finite-difference analogue to the dispersion relation associated with the original continuous problem. The discrete-dispersion relation is obtained by substituting a traveling wave solution of the form

$$\phi_j^n = e^{i(kj\Delta x - \omega n \Delta t)} \tag{2.89}$$

into the finite-difference formula and solving for ω. If the frequency is separated into its real and imaginary parts ($\omega_r + i\omega_i$), (2.89) becomes

$$\phi_j^n = e^{\omega_i n \Delta t} e^{i(kj\Delta x - \omega_r n \Delta t)} = |A|^n e^{i(kj\Delta x - \omega_r n \Delta t)}. \tag{2.90}$$

The determination of the imaginary part of ω is tantamount to a Von Neumann stability analysis, since ω_i determines the amplification factor and governs the rate of numerical dissipation. Information about the phase-speed error can be obtained from ω_r.

Suppose that the advection equation is approximated with leapfrog-time and second-order centered-space differencing such that

$$\frac{\phi_j^{n+1} - \phi_j^{n-1}}{2\Delta t} + c\frac{\phi_{j+1}^n - \phi_{j-1}^n}{2\Delta x} = 0. \tag{2.91}$$

Substitution of (2.89) into this finite-difference scheme gives

$$\left(\frac{e^{-i\omega\Delta t} - e^{i\omega\Delta t}}{2\Delta t}\right)\phi_j^n = -c\left(\frac{e^{ik\Delta x} - e^{-ik\Delta x}}{2\Delta x}\right)\phi_j^n,$$

or, equivalently,

$$\sin\omega\Delta t = \mu\sin k\Delta x, \tag{2.92}$$

where $\mu = c\Delta t/\Delta x$. Inspection of (2.92) demonstrates that if $|\mu| < 1$, ω will be real and the scheme will be neutral. The scheme also appears to be neutral when $|\mu| = 1$, but this is a special case. When $\mu = 1$, (2.92) reduces to

$$c^* = \frac{\Delta x}{\Delta t} = c,$$

showing that the numerical solution propagates at the correct phase speed. Although there are no phase-speed errors when $|\mu| = 1$, the two roots of (2.92) become identical if $k\Delta x = \pi/2$, and as a consequence of this double root, the scheme admits a weakly unstable $4\Delta x$ wave. When $\mu = 1$, the weakly growing mode has the form

$$\phi_j^n = n\cos[\pi(j - n)/2]. \tag{2.93}$$

The distinction between the sufficient condition for stability $|\mu| < 1$ and the more easily derived necessary condition $|\mu| \leq 1$ is, however, of little practical significance because uncertainties about the magnitudes of the spatially and temporally varying velocities in real-world applications usually make it impossible to choose a time step such that $|\mu| = 1$.

The frequencies resolvable in the discretized time domain lie in the interval $0 \leq \omega_r \leq \pi/\Delta t$. Except for the special case just considered when $|\mu| = 1$ and $k\Delta x = \pi/2$, there are two resolvable frequencies that satisfy (2.92). Dividing these frequencies by k gives the phase speed of the physical and computational modes

$$c_{phys}^* \equiv \frac{\omega_{phys}}{k} = \frac{1}{k\Delta t}\arcsin(\mu\sin k\Delta x)$$

and

$$c_{comp}^* \equiv \frac{\omega_{comp}}{k} = \frac{1}{k\Delta t}[\pi - \arcsin(\mu\sin k\Delta x)].$$

As in the differential–difference problem, the $2\Delta x$ physical mode does not propagate. The $2\Delta x$ computational mode flips sign each time step, or equivalently, it moves at the speed $\Delta x/\Delta t$. In the limit of good spatial resolution ($k\Delta x \to 0$), the Taylor series approximations

$$\sin x \approx x - \tfrac{1}{6}x^3 \quad \text{and} \quad \arcsin x \approx x + \tfrac{1}{6}x^3$$

can be used to obtain

$$c^*_{\text{phys}} \approx c\left(1 - \frac{k^2 \Delta x^2}{6}(1 - \mu^2)\right). \tag{2.94}$$

If the time step is chosen to ensure stability, then $\mu^2 < 1$, $|c^*| < |c|$, and the decelerating effect of centered spatial differencing dominates the accelerating effects of leapfrog-time differencing. As suggested by (2.94), in practical computations the most accurate results are obtained using a time step such that the maximum value of $|\mu|$ is slightly less than one.

Now consider the forward-time one-sided space scheme

$$\frac{\phi_j^{n+1} - \phi_j^n}{\Delta t} + c\frac{\phi_j^n - \phi_{j-1}^n}{\Delta x} = 0, \tag{2.95}$$

sometimes referred to as the *donor-cell* scheme. Substitution of (2.89) into (2.95) gives

$$e^{-i\omega\Delta t} - 1 = \mu\left(e^{-ik\Delta x} - 1\right). \tag{2.96}$$

It follows that the exact dispersion relation and the exact solution are obtained in the special case when $\mu = 1$. Further analysis is facilitated by separating (2.96) into its real and imaginary parts

$$|A|\cos\omega_r\Delta t - 1 = \mu(\cos k\Delta x - 1) \tag{2.97}$$

and

$$|A|\sin\omega_r\Delta t = \mu\sin k\Delta x, \tag{2.98}$$

where, $\omega = \omega_r + i\omega_i$, and $|A| \equiv e^{\omega_i\Delta t}$ is the modulus of the amplification factor. Squaring both sides of (2.97) and (2.98) and adding yields

$$|A|^2 = 1 - 2\mu(1 - \mu)(1 - \cos k\Delta x),$$

which implies that the donor-cell scheme is stable and damping for $0 \le \mu \le 1$, and that the maximum damping per time step occurs at $\mu = \tfrac{1}{2}$.[8]

The discrete dispersion relation

$$\omega_r = \frac{1}{\Delta t}\arctan\left(\frac{\mu\sin k\Delta x}{1 + \mu(\cos k\Delta x - 1)}\right)$$

---

[8]This stability condition is identical to that obtained via the standard Von Neumann stability analysis in Section 2.3.3.

may be obtained after dividing (2.98) by (2.97). The function $\arctan \omega_r \Delta t$ is single-valued over the range of resolvable frequencies $0 \le \omega_r \le \pi/\Delta t$, so as expected for a two-time-level scheme, there is no computational mode. In the limit of good numerical resolution,

$$c^* \equiv \frac{\omega_r}{k} \approx c \left[1 - \frac{(k \Delta x)^2}{6} (1 - \mu)(1 - 2\mu) \right],$$

showing that phase-speed error is minimized by choosing either $\mu = 1$ or $\mu = \frac{1}{2}$. The donor cell scheme is decelerating for $0 < \mu < \frac{1}{2}$, and accelerating for $\frac{1}{2} < \mu < 1$. The phase-speed error in the donor-cell scheme may be minimized by choosing a time step such that $\mu_{avg} \approx \frac{1}{2}$. Under such circumstances, the donor-cell method will generate less phase-speed error than the leapfrog centered-space scheme. Unfortunately, the good phase-speed characteristics of the donor-cell method are overshadowed by its large dissipation.

It is somewhat surprising that there are values of μ for which the donor cell scheme is accelerating, since forward time-differencing is decelerating and one-sided spatial differencing reduces the phase speed of solutions to the differential–difference advection equation. This example illustrates the danger of relying too heavily on results obtained through the independent analysis of space- and time-truncation error.

2.5.2   The Modified Equation

As an alternative to the discrete dispersion equation, numerical dissipation and dispersion can be analyzed by examining a "modified" partial differential equation whose solution satisfies the finite-difference equation to a higher order of accuracy than the solution to the original partial differential equation. This technique is similar to that described in Section 2.4.2 except that since the truncation error includes derivatives with respect to both space and time, all the time derivatives must be expressed as spatial derivatives in order to isolate those terms responsible for numerical dissipation and dispersion. As an example, consider (2.95), the upstream approximation to the constant-wind-speed advection equation, which is a third-order accurate approximation to the modified equation

$$\frac{\partial \psi}{\partial t} + c \frac{\partial \psi}{\partial x} = \frac{c \Delta x}{2} (1 - \mu) \frac{\partial^2 \psi}{\partial x^2} - \frac{c (\Delta x)^2}{6} (1 - \mu)(1 - 2\mu) \frac{\partial^3 \psi}{\partial x^3}. \tag{2.99}$$

Examination of this equation shows that upstream differencing generates numerical dissipation of $O\left[(\Delta x)^2\right]$ and numerical dispersion of $O\left[(\Delta x)^3\right]$. Both the dissipation and dispersion are minimized as $\mu \to 1$, and the dispersion is also eliminated when $\mu = \frac{1}{2}$.

In deriving the modified equation, the original partial differential equation cannot be used to express all the higher-order time derivatives as spatial derivatives because the finite-difference scheme must approximate the modified equation more accurately than the original partial differential equation (Warming and Hyett 1974). The upstream method (2.99) provides a first-order approximation to the advection equation (2.86), a second-order approximation to

$$\frac{\partial \psi}{\partial t} = -c\frac{\partial \psi}{\partial x} - \frac{\Delta t}{2}\frac{\partial^2 \psi}{\partial t^2} + c\frac{\Delta x}{2}\frac{\partial^2 \psi}{\partial \psi^2}, \qquad (2.100)$$

and a third-order approximation to

$$\frac{\partial \psi}{\partial t} = -c\frac{\partial \psi}{\partial x} - \frac{\Delta t}{2}\frac{\partial^2 \psi}{\partial t^2} - \frac{(\Delta t)^2}{6}\frac{\partial^3 \psi}{\partial t^3} + c\frac{\Delta x}{2}\frac{\partial^2 \psi}{\partial x^2} - c\frac{(\Delta x)^2}{6}\frac{\partial^3 \psi}{\partial x^3}. \qquad (2.101)$$

The third-order accurate modified equation (2.99) is obtained by repeatedly substituting derivatives of (2.100) into (2.101) until all the first-order terms involving time derivatives are eliminated. The time derivatives in the remaining second-order terms can then be eliminated using the first-order-accurate relation (2.86).

2.5.3   The Lax–Wendroff Method

None of the schemes considered previously achieves $O\left[(\Delta t)^2\right]$ accuracy without multistage computation or implicitness or the use of data from two or more previous time levels. Lax and Wendroff (1960) proposed a general method for creating $O\left[(\Delta t)^2\right]$ schemes in which the time derivative is approximated by forward differencing and the $O(\Delta t)$ truncation error generated by that forward difference is canceled by terms involving finite-difference approximations to spatial derivatives. Needless to say, it is impossible to analyze the behavior of a Lax–Wendroff method properly without considering the combined effects of space- and time-differencing.

One important example of a Lax–Wendroff scheme is the following approximation to the advection equation (2.86):

$$\frac{\phi_j^{n+1} - \phi_j^n}{\Delta t} + c\left(\frac{\phi_{j+1}^n - \phi_{j-1}^n}{2\Delta x}\right) = \frac{c^2 \Delta t}{2}\left(\frac{\phi_{j+1}^n - 2\phi_j^n + \phi_{j-1}^n}{(\Delta x)^2}\right). \qquad (2.102)$$

The lowest-order truncation error in the first term of (2.102), the forward time difference, is

$$\frac{\Delta t}{2}\frac{\partial^2 \psi}{\partial t^2}.$$

However, since ψ is the exact solution to the continuous problem (2.86),

$$\frac{\Delta t}{2}\frac{\partial^2 \psi}{\partial t^2} = \frac{\Delta t}{2}\frac{\partial}{\partial t}\left(-c\frac{\partial \psi}{\partial x}\right) = \frac{c^2 \Delta t}{2}\frac{\partial^2 \psi}{\partial x^2}.$$

The term on the right side of (2.102) will therefore cancel the $O(\Delta t)$ truncation error in the forward time difference to within $O\left[\Delta t(\Delta x)^2\right]$, and as a consequence, the entire scheme is $O\left[(\Delta t)^2\right] + O\left[(\Delta x)^2\right]$ accurate.

The second-order nature of (2.102) may also be demonstrated by expressing it as a two-step formula in which each individual step is centered in space and time. In the first step, intermediate values staggered in space and time are calculated from the relations

$$\frac{\phi_{j+\frac{1}{2}}^{n+\frac{1}{2}} - \frac{1}{2}\left(\phi_{j+1}^n + \phi_j^n\right)}{\frac{1}{2}\Delta t} = -c\left(\frac{\phi_{j+1}^n - \phi_j^n}{\Delta x}\right), \tag{2.103}$$

$$\frac{\phi_{j-\frac{1}{2}}^{n+\frac{1}{2}} - \frac{1}{2}\left(\phi_j^n + \phi_{j-1}^n\right)}{\frac{1}{2}\Delta t} = -c\left(\frac{\phi_j^n - \phi_{j-1}^n}{\Delta x}\right). \tag{2.104}$$

In the second step, ϕ_j^{n+1} is computed from

$$\frac{\phi_j^{n+1} - \phi_j^n}{\Delta t} = -c\left(\frac{\phi_{j+\frac{1}{2}}^{n+\frac{1}{2}} - \phi_{j-\frac{1}{2}}^{n+\frac{1}{2}}}{\Delta x}\right). \tag{2.105}$$

The single-step formula (2.102) may be recovered by using (2.103) and (2.104) to eliminate $\phi^{n+\frac{1}{2}}$ and $\phi^{n-\frac{1}{2}}$ from (2.105). One advantage of the two-step formulation is that its extension to more complex problems can be immediately apparent. For example, if the wind speed is a function of the spatial coordinate, c is replaced by $c_{j+\frac{1}{2}}$, $c_{j-\frac{1}{2}}$, and c_j in (2.103), (2.104), and (2.105), respectively. In contrast, the equivalent modification of the single-step formula (see (2.107)) is slightly less obvious.

The amplitude and phase-speed errors of the Lax–Wendroff approximation to the constant-wind-speed advection equation may be examined by substituting a solution of the form (2.90) into (2.102), which yields

$$|A|(\cos \omega_r \Delta t - i \sin \omega_r \Delta t) = 1 + \mu^2(\cos k\Delta x - 1) - i\mu \sin k\Delta x. \tag{2.106}$$

Equating the real and imaginary parts of the preceding equation, and then eliminating $|A|$, one obtains the discrete-dispersion relation

$$\omega_r = \frac{1}{\Delta t} \arctan\left(\frac{\mu \sin k\Delta x}{1 + \mu^2(\cos k\Delta x - 1)}\right).$$

In the limit $k\Delta x \ll 1$, ω_r/k reduces to (2.94), showing that for well-resolved waves, the phase-speed error of the Lax–Wendroff method is identical to that of the leapfrog centered-space scheme. Eliminating ω_r from the real and imaginary parts of (2.106), one obtains

$$|A|^2 = 1 - \mu^2(1 - \mu^2)(1 - \cos k\Delta x)^2,$$

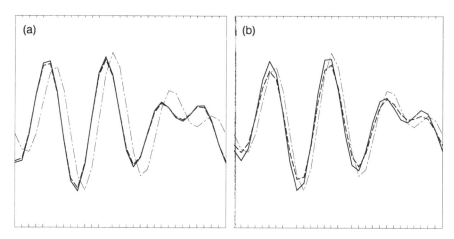

FIGURE 2.19. Leapfrog, second-order space (solid), Lax–Wendroff (dashed), and exact solution (dot-dashed) for the advection of the sum of equal-amplitude $7.5\Delta x$ and $10\Delta x$ sine waves over a distance of twelve grid points using a Courant number of (a) 0.1, and (b) 0.75.

from which it follows that the Lax Wendroff scheme is stable for $\mu^2 \leq 1$. Short wavelengths are damped most rapidly; the $2\Delta x$ wave is completely eliminated in a single time step if $|\mu| = 1/\sqrt{2}$. Since the shortest wavelengths are seriously in error—once again the phase speed of the $2\Delta x$ wave is zero—this scale-selective damping can be advantageous. Indeed, the scale-selectivity of the dissipation in the Lax–Wendroff scheme is the same as that of a fourth-order spatial filter. Unfortunately, the numerical analyst has little control over the actual magnitude of the dissipation because it is a function of the Courant number, and in most practical problems, μ will vary throughout the computational domain. The dependence of the damping on the Courant number is illustrated in Fig. 2.19, which compares solutions generated by the Lax–Wendroff method and the leapfrog scheme (2.91) using Courant numbers of 0.75 and 0.1. When $\mu = 0.1$, the leapfrog and Lax–Wendroff schemes give essentially the same result, but when μ is increased to 0.75, the damping of the Lax–Wendroff solution relative to the leapfrog scheme is clearly evident. Fig. 2.19 also demonstrates how the phase-speed error in both numerical solutions is reduced as the Courant number increases toward unity.

The term that cancels the $O(\Delta t)$ truncation error in a Lax–Wendroff scheme must be specifically reformulated for each new problem. The following three examples illustrate the general approach. If the flow velocity in (2.102) is a function of x, then

$$\frac{\partial^2 \psi}{\partial t^2} = c\frac{\partial}{\partial x}\left(c\frac{\partial \psi}{\partial x}\right),$$

and the right side of (2.102) becomes

$$\frac{c_j \Delta t}{2} \left(\frac{c_{j+\frac{1}{2}} (\phi^n_{j+1} - \phi^n_j) - c_{j-\frac{1}{2}} (\phi^n_j - \phi^n_{j-1})}{(\Delta x)^2} \right). \qquad (2.107)$$

If the flow is two-dimensional, the advection problem becomes

$$\frac{\partial \psi}{\partial t} + u \frac{\partial \psi}{\partial x} + v \frac{\partial \psi}{\partial y} = 0,$$

and if u and v are constant,

$$\frac{\partial^2 \psi}{\partial t^2} = u \frac{\partial^2 \psi}{\partial x^2} + v \frac{\partial^2 \psi}{\partial y^2} + 2uv \frac{\partial^2 \psi}{\partial x \partial y},$$

which must be approximated by a second-order spatial difference. Finally, consider a general system of "conservation laws" of the form

$$\frac{\partial \mathbf{v}}{\partial t} + \frac{\partial}{\partial x} \mathbf{F}(\mathbf{v}) = \mathbf{0},$$

where $\mathbf{v}$ and $\mathbf{F}$ are column vectors. Then

$$\frac{\partial^2 \mathbf{v}}{\partial t^2} = \frac{\partial}{\partial x} \left(\mathbf{J} \frac{\partial \mathbf{F}}{\partial x} \right), \qquad (2.108)$$

where $\mathbf{J}$ is the Jacobian matrix whose ijth element is $\partial F_i / \partial v_j$. Once again, this matrix operator must be approximated by second-order spatial differences.

In many applications the Lax–Wendroff method can be implemented more easily and more efficiently using the two-step method (2.103)–(2.105), or the following variant of the two-step method suggested by MacCormack (1969):

$$\tilde{\mathbf{v}}_j = \mathbf{v}^n_j - \frac{\Delta t}{\Delta x} \left[\mathbf{F}(\mathbf{v}^n_j) - \mathbf{F}(\mathbf{v}^n_{j-1}) \right],$$

$$\tilde{\tilde{\mathbf{v}}}_j = \tilde{\mathbf{v}}_j - \frac{\Delta t}{\Delta x} \left[\mathbf{F}(\tilde{\mathbf{v}}_{j+1}) - \mathbf{F}(\tilde{\mathbf{v}}_j) \right],$$

$$\mathbf{v}^{n+1}_j = \frac{1}{2} \left(\tilde{\mathbf{v}}_j + \tilde{\tilde{\mathbf{v}}}_j \right).$$

These two-step methods generate numerical approximations to the higher-order spatial derivatives required to cancel the $O(\Delta t)$ truncation error in the forward time difference without requiring the user to explicitly evaluate complex expressions like (2.108). The MacCormack method is particularly useful, since it easily generalizes to problems in two or more spatial dimensions.

In the classical Lax–Wendroff method, the spatial derivatives are approximated using centered differences, but other approximations are also possible. If the spatial dependence of ψ is not discretized, the Lax–Wendroff approximation to the advection equation (2.102) may be written

$$\frac{\phi^{n+1} - \phi^n}{\Delta t} + c\frac{\partial \phi^n}{\partial x} = \frac{c^2 \Delta t}{2}\frac{\partial \phi^n}{\partial x^2}.$$

Warming and Beam (1976) proposed the following upwind approximation to the preceding:

$$\phi_j^{n+1} = \phi_j^n - \mu\left(\phi_j^n - \phi_{j-1}^n\right) - \frac{\mu}{2}(1 - \mu)\left(\left(\phi_j^n - 2\phi_{j-1}^n + \phi_{j-2}^n\right)\right), \quad (2.109)$$

which is $O\left[(\Delta t)^2\right] + O\left[(\Delta x)^2\right]$ accurate and is stable for $0 \le \mu \le 2$.

2.6 Summary Discussion of Elementary Methods

In this chapter we have investigated the performance of schemes for approximating the constant-wind-speed advection equation. Let us now recapitulate the better methods discussed in this chapter and briefly summarize the conditions under which they might be expected to yield good results in more complicated problems. Further analysis of the performance of these schemes in more complex situations will be presented in the following chapters of this book.

First consider the relatively atypical class of problems in which the solution is sufficiently smooth that it can always be properly resolved on the numerical mesh.[9] Under these circumstances any stable method can be expected to converge to the correct result as the space–time grid is refined. Higher-order schemes will converge to *smooth* solutions more rapidly than low-order methods as the mesh size is decreased. Thus, even though higher-order methods require more computations per grid point per time step, genuinely high accuracy (i.e., several significant digits) can usually be achieved more efficiently by using a high-order scheme on a relatively coarse mesh than by using a low-order scheme on a finer mesh. Suitable high-order time differences include the third-order Adams–Bashforth method and the third- and fourth-order Runge–Kutta methods. Spatial differences might be computed using either explicit or compact fourth- or sixth-order differences; however, spectral methods, which will be discussed in Chapter 4, can be a better choice when very high accuracy is desired.

Most low-viscosity flows do not remain completely smooth. Instead, they develop at least some features with spatial scales shorter than or equal to that of an individual grid cell. Such small-scale features cannot be accurately captured by any numerical scheme, and the unavoidable errors in these small scales can feed back on the larger-scale flow and thereby exert a significant influence on the overall solution. In such circumstances there is no hope of computing an approximation to the correct solution that is accurate to several significant digits.

---

[9] An example of this type is provided by the barotropic vorticity equation, which will be discussed in Section 3.6.2.

Although the larger-scale features may be approximated with considerable quantitative accuracy, generally one must either be content with a qualitatively correct representation of the shortest-scale features or must remove these features with some type of numerical smoothing. Since it is not realistic to expect convergence to the correct solution in such problems, it is not particularly important to use high-order methods. Instead, one generally employs the finest possible numerical grid, selects a method that captures the behavior of moderately resolved waves with reasonable fidelity, and ensures that any spurious poorly resolved waves are eliminated by either explicit or implicit numerical dissipation. The numerical dissipation associated with all the schemes considered in this chapter is applied throughout the entire numerical domain. An alternative approach will be considered in Chapter 5, in which the implicit dissipation is primarily limited to those regions where the approximate solution is discontinuous or very poorly resolved.

Given that some degree of dissipation must generally be included to generalize the methods described in this chapter to practical problems involving low-viscosity flow, the neutral amplification factors associated with leapfrog time differencing and centered spatial differences are less advantageous than they may first appear. The difficulties associated with time splitting that can arise in nonlinear problems make the leapfrog scheme relatively unattractive in comparison with the third-order Adams–Bashforth or third- or fourth-order Runge–Kutta methods. The advantages of these relatively high-order methods are not primarily associated with their small truncation error (since some features will be poorly resolved) but arise from their stability and relative efficiency. The second-order Magazenkov and leapfrog–trapezoidal methods are also possible alternatives to the leapfrog scheme. Even forward differencing is a possibility, provided that it is used in a Lax–Wendroff method and that the implicit diffusion in the Lax–Wendroff scheme is limited by using a sufficiently small time step.

Now consider the choice of spatial difference approximations. Approximations based on centered spatial differences typically require the use of an explicit fourth- or sixth-derivative dissipative filter and are therefore less efficient than a third-order upstream approximation. This lack of efficiency is compensated by two practical advantages. First, it is not necessary to determine the upstream direction at each grid point when formulating the computer algorithm to evaluate a centered spatial difference. The determination of the upstream direction is not particularly difficult in advection problems where all signal propagation is directed along a clearly defined flow, but it can be far more difficult in problems admitting wave solutions that propagate both to the right and to the left. The second advantage of a centered difference used in conjunction with a spatial filter is that one can explicitly control the magnitude of the artificial dissipation, whereas the magnitude of the numerical dissipation associated with an upstream difference is implicitly determined by the local wind speed. The compact schemes appear to provide particularly good formulae for the evaluation of centered spatial differences because they remain accurate at relatively short wavelengths ($3\Delta x$ or $4\Delta x$) and use information at a minimum number of spatial grid points, which reduces the amount of special coding required near the boundaries of the spatial domain.

Problems

1. Suppose that $f(x)$ is to be represented at discrete points x_j on an *uneven* mesh and that $\Delta_{j-\frac{1}{2}} = x_j - x_{j-1}$. Use Taylor series expansions to derive a second-order finite-difference approximation to df/dx using a three-point stencil of the form

$$\alpha f_{j+1} + \beta f_j + \gamma f_{j-1}.$$

Hint: the result may be written in the form

$$\left(\frac{\Delta_{j-\frac{1}{2}}}{\Delta_{j+\frac{1}{2}} + \Delta_{j-\frac{1}{2}}} \right) \left(\frac{f_{j+1} - f_j}{\Delta_{j+\frac{1}{2}}} \right) + \left(\frac{\Delta_{j+\frac{1}{2}}}{\Delta_{j+\frac{1}{2}} + \Delta_{j-\frac{1}{2}}} \right) \left(\frac{f_j - f_{j-1}}{\Delta_{j-\frac{1}{2}}} \right).$$

2. Determine an $O(\Delta x)^2$-accurate *one-sided* finite-difference approximation to the first derivative $\partial \psi / \partial x$. Use the minimum number of points. Suppose that the numerical solution ϕ_j is available at points x_j, and that the derivative will be calculated using points to the right of x_j (i.e., x_j, x_{j+1}, ...). Assume constant grid spacing. How does the magnitude of the leading-order term in the truncation error of this one-sided approximation compare with that for the centered difference

$$\frac{\phi_{j+1} - \phi_{j-1}}{2\Delta x} ?$$

3. Determine those regions of the x-t plane in which the solution of

$$\left(\frac{\partial}{\partial t} + U \frac{\partial}{\partial x} \right)^2 \psi - c^2 \frac{\partial^2 \psi}{\partial x^2} = 0,$$

depends on ψ at some fixed point (x_0, t_0). Assuming that U and c are non-negative constants, schematically plot these regions and label them as either the "domain of influence" or the "domain of dependence." Draw a plot for the case $U > c$ and a plot for $U < c$.

4. Explain how the unconditional stability of the trapezoidally time-differenced one-dimensional advection equation

$$\frac{\phi_j^{n+1} - \phi_j^n}{\Delta t} + \frac{c}{2\Delta x} \left(\frac{\phi_{j+1}^{n+1} + \phi_{j+1}^n}{2} - \frac{\phi_{j-1}^{n+1} + \phi_{j-1}^n}{2} \right) = 0$$

is consistent with the CFL stability condition.

5. Consider the Lax–Fredrichs approximation to the scalar advection equation

$$\frac{\phi_j^{n+1} - \frac{1}{2}(\phi_{j+1}^n + \phi_{j-1}^n)}{\Delta t} + c\frac{\phi_{j+1}^n - \phi_{j-1}^n}{2\Delta x} = 0.$$

(a) Determine the truncation error for this scheme. Under what conditions does this scheme provide a consistent approximation to the advection equation? Would the condition required for consistency be difficult to satisfy in a series of simulations in which Δx is repeatedly halved?

(b) Determine the values of $c\Delta t/\Delta x$ for which this scheme is stable.

6. Suppose that the one-dimensional diffusion equation

$$\frac{\partial \phi}{\partial t} = K\frac{\partial^2 \phi}{\partial x^2} \tag{1}$$

is approximated using the Dufort–Frankel method

$$\frac{\phi_j^{n+1} - \phi_j^{n-1}}{2\Delta t} = K\left(\frac{\phi_{j+1}^n - (\phi_j^{n+1} + \phi_j^{n-1}) + \phi_{j-1}^n}{\Delta x^2}\right).$$

(a) Determine the truncation error associated with this approximation. Under what conditions does this scheme provide a consistent approximation to the diffusion equation? Would the condition required for consistency be difficult to satisfy in a series of simulations in which Δx is repeatedly halved?

(b) The advantage of the Dufort–Frankel scheme is that it is both explicit and unconditionally stable. Show that the scheme is indeed stable for all Δt.

7. When applied to the oscillation equation, Matsuno time differencing preferentially damps the higher frequencies (provided that $\kappa_{max}\Delta t < 1/\sqrt{2}$). Yet, if we turn our attention to the constant-wind-speed advection equation, the Lax–Wendroff scheme (2.102) damps $2\Delta x$ waves much more rapidly than does the following combination of Matsuno time differencing and centered space differencing:

$$\tilde{\phi}^n = \phi^n - c\Delta t\,\delta_{2x}\phi^n, \qquad \phi^{n+1} = \phi^n - c\Delta t\,\delta_{2x}\tilde{\phi}^n.$$

Explain why. Consider only those time steps for which $c\Delta t$ times the effective horizontal wave number is less than $1/\sqrt{2}$.

8. Consider the shallow-water equations, linearized about a state at rest,

$$\frac{\partial u}{\partial t} + g\frac{\partial \eta}{\partial x} = 0, \qquad \frac{\partial \eta}{\partial t} + H\frac{\partial u}{\partial x} = 0.$$

Prove, *without* doing a Von Neumann stability analysis, that the following finite-difference approximation to the preceding system must be unstable:

$$\frac{u_j^{n+1} - u_j^n}{\Delta t} + g\left(\frac{\eta_j^n - \eta_{j-1}^n}{\Delta x}\right) = 0,$$

$$\frac{\eta_j^{n+1} - \eta_j^n}{\Delta t} + H\left(\frac{u_j^n - u_{j-1}^n}{\Delta x}\right) = 0.$$

9. The most general form for an explicit, noniterative, three-time-level method is given by (2.41). Constraining the three-time-level method to be of at least second-order requires satisfaction of (2.42). Suppose that the free parameter α_2 is chosen to minimize the truncation error.

   (a) Determine the coefficients α_1, α_2, β_1, and β_2 for this scheme and give the resulting finite-difference formula. What is the order of accuracy of this scheme?

   (b) Prove that this is not a useful scheme for the numerical integration of the oscillation equation.

10. Compare and contrast the instabilities that arise when the oscillation equation is integrated using either forward (Euler) differencing or the time-differencing scheme given in Problem 9. If the integration is to be terminated at a fixed time t_f, can either scheme be used to obtain a numerical solution that converges to $\psi(t_f)$ as Δt is repeatedly decreased?

11. Determine the truncation error in the following approximation to the one-dimensional advection equation:

$$\frac{\phi_j^{n+1} - \phi_j^n}{\Delta t} + c\left(\frac{3\phi_j^n - 4\phi_{j-1}^n + \phi_{j-2}^n}{2\Delta x}\right) = 0.$$

Also determine the range of Δt over which the scheme is stable.

12. Suppose we try to reduce the phase-speed errors in the Lax–Wendroff scheme (2.102) by using the following approximation to the constant-wind-speed advection equation:

$$\frac{\phi_j^{n+1} - \phi_j^n}{\Delta t} + c\left(\frac{4}{3}\delta_{2x}\phi_j^n - \frac{1}{3}\delta_{4x}\phi_j^n\right) = \frac{c^2\Delta t}{2}\delta_x^2\phi_j^n.$$

Evaluate the truncation errors in this scheme to determine the leading-order dissipation and dispersion errors, and determine the condition (if any) under which the scheme is stable. Compare your results with those for (2.102) and for the leapfrog-time fourth-order-space scheme

$$\delta_{2t}\phi_j^n + c\left(\frac{4}{3}\delta_{2x}\phi_j^n - \frac{1}{3}\delta_{4x}\phi_j^n\right) = 0.$$

13. Determine the order of accuracy and the stability properties of the "slant-derivative" approximation to the constant-wind-speed advection equation

$$\frac{\phi_j^{n+1} - \phi_j^n}{\Delta t} + \frac{c}{2}\left(\frac{\phi_j^{n+1} - \phi_{j-1}^{n+1}}{\Delta x} + \frac{\phi_{j+1}^n - \phi_j^n}{\Delta x}\right) = 0.$$

14. Show that the upwind method of Warming and Beam (2.109) is second-order accurate with truncation error $O(\Delta t^2 + \Delta t \Delta x + \Delta x^2)$, and that it is stable for $0 \le \mu \le 2$.

15. Suppose that the advection equation is approximated by a second-order Runge–Kutta time difference and a centered second-order spatial differ-ence. Show that the auxiliary condition $O(\Delta t) \le O\left[(\Delta x)^{4/3}\right]$ is a neces-sary condition for this scheme to converge to the true solution in the limit $\Delta t, \Delta x \to 0$.

16. Suppose that the time derivative in the differential–difference equation (2.60) is approximated using a fourth-order Runge–Kutta Scheme. Determine the maximum value of $c\Delta t/\Delta x$ for which this scheme will be stable using the stability criteria for the oscillation equation given in Table 2.2. Explain how this value can exceed unity without violating the CFL condition.

17. Derive the modified equation that is approximated through order three by the leapfrog-time centered-space scheme (2.91). Compare this with the mod-ified equation for the Lax–Wendroff scheme

$$\frac{\partial \phi}{\partial t} + c\frac{\partial \phi}{\partial x} = -(1 - \mu^2)\frac{c(\Delta x)^2}{6}\frac{\partial^3 \phi}{\partial x^3} - \mu(1 - \mu^2)\frac{c(\Delta x)^3}{8}\frac{\partial^4 \phi}{\partial x^4},$$

where $\mu = c\Delta t/\Delta x$. Discuss whether the behavior of these two schemes, as illustrated in Fig. 2.19, is consistent with the leading-order error terms in each scheme's modified equation.

18. Derive an expression for the upstream approximation to the constant-wind-speed advection equation that remains upstream independent of the sign of the velocity field. Express the upstream spatial derivative as the combina-tion of a centered-space derivative and a diffusive smoother in a manner similar to that in (2.78).

19. *Evaluate the performance of several numerical schemes for approximating the two-component system of ordinary differential equations

$$\frac{du}{dt} = fv, \qquad \frac{dv}{dt} = -fu,$$

subject to the initial conditions $u(0) = 1$, $v(0) = 0$. Set $f = \pi$.

(a) Compare the exact solution with those obtained using the following schemes: (1) forward, (2) backward, (3) trapezoidal, (4) Matsuno, (5) Huen variant of second-order Runge–Kutta, (6) Leapfrog, and (7) second-order Adams–Bashforth. Initialize the leapfrog and second-order Adams–Bashforth schemes by taking one forward time step and set $f\Delta t = \pi/6$. Submit plots of u as a function of t comparing the various methods with the exact solution over the interval $0 \le t \le 6$. Set the vertical scale to $-2 \le u \le 2$ and terminate the curve for wildly unstable schemes when u exceeds these limits.

(b) Compare the average damping or amplification and the average phase-speed error per time step in your solution with the theoretical value *for small $f\Delta t$*. Choose $f\Delta t = 0.2$ for this comparison and present your results in a table. The table should contain the amplification per time step as predicted by theory (in the limit of good numerical resolution) and as determined from the numerical simulation. The table should also contain the phase-speed error as predicted by theory and as determined by the numerical solution. In gathering data for the table, run the simulations long enough to get reasonable estimates for each numerical scheme.

20. *Compare the performance of two strategies for initializing the leapfrog approximation to the equations in Problem 19. As the first strategy initialize the leapfrog scheme with a single forward time step. As the second strategy take a single forward step of length $\Delta t/2$ followed by a leapfrog step of length $\Delta t/2$ to obtain the unknowns at time Δt. This second approach is equivalent to a second-order Runge–Kutta midpoint method. As before, let $u(0) = 1$, $v(0) = 0$, and determine the error in $v(t = 4)$ generated by each scheme when $f\Delta t$ is $0.1, 0.3, \ldots, 0.9$.

21. *Compute solutions to the constant-wind-speed advection equation on the periodic domain $0 \le x \le 1$ subject to the initial condition $\psi(x, 0) = \sin^6(2\pi x)$. Use centered-space differencing $\partial\psi/\partial x \approx \delta_{2x}\phi$ and set $c = 0.1$.

(a) Compare the exact solution with numerical solutions obtained using forward and leapfrog differencing. Use a Courant number $c\Delta t/\Delta x = 0.1$ and plot your solutions at $t = 50$ using a vertical scale that includes $-40 \le \phi \le 40$. Compare and explain the results obtained using $\Delta x = 1/20$, $\Delta x = 1/40$ and $\Delta x = 1/80$. Use a single forward time step to initialize the leapfrog integration.

(b) Compare the exact solution with numerical solutions obtained using Heun (second-order Runge–Kutta) and leapfrog differencing. Use a Courant number $c\Delta t/\Delta x = 0.5$ and plot your solutions at $t = 60$ using a vertical scale that includes $-2 \le \phi \le 2$. Compare and explain the results obtained using $\Delta x = 1/20$, $1/40$, $1/80$, and $1/160$.

(c) Repeat the simulation in (b) with $\Delta x = 1/160$ but use a Courant number of 1.2 and integrate to $t = 2.1$. Compare and contrast the nature of the instabilities exhibited by the forward, leapfrog, and Heun methods in the simulations in (a), (b), and (c).

22. *Find solutions to the advection equation

$$\frac{\partial \psi}{\partial t} + c\frac{\partial \psi}{\partial x} = 0$$

in a periodic domain $0 \le x \le 1$. Suppose that $c = 0.2 \text{ ms}^{-1}$ and

$$\psi(x, 0) = \begin{cases} 9^4 \left[(x - \frac{5}{6})^2 - (\frac{1}{9})^2\right]^2, & \text{if } |x - \frac{5}{6}| \le \frac{1}{9}; \\ 0, & \text{otherwise}. \end{cases}$$

Obtain solutions using (1) leapfrog time differencing and centered second-order spatial differencing, (2) upstream (or donor cell) differencing and (3) the Lax–Wendroff method. Choose $\Delta x = 1/36$. Examine the sensitivity of the numerical solutions to the Courant number $(c\Delta t / \Delta x)$. Try Courant numbers of 0.1, 0.5, and 0.9. For each Courant number, submit a plot of the three numerical solutions and the exact solution at time $t = 5$. Scale the vertical axis on the plot to the range $-0.6 \le \psi \le 1.6$. Discuss the relative quality of the solutions and their dependence on Courant number. Is the dependence of the solutions on the Courant number consistent with the modified equation (7.23) and the results obtained in Problem 17?

23. *Consider the leapfrog-time, fourth-order space approximation to the constant-wind-speed advection equation

$$\delta_{2t}\phi_j^n + c\left(\frac{4}{3}\delta_{2x}\phi_j^n - \frac{1}{3}\delta_{4x}\phi_j^n\right) = 0.$$

(a) Determine the maximum Courant number $(c\Delta t / \Delta x)$ for which this scheme is stable.

(b) Repeat the comparison in Problem 22 including results from this fourth-order scheme and the second-order leapfrog scheme on each plot. Use Courant numbers 0.1, 0.5, and 0.72. Does the accuracy of all three approximate solutions improve as the Courant number approaches its maximum stable value? Why or why not?

3
Beyond the One-Way Wave Equation

The basic properties of finite-difference methods were explored in Chapter 2 by applying each scheme to a simple prototype problem: the one-way wave equation (or, equivalently, the one-dimensional constant-wind-speed advection equation). The equations governing wave-like geophysical flows include additional complexities. In particular, the flow may depend on several unknown functions that are related by a system of partial differential equations, the unknowns may be functions of more than two independent variables, and the equations may be nonlinear. It may also be necessary to account for weak dissipation, sources, and sinks. In this chapter we will examine some of the additional considerations that arise in the design and analysis of finite-difference schemes for the approximation of these more general problems.

3.1 Systems of Equations

Suppose that the problem of interest involves several unknown functions of x and t and that the governing equations for the system are linear with constant coefficients. An example of this type is the linearized one-dimensional shallow-water system

$$\frac{\partial u}{\partial t} + U \frac{\partial u}{\partial x} + g \frac{\partial h}{\partial x} = 0, \tag{3.1}$$

$$\frac{\partial h}{\partial t} + U \frac{\partial h}{\partial x} + H \frac{\partial u}{\partial x} = 0, \tag{3.2}$$

in which U and $u(x, t)$ represent the mean and perturbation fluid velocity, H and $h(x, t)$ are the mean and perturbation fluid depth, and g is the gravitational acceleration. The procedure for determining the truncation error, consistency, and order of accuracy of finite-difference approximations to a system such as (3.1) and (3.2) is identical to that discussed in Section 2.1. Taylor series expansions for the exact solution at the various grid points $(x_0, x_0 \pm \Delta x, \ldots)$ are substituted into each finite difference, and the order of accuracy of the overall scheme is determined by the lowest powers of Δx and Δt appearing in the truncation error. The stability analysis for finite-difference approximations to systems of partial differential equations is, on the other hand, more complex than that for a single equation.

3.1.1   Stability

Recall that a Von Neumann stability analysis of the finite-difference approximation to a linear constant-coefficient scalar equation is performed by determining the magnitude of the amplification factor A_k. Here, as in Section 2.2.2, the amplification factor for a two-time-level scheme is defined such that a single step of the finite-difference integration maps the Fourier component e^{ikx} to $A_k e^{ikx}$. However, when the governing equations are approximated by a linear constant-coefficient system of finite-difference equations, the kth Fourier component of the solution is represented by the vector $\mathbf{v}_k$, and the amplification factor becomes an *amplification matrix* $\mathbf{A}_k$. For example, in the shallow-water system (3.1) and (3.2), the vector representing the kth Fourier mode is

$$\mathbf{v}_k = \begin{pmatrix} u_k \\ h_k \end{pmatrix} e^{ikx}.$$

If the true solution does not grow with time, an appropriate stability condition is that

$$\|\mathbf{v}_k^n\| = \|\mathbf{A}_k^n \mathbf{v}_k^0\| \leq \|\mathbf{v}_k^0\|,$$

for all n and all wave numbers k resolved on the numerical mesh. For a single scalar equation, this condition reduces to (2.16). If the true solution grows with time, or if one is interested only in establishing sufficient conditions for the convergence of a consistent finite-difference scheme, the preceding should be relaxed to

$$\|\mathbf{v}_k^n\| = \|\mathbf{A}_k^n \mathbf{v}_k^0\| \leq C_T \|\mathbf{v}_k^0\| \quad \text{for all} \quad n\Delta t \leq T$$

and all sufficiently small values of Δt and Δx. Here C_T may depend on T, the time period over which the equations are integrated, but not on Δt or Δx. In the case of a single scalar equation, the preceding reduces to (2.15). Possible vector norms for use in these inequalities include $\| \ \|_\infty$ and $\| \ \|_2$ (defined by (2.13) and (2.14)).

Power Bounds on Matrices

The preceding stability conditions may be expressed in terms of the amplification matrix after introducing the concept of matrix norms. The norm of a matrix is defined in terms of the more familiar vector norm such that if $\mathbf{B}$ is an $M \times N$ matrix and $\mathbf{z}$ a column vector of length N, then

$$\|\mathbf{B}\| = \sup_{\|\mathbf{z}\|=1} \|\mathbf{Bz}\| = \sup_{\|\mathbf{z}\|\neq 0} \frac{\|\mathbf{Bz}\|}{\|\mathbf{z}\|}.$$

In particular, if b_{ij} is the element in the ith row and jth column of $\mathbf{B}$, then

$$\|B\|_\infty = \max_{1 \leq i \leq M} \sum_{j=1}^{N} |b_{ij}|$$

and

$$\|B\|_2 = \rho(\mathbf{B}^*\mathbf{B})^{1/2},$$

where $\mathbf{B}^*$ is the conjugate transpose of $\mathbf{B}$, and ρ is the *spectral radius*, defined as the maximum in absolute value of the eigenvalues of a square matrix.

Necessary and sufficient conditions for the stability of a constant-coefficient linear system may be expressed using this norm notation as

$$\|\mathbf{A}_k^n\| \leq 1 \tag{3.3}$$

for nongrowing solutions, and as

$$\|\mathbf{A}_k^n\| \leq C_T \tag{3.4}$$

in cases where the true solution grows with time or where the interest is only in ensuring that the numerical solution will be sufficiently stable to converge in the limit of $\Delta x, \Delta t \to 0$. (Once again, C_T depends on time, but not on Δx and Δt.)

Up to this point, the stability analysis for the single scalar equation and the system are essentially the same. The difference between the two arises when one attempts to reduce the preceding conditions on $\|\mathbf{A}_k^n\|$ to a constraint on $\|\mathbf{A}_k\|$. In the scalar case the necessary and sufficient condition that $|A_k^n| \leq 1$ is just $|A_k| \leq 1$. On the other hand, when the amplification factor is a matrix, the necessary and sufficient conditions for $\mathbf{A}_k$ to be "power bounded" are rather complex. Since $\|\mathbf{A}_k^n\| \leq \|\mathbf{A}_k\|^n$ (by the fundamental properties of any norm), the condition $\|\mathbf{A}_k\| \leq 1$ will ensure stability. This condition is not, however, necessary for stability, as may be seen by considering the matrix

$$\mathbf{E} = \begin{pmatrix} 1 & -1 \\ 0 & -1 \end{pmatrix},$$

for which $\|\mathbf{E}\|_\infty = 2$, $\|\mathbf{E}\|_2 = (3 + \sqrt{5})/2$, and yet for all positive integers m, $\mathbf{E}^{2m}$ is the identity matrix, whose norm is unity.

The precise necessary and sufficient conditions for an arbitrary matrix to be power bounded are given by the Kreiss matrix theorem (Kreiss 1962; Strikwerda 1989, p. 188) and are relatively complicated. Necessary conditions for the boundedness of $\|\mathbf{A}_k^n\|$ can, however, be expressed quite simply. In order to have $\|\mathbf{A}_k^n\| \leq 1$, i.e., to have a nongrowing numerical solution, it is necessary that

$$\rho(\mathbf{A}_k) \leq 1. \tag{3.5}$$

In order to satisfy the bound on the amplification matrix for growing solutions (3.4) it is necessary that

$$\rho(\mathbf{A}_k) \leq 1 + \gamma \Delta t, \tag{3.6}$$

where γ is a constant independent of Δx and Δt. The fact that the preceding are not sufficient conditions for stability is illustrated by the matrix

$$\mathbf{F} = \begin{pmatrix} 1 & 1 \\ 0 & 1 \end{pmatrix},$$

for which $\rho(\mathbf{F}) = 1$, but

$$\mathbf{F}^n = \begin{pmatrix} 1 & n \\ 0 & 1 \end{pmatrix},$$

so $\|\mathbf{F}^n\|$ grows linearly with n. This linear growth is, however, much weaker than the geometric growth in $\|\mathbf{F}^n\|$ that would occur if the spectral radius of $\mathbf{F}$ were bigger than one. The fact that the condition $\rho(\mathbf{A}_k) \leq 1$ is capable of eliminating all highly unstable cases with geometrically growing solutions is an indication that the spectral radius criteria (3.5) and (3.6) are "almost" strong enough to ensure stability. Indeed, if $\mathbf{A}_k$ can be transformed to a diagonal matrix, which is frequently the situation when hyperbolic partial differential equations are approximated by finite differences, (3.5) and (3.6) are both necessary and sufficient conditions for stability. Even if $\mathbf{A}_k$ cannot be transformed to a diagonal matrix, (3.5) will be sufficient to ensure nongrowing solutions, provided that the moduli of all but one of the eigenvalues of $\mathbf{A}_k$ are strictly less than unity.

In most practical applications, the governing equations will contain either nonlinear terms or linear terms with variable coefficients, and in order to perform a Von Neumann stability analysis, one must first approximate the full equations with a frozen-coefficient linearized system. Subsequent analysis of the frozen-coefficient linearized system yields necessary, but not sufficient, conditions for the stability of the numerical solution to the original problem. It is therefore often not profitable to exert great effort to determine sufficient conditions for the stability of the frozen-coefficient linearized system. Instead, it is common practice to evaluate condition (3.5) or (3.6) with the understanding that they provide necessary conditions for the stability of both the original problem and the associated frozen-coefficient linear system, but do not guarantee stability in either case.

Reanalysis of Leapfrog Time-Differencing

A simple system of finite-difference equations illustrating the preceding concepts can be obtained by writing the leapfrog time-differenced approximation to the oscillation equation (2.30) as

$$\phi^{n+1} = \chi^n + 2i\kappa\,\Delta t\,\phi^n, \tag{3.7}$$
$$\chi^{n+1} = \phi^n. \tag{3.8}$$

Although in this example the original problem is governed by a single ordinary differential equation rather than a system of partial differential equations, the stability analysis of the finite-difference system proceeds as if (3.7) and (3.8) had been obtained directly from a more complicated problem. Let $\mathbf{A}$ denote the amplification matrix obtained by writing the system in the matrix form

$$\begin{pmatrix} 2i\kappa\,\Delta t & 1 \\ 1 & 0 \end{pmatrix} \begin{pmatrix} \phi \\ \chi \end{pmatrix}^n = \begin{pmatrix} \phi \\ \chi \end{pmatrix}^{n+1}.$$

The eigenvalues of the amplification matrix satisfy

$$\lambda^2 - 2i\kappa\,\Delta t\,\lambda - 1 = 0,$$

which is the same quadratic equation obtained for the amplification factor in the analysis of the leapfrog scheme in Section 2.3.4, where it was shown that $|\lambda_\pm| = 1$ if and only if $|\kappa\,\Delta t| \leq 1$. Thus the necessary condition for stability $\rho(\mathbf{A}) \leq 1$ is satisfied when $|\kappa\,\Delta t| \leq 1$.

Sufficient conditions for stability are easy to obtain if the amplification matrix is diagonalizable, since $\rho(\mathbf{A}) \leq 1$ is both a necessary and sufficient condition for stability if there exist bounded matrices $\mathbf{T}$ and $\mathbf{T}^{-1}$ such that $\mathbf{T}^{-1}\mathbf{A}\mathbf{T}$ is a diagonal matrix. Any matrix can be transformed to a diagonal matrix if it has a complete set of linearly independent eigenvectors. The eigenvectors of the leapfrog amplification matrix

$$\begin{pmatrix} i\kappa\,\Delta t + [1 - (\kappa\,\Delta t)^2]^{1/2} \\ 1 \end{pmatrix} \quad \text{and} \quad \begin{pmatrix} i\kappa\,\Delta t - [1 - (\kappa\,\Delta t)^2]^{1/2} \\ 1 \end{pmatrix}$$

are linearly independent for $\kappa\,\Delta t \neq \pm 1$. The leapfrog scheme must therefore be stable when $|\kappa\,\Delta t| < 1$. When $\kappa\,\Delta t = 1$, however, the eigenvectors of the leapfrog amplification matrix are not linearly independent, and the matrix is not diagonalizable. In this case,

$$\mathbf{A} = \begin{pmatrix} 2i & 1 \\ 1 & 0 \end{pmatrix} \quad \text{and} \quad \mathbf{A}^n = i^n \begin{pmatrix} n+1 & -in \\ -in & 1-n \end{pmatrix}.$$

Since $\|\mathbf{A}^n\|$ grows linearly with n, the leapfrog scheme is not stable for $\kappa\,\Delta t = 1$. Similar reasoning shows that the choice $\kappa\,\Delta t = -1$ is also unstable. The overall conclusion, that the leapfrog differencing is stable for $|\kappa\,\Delta t| < 1$, is identical to that obtained in Section 2.3.4.

The Discrete Dispersion Relation

One useful way to obtain necessary conditions for the stability of wave-like solutions of systems of partial differential equations is to evaluate the discrete dispersion relation. The discrete dispersion relation is particularly simple in instances where the original system is approximated by finite differences that are centered in space and time. For example, suppose the one-dimensional shallow-water system (3.1) and (3.2) is approximated using leapfrog-time centered-second-order space differencing as

$$\delta_{2t}u + U\delta_{2x}u + g\delta_{2x}h = 0, \tag{3.9}$$
$$\delta_{2t}h + U\delta_{2x}h + H\delta_{2x}u = 0. \tag{3.10}$$

(The finite-difference operator δ_{nx} is defined in the Appendix by (A.1).) As in the case of the scalar advection equation discussed in Section 2.5.1, the construction of the discrete dispersion relation mimics the procedure used to obtain the dispersion relation for the continuous system. Wave solutions to the discretized shallow-water equations are sought in the form

$$u_j^n = u_0 e^{i(kj\Delta x - \omega n \Delta t)}, \qquad h_j^n = h_0 e^{i(kj\Delta x - \omega n \Delta t)}, \tag{3.11}$$

where u_0 and h_0 are complex constants determining the wave amplitude, and the physically relevant portion of the solution is the real part of u_j^n and h_j^n. Substitution of (3.11) into the finite-differenced governing equations (3.9) and (3.10) yields

$$\left(-\frac{\sin \omega \Delta t}{\Delta t} + U\frac{\sin k\Delta x}{\Delta x}\right)u_0 + g\frac{\sin k\Delta x}{\Delta x}h_0 = 0,$$
$$\left(-\frac{\sin \omega \Delta t}{\Delta t} + U\frac{\sin k\Delta x}{\Delta x}\right)h_0 + H\frac{\sin k\Delta x}{\Delta x}u_0 = 0.$$

Nontrivial values of u_0 and h_0 will satisfy the preceding pair of homogeneous equations when the determinant of the coefficients of u_0 and h_0 is zero, which requires

$$\left(\frac{\sin \omega \Delta t}{\Delta t} - U\frac{\sin k\Delta x}{\Delta x}\right)^2 = gH\left(\frac{\sin k\Delta x}{\Delta x}\right)^2,$$

or defining $c = \sqrt{gH}$,

$$\sin \omega \Delta t = \frac{\Delta t}{\Delta x}(U \pm c)\sin k\Delta x. \tag{3.12}$$

In the limit of $\Delta t, \Delta x \to 0$, this discrete-dispersion relation approaches the dispersion relation for the continuous problem $\omega = (U \pm c)k$.

The discrete-dispersion relation for the linearized shallow-water system is identical to that for the scalar advection equation (2.92) except that it supports two physical modes moving-at velocities $U + c$ and $U - c$. The amplitude and phase-speed error in each wave may be analyzed in the same manner as that in the scalar advection problem (see Section 2.5.1). The analysis of amplitude error, for example, proceeds by examining the amplification factor $e^{\Im(\omega)\Delta t}$ by which the waves (3.11) grow or decay during each time step. Since the horizontal wave number (k) of periodic waves is real, the imaginary part of ω will be zero unless the right side

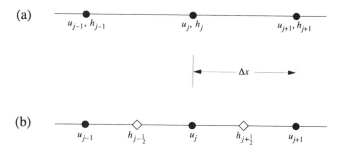

FIGURE 3.1. Distribution of u and h on (a) an unstaggered and (b) a staggered mesh.

of (3.12) exceeds unity. When the right side is greater than unity, one of the ω satisfying (3.12) has a positive imaginary part, and the numerical solution grows at each time step. Thus, a necessary condition for the stability of the finite-difference scheme is that

$$\left| \frac{\Delta t}{\Delta x} (U \pm c) \sin k \Delta x \right| \leq 1$$

for all k resolvable on the numerical mesh, or since $k = \pi/(2\Delta x)$ is a resolvable wave,

$$(|U| + c) \frac{\Delta t}{\Delta x} \leq 1. \tag{3.13}$$

This condition is not quite sufficient to guarantee stability; since the time differencing is leapfrog, sufficient conditions for stability require strict inequality in (3.13).

3.1.2   Staggered Meshes

When simulating a system of equations with several unknowns, it not necessary to define all the unknown variables at the same grid points. Significant improvements in the accuracy of the short-wavelength components of the solution can sometimes be obtained by the use of *staggered* meshes.

Spatial Staggering

Consider, once again, the linearized shallow-water system (3.1) and (3.2) and in order to reveal the benefits of staggering more clearly, suppose that $U = 0$. The finite-difference approximations (3.9) and (3.10) assume that the perturbation velocity (u) and depth (h) are defined at the same grid points, as shown schematically in Fig. 3.1a. An alternative arrangement is shown in Fig. 3.1b, in which the grid points where h is defined are shifted $\Delta x/2$ to the right (or left) of u grid points. A centered-difference approximation to the linearized shallow-water system with $U = 0$ may be written for the staggered mesh as

$$\delta_{2t} u_j + g \delta_x h_j = 0, \tag{3.14}$$

$$\delta_{2t} h_{j+\frac{1}{2}} + H \delta_x u_{j+\frac{1}{2}} = 0. \tag{3.15}$$

The discrete-dispersion relation for the solution to these equations is

$$\sin \omega \Delta t = \pm \frac{2c \Delta t}{\Delta x} \sin \left(\frac{k \Delta x}{2} \right), \tag{3.16}$$

from which it follows that (3.14) and (3.15) are stable when $|c \Delta t / \Delta x| < \frac{1}{2}$. The maximum time step available for integrations on the staggered mesh is only one-half that which may be used on the unstaggered mesh. The more stringent restriction on the time step is, however, not entirely bad, because shorter time steps are generally required by spatial differencing schemes that more faithfully capture the high-frequency components of the solution. Analysis of the truncation error shows that both the staggered and the unstaggered schemes are $O\left[(\Delta x)^2\right]$ and that the leading-order truncation error is smaller for the staggered scheme.

A more revealing comparison of the accuracy of each scheme is provided by examining their discrete dispersion relations in the limit of good temporal resolution ($\omega \Delta t \to 0$). Let c_u and c_s denote the phase speeds of the numerical solutions on the unstaggered and staggered meshes, respectively, then

$$c_u = \frac{c}{k \Delta x} \sin k \Delta x \qquad \text{and} \qquad c_s = \frac{2c}{k \Delta x} \sin \left(\frac{k \Delta x}{2} \right).$$

Curves showing c_u and c_s are plotted as a function of spatial resolution in Fig. 3.2. Also plotted in Fig. 3.2 is the phase-speed obtained when the explicit fourth-order difference (2.6) is used to approximate the spatial derivatives on the unstaggered mesh. As evident in Fig. 3.2, the phase-speed error in the poorly resolved waves is greatly reduced on the staggered mesh. In particular, the $2\Delta x$ wave propagates at 64% of the correct speed on the staggered mesh but remains stationary on the unstaggered mesh.

Substantial improvements in the group velocity of the shortest waves are also achieved using the staggered mesh. Assuming good temporal resolution, the group velocities of the right-moving wave for the second-order schemes on the unstaggered and staggered meshes are, respectively,

$$\left(\frac{\partial \omega}{\partial k} \right)_u = c \cos k \Delta x \qquad \text{and} \qquad \left(\frac{\partial \omega}{\partial k} \right)_s = c \cos \left(\frac{k \Delta x}{2} \right).$$

The group velocity of a $2\Delta x$ wave is $-c$ on the unstaggered mesh and zero on the staggered mesh. Since the correct group velocity is c, both schemes generate serious error, but the error on the unstaggered mesh is twice as large.

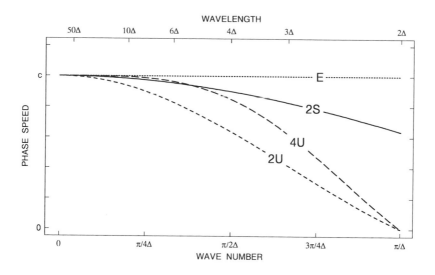

FIGURE 3.2. Phase speed as a function of spatial resolution for the exact solution (E), for second- (2U) and fourth-order (4U) spatial derivatives on an unstaggered mesh, and for second-order spatial derivatives on a staggered mesh (2S).

Temporal Staggering and Forward–Backward Differencing

Just as every unknown variable need not be defined at every spatial grid point, it is also not necessary to define all the unknown variables at each time level. For example, the finite-difference scheme (3.14) and (3.15) could be stepped forward in time using only the values of u at the odd time levels ($[2n+1]\Delta t$) and h at the even time levels ($2n\Delta t$). Staggering u and h in time could, therefore, halve the total computation required for a given simulation. Unfortunately, time-staggering can be difficult to program and may be incompatible with the time-differencing used to integrate other terms in the governing equations, such as those representing advection by the mean wind in (3.1) and (3.2).

Sometimes the advantages of time-staggering can be achieved without actually staggering the unknowns in time by evaluating the various terms in the governing equations at different time levels. In the case of the shallow-water system, the benefits of true time-staggering can be obtained using forward-backward differencing. One possible forward-backward formulation of the spatially staggered finite-difference approximation to the linearized shallow-water system (3.14) and (3.15) is

$$\delta_t u_j^{n+\frac{1}{2}} + g\delta_x h_j^n = 0, \tag{3.17}$$

$$\delta_t h_{j+\frac{1}{2}}^{n+\frac{1}{2}} + H\delta_x u_{j+\frac{1}{2}}^{n+1} = 0. \tag{3.18}$$

In the above, the momentum equation (3.17) is first updated using forward differencing, and then the continuity equation (3.18) is integrated using backward differencing. The backward difference does not introduce an implicit coupling between the unknowns in the forward-backward scheme because u^{n+1} is computed in (3.17) before it is required in (3.18). The overall stability and accuracy of the forward-backward scheme is independent of which equation is updated first; the continuity equation could be integrated first with a forward difference and then the momentum equation could be advanced using a backward difference.

The discrete dispersion relation associated with the forward-backward approximation on the spatially staggered mesh, (3.17) and (3.18), is

$$\sin\left(\frac{\omega\Delta t}{2}\right) = \pm\frac{c\Delta t}{\Delta x}\sin\left(\frac{k\Delta x}{2}\right). \tag{3.19}$$

If $|c\Delta t/\Delta x| < 1$, there will be real-valued ω that satisfy (3.19) and no weakly amplifying double-root solutions, and the scheme will be stable. The time-step restriction introduced by spatial staggering can therefore be avoided if leapfrog differencing is replaced by the forward-backward scheme. In addition, forward-backward differencing involves only two time levels and thereby avoids the introduction of computational modes.

In the case of the linearized shallow-water system with $U = 0$, forward-backward differencing on a spatially staggered mesh is clearly superior to the leapfrog spatially unstaggered scheme. However, in applications where several different terms appear in each governing equation, it is often impossible to choose a single staggering that improves the accuracy of every term. In such situations the advantages of staggering can be substantially reduced. As an example, suppose that the preceding forward-backward spatially staggered approximation is to be extended to shallow-water problems with nonzero mean flow. The simplest $O\left[(\Delta x)^2\right]$ approximation to the spatial derivatives in the advection terms is the same centered difference used in the unstaggered equations (3.9) and (3.10). The staggering of u with respect to h does not interfere with the construction of these centered differences, but it does not improve their accuracy either. The incorporation of the advection terms in the forward-backward time difference poses more of a problem, since a forward-difference approximation to the advection equation (see Section 2.3.2) is unstable. One possible approach is to perform the forward-backward differencing over an interval of $2\Delta t$ and to use the intermediate time level to evaluate the advection terms with leapfrog differencing as follows:

$$\delta_{2t}u_j^n + U\delta_{2x}u_j^n + g\delta_x h_j^{n-1} = 0,$$

$$\delta_{2t}h_{j+\frac{1}{2}}^n + U\delta_{2x}h_{j+\frac{1}{2}}^n + H\delta_x u_{j+\frac{1}{2}}^{n+1} = 0.$$

The discrete dispersion relation for this system is

$$\sin\omega\Delta t = \frac{\Delta t}{\Delta x}\left(U\sin k\Delta x \pm 2c\sin\left(\frac{k\Delta x}{2}\right)\right),$$

and is identical to that which would be obtained if leapfrog time-differencing was used to integrate every term. The benefits of forward-backward time-differencing have been lost. Once again, the numerical scheme includes computational modes, and for $c \gg |U|$ the maximum stable time step is one-half that allowed in the spatially unstaggered scheme. The benefits of spatial staggering are retained, but apply only to that portion of the total velocity of propagation that is produced by the pressure gradient and divergence terms (i.e., by the mechanisms that remain active in the limit $U \to 0$). In situations where $c \gg |U|$, spatial staggering yields substantial improvement, but in those cases where $|U| \gg c$, the errors in the $2\Delta x$ waves introduced by the advection terms dominate the total solution and mask the benefits of spatial staggering. One way to improve accuracy when $|U| \geq c$ is to use fourth-order centered differencing for the advection terms. Fourth-order differencing is not used to obtain high accuracy in the well-resolved waves, but rather to reduce the phase-speed error in the moderately resolved waves to a value comparable to that generated by second-order staggered differencing (see Fig. 3.2).

3.2   Three or More Independent Variables

In most time-dependent problems of practical interest, the unknowns are functions of three or four independent variables (i.e., time and two or three spatial coordinates). The accuracy, consistency, and stability of finite-difference approximations to higher-dimensional equations are determined using essentially the same procedures described in Chapter 2. Two specific examples will be considered in the following section: scalar advection in two dimensions and the Boussinesq equations.

3.2.1   Scalar Advection in Two Dimensions

The advection equation for two-dimensional flow can be approximated using leapfrog-time centered-space schemes that are obvious generalizations of the finite-difference approximations employed in the one-dimensional problem. New considerations involving the incorporation of mixed spatial derivatives do, however, arise in designing accurate and efficient forward-in-time approximations. These considerations will be explored after first examining schemes that are centered in space and time.

Centered-in-Time Schemes

When explicit finite-difference schemes for the integration of one-dimensional problems are extended to two or more spatial dimensions, the stability criteria for the multidimensional problems are often more stringent than those for the one-dimensional formulation. As an example, consider the two-dimensional advection equation

$$\frac{\partial \psi}{\partial t} + U \frac{\partial \psi}{\partial x} + V \frac{\partial \psi}{\partial y} = 0, \tag{3.20}$$

which may be approximated by leapfrog-time, centered second-order space differencing as

$$\delta_{2t}\phi + U\delta_{2x}\phi + V\delta_{2y}\phi = 0. \tag{3.21}$$

Let $\mu = U \Delta t / \Delta x$ and $\nu = V \Delta t / \Delta y$ be the Courant numbers for flow parallel to the x- and y-axes. The finite-difference equation (3.21) has discrete solutions of the form

$$\phi_{m,n}^{j} = e^{i(km\Delta x + \ell n \Delta y - \omega j \Delta t)},$$

provided that ω, k, and ℓ satisfy the discrete dispersion relation

$$\sin(\omega \Delta t) = \mu \sin(k \Delta x) + \nu \sin(\ell \Delta y). \tag{3.22}$$

A necessary condition for stability is that ω be real, or equivalently, that

$$|\mu| + |\nu| \le 1. \tag{3.23}$$

As discussed in connection with (2.92), the sufficient condition for stability actually requires strict inequality in (3.23) in order to avoid weakly growing modes such as

$$\phi_{m,n}^{j} = j \cos \left[\pi (m + n - j)/2 \right],$$

which is a solution to (3.21) when $\mu = \nu = \frac{1}{2}$. The distinction between strict inequality and the condition given in (3.23) is, however, of little practical significance.

In order to better compare this stability condition with that for one-dimensional advection, suppose that $\Delta x = \Delta y = \Delta s$ and express the wind components in terms of wind speed c and direction θ such that $U = c \cos \theta$ and $V = c \sin \theta$. Then the stability condition may be written

$$c\left(|\cos \theta| + |\sin \theta| \right) \frac{\Delta t}{\Delta s} < 1.$$

The left side of the preceding inequality is maximized when the wind blows diagonally across the mesh. If C denotes a bound on the magnitude of the two-dimensional velocity vector, the stability condition becomes $C \Delta t / \Delta s < 1/\sqrt{2}$. Comparing this with the corresponding result for one-dimensional flow, it is apparent that the maximum stable time step in the two-dimensional case is decreased by a factor of $1/\sqrt{2}$.

The stability condition for two-dimensional flow is more restrictive than that for the one-dimensional case because shorter-wavelength disturbances are present on the two-dimensional mesh. The manner in which two-dimensional grids can support wavelengths shorter than $2\Delta x$ is illustrated in Fig. 3.3. In the case shown in Fig. 3.3, $\Delta x = \Delta y = \Delta s$. Grid points beneath a wave crest are indicated by solid circles, those beneath a trough by open circles. The apparent wavelength parallel

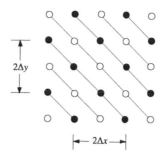

FIGURE 3.3. Distribution of wave crest (solid circles) and wave troughs (open circles) in the shortest-wavelength disturbance resolvable on a square mesh in which $\Delta x = \Delta y$.

to both the x- and y-axes is $2\Delta s$; however, the true wavelength measured along the line $x = y$ is $\sqrt{2}\Delta s$. The maximum stable time step is inversely proportional to the highest frequency resolvable by the numerical scheme, and in the case of the advection equation, the frequency is proportional to the wave number times the wind speed. Since the wave number of a diagonally propagating wave exceeds the apparent wave numbers in the x and y directions by a factor of $\sqrt{2}$, the maximum resolvable frequency is increased by the same factor, and the maximum stable time step is reduced by $1/\sqrt{2}$.

One way to avoid this restriction on the maximum stable time step is to average each spatial derivative as follows (Abarbanel and Gottlieb 1976):

$$\delta_{2t}\phi + U \langle \delta_{2x}\phi \rangle^{2y} + V \langle \delta_{2y}\phi \rangle^{2x} = 0, \tag{3.24}$$

where $\langle \ \rangle^x$ is an averaging operator defined by

$$\langle f(x) \rangle^{nx} = \left[\frac{f(x + n\Delta x/2) + f(x - n\Delta x/2)}{2} \right]. \tag{3.25}$$

The discrete dispersion relation for this "averaging" scheme is

$$\sin(\omega\Delta t) = \mu \sin(k\Delta x)\cos(\ell\Delta y) + \nu \sin(\ell\Delta y)\cos(k\Delta x). \tag{3.26}$$

Let $\xi = k\Delta x$ and $\zeta = \ell\Delta y$, and note that Schwarz's inequality,[1] implies

$$|\sin\xi||\cos\zeta| + |\sin\zeta||\cos\xi| \le \left(\sin^2\xi + \cos^2\xi \right)^{1/2} \left(\sin^2\zeta + \cos^2\zeta \right)^{1/2}.$$

Since

$$|\sin(\omega\Delta t)| = |\mu \sin\xi \cos\zeta + \nu \sin\zeta \cos\xi|$$

$$\le \max\{|\mu|, |\nu|\} \left(|\sin\xi||\cos\zeta| + |\sin\zeta||\cos\xi| \right),$$

---

[1] $\sum_j a_j b_j \le \left(\sum_j a_j^2 \right)^{1/2} \left(\sum_j b_j^2 \right)^{1/2}.$

real values of ω are obtained whenever

$$\max \{|\mu|, |\nu|\} \leq 1. \tag{3.27}$$

As before, suppose that $U = c \cos \theta$, $V = c \sin \theta$, $\Delta x = \Delta y = \Delta s$, and that C is a bound on $|c|$; then requiring strict inequality in (3.27) to guarantee that the leapfrog time difference does not admit weakly unstable modes, the stability condition becomes $C \Delta t / \Delta s < 1$, which is identical to that for the one-dimensional case.

Although the averaging scheme is potentially more efficient because it permits longer time steps, it is also less accurate. This loss of accuracy is not clearly reflected in the truncation error, which is $O\left[(\Delta x)^2\right] + O\left[(\Delta y)^2\right]$ for both the nonaveraged method (3.21) and averaging scheme (3.24). The problems with the averaging scheme appear in the representation of the poorly resolved waves. As discussed in connection with Fig. 3.3, shorter waves are resolvable on a two-dimensional grid, and if properly represented by the spatial differencing, they *should* generate higher-frequency oscillations and reduce the maximum stable time step. The averaging scheme avoids such time-step reduction by artificially reducing the phase speeds of the diagonally propagating waves.

The phase-speed errors introduced by the spatial differencing in both methods may be examined by a generalization of the one-dimensional approach discussed in Section 2.4.1. First consider the propagation of two-dimensional waves in the nondiscretized problem. Waves of the form

$$\psi(x, y, t) = e^{i(kx + \ell y - \omega t)}$$

satisfy the two-dimensional advection equation (3.20), provided that

$$\omega = \mathbf{v} \cdot \mathbf{k}, \tag{3.28}$$

where $\mathbf{v}$ is the velocity vector and $\mathbf{k}$ is the wave number vector with components (k, ℓ). If K denotes the magnitude of $\mathbf{k}$, then the x and y components of the wave number vector may be expressed as

$$k = K \cos \theta \quad \text{and} \quad \ell = K \sin \theta, \tag{3.29}$$

where θ is the angle between the wave number vector and the x-axis. The dispersion relation (3.28) implies that all apparent wave propagation is parallel to the wave number vector. Consider, therefore, the case in which the velocity vector is parallel to the wave number vector. Then if c is the wind speed,

$$U = c \cos \theta \quad \text{and} \quad V = c \sin \theta. \tag{3.30}$$

Substituting (3.29) and (3.30) into the dispersion relation demonstrates that $c = \omega / K$, implying that the phase speed is equal to the wind speed.

As in the one-dimensional case, the phase speeds of the waves generated by the finite-difference approximations (3.21) and (3.24) are defined as $c^* = \omega^* / K$,

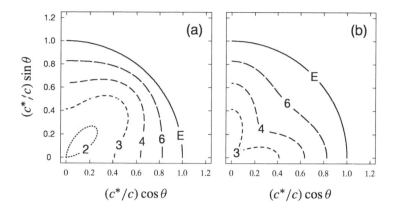

FIGURE 3.4. Polar plot of the relative phase speeds of $2\Delta s$ (shortest dashed line), $3\Delta s$, $4\Delta s$, and $6\Delta s$ (longest dashed line) waves generated by (a) the nonaveraged finite-difference formula and (b) the averaging scheme. Also plotted is the curve for perfect propagation (labeled E), which is independent of the wavelength and appears as a circular arc of radius unity.

where ω^* is the frequency satisfying the discrete-dispersion relations (3.22) or (3.26). In the limit of good time resolution, the phase speed for the nonaveraging scheme is

$$c^*_{na} = \frac{1}{K}\left(U\,\frac{\sin(k\Delta x)}{\Delta x} + V\,\frac{\sin(\ell\Delta y)}{\Delta y}\right),$$

and that for the averaging scheme is

$$c^*_a - \frac{1}{K}\left(U\,\frac{\sin(k\Delta x)}{\Delta x}\cos(\ell\Delta y) + V\,\frac{\sin(\ell\Delta y)}{\Delta y}\cos(k\Delta x)\right).$$

Suppose $\Delta x = \Delta y = \Delta s$ and define $\beta = K\,\Delta s$; then using (3.29) and (3.30) to evaluate the velocity and wave number components in the preceding expressions, the relative phase speed for each scheme becomes

$$\frac{c^*_{na}}{c} = \frac{\cos\theta\,\sin(\beta\cos\theta) + \sin\theta\,\sin(\beta\sin\theta)}{\beta}$$

and

$$\frac{c^*_a}{c} = \frac{\cos\theta\,\sin(\beta\cos\theta)\cos(\beta\sin\theta) + \sin\theta\,\sin(\beta\sin\theta)\cos(\beta\cos\theta)}{\beta}.$$

These expressions for the relative phase speed were evaluated for wavelengths $2\pi/K = 2\Delta s$, $3\Delta s$, $4\Delta s$, and $6\Delta s$ and plotted as a function of θ in Fig. 3.4. Figure 3.4 is a polar plot in which the relative phase speed of a $3\Delta s$ wave propagating along a ray extending outward from the origin at an angle θ with respect to the x-axis is plotted as the radial distance between the origin and the point where that ray intersects the dashed curve labeled "3." As indicated in Fig. 3.4a, the nonaveraging scheme does not resolve the propagation of $2\Delta s$ waves parallel to either the x- or the y-axis, but $2\Delta s$ waves can move at greatly reduced speed along the diagonal line $x = y$ ($\theta = \pi/4$). The phase-speed error diminishes as the wavelength increases, with the maximum error in $6\Delta s$ waves being no larger than 20%. These results can be compared with the relative phase speed curves for the averaging scheme plotted in Fig. 3.4b. The averaging scheme generates substantial errors in the phase speed of waves moving diagonally along the line $x = y$; the $2\Delta s$ wave does not propagate at all, and even the $6\Delta s$ wave is significantly retarded. These reduced phase speeds allow the averaging scheme to remain stable for large time steps, but as is apparent in Fig. 3.4, the enhanced stability is obtained at the cost of increased phase-speed errors in the poor and moderately resolved waves.

Forward-in-Time Schemes

Assuming that $U \geq 0$ and $V \geq 0$, one generalization of the upstream method to the two-dimensional advection equation (3.20) is

$$\delta_t \phi_{m,n}^{j+\frac{1}{2}} + U \delta_x \phi_{m-\frac{1}{2},n}^{j} + V \delta_y \phi_{m,n-\frac{1}{2}}^{j} = 0. \tag{3.31}$$

The stability of this scheme may be investigated using the standard Von Neumann method. Let

$$\phi_{m,n}^{j} = A^j e^{i(km\Delta x + \ell n \Delta y)}. \tag{3.32}$$

Then

$$A = 1 - \mu(1 - e^{-i\xi}) - \nu(1 - e^{-i\zeta}), \tag{3.33}$$

where as before, $\mu = U\Delta t/\Delta x$, $\nu = V\Delta t/\Delta y$, $\xi = k\Delta x$, and $\zeta = \ell \Delta y$. Necessary and sufficient conditions for stability are

$$0 \leq \mu, \qquad 0 \leq \nu, \qquad \text{and} \qquad \mu + \nu \leq 1. \tag{3.34}$$

The necessity of the preceding may be established by considering the three cases $\xi = 0$, $\zeta = 0$, and $\xi = \zeta$, for each of which the dependence of the amplification factor on the wave number reduces to an expression of the same form as in the one-dimensional case (2.25). The sufficiency of (3.34) follows from

$$|A| \leq |1 - \mu - \nu| + |\mu e^{-i\xi}| + |\nu e^{-i\zeta}|$$
$$= |1 - \mu - \nu| + |\mu| + |\nu|,$$

which implies that $|A| \leq 1$ whenever μ and ν satisfy (3.34).

Suppose that $\Delta x = \Delta y = \Delta s$ and that C is a bound on the magnitude of the two-dimensional velocity vector; then provided that the spatial differences are evaluated in the upstream direction, the stability condition is $C\Delta t/\Delta s \leq 1/\sqrt{2}$. As was the case with the leapfrog approximation (3.21), the maximum stable time step is approximately 30% less than that in the analogous one-dimensional problem. In contrast to the situation with centered-in-time schemes, it is, however, possible to improve the stability of forward-in-time approximations to the two-dimensional advection equation while simultaneously improving at least some aspects of their accuracy.

One natural way to derive the upstream approximation to the one-dimensional advection equation is through the method of characteristics (Courant et al. 1952). The true solution of the one-dimensional advection equation is constant along characteristic curves whose slopes are $dx/dt = U$. The characteristic curve passing through the point $[m\Delta x, (j + 1)\Delta t]$ also passes through $[(m - \mu)\Delta x, j\Delta t]$, and as discussed in Section 6.1.1, the upstream scheme

$$\phi_m^{j+1} = (1 - \mu)\phi_m^j + \mu\phi_{m-1}^j$$

is obtained if the value of ϕ^j at $(m - \mu)\Delta x$ is estimated from ϕ_m^j and ϕ_{m-1}^j by linear interpolation. The method of characteristics is naturally extended to the two-dimensional advection problem using bilinear interpolation, in which case

$$\phi_{m,n}^{j+1} = (1 - \mu)\left[(1 - \nu)\phi_{m,n}^j + \nu\phi_{m,n-1}^j\right]$$
$$+\mu\left[(1 - \nu)\phi_{m-1,n}^j + \nu\phi_{m-1,n-1}^j\right] \tag{3.35}$$

(Bates and McDonald 1982). Colella (1990), who derived the same scheme using a finite-volume argument (see Section 5.7.2), has referred to this scheme as the CTU (corner transport upstream) method.

The CTU method may be expressed in the alternative form

$$\delta_t\phi_{m,n}^{j+\frac{1}{2}} + U\delta_x\phi_{m-\frac{1}{2},n}^j + V\delta_y\phi_{m,n-\frac{1}{2}}^j = UV\Delta t\delta_x\delta_y\phi_{m-\frac{1}{2},n-\frac{1}{2}}^j, \tag{3.36}$$

which shows that it differs from (3.31) by a term that is a finite-difference approximation to

$$UV\Delta t\frac{\partial^2\psi}{\partial x\partial y}.$$

The addition of this cross-derivative term improves the stability of the CTU scheme relative to (3.31). Substituting (3.32) into (3.36) yields

$$A = \left(1 - \mu + \mu e^{-i\xi}\right)\left(1 - \nu + \nu e^{-i\zeta}\right).$$

Each factor in the preceding has the same form as (2.25), so the magnitude of each factor will be less than one, and the CTU scheme will be stable if $0 \le \mu \le 1$ and $0 \le \nu \le 1$. If the computational mesh is uniform and the wind speed is bounded by C, the stability condition becomes $C\Delta t/\Delta s \le 1$, which is identical to that for the upstream approximation to the one-dimensional problem.

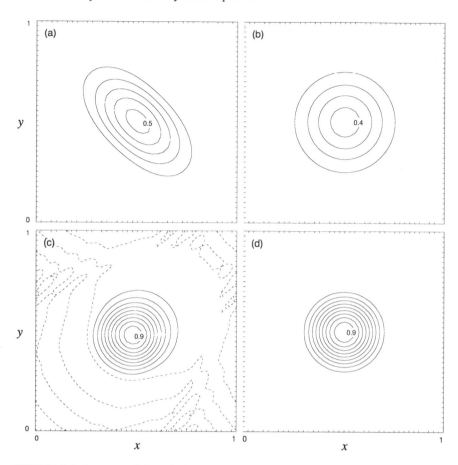

FIGURE 3.5. Contours of the solution to the constant-wind-speed advection equation on a two-dimensional periodic mesh obtained using: (a) the two-dimensional upstream method, (b) the CTU method, (c) the upstream-biased Lax-Wendroff method and (d) the true solution. The contour interval is 0.1, and the zero contour is dashed.

Solutions generated by the CTU method are compared with those obtained using the two-dimensional upstream scheme (3.31) in Fig. 3.5. The spatial domain is $0 \le x \le 1$, $0 \le y \le 1$ and is discretized using a square mesh with $\Delta x = \Delta y = 0.025$. The lateral boundary conditions are periodic, and the initial condition is

$$\phi(x, y, 0) = \frac{1}{2}[1 + \cos(\pi r)],$$

where

$$r(x, y) = \min\left(1, 4\sqrt{(x - 1/2)^2 + (y - 1/2)^2}\right).$$

The wind is directed diagonally across the mesh with $U = V = 1$. The time step is chosen such that $\mu = \nu = 0.5$. The results are displayed at $t = 1$, at which time the flow has made exactly one circuit around the domain. The solution obtained using (3.31) is shown in Fig. 3.5a; that obtained using the CTU method appears in Fig. 3.5b, and the true solution is plotted in Fig. 3.5d. Since they are first-order methods, both upstream solutions are heavily damped. The solution generated by the two-dimensional upstream method has also developed a pronounced asymmetry, whereas that produced by the CTU method appears axisymmetric. The CTU solution is, however, damped slightly more than the two-dimensional upstream solution.

The tendency of the two-dimensional upstream scheme to distort the solution as shown in Fig. 3.5a can be understood by noting that (3.31) is a second-order approximation to the modified equation

$$\frac{\partial \psi}{\partial t} + U \frac{\partial \psi}{\partial x} + V \frac{\partial \psi}{\partial y} = \frac{U \Delta x}{2} (1 - \mu) \frac{\partial^2 \psi}{\partial x^2} + \frac{V \Delta y}{2} (1 - \nu) \frac{\partial^2 \psi}{\partial y^2} - U V \Delta t \frac{\partial^2 \psi}{\partial x \partial y}.$$

The CTU method, on the other hand, is a second-order approximation to

$$\frac{\partial \psi}{\partial t} + U \frac{\partial \psi}{\partial x} + V \frac{\partial \psi}{\partial y} = \frac{U \Delta x}{2} (1 - \mu) \frac{\partial^2 \psi}{\partial x^2} + \frac{V \Delta y}{2} (1 - \nu) \frac{\partial^2 \psi}{\partial y^2};$$

the mixed spatial derivative does not appear because it is canceled (to second order) by the finite difference on the right side of (3.36). The influence of the mixed spatial derivative on the error in the two-dimensional upstream scheme can be isolated by considering the simplified equation

$$\frac{\partial \varphi}{\partial t} = -\frac{\partial^2 \varphi}{\partial x \partial y}.$$

Expressing the preceding in a coordinate system rotated by 45°, so that the new independent variables are $r = x + y$ and $s = x - y$, yields

$$\frac{\partial \varphi}{\partial t} = \frac{\partial^2 \varphi}{\partial s^2} - \frac{\partial^2 \varphi}{\partial r^2}.$$

Thus, perturbations in ψ diffuse along lines of constant r and "anti-diffuse" along lines of constant s. Whenever $UV > 0$, this process of diffusion and anti-diffusion tends to distort the solution as shown in Fig. 3.5a. In contrast, the leading-order error in the CTU method is purely isotropic when $U = V$ and $\Delta x = \Delta y$.

Second-order forward-in-time approximations can be obtained using the Lax–Wendroff method. The scheme

$$\delta_t \phi^{j+\frac{1}{2}} + U \delta_{2x} \phi^j + V \delta_{2y} \phi^j = \frac{U^2 \Delta t}{2} \delta_x^2 \phi^j + \frac{V^2 \Delta t}{2} \delta_y^2 \phi^j \tag{3.37}$$

has sometimes been proposed as a generalization of the one-dimensional Lax–Wendroff method for constant-wind-speed advection in two dimensions, but this scheme is not second-order accurate because the right side is not a second-order approximation to the leading-order truncation error in the forward time difference,

$$\frac{\Delta t}{2}\frac{\partial^2 \psi}{\partial t^2} = \frac{\Delta t}{2}\left(u\frac{\partial^2 \psi}{\partial x^2} + v\frac{\partial^2 \psi}{\partial y^2} + 2uv\frac{\partial^2 \psi}{\partial x \partial y}\right).$$

In addition to its lack of accuracy, (3.37) is unstable for all Δt.

A stable second-order Lax–Wendroff approximation to the two-dimensional constant-wind-speed advection equation may be written in the form

$$\delta_t \phi^{j+\frac{1}{2}} + U\delta_{2x}\phi^j + V\delta_{2y}\phi^j = \frac{U^2 \Delta t}{2}\delta_x^2 \phi^j + \frac{V^2 \Delta t}{2}\delta_y^2 \phi^j + UV\Delta t \delta_{2x}\delta_{2y}\phi^j.$$

Necessary and sufficient conditions for the stability of this method are that

$$\mu^{2/3} + \nu^{2/3} \leq 1$$

(Turkel 1977). If C is a bound on the magnitude of the two-dimensional wind vector and $\Delta x = \Delta y = \Delta s$, the stability condition becomes $C\Delta t/\Delta s \leq \frac{1}{2}$, which is more restrictive than that for the two-dimensional leapfrog and upstream schemes, and much more restrictive than the stability condition for the CTU method.

The stability of the two-dimensional Lax–Wendroff approximation can be greatly improved using an upstream finite-difference approximation to the mixed spatial derivative (Leonard et al. 1993). If $U \geq 0$ and $V \geq 0$, the resulting scheme is

$$\delta_t \phi_{m,n}^{j+\frac{1}{2}} + U\delta_{2x}\phi_{m,n}^j + V\delta_{2y}\phi_{m,n}^j$$

$$= \frac{U^2 \Delta t}{2}\delta_x^2 \phi_{m,n}^j + \frac{V^2 \Delta t}{2}\delta_y^2 \phi_{m,n}^j + UV\Delta t \delta_x \delta_y \phi_{m-\frac{1}{2},n-\frac{1}{2}}^j. \quad (3.38)$$

The stability condition for this scheme is identical to that for the CTU method, $0 \leq \mu \leq 1$ and $0 \leq \nu \leq 1$ (Hong et al. 1997). If the mixed spatial derivative is calculated in the upstream direction and the magnitude of the two-dimensional wind vector is bounded by C, the stability condition for an isotropic mesh may be expressed as $C\Delta t/\Delta s \leq 1$.

Numerical solutions computed using (3.38) appear in Fig. 3.5c. The initial condition and the physical and numerical parameters are identical to those used to obtain the two-dimensional upstream and CTU solutions. As might be expected in problems where there is adequate numerical resolution, the amplitude error in the second-order solution is far less than that in either first-order solution. The leading-order dispersive error in the second-order method does, however, generate regions where ϕ is slightly negative. Techniques for minimizing or eliminating these spurious negative values will be discussed in Chapter 5.

3.2.2   Systems of Equations in Several Dimensions

Although the stability analysis of systems of linear finite-difference equations in several spatial dimensions is conceptually straightforward, in practice it can be

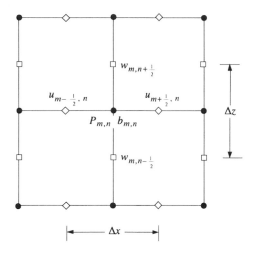

FIGURE 3.6. Distribution of the dependent variables on a staggered mesh for the finite-difference approximation of the two-dimensional Boussinesq system.

somewhat tedious. The easiest way to obtain necessary conditions for the stability of linear centered-difference approximations to problems involving wave-like flow is to examine the discrete dispersion relation. As an example, consider the linearized Boussinesq equations governing the incompressible flow of a continuously stratified fluid in the x-z plane. If U and 0 are the constant basic-state horizontal and vertical wind speeds, the linearized versions of (1.61)–(1.63) become

$$\frac{\partial u}{\partial t} + U\frac{\partial u}{\partial x} + \frac{\partial P}{\partial x} = 0, \tag{3.39}$$

$$\frac{\partial w}{\partial t} + U\frac{\partial w}{\partial x} + \frac{\partial P}{\partial z} = b \tag{3.40}$$

$$\frac{\partial b}{\partial t} + U\frac{\partial b}{\partial x} + N^2 w = 0, \tag{3.41}$$

$$\frac{\partial u}{\partial x} + \frac{\partial w}{\partial z} = 0, \tag{3.42}$$

where as before, P is the perturbation pressure divided by ρ_0, b is the buoyancy, and N^2 is the Boussinesq approximation to the Brunt–Väisälä frequency.

This system is often discretized using the staggered mesh shown in Fig. 3.6, which is sometimes referred to as the Arakawa "C" grid (Arakawa and Lamb 1977). One important property of the C-grid is that it allows an accurate computation of the pressure gradient and velocity divergence using a compact stencil on the staggered mesh, as in the following finite-difference approximation to (3.39)–(3.42)

$$\delta_{2t}u_{m-\frac{1}{2},n} + U\delta_{2x}u_{m-\frac{1}{2},n} + \delta_x P_{m-\frac{1}{2},n} = 0, \tag{3.43}$$

$$\delta_{2t} w_{m,n-\frac{1}{2}} + U \delta_{2x} w_{m,n-\frac{1}{2}} + \delta_z P_{m,n-\frac{1}{2}} = \left(b_{m,n-\frac{1}{2}} \right)^z, \tag{3.44}$$

$$\delta_{2t} b_{m,n} + U \delta_{2x} b_{m,n} + N^2 \left(w_{m,n} \right)^z = 0, \tag{3.45}$$

$$\delta_x u_{m,n} + \delta_z w_{m,n} = 0. \tag{3.46}$$

This system of finite-difference equations does not provide a complete algorithm for the time integration of the Boussinesq system because it does not include an equation for P^{j+1}. There is no equation for P^{j+1} because the Boussinesq system does not have a prognostic pressure equation. Techniques for determining the time tendency of the pressure field will be discussed in Section 7.1. For the present, it is assumed that the pressure is determined in some unspecified way that guarantees satisfaction of the finite-difference system (3.43)–(3.46).

Substituting solutions of the form

$$\begin{pmatrix} u \\ w \\ b \\ P \end{pmatrix}_{m,n}^{j} = \begin{pmatrix} u_0 \\ w_0 \\ b_0 \\ P_0 \end{pmatrix} e^{i(km\Delta x + \ell n \Delta z - \omega j \Delta t)}$$

into the finite-difference system (3.43)–(3.46) and setting the determinant of the coefficients of u_0, w_0, b_0, and P_0 to zero, one obtains the discrete dispersion relation

$$\tilde{\omega} = U \tilde{k}_2 \pm \frac{\tilde{N} \tilde{k}_1}{(\tilde{k}_1^2 + \tilde{\ell}^2)^{1/2}}, \tag{3.47}$$

in which

$$\tilde{k}_1 = \frac{\sin(k\Delta x/2)}{\Delta x/2}, \qquad \tilde{k}_2 = \frac{\sin(k\Delta x)}{\Delta x},$$

$$\tilde{\omega} = \frac{\sin \omega \Delta t}{\Delta t}, \qquad \tilde{\ell} = \frac{\sin(\ell \Delta z/2)}{\Delta z/2}, \qquad \tilde{N} = N \cos(\ell \Delta z/2).$$

The preceding is identical to the dispersion relation for the continuous problem except that the true frequencies and wave numbers are replaced by their numerical approximations. Note that there are two different approximations to the horizontal wave number: $\tilde{k}_1$ arises from a centered finite-difference approximation to the horizontal derivative on a Δx-wide stencil, and $\tilde{k}_2$ is associated with finite differences on a $2\Delta x$-wide stencil. The factor $\tilde{k}_1$ is associated with the discretized pressure-gradient and divergence operators, whereas $\tilde{k}_2$ is associated with the advection operator.

The numerical solution should not grow with time because linear-wave solutions to Boussinesq equations are nonamplifying. A necessary condition for the absence of amplifying waves in the discretized solution is that $\tilde{\omega}$ be real or, equivalently, that the magnitude of the right side of (3.47) be less than one. Since the factor multiplying N in (3.47) is bounded by unity, this stability condition is

$$\left(\frac{|U|}{\Delta x} + N \right) \Delta t \le 1.$$

3.3   Splitting into Fractional Steps

More efficient integration schemes can often be obtained by splitting complex finite-difference formulae into a series of fractional steps. As an example, the two-dimensional advection equation (3.20) might be approximated by the scheme

$$\phi^s = \phi^n - \frac{U\,\Delta t}{2}\delta_{2x}(\phi^n + \phi^s), \tag{3.48}$$

$$\phi^{n+1} = \phi^s - \frac{V\,\Delta t}{2}\delta_{2y}(\phi^s + \phi^{n+1}), \tag{3.49}$$

where ϕ^s is a temporary quantity computed during the first fractional step. (Note that ϕ^s is not a consistent approximation to the true solution at any particular time level, which complicates the specification of boundary conditions for ϕ^s and makes it difficult to use multilevel time differencing in time-split methods.) If the computational domain contains $N_x \times N_y$ grid points, each integration step of (3.48) and (3.49) requires the solution of $N_x + N_y$ tridiagonal systems. In contrast, a single integration step of the corresponding unsplit formula,

$$\phi^{n+1} = \phi^n - \frac{U\,\Delta t}{2}\delta_{2x}(\phi^n + \phi^{n+1}) - \frac{V\,\Delta t}{2}\delta_{2y}(\phi^n + \phi^{n+1}), \tag{3.50}$$

requires the solution of a linear system with an $N_x N_y \times N_x N_y$ coefficient matrix whose bandwidth is the smaller of $2N_x + 1$ and $2N_y + 1$. The fractional step approach is more efficient because fewer computations are required to solve the $N_x + N_y$ tridiagonal problems than to solve the single linear system associated with the band matrix. Some loss of accuracy may, however, be introduced when the problem is split into fractional steps.

In order to examine the accuracy and stability of fractional-step splittings in a general context, consider the class of partial differential equations of the form

$$\frac{\partial \psi}{\partial t} = \mathcal{L}\psi = \mathcal{L}_1\psi + \mathcal{L}_2\psi, \tag{3.51}$$

where $\mathcal{L}$ is the linear operator formed by the sum of two time-independent linear operators $\mathcal{L}_1$ and $\mathcal{L}_2$. In the preceding case of two-dimensional advection,

$$\mathcal{L}_1 = U\frac{\partial}{\partial x} \quad \text{and} \quad \mathcal{L}_2 = V\frac{\partial}{\partial y}, \tag{3.52}$$

and $\mathcal{L}$ is split into operators involving derivatives parallel to each spatial coordinate. In other applications, the governing equations might be split into subproblems representing different physical processes. In a simulation of chemically reacting flow, for example, the terms representing advection might be grouped together into $\mathcal{L}_1$, while terms describing chemistry might appear in $\mathcal{L}_2$.

Since $\mathcal{L}$ is assumed to be time-independent, the exact solution to (3.51) may be written in the form

$$\psi(t) = \exp(t\mathcal{L})\psi(0),$$

where the exponential of the operator $\mathcal{L}$ is defined by the infinite series

$$\exp(t\mathcal{L}) = I + t\mathcal{L} + \frac{t^2}{2}\mathcal{L}^2 + \frac{t^3}{6}\mathcal{L}^3 + \cdots,$$

and I is the identity operator. The change in ψ over one time step is therefore

$$\psi(t + \Delta t) = \exp[(\Delta t + t)\mathcal{L}] = \exp(\Delta t \mathcal{L})\exp(t\mathcal{L}) = \exp(\Delta t \mathcal{L})\psi(t).$$

Suppose that a numerical approximation to the preceding has the form

$$\phi^{n+1} = \mathcal{F}(\Delta t)\phi^n. \tag{3.53}$$

If the global truncation error in this approximation is $O\left[(\Delta t)^n\right]$, the local trunca-tion error[2] is $O\left[(\Delta t)^{n+1}\right]$ and

$$\mathcal{F}(\Delta t) = \exp(\Delta t \mathcal{L}) + O\left[(\Delta t)^{n+1}\right]. \tag{3.54}$$

In practice, $\mathcal{F}$ may involve approximations to spatial derivatives, but the funda-mental properties of the fractional-step method can be explored without explicitly considering the discretization of the spatial derivatives.

3.3.1   Split Explicit Schemes

The unsplit forward-difference approximation

$$\mathcal{F}(\Delta t) = (I + \Delta t \mathcal{L}_1 + \Delta t \mathcal{L}_2)$$

satisfies (3.54) with $n = 1$, as would be expected, since forward differencing is $O(\Delta t)$ accurate. It is easy to achieve the same level of accuracy using $O(\Delta t)$-accurate fractional steps. For example, the split scheme consisting of the two for-ward steps

$$\phi^s = (I + \Delta t \mathcal{L}_1)\phi^n, \tag{3.55}$$
$$\phi^{n+1} = (I + \Delta t \mathcal{L}_2)\phi^s \tag{3.56}$$

generates the approximate finite-difference operator

$$(I + \Delta t \mathcal{L}_2)(I + \Delta t \mathcal{L}_1) = I + \Delta t \mathcal{L}_1 + \Delta t \mathcal{L}_2 + (\Delta t)^2 \mathcal{L}_2 \mathcal{L}_1,$$

and is therefore $O(\Delta t)$ accurate.

It is more difficult to design split schemes that are $O\left[(\Delta t)^2\right]$ accurate unless the operators $\mathcal{L}_1$ and $\mathcal{L}_2$ commute. Suppose the forward differences in (3.55) and (3.56) are replaced by second-order numerical operators $\mathcal{F}_1$ and $\mathcal{F}_2$. One possible choice for $\mathcal{F}_1$ and $\mathcal{F}_2$ is the second-order Runge–Kutta method, in which (3.53)

---

[2] See Section 2.3.2 for the definition of local and global truncation error.

would be evaluated in the two-stages

$$\phi^* = \phi^n + \frac{\Delta t}{2\beta}\mathcal{L}\phi^n,$$

$$\phi^{n+1} = \phi^n + \beta\Delta t\mathcal{L}\phi^* + (1-\beta)\Delta t\mathcal{L}\phi^n,$$

and β is a free parameter (see Section 2.3.3). A second possibility is the Lax–Wendroff method, in which $\mathcal{L}$ and $\mathcal{L}^2$ are both evaluated during a single forward time step such that

$$\phi^{n+1} = \phi^n + \Delta t\mathcal{L}\phi^n + \frac{(\Delta t)^2}{2}\mathcal{L}^2\phi^n.$$

Whatever the exact formulation of $\mathcal{F}_1$ and $\mathcal{F}_2$, since they are of second order,

$$\mathcal{F}_1(\Delta t) = I + \Delta t\mathcal{L}_1 + \frac{(\Delta t)^2}{2}\mathcal{L}_1^2 + O\left[(\Delta t)^3\right],$$

$$\mathcal{F}_2(\Delta t) = I + \Delta t\mathcal{L}_2 + \frac{(\Delta t)^2}{2}\mathcal{L}_2^2 + O\left[(\Delta t)^3\right],$$

and the composite operator for a complete integration step is

$$[\mathcal{F}_2(\Delta t)][\mathcal{F}_1(\Delta t)] =$$

$$I + \Delta t(\mathcal{L}_2 + \mathcal{L}_1) + \frac{(\Delta t)^2}{2}(\mathcal{L}_2^2 + 2\mathcal{L}_2\mathcal{L}_1 + \mathcal{L}_1^2) + O\left[(\Delta t)^3\right].$$

The preceding will not be a second-order approximation to the exact operator

$$\exp(\Delta t\mathcal{L}) = I + \Delta t(\mathcal{L}_1 + \mathcal{L}_2) + \frac{(\Delta t)^2}{2}(\mathcal{L}_1^2 + \mathcal{L}_1\mathcal{L}_2 + \mathcal{L}_2\mathcal{L}_1 + \mathcal{L}_2^2) + O\left[(\Delta t)^3\right],$$

unless $\mathcal{L}_1\mathcal{L}_2 = \mathcal{L}_2\mathcal{L}_1$. Unfortunately, in many practical applications $\mathcal{L}_1$ and $\mathcal{L}_2$ do not commute. For example, if $\mathcal{L}_1$ and $\mathcal{L}_2$ are the one-dimensional advection operators defined by (3.52) and U and V are functions of x and y, then

$$\mathcal{L}_1\mathcal{L}_2 = U\frac{\partial V}{\partial x}\frac{\partial}{\partial y} + UV\frac{\partial^2}{\partial x\partial y},$$

$$\mathcal{L}_2\mathcal{L}_1 = V\frac{\partial U}{\partial y}\frac{\partial}{\partial x} + UV\frac{\partial^2}{\partial x\partial y},$$

and $\mathcal{L}_1\mathcal{L}_2 \neq \mathcal{L}_2\mathcal{L}_1$ unless

$$U\frac{\partial V}{\partial x} = V\frac{\partial U}{\partial y} = 0.$$

Strang (1968) noted that even if $\mathcal{L}_1$ and $\mathcal{L}_2$ don't commute, $\mathcal{F}_1$ and $\mathcal{F}_2$ can still be used to construct the following $O\left[(\Delta t)^2\right]$ operator:

$$[\mathcal{F}_1(\Delta t/2)][\mathcal{F}_2(\Delta t)][\mathcal{F}_1(\Delta t/2)]. \tag{3.57}$$

It might appear that this splitting requires 50% more computation than the binary products considered previously. However, since

$$[\mathcal{F}_1(\Delta t/2)][\mathcal{F}_1(\Delta t/2)] = [\mathcal{F}_1(\Delta t)] + O\left[(\Delta t)^3\right],$$

two consecutive $\mathcal{F}_1$-integration steps of length $\Delta t/2$ may be combined into a single $\mathcal{F}_1$-integration of length Δt without sacrificing second-order accuracy. Thus, if the physical solution is not required at every time step, a series of consecutive steps involving the operator (3.57) can be consolidated as

$$[\mathcal{F}_1(\Delta t/2)][\mathcal{F}_2(\Delta t)][\mathcal{F}_1(\Delta t)] \cdots [\mathcal{F}_2(\Delta t)][\mathcal{F}_1(\Delta t/2)],$$

where a single half step has been performed at the beginning and the end of the interval and all other steps are full steps of length Δt. Such consolidation can greatly improve efficiency in problems where the approximate solution is not needed at every time step (e.g., if the solution is required only once every 100 time steps for output to a plotting program).

Problems can be split into more than two subproblems, although if the individual operators do not commute, the effort required to evaluate an $O\left[(\Delta t)^2\right]$ split can be substantial. If the original problem is approximated by a series of numerical operators $\mathcal{F}_1, \mathcal{F}_2, \ldots, \mathcal{F}_N$ and the least accurate of the $\mathcal{F}_j$ is $O\left[(\Delta t)^n\right]$, then the simplest fractional step splitting

$$[\mathcal{F}_1(\Delta t)][\mathcal{F}_2(\Delta t)] \cdots [\mathcal{F}_N(\Delta t)]$$

is $O(\Delta t)$ unless all the individual operators commute, in which case the accuracy is $O\left[(\Delta t)^n\right]$. When $\mathcal{F}_1, \mathcal{F}_2, \ldots, \mathcal{F}_N$ don't commute, $O\left[(\Delta t)^2\right]$ accuracy can be obtained using higher-dimensional forms of the Strang splitting (3.57). In the case of three operators, the Strang splitting is

$$[\mathcal{F}_1(\Delta t/2)][\mathcal{F}_2(\Delta t/2)][\mathcal{F}_3(\Delta t)][\mathcal{F}_2(\Delta t/2)][\mathcal{F}_1(\Delta t/2)].$$

3.3.2   Split Implicit Schemes

Although Strang splitting can be used to obtain second-order accuracy when explicit time-differencing is used in the individual fractional steps, other techniques are required when the time-differencing is implicit. The trapezoidal scheme is the most important second-order implicit time difference used in split schemes. The trapezoidal approximation to the general partial differential equation (3.51) may be expressed using the preceding operator notation as

$$\left[I - \frac{\Delta t}{2}\mathcal{L}\right]\phi^{n+1} = \left[I + \frac{\Delta t}{2}\mathcal{L}\right]\phi^n,$$

or

$$\phi^{n+1} = \left[I - \frac{\Delta t}{2}\mathcal{L}\right]^{-1}\left[I + \frac{\Delta t}{2}\mathcal{L}\right]\phi^n.$$

If trapezoidal time-differencing is employed in two successive fractional steps, as in (3.48) and (3.49), the composite operator is

$$\mathcal{F}(\Delta t) = \left[I - \frac{\Delta t}{2}\mathcal{L}_2\right]^{-1}\left[I + \frac{\Delta t}{2}\mathcal{L}_2\right]\left[I - \frac{\Delta t}{2}\mathcal{L}_1\right]^{-1}\left[I + \frac{\Delta t}{2}\mathcal{L}_1\right]. \quad (3.58)$$

This operator may be expanded using the formula for the sum of a geometric series,

$$(1 - x)^{-1} = 1 + x + x^2 + x^3 + \cdots,$$

to yield

$$\mathcal{F}(\Delta t) = I + \Delta t\,(\mathcal{L}_2 + \mathcal{L}_1) + \frac{(\Delta t)^2}{2}(\mathcal{L}_2^2 + 2\mathcal{L}_2\mathcal{L}_1 + \mathcal{L}_1^2) + O\left[(\Delta t)^3\right].$$

This is the same expression obtained using second-order explicit differences in each fractional step, and as before, it will not agree with $\exp(\Delta t \mathcal{L}_1 + \Delta t \mathcal{L}_2)$ through $O\left[(\Delta t)^2\right]$ unless $\mathcal{L}_1$ and $\mathcal{L}_2$ commute.

Even if $\mathcal{L}_1$ and $\mathcal{L}_2$ don't commute, an $O\left[(\Delta t)^2\right]$ approximation can be achieved by the following permutation of the operators in (3.58):

$$\mathcal{F}(\Delta t) = \left[I - \frac{\Delta t}{2}\mathcal{L}_2\right]^{-1}\left[I - \frac{\Delta t}{2}\mathcal{L}_1\right]^{-1}\left[I + \frac{\Delta t}{2}\mathcal{L}_1\right]\left[I + \frac{\Delta t}{2}\mathcal{L}_2\right].$$

The resulting scheme,

$$\left[I - \frac{\Delta t}{2}\mathcal{L}_1\right]\left[I - \frac{\Delta t}{2}\mathcal{L}_2\right]\phi^{n+1} = \left[I + \frac{\Delta t}{2}\mathcal{L}_1\right]\left[I + \frac{\Delta t}{2}\mathcal{L}_2\right]\phi^n, \quad (3.59)$$

may be efficiently implemented using the Peaceman–Rachford *alternating direction* algorithm

$$\left[I - \frac{\Delta t}{2}\mathcal{L}_1\right]\phi^s = \left[I + \frac{\Delta t}{2}\mathcal{L}_2\right]\phi^n, \quad (3.60)$$

$$\left[I - \frac{\Delta t}{2}\mathcal{L}_2\right]\phi^{n+1} = \left[I + \frac{\Delta t}{2}\mathcal{L}_1\right]\phi^s. \quad (3.61)$$

In order to demonstrate the equivalence of (3.59) and the Peaceman–Rachford formulation, apply $I - \frac{\Delta t}{2}\mathcal{L}_1$ to each side of (3.61) and observe that

$$\left[I - \frac{\Delta t}{2}\mathcal{L}_1\right]\left[I - \frac{\Delta t}{2}\mathcal{L}_2\right]\phi^{n+1} = \left[I - \frac{\Delta t}{2}\mathcal{L}_1\right]\left[I + \frac{\Delta t}{2}\mathcal{L}_1\right]\phi^s$$

$$= \left[I + \frac{\Delta t}{2}\mathcal{L}_1\right]\left[I - \frac{\Delta t}{2}\mathcal{L}_1\right]\phi^s$$

$$= \left[I + \frac{\Delta t}{2}\mathcal{L}_1\right]\left[I + \frac{\Delta t}{2}\mathcal{L}_2\right]\phi^n,$$

where the second equality is obtained because $\mathcal{L}_1$ commutes with itself, and substitution from (3.60) is used to form the final equality.

3.3.3 Stability of Split Schemes

When the numerical operators $\mathcal{F}_1$ and $\mathcal{F}_2$ commute, the stability of the split scheme $\mathcal{F}_1\mathcal{F}_2$ is guaranteed by the stability of the individual operators. In order to demonstrate this, suppose $\mathbf{A}_1$ and $\mathbf{A}_2$ are the amplification matrices associated with $\mathcal{F}_1$ and $\mathcal{F}_2$, and note that if $\mathcal{F}_1$ and $\mathcal{F}_2$ commute, their amplification matrices also commute. The amplification matrix for the split scheme is $\mathbf{A}_1\mathbf{A}_2$, and

$$\|(\mathbf{A}_1\mathbf{A}_2)^n\| = \|\mathbf{A}_1\mathbf{A}_2\mathbf{A}_1\mathbf{A}_2\cdots\mathbf{A}_1\mathbf{A}_2\|$$
$$= \|(\mathbf{A}_1)^n(\mathbf{A}_2)^n\| \tag{3.62}$$
$$\leq \|(\mathbf{A}_1)^n\|\,\|(\mathbf{A}_2)^n\|, \tag{3.63}$$

and it is apparent that the split scheme inherits the stability properties of the individual operators.

If $\mathcal{F}_1$ and $\mathcal{F}_2$ don't commute, equality (3.62) does not hold, and

$$\|(\mathbf{A}_1\mathbf{A}_2)^n\| \leq \|\mathbf{A}_1\|^n\|\mathbf{A}_2\|^n$$

is the best bound that can be obtained without specific knowledge of $\mathbf{A}_1$ and $\mathbf{A}_2$. The preceding guarantees the stability of the split scheme when $\|\mathbf{A}_1\|$ and $\|\mathbf{A}_2\|$ are less than or equal to unity; however, as discussed in Section 3.1.1, $\|\mathbf{A}_1\| \leq 1$ is not necessary for the stability of $\mathcal{F}_1$.

As an illustration of the preceding, consider the system of ordinary differential equations

$$\frac{du}{dt} = icv + ibu, \tag{3.64}$$

$$\frac{dv}{dt} = icu, \tag{3.65}$$

where b and c are real constants. Wave solutions to the preceding problem exist of the form

$$\begin{pmatrix} u \\ v \end{pmatrix} = \begin{pmatrix} 1 \\ -c/\omega \end{pmatrix} Ae^{-i\omega t},$$

where A is an arbitrary amplitude and ω is one of the two real roots to the dispersion relation

$$\omega^2 + b\omega - c^2 = 0.$$

A split scheme in which each step is stable but the composite scheme is unconditionally unstable can be obtained by constructing the following finite-difference approximation to (3.64) and (3.65). In the first step, integrate the terms involving c using forward-backward differencing:

$$\frac{u^s - u^n}{\Delta t} = icv^n,$$

$$\frac{v^s - v^n}{\Delta t} = icv^s,$$

and then integrate the term involving b using trapezoidal differencing:

$$\frac{u^{n+1} - u^s}{\Delta t} = ib\left(\frac{u^{n+1} + u^s}{2}\right),$$

$$v^{n+1} = v^s.$$

Letting $\hat{c} = c\Delta t$, the first step may be written in matrix form as

$$\begin{pmatrix} 1 & 0 \\ -i\hat{c} & 1 \end{pmatrix} \begin{pmatrix} u^s \\ v^s \end{pmatrix} = \begin{pmatrix} 1 & i\hat{c} \\ 0 & 1 \end{pmatrix} \begin{pmatrix} u^n \\ v^n \end{pmatrix},$$

or

$$\begin{pmatrix} u^s \\ v^s \end{pmatrix} = \begin{pmatrix} 1 & i\hat{c} \\ i\hat{c} & 1 - \hat{c}^2 \end{pmatrix} \begin{pmatrix} u^n \\ v^n \end{pmatrix}. \tag{3.66}$$

Denote the matrix in (3.66) by $\mathbf{A}_1$. The eigenvalues of $\mathbf{A}_1$ are

$$1 - \frac{\hat{c}^2}{2} \pm \frac{i\hat{c}}{2}(4 - \hat{c}^2)^{1/2}.$$

For $|\hat{c}| \le 2$, the magnitude of both eigenvalues is unity, and since $\mathbf{A}_1$ is symmetric, the scheme is stable. Observe, however, that the norm of $\mathbf{A}_1$,

$$\|\mathbf{A}_1\|_2 = \left(1 + \frac{\hat{c}^4}{2} + \frac{\hat{c}^2}{2}(4 + \hat{c}^4)^{1/2}\right)^{1/2},$$

exceeds unity for all nonzero Δt.

If $\hat{b} = b\Delta t$, the second fractional step may be written as

$$\begin{pmatrix} u^{n+1} \\ v^{n+1} \end{pmatrix} = \begin{pmatrix} \dfrac{2 + i\hat{b}}{2 - i\hat{b}} & 0 \\ 0 & 1 \end{pmatrix} \begin{pmatrix} u^s \\ v^s \end{pmatrix}. \tag{3.67}$$

Let $\mathbf{A}_2$ represent the amplification matrix in (3.67). One can easily show that $\|\mathbf{A}_2\| = 1$, so this scheme is also stable.

The amplification matrix for the composite scheme is

$$\mathbf{A}_2\mathbf{A}_1 = \begin{pmatrix} \dfrac{2 + i\hat{b}}{2 - i\hat{b}} & \dfrac{2 + i\hat{b}}{2 - i\hat{b}}i\hat{c} \\ i\hat{c} & 1 - \hat{c}^2 \end{pmatrix}.$$

Since $\mathbf{A}_2\mathbf{A}_1 \ne \mathbf{A}_1\mathbf{A}_2$, the stability of the individual steps does not guarantee the stability of the composite scheme. Moreover, the inequality

$$\|\mathbf{A}_2\mathbf{A}_1\| \le \|\mathbf{A}_2\|\|\mathbf{A}_1\|$$

cannot be used to show stability, since $\|\mathbf{A}_1\| > 1$. In fact, numerical calculations show that the magnitude of the largest eigenvalue of $\mathbf{A}_2\mathbf{A}_1$ is greater than unity for all $|\hat{b}|, |\hat{c}| > 0$, so the composite scheme is unconditionally unstable.

3.4   Diffusion, Sources, and Sinks

Some background dissipation is present in most physical systems, and even when the phenomena of interest are governed by essentially inviscid processes, it is often necessary to incorporate the effects of this dissipation in a numerical model. Sources and sinks also may need to be included. In the following, we will consider finite-difference methods for the incorporation of diffusive and Rayleigh-damping processes. Numerical approximations to the pure diffusion equation

$$\frac{\partial \psi}{\partial t} = M \frac{\partial^2 \psi}{\partial x^2} \tag{3.68}$$

(where $M > 0$ is a molecular diffusivity) will be examined before considering the combined effects of advection and diffusion or Rayleigh damping.

3.4.1   Pure Diffusion

Suppose that (3.68) is approximated using the forward-time centered-space scheme

$$\frac{\phi_j^{n+1} - \phi_j^n}{\Delta t} = M \left(\frac{\phi_{j+1}^n - 2\phi_j^n + \phi_{j-1}^n}{(\Delta x)^2} \right).$$

The standard Von Neumann stability analysis yields an amplification factor for this scheme of

$$A_k = 1 - 2\nu(1 - \cos k\Delta x), \tag{3.69}$$

where $\nu = M\Delta t/(\Delta x)^2$. The amplification factor is maximized for $k = \pi/\Delta x$ (the $2\Delta x$ mode), and thus $|A_k| \leq 1$ and the numerical solution decays with time, provided that $0 \leq \nu \leq \frac{1}{2}$. Note that in contrast to the conditional stability criteria obtained for finite-difference approximations to the advection equation, this scheme does not remain stable as $\Delta t, \Delta x \to 0$ unless Δt decreases much more rapidly than Δx, i.e., unless $\Delta t/\Delta x < O(\Delta x)$. This makes the preceding scheme very inefficient at high spatial resolution.

Although it guarantees that the solution will not blow up, the criterion $0 \leq \nu \leq \frac{1}{2}$ is not adequate to ensure a qualitatively correct simulation of the $2\Delta x$ mode. The amplitude b_k of the kth Fourier mode of the exact solution to (3.68) satisfies

$$\frac{db_k}{dt} = -Mk^2 b_k, \tag{3.70}$$

implying that the correct amplification factor for the kth mode is

$$\frac{b_k(t + \Delta t)}{b_k(t)} = e^{-Mk^2 \Delta t},$$

which (for $M > 0$) is a real number between 0 and 1. However, if $\frac{1}{4} < \nu \le \frac{1}{2}$, the numerical amplification factor for the $2\Delta x$ mode lies in the interval $[-1, 0)$, and as a consequence, the sign of the $2\Delta x$ mode flips every time step as it gradually damps toward zero. If one wishes to avoid "over damping" the poorly resolved modes, Δt must satisfy the more restrictive criterion that $0 \le \nu \le \frac{1}{4}$.

As is the case with numerical approximations to the advection equation, the stability of an explicit finite-difference approximation to the diffusion equation is limited by the $2\Delta x$ mode. The behavior of the $2\Delta x$ mode relative to the longer modes in the advection problem is, however, quite different from that in the diffusion problem. In the advection problem the shortest waves translate without loss of amplitude (and may even amplify as the result of deformation in the wind field or nonlinear processes), so any errors in the simulation of the short waves can have a serious impact on the accuracy of the overall solution. On the other hand, as implied by (3.70), diffusion preferentially damps the shortest modes, and after a brief time the amplitude in these modes becomes negligible relative to that of the total solution. Since the accuracy with which the short waves are simulated is irrelevant once those waves have dissipated, an acceptable approximation to the overall solution can often be obtained without accurately simulating the transient decay of the most poorly resolved initial perturbations. It can therefore be very advantageous to approximate the diffusion equation using unconditionally stable schemes like the trapezoidal method, for which the time step is limited only by accuracy considerations. In particular, the time step can be chosen to accurately simulate the transient decay of the physical scales of primary interest, while any inaccuracies generated by this time step in the poorly resolved modes are hidden by their rapid decay.

If the time-differencing in the finite-difference approximation to the diffusion equation is trapezoidal, the resulting scheme

$$\frac{\phi_j^{n+1} - \phi_j^n}{\Delta t} = \frac{M}{2} \left(\delta_x^2 \phi_j^{n+1} + \delta_x^2 \phi_j^n \right) \tag{3.71}$$

is known as the *Crank–Nicolson* method. The amplification factor for this scheme is

$$A_k = \frac{1 - \nu(1 - \cos k\Delta x)}{1 + \nu(1 - \cos k\Delta x)},$$

and since $M > 0$, it follows that $|A_k| \le 1$ and the scheme is stable for all Δt. Assuming that boundary conditions are specified at the edges of the spatial domain, (3.71) constitutes a tridiagonal linear system for the unknown ϕ_j^{n+1}, which can be solved with minimal computational effort as discussed in the Appendix. Although the Crank–Nicolson approximation to the one-dimensional diffusion equation yields a set of algebraic equations that can be solved very efficiently, when the same method is used to approximate the three-dimensional diffusion equation, the matrix associated with the resulting algebraic system has a much wider bandwidth, and a considerable increase in computational effort is required to obtain its solution. Nevertheless, in comparison with explicit finite-difference methods, the extra work per time step required by the trapezoidal scheme can usually be more than offset by using a much larger time step.

3.4.2 Advection and Diffusion

Now consider the combined advection–diffusion problem, which in one dimension is governed by the equation

$$\frac{\partial \psi}{\partial t} + c \frac{\partial \psi}{\partial x} = M \frac{\partial^2 \psi}{\partial x^2}. \tag{3.72}$$

Before discussing how to approximate the time derivative in the preceding, we will examine accuracy issues that arise solely from the approximation of the spatial derivatives. If the first spatial derivative is approximated by an upstream difference (with $c > 0$) and the second derivative is approximated by the standard three-point stencil, the resulting differential–difference approximation to (3.72) is

$$\frac{d\phi_j}{dt} + c \left(\frac{\phi_j - \phi_{j-1}}{\Delta x} \right) = M \left(\frac{\phi_{j+1} - 2\phi_j + \phi_{j-1}}{(\Delta x)^2} \right). \tag{3.73}$$

Evaluating the truncation error in the preceding shows that it is an $O\left[(\Delta x)^2\right]$-accurate approximation to the modified equation

$$\frac{\partial \psi}{\partial t} + c \frac{\partial \psi}{\partial x} = M \left(1 + \frac{Pe}{2} \right) \frac{\partial^2 \psi}{\partial x^2}, \tag{3.74}$$

where $Pe = c\Delta x / M$ is the numerical Péclet number. The Péclet number is a nondimensional parameter classically defined as the ratio of the strength of thermal advection to the strength of thermal diffusion.[3] Since the length scale in the numerical Péclet number is the grid spacing, Pe is a measure of the relative strengths of advection and diffusion at the smallest spatial scales resolved on the numerical mesh. A comparison of the modified equation (3.74) with the original advection–diffusion equation (3.72) shows that the differential–difference approximation (3.73) generates an inaccurate approximation to the diffusion term unless $Pe \ll 1$, i.e., unless diffusion dominates advective transport on the shortest resolvable scales. This difficulty arises because the total diffusion is dominated

---

[3]The Péclet number is completely analogous to the more familiar Reynolds number, which is the ratio of momentum advection to momentum diffusion. The difference between the Péclet and Reynolds numbers is due to the difference in the diffusivities of heat and momentum. In particular, the ratio of the Péclet number to the Reynolds number is equal to the Prandtl number, which is the ratio of the kinematic viscosity (or momentum diffusivity) to the thermal diffusivity.

by numerical diffusion unless the molecular diffusivity is very large or the grid resolution is very fine. In order to accurately represent the diffusion term in low-viscosity flow, it is generally necessary to use a less diffusive approximation to $\partial \psi / \partial x$, such as a centered difference or a higher-order one-sided difference.

As the horizontal resolution increases, the numerical Péclet number decreases, and in principle, there is some grid size at which diffusive transport dominates advective transport in the shortest resolvable modes. Nevertheless, in many problems involving low-viscosity flow this grid size may be several orders of magnitude smaller than the physical scales of primary interest, so that there is no possibility of resolving the scales at which molecular viscosity dominates numerical diffusion without exceeding the resources of the most advanced computers. Even when molecular diffusion has no direct influence on the resolved-scale fields, an essentially inviscid transport by sub-grid-scale eddies may produce turbulent mixing whose influence on the resolved-scale fields is often parametrized by a diffusion term in which the true molecular diffusivity M is replaced by an "eddy" diffusivity $\tilde{M}$ (Yih 1977, p. 572). The eddy diffusivity is generally parametrized such that $\tilde{M}$ is proportional to the mesh size, in which case the numerical Péclet number $c\Delta x / \tilde{M}$ does not decrease as the grid is refined, and it is not possible to make the parametrized eddy diffusion dominate the numerical diffusion by reducing the mesh size.

Now consider the effects of time-differencing on the stability of finite-difference approximations to the advection–diffusion equation. In order to isolate the role of time-differencing on the numerical solution, (3.72) is Fourier transformed with respect to the x coordinate to obtain

$$\frac{d\hat{\psi}}{dt} + ick\hat{\psi} = -Mk^2\hat{\psi}.$$

The coefficients in the preceding ordinary differential equation can be further simplified to yield the following prototype equation for the investigation of time-differences in the advection–diffusion problem:

$$\frac{d\psi}{dt} = i\omega\psi + \lambda\psi, \tag{3.75}$$

where ω and λ are real. Solutions to (3.75) satisfy $|\psi(t)| \leq |\psi(0)|$, provided that $\lambda \leq 0$. Practically useful numerical approximations to (3.75) must satisfy the analogous stability condition that $|\phi^n| \leq |\phi^0|$. Numerical approximations to (3.75) computed with a specific value of the parameters $\omega\Delta t$ and $\lambda\Delta t$ are defined to be *absolutely stable* if $|\phi^n| \leq |\phi^0|$ for all n.

The amplification factor associated with a forward-difference approximation to (3.75) is

$$A_f = 1 + \tilde{\lambda} + i\tilde{\omega},$$

where $\tilde{\omega} = \omega\Delta t$ and $\tilde{\lambda} = \lambda\Delta t$. Thus the forward-difference approximation to (3.75) will be absolutely stable when

$$(1 + \tilde{\lambda})^2 + \tilde{\omega}^2 \leq 1. \tag{3.76}$$

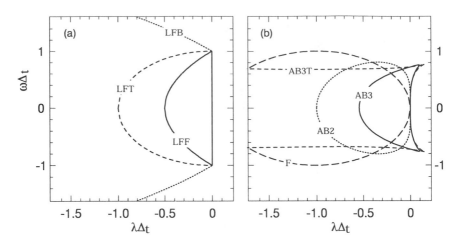

FIGURE 3.7. (a) Regions of useful stability in the $\tilde{\lambda}$-$\tilde{\omega}$ plane if the oscillatory forcing is integrated using leapfrog differencing and the damping is integrated using the forward (LFF), trapezoidal (LFT), or backward (LFB) methods. (b) Region of absolute stability when both the oscillatory forcing and the damping are both integrated using the forward (F), the second-order (AB2), or the third-order (AB3) Adams–Bashforth scheme. Also shown in (b) is the region of absolute stability when the integration of the oscillatory forcing is third-order Adams–Bashforth and the damping is integrated using the trapezoidal method (AB3T). Note the difference in the horizontal and vertical scales.

The portion of the $\tilde{\lambda}$-$\tilde{\omega}$ plane satisfying this inequality is the circle of unit radius centered at $(-1, 0)$ plotted in Fig. 3.7b. As discussed in Section 2.3, forward differencing generates unstable solutions to the pure advection problem. This can also be deduced from Fig. 3.7b by noting that the region of absolute stability does not include a finite segment of the $\tilde{\omega}$ axis. Forward differencing will therefore fail to generate a stable numerical solution to the advection–diffusion problem unless the diffusion is relatively large.

One might attempt to improve the stability of the numerical solution to the low-viscosity advection–diffusion problem by using leapfrog time-differencing, which as demonstrated in Section 2.3.4 is stable in the zero-viscosity limit. The two amplification factors associated with the leapfrog approximation to (3.75) are

$$A_{\text{lf}} = i\tilde{\omega} + \tilde{\lambda} \pm \left[(i\tilde{\omega} + \tilde{\lambda})^2 + 1 \right]^{1/2}.$$

The conditions under which both the computational and physical modes are stable can be most easily established by defining a complex number ζ such that

$$i \cos \zeta = i\tilde{\omega} + \tilde{\lambda}.$$

Then

$$A_{\text{lf}} = i \cos \zeta \pm \sin \zeta = i e^{\mp i\zeta},$$

and thus the two amplification factors associated with the leapfrog scheme have magnitudes $|e^{-i\zeta}|$ and $|e^{i\zeta}|$. One of these will exceed unity unless ζ is real, and since

$$\zeta = \cos^{-1}(\tilde{\omega} - i\tilde{\lambda}),$$

the condition for absolute stability[4] reduces to

$$\tilde{\lambda} = 0 \quad \text{and} \quad |\tilde{\omega}| < 1.$$

The region of the $\tilde{\lambda}$-$\tilde{\omega}$ plane within which the leapfrog scheme is stable is just a line segment along the $\tilde{\omega}$-axis. Any amount of diffusion destabilizes the leapfrog scheme.

A stable approximation to (3.75) that uses leapfrog time differencing for the oscillatory forcing may be obtained by evaluating the diffusion term at time level $n-1$ such that

$$\frac{\phi^{n+1} - \phi^{n-1}}{2\Delta t} = i\omega\phi^n + \lambda\phi^{n-1}. \tag{3.77}$$

Note that although the standard leapfrog method is accurate to $O\left[(\Delta t)^2\right]$, this approach introduces an $O(\Delta t)$ truncation error in the approximation of the diffusion term. The amplification factor for the leapfrog-forward scheme (3.77) is

$$A_{\text{lff}} = i\tilde{\omega} \pm (-\tilde{\omega}^2 + 1 + 2\tilde{\lambda})^{1/2}.$$

When $1 + 2\tilde{\lambda} > \tilde{\omega}^2$, the square root is real; both amplification factors have the same magnitude, and

$$|A_{\text{lff}}|^2 = 1 + 2\tilde{\lambda}.$$

It follows that the leapfrog-forward scheme will be stable when $1 + 2\tilde{\lambda} > \tilde{\omega}^2$ and $\tilde{\lambda} \leq 0$, or

$$\frac{\tilde{\omega}^2 - 1}{2} < \tilde{\lambda} \leq 0.$$

The region of the $\tilde{\lambda}$-$\tilde{\omega}$ plane that satisfies this inequality lies inside the curve labeled LFF in Fig. 3.7a. Consideration of the case $1 + 2\tilde{\lambda} \leq \tilde{\omega}^2$ shows that the actual region of the $\tilde{\lambda}$-$\tilde{\omega}$ plane throughout which $|A_{\text{lff}}| \leq 1$ is a larger triangular region with vertices at $(\tilde{\lambda}, \tilde{\omega})$ equal to $(0, 1)$, $(-1, 0)$, and $(0, -1)$. This larger region of formal stability is of no practical use, however, since A_{lff} is pure imaginary when $1 + 2\tilde{\lambda} \leq \tilde{\omega}^2$, and as a consequence the numerical solution undergoes a $4\Delta t$ oscillation independent of the actual value of $\omega\Delta t$. Because the region of useful stability for the leapfrog-forward scheme is relatively small, this scheme is appropriate only for problems with very low viscosity. Even when the value of $\lambda\Delta t$ is as small as 0.3, stability considerations require a significant reduction in $\omega\Delta t$ relative to that which would be stable in the inviscid limit.

Much better stability properties can be obtained using A-stable finite-difference schemes. An *A-stable* finite-difference approximation to (3.75) is absolutely stable for all $\lambda\Delta t \leq 0$. An A-stable method generates a bounded numerical solu-

---

[4]As discussed in Section 2.3.4, this scheme is subject to a weak instability when $\tilde{\omega} = \pm 1$ and $\tilde{\lambda} = 0$.

tion to (3.75) whenever the true solution is bounded. An ideal scheme for the solution of (3.75) would be A-stable, accurate, and efficient. Both the backward and trapezoidal time-difference approximations to (3.75) are A-stable. The backward and trapezoidal methods are also implicit; there are no A-stable explicit schemes. Although the trapezoidal method is stable and accurate, it can be inefficient to approximate every term in the governing equations using trapezoidal time-differencing. Some of the terms that generate oscillatory forcing in practical applications are nonlinear (e.g., the nonlinear advection operator $\mathbf{v} \cdot \nabla \mathbf{v}$ in the momentum equations (1.31)), and in order to avoid solving nonlinear implicit algebraic equations on every time step, the nonlinear advection terms are usually approximated using an explicit finite difference. Consider, therefore, a pair of hybrid schemes in which the oscillatory forcing in (3.75) is integrated using leapfrog differencing, but the dissipation term is approximated using either backward or trapezoidal differencing.

If (3.75) is approximated using the leapfrog-backward scheme

$$\frac{\phi^{n+1} - \phi^{n-1}}{2\Delta t} = i\omega\phi^n + \lambda\phi^{n+1}, \tag{3.78}$$

the combinations of $\tilde{\lambda}$ and $\tilde{\omega}$ that yield stable physically reasonable solutions satisfy

$$\tilde{\omega}^2 < 1 - 2\tilde{\lambda} \quad \text{and} \quad \tilde{\lambda} \le 0.$$

This region lies within and to the left of the curves labeled LFB in Fig. 3.7a. The region of useful stability for the leapfrog-trapezoidal method,

$$\frac{\phi^{n+1} - \phi^{n-1}}{2\Delta t} = i\omega\phi^n + \frac{\lambda}{2}\left(\phi^{n+1} + \phi^{n-1}\right), \tag{3.79}$$

is

$$\tilde{\omega}^2 + \tilde{\lambda}^2 < 1 \quad \text{and} \quad \tilde{\lambda} \le 0,$$

and lies within the curve labeled LFT in Fig. 3.7a. Since damping reduces the amplitude of the true solution to (3.75), and since the damping term in both (3.78) and (3.79) is approximated by an A-stable scheme, one might hope that the stability condition for the leapfrog-differenced purely oscillatory problem would be a sufficient condition for the stability of the full hybrid scheme. This is the case for the leapfrog-backward method, which is stable whenever $|\omega\Delta t| < 1$. On the other hand, in order to maintain the useful stability of the leapfrog-trapezoidal method, the time step must be reduced as the damping rate $|\lambda|$ increases, and whenever $\lambda\Delta t < -1$ the method is either unstable or exhibits $4\Delta t$ oscillations independent of the actual value of $\omega\Delta t$.

The leapfrog-backward scheme is not a particularly attractive method because it is only first-order accurate, yet being implicit, it requires essentially the same computational overhead as the more accurate trapezoidal method. The stability and accuracy potentially available through a trapezoidal approximation to the damping term can be better realized if the leapfrog approximation to the oscillatory term is replaced by the third-order Adams–Bashforth method. Before examining the combined Adams–Bashforth-trapezoidal scheme, consider the simpler third-order Adams–Bashforth approximation to the entire advection–diffusion problem

$$\frac{\phi^{n+1} - \phi^n}{\Delta t} = \frac{(i\omega + \lambda)}{12} \left(23\phi^n - 16\phi^{n-1} + 5\phi^{n-2} \right). \tag{3.80}$$

The region of the $\tilde{\lambda}$-$\tilde{\omega}$ plane for which solutions obtained using (3.80) are absolutely stable can be determined by numerically solving a cubic equation for the amplification factor. The stable region for this third-order method, which is plotted in Fig. 3.7b, is only slightly smaller than that for the first-order leapfrog-forward scheme shown in Fig. 3.7a. Also shown in Fig. 3.7b is the region of absolute stability for the second-order Adams–Bashforth solution (2.46). One might suppose that the second-order Adams–Bashforth method would be a more economical alternative to the third-order scheme. If the flow is sufficiently viscous, this can be the case. However, as shown in Fig. 3.7b, the second-order Adams–Bashforth method does not generate absolutely stable solutions to the undamped ($\lambda = 0$) problem. On the other hand, when $\lambda = 0$, the third-order Adams–Bashforth method is absolutely stable for $|\omega \Delta t| < 0.72$.

If the damping term in (3.80) is approximated using trapezoidal time-differencing such that

$$\frac{\phi^{n+1} - \phi^n}{\Delta t} = \frac{i\omega}{12} \left(23\phi^n - 16\phi^{n-1} + 5\phi^{n-2} \right) + \frac{\lambda}{2} \left(\phi^{n+1} + \phi^n \right), \tag{3.81}$$

the region of absolute stability expands, as indicated in Fig. 3.7b, so that it is determined almost entirely by the value of $|\omega \Delta t|$. If Δt is small enough to yield stable solutions to the inviscid problem, then the full advection–diffusion problem will be stable for almost all values of $\lambda < 0$. Therefore, the Adams–Bashforth-trapezoidal approximation appears to be the best scheme for simulating advection–diffusion in situations where there are both regions of zero viscosity and patches of high diffusivity. Patches of high eddy diffusivity may appear in a nominally inviscid fluid in localized regions where the flow is dynamically unstable to small-scale perturbations. The high eddy diffusion in these isolated regions can have a severe impact on the time step of the overall numerical integration unless an artificial cap is imposed on the maximum eddy diffusivity or the time-differencing is approximated with a very stable method like the Adams–Bashforth-trapezoidal scheme.

The preceding analyses of the prototype ordinary differential equation (3.75) are sufficient to characterize the stability of the various time differencing schemes qualitatively, but they do not yield the precise stability limits of the complete finite-difference equation obtained when the spatial derivatives in the advection–diffusion equation are approximated by finite differences. As an example, consider the forward-time centered-space approximation

$$\frac{\phi_j^{n+1} - \phi_j^n}{\Delta t} + c\delta_{2x}\phi_j^n = M\delta_x^2\phi_j^n.$$

The standard Von Neumann analysis shows that the numerical solution will be nonamplifying when

$$|A_k|^2 = [1 - 2\nu(1 - \cos k\Delta x)]^2 + \mu^2 \sin^2 k\Delta x \le 1,$$

for all k in the interval $[0, \pi/\Delta x]$. Here, as before, $\mu = c\Delta t/\Delta x$ and $\nu = M\Delta t/(\Delta x)^2$. Necessary and sufficient conditions for the stability of this scheme are

$$0 \le \nu \le \frac{1}{2} \quad \text{and} \quad \mu^2 \le 2\nu. \tag{3.82}$$

In order to establish the necessity of these conditions note that the first condition is required for stability when $k\Delta x = \pi$, and the second condition is required for stability in the limit $k\Delta x \to 0$, in which case

$$|A_k|^2 \to 1 - (2\nu - \mu^2)(k\Delta x)^2.$$

In order to establish that (3.82) is sufficient for stability, suppose that $\mu^2 \le 2\nu$. Then

$$|A_k|^2 \le (1 - 2\nu(1 - \cos k\Delta x))^2 + 2\nu \sin^2 k\Delta x$$
$$= 1 - 2\nu(1 - 2\nu)(1 - \cos k\Delta x)^2,$$

which is less than unity whenever $0 \le \nu \le \frac{1}{2}$. Note that the stable region of the ν-μ plane defined by (3.82) is a portion of a parabola, whereas the stable region of the $\tilde{\lambda}$-$\tilde{\omega}$ plane defined by (3.76) is a circle.

3.4.3 Advection with Sources and Sinks

Sources and sinks typically appear as functions of the temporal and spatial coordinates and the undifferentiated unknown variables. The evolution of the unknown variables in the pure source–sink problem is therefore governed by ordinary differential equations. Elementary numerical methods for the solution of ordinary differential equations have been discussed in Section 2.3 and in the preceding Section. Further details may be found in standard texts such as Gear (1971) or Iserles (1996). In the following we will consider the combined effects of advection and sources or sinks in two prototypical cases.

First, suppose that the source or sink is a function only of the coordinate variables, in which case the advection–source–sink equation is

$$\frac{\partial \psi}{\partial t} + c\frac{\partial \psi}{\partial x} = s(x, t).$$

Almost any finite-difference scheme suitable for the approximation of the pure advection problem can be trivially modified to approximate the preceding equation. The only subtlety involves the numerical specification of $s(x, t)$. It is natural to specify $s(x, t)$ at the finest spatial and temporal scales resolvable on the space–

time grid, but this may generate excessive noise in the numerical solution. One should make a distinction between the shortest scale present on the numerical mesh and the shortest scale at which the finite-difference scheme can be expected to yield physically meaningful results. The accuracy of the solution is generally not improved by applying forcing at wavelengths too short to be adequately simulated by the numerical scheme. Since almost all numerical methods do a very poor job of simulating $2\Delta x$ waves and $2\Delta t$ oscillations, it is usually unwise to include spatial and temporal scales in $s(x, t)$ corresponding to wavelengths shorter than about $3\Delta x$ or periods shorter than about $3\Delta t$.[5] The optimal cutoff depends on the numerical scheme and the nature of the problem being approximated. See Lander and Hoskins (1997) for an example in which a cutoff wave number is determined for external forcing in a spectral model of the Earth's atmosphere.

Now suppose that the sink is a linear function of ψ, so that the advection–source–sink equation is

$$\frac{\partial \psi}{\partial t} + c\frac{\partial \psi}{\partial x} = -r\psi, \tag{3.83}$$

where positive values of r represent sinks and negative values represent sources. Confusion can arise in assessing the stability of numerical approximations to (3.83). Consider the stability of the upstream approximation

$$\frac{\phi_j^{n+1} - \phi_j^n}{\Delta t} + c\left(\frac{\phi_j^n - \phi_{j-1}^n}{\Delta x}\right) = -r\phi_j^n. \tag{3.84}$$

The amplification factor for this scheme is

$$A_k = (1 - \mu - \lambda) + \mu e^{-ik\Delta x}, \tag{3.85}$$

where $\lambda = r\Delta t$. If $r < 0$, the true solution grows with time, and it is clearly inappropriate to require $|A_k| \le 1$. In this case, all that is required is that the scheme be sufficiently stable to converge in the limit $\Delta t \to 0$, $\Delta x \to 0$, for which the Von Neumann condition is

$$|A_k| \le 1 + \gamma \Delta t, \tag{3.86}$$

where γ is a constant independent of Δt and Δx. In order to establish criteria guaranteeing satisfaction of (3.86) let

$$\tilde{A}_k = (1 - \mu) + \mu e^{-ik\Delta x},$$

which is just the amplification factor for upstream differencing. Since $A_k = \tilde{A}_k - r\Delta t$,

$$|A_k| \le |\tilde{A}_k| + |r\Delta t|.$$

---

[5] For similar reasons, it is often unwise to include $2\Delta x$ features in the initial data.

If $0 \le \mu \le 1$, then as demonstrated in Section 2.2.2, $|\tilde{A}_k| \le 1$ and

$$|A_k| \le 1 + |r| \Delta t,$$

which satisfies the stability criterion (3.86). The finite-difference approximation (3.84) to the advection–source equation is therefore stable whenever the associated approximation to the pure advection problem is stable.

Clearly, the methodology used in the preceding stability analysis can be generalized to a wider class of problems. Let $L(\psi)$ be a linear operator involving partial derivatives of ψ and consider the family of partial differential equations of the form

$$\frac{\partial \psi}{\partial t} + L(\psi) + r\psi = 0.$$

As demonstrated by Strang (1964), the range of Δt for which any explicit two-time-level approximation to the preceding partial differential equation satisfies the stability condition (3.86) is independent of the value of r. Unfortunately, it is rather easy to misinterpret this result. The Strang perturbation theorem guarantees only that the value of r has no influence on the ability of consistent finite-difference approximations to converge to the correct solution in the limit of Δt, $\Delta x \to 0$. The value of r does affect the boundedness of numerical solutions computed with finite values of Δt and Δx.

The solution to the advection–sink problem (3.83) is bounded whenever $r > 0$, and in such circumstances the numerical approximation obtained using finite Δx and Δt should satisfy the more strict stability condition $|A_k| \le 1$. The conditions on $r \Delta t$ required to guarantee a nongrowing solution may be determined as follows. Using (3.85),

$$|A_k|^2 = (1 - \lambda)^2 - 2\mu(1 - \mu - \lambda)(1 - \cos k \Delta x).$$

Now consider two cases. First, if $\mu(1 - \mu - \lambda) \ge 0$, the largest amplification factor occurs when $k = 0$ and

$$|A_0| = (1 - \lambda)^2.$$

In this case all $|A_k|$ are less than unity when $(1 - \lambda)^2 \le 1$ or $0 \le \lambda \le 2$. This inequality is always satisfied, since $r > 0$, $c > 0$, and by assumption $\mu(1 - \mu - \lambda) \ge 0$. All the stability restrictions on Δt arise therefore from the the second case, for which $\mu(1 - \mu - \lambda) < 0$. In this case the largest amplification factor occurs for $k \Delta x = \pi$ and

$$|A_{\pi \Delta x}|^2 = (1 - \lambda)^2 - 4\mu(1 - \mu - \lambda) = (1 - \lambda - 2\mu)^2.$$

Thus all $|A_k|$ are less than unity when $(1 - \lambda - 2\mu)^2 \le 1$ or $0 \le \lambda + 2\mu \le 2$, or equivalently,

$$0 \le \frac{c\Delta t}{\Delta x}\left(1 + \frac{r\Delta x}{2c}\right) \le 1.$$

The last expression shows that the value of r ceases to restrict the maximum stable time step as $\Delta x \to 0$. This is consistent with the implication of the Strang perturbation theorem that the stability condition sufficient to guarantee convergence cannot depend on r. The value of r may, nevertheless, have a dramatic impact on the maximum stable time step when Δx is finite.

3.5   Linear Equations with Variable Coefficients

Some of the simplest equations of practical interest are linear equations with variable coefficients. Consider, for example, one-dimensional advection by a spatially varying wind speed, which is governed by the partial differential equation

$$\frac{\partial \psi}{\partial t} + c(x)\frac{\partial \psi}{\partial x} = 0. \tag{3.87}$$

Suppose $\phi_j(t)$ is the numerical approximation to $\psi(t, j\Delta x)$ and that c is available at the same set of spatial grid points. One obvious differential–difference approximation to the preceding is

$$\frac{d\phi_j}{dt} + c_j\delta_{2x}\phi_j = 0. \tag{3.88}$$

The stability of this scheme is often assessed by "freezing" $c(x)$ at some constant value c_0 and studying the stability of the family of frozen-coefficient problems obtained by varying c_0 over the range of all possible $c(x)$. It should be noted, however, that in some pathological examples there is no relation between the stability of the variable-coefficient problem and the corresponding family of frozen-coefficient problems (see Problem 10).

Suppose that the time derivative in (3.88) is replaced by leapfrog time-differencing; then a necessary condition for the stability of the resulting scheme is

$$\max_x |c(x)| \frac{\Delta t}{\Delta x} < 1.$$

If this stability condition is violated in some small region of the flow, the instability will initially be confined to the same region and will appear as a packet of rapidly amplifying short waves typically having wavelengths between $2\Delta x$ and $4\Delta x$. If the numerical solution and the variable coefficients remain smooth and well-resolved, the frozen-coefficient analysis can also yield sufficient conditions for stability. In order to guarantee stability via a frozen-coefficient analysis, the numerical scheme must include some dissipative smoothing (Gustafsson et al. 1995, p. 235). The stability of some completely nondissipative methods can, nevertheless, be established by the energy method (see Section 3.5.2).

A second reasonable differential–difference approximation to (3.87) may be written in the form

$$\frac{d\phi_j}{dt} + \left(\langle c_j\rangle^x \delta_x\phi_j\right)^x = 0, \tag{3.89}$$

where the averaging operator $\langle\ \rangle^x$ is defined in (A.2) of the Appendix. When c is a constant, the preceding is identical to (3.88). If identical time differences are employed to solve (3.88) and (3.89), a frozen-coefficient stability analysis will yield the same stability condition for each scheme. An analysis of truncation error, performed by substituting Taylor series expansions for c and ψ into the differential–difference equations, shows that both schemes are accurate to $O\left[(\Delta x)^2\right]$. Is there any practical difference between (3.88) and (3.89)? There is, but the difference is not obvious unless one considers problems in which c and ϕ are poorly resolved on the numerical mesh or situations where the true solution has additional conservation properties (such as advection in a nondivergent flow) that are not automatically retained by finite-difference approximations.

First consider the problems that can arise when there are large-amplitude poorly resolved perturbations in ϕ or c. Under such circumstances, both of the preceding numerical approximations can exhibit serious instabilities. The structure and growth rates of the unstable perturbations generated by each scheme can, however, be very different. Perhaps the most useful way to explore these instabilities is to examine the aliasing error produced by (3.88) and (3.89).

3.5.1 Aliasing Error

Aliasing error occurs when a short-wavelength fluctuation is sampled at discrete intervals and misinterpreted as a longer-wavelength oscillation. The shortest wavelength that can be represented on a numerical grid is twice the grid interval; all shorter wavelengths will be aliased. Figure 3.8 illustrates the aliasing of a $4\Delta x/3$ wave into a $4\Delta x$ wave. The apparent equivalence of the $4\Delta x/3$ and $4\Delta x$ waves follows from the fact that for all integers n, the relation

$$e^{ikj\Delta x} = \left[e^{i2n\pi}\right]^j e^{ikj\Delta x} = e^{i(k+2n\pi/\Delta x)j\Delta x}, \tag{3.90}$$

is satisfied at all spatial locations $j\Delta x$ on the discrete mesh. In the case shown in Fig. 3.8, the wave number of the aliased wave is $k = (2\pi)/(4\Delta x/3) = 3\pi/(2\Delta x)$, and the wave number of the resolved wave is $-\pi/(2\Delta x)$, so that (3.90) applies with $n = -1$. The change in the sign of the wave number during aliasing is visible in Fig. 3.8 as the $180°$ phase shift between the original and the aliased wave.

Aliasing error may occur when the initial data are represented on a discrete grid or projected onto a truncated series of Fourier expansion functions. Aliasing error can also occur during the computation of the product $c\partial\psi/\partial x$ on a finite-resolution numerical grid. In order to illustrate how the product of two spatially varying functions may introduce aliasing error, suppose that the product $\phi(x)\chi(x)$ is computed at a discrete set of grid points. If $\phi = e^{ik_1 x}$ and $\chi = e^{ik_2 x}$, then $\phi\chi = e^{i(k_1+k_2)x}$. Since ϕ and χ were representable on the numerical grid, $|k_1|$, $|k_2| \leq \pi/\Delta x$. It is possible, however, that the wave number of their product lies in the range $\pi/\Delta x < |k_1 + k_2| \leq 2\pi/\Delta x$, in which case the product cannot be resolved on the numerical mesh and will misrepresented as a longer wave. The wave number $\tilde{k}$ into which a binary product is aliased is determined by the

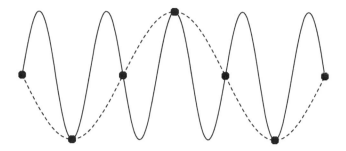

FIGURE 3.8. Misrepresentation of a $4\Delta x/3$ wave as a $4\Delta x$ wave when sampled on a discrete mesh.

relation

$$\tilde{k} = \begin{cases} k_1 + k_2 - 2\pi/\Delta x, & \text{if } k_1 + k_2 > \pi/\Delta x, \\ k_1 + k_2 + 2\pi/\Delta x, & \text{if } k_1 + k_2 < -\pi/\Delta x. \end{cases} \tag{3.91}$$

There is no aliasing when $|k_1 + k_2| \leq \pi/\Delta x$. In particular, if both ϕ and χ are $4\Delta x$ waves or longer, their product will not be aliased. A graphical diagram of the aliasing process may be created by plotting $k_1 + k_2$ and the cutoff wave number $\pi/\Delta x$ on a number line extending from zero to $2\pi/\Delta x$. Since $(k_1 + k_2 + |\tilde{k}|)/2 = \pi/\Delta x$, $|\tilde{k}|$ appears as the reflection of $k_1 + k_2$ about the cutoff wave number, as illustrated in Fig. 3.9. Note that the product of two extremely short waves is aliased into a relatively long wave. For example, the product of a $2\Delta x$ wave and a $2.5\Delta x$ wave appears as a $10\Delta x$ wave.

Unstable Growth Through Aliasing Error

Let us now consider the effects of aliasing error on stability.[6] Suppose that (3.87) is approximated as the differential–difference equation (3.88) and that a solution is sought over the periodic domain $[-\pi, \pi]$. Let the spatial domain be discretized-

FIGURE 3.9. Aliasing of $k_1 + k_2$ into $\tilde{k}$ such that $|\tilde{k}| = 2\pi/\Delta x - (k_1 + k_2)$ appears to be the symmetric reflection of $k_1 + k_2$ about the cutoff wave number $\pi/\Delta x$.

---

[6]The following example was anticipated by Miyakoda (1962) and Gary (1979).

into the $2N + 1$ points

$$x_j = \frac{\pi j}{N}, \quad j = -N, \dots, N$$

and suppose that the initial data are representable as the sum of the Fourier modes in the set $\{e^0, e^{iNx/2}, e^{-iNx/2}, e^{iNx}\}$. This set of four modes is closed under multiplication on the discrete mesh due to aliasing error; for example,

$$e^{iNx_j/2} \times e^{iNx_j} = e^{i\pi j/2} \times e^{i\pi j} = e^{i2\pi j}e^{-i\pi j/2} = e^{-iNx_j/2}.$$

Let c and ϕ be arbitrary combinations of these four Fourier modes. Under the assumption that c is real, the velocity

$$c(x_j) = c_0 + c_{n/2}e^{i\pi j/2} + c_{-n/2}e^{-i\pi j/2} + c_n e^{i\pi j}$$

may be alternatively expressed as

$$c(x_j) = c_0 + (c_r + ic_i)e^{i\pi j/2} + (c_r - ic_i)e^{-i\pi j/2} + c_n e^{i\pi j}, \tag{3.92}$$

where the coefficients c_0, c_r, c_i, and c_n are all real. Assuming that ϕ is also real, it may be written in the similar form

$$\phi(x_j) = a_0 + (a_r + ia_i)e^{i\pi j/2} + (a_r - ia_i)e^{-i\pi j/2} + a_n e^{i\pi j}. \tag{3.93}$$

Substituting (3.92) and (3.93) into the nonaveraging scheme (3.88) and requiring the linearly dependent terms to sum to zero yields

$$\dot{a}_0 = 2(a_i c_r - a_r c_i)/\Delta x,$$
$$\dot{a}_n = 2(a_i c_r + a_r c_i)/\Delta x,$$
$$\dot{a}_r = a_i(c_n + c_0)/\Delta x, \tag{3.94}$$
$$\dot{a}_i = a_r(c_n - c_0)/\Delta x, \tag{3.95}$$

where the dot denotes differentiation with respect to time. Eliminating a_i between (3.94) and (3.95), one obtains

$$\ddot{a}_r = \frac{c_n^2 - c_0^2}{(\Delta x)^2} a_r. \tag{3.96}$$

A similar equation holds for a_i. According to (3.96), the behavior of the $4\Delta x$ wave in ϕ is determined by the relative magnitudes of the mean wind speed and the $2\Delta x$ wind-speed perturbation. If the mean wind is stronger than the $2\Delta x$ perturbation, a_r oscillates sinusoidally. On the other hand, if c_n exceeds c_0, the $4\Delta x$ component in ϕ grows exponentially. The growth criterion $c_n > c_0$ is particularly simple in the special case $c_r = c_i = 0$. Then growth will occur whenever the wind speed changes sign. This exponential growth is clearly a nonphysical instability, since the true solution is constant along the characteristic curves $dx/dt = c(x)$ and therefore bounded between the maximum and minimum initial values of ϕ.

FIGURE 3.10. Comparison of the two differential–difference solutions to the one-dimensional advection equation in a periodically reversing flow field. (a) Initial condition (dot-dashed line) and the time-invariant velocity (solid line). (b) Solution to a high-resolution simulation at $t = 0.5$s. Averaging and non-averaging differential-difference solutions at (c) $t = 1.0$s and (d) 2.2s.

The preceding demonstrates that the nonaveraging scheme (3.88) can be unstable in a rather pathological case. Both (3.88) and (3.89) can produce unstable growth in less pathological examples, provided that the wind field forces the development of unresolvable short-wavelength perturbations in ϕ. An example of this type is illustrated in Fig. 3.10, in which the initial distributions of c and ϕ are smooth and well-resolved but ϕ develops unresolvable perturbations as a result of 180° changes in the wind direction. In this example the wind speed and the initial condition on ψ are

$$c(x) = 0.5\cos(4\pi x) \qquad \text{and} \qquad \psi(x, 0) = \sin(2\pi x).$$

The computational domain is $0 \leq x \leq 1$ and $\Delta x = 1/80$. The time derivatives in (3.88) and (3.89) were integrated using a fourth-order Runge–Kutta method and a small Courant number. The velocity (solid line) and the initial condition ϕ^0 (dot-dashed line) are plotted together in Fig. 3.10a. The velocity field is convergent in the portions of the spatial domain labeled "Con" along the bottom of Fig. 3.10a and is divergent in the regions labeled "Div." The character of the true solution is illustrated in Fig. 3.10b, which shows a very-high-resolution simulation at $t = 0.5$ s. Observe that the true solution is tending toward a square wave of amplitude $\sqrt{2}/2$ with a unit-amplitude spike at the left edge of each plateau. The spikes are located at the nodal points in the velocity field where the flow is convergent. Away from the spike the solution is tending toward the initial value of ψ at the divergent node. A comparison of the two differential–difference solutions is shown at $t = 1.0$ s in Fig. 3.10c and at $t = 2.2$ s in Fig. 3.10d. In both Fig. 3.10c and Fig. 3.10d the solution obtained with the nonaveraging scheme (3.88) is dominated by a growing $2\Delta x$ component. In contrast, the solution calculated with the averaging scheme (3.89) never develops a large-amplitude $2\Delta x$ component. At the earlier time (Fig. 3.10c) the averaging scheme generates a solution that is reasonably accurate and far superior to the nonaveraging result. This superiority is lost, however, by the later time (Fig. 3.10d), at which the averaging scheme has generated a longer-wavelength disturbance that rapidly amplifies and dominates the solution.

Comparison of the Aliasing Error in Two Schemes

As suggested by the preceding example, one important difference between the nonaveraging scheme (3.88) and the averaging method (3.89) lies in the nature of the aliasing error generated by each approximation. Let us examine the aliasing error produced by each formula in a more general context. Suppose that numerical solutions are sought to the one-dimensional advection equation (3.87), and that at some instant in time, c and ϕ are expanded into Fourier modes. Consider the interaction of the individual pair of modes:

$$c = c_{k_1} e^{i k_1 x} \quad \text{and} \quad \phi = a_{k_2} e^{i k_2 x}. \tag{3.97}$$

If the unapproximated product of c and $\partial \phi / \partial x$ is evaluated at grid points $j \Delta x$ on a discrete mesh, one obtains

$$c \frac{\partial \phi}{\partial x} = i c_{k_1} a_{k_2} k_2 e^{i(k_1 + k_2) j \Delta x}. \tag{3.98}$$

Evaluating c times the nonaveraging spatial-difference operator $\delta_{2x}\phi$ on the same mesh gives

$$c_j \delta_{2x} \phi_j = \frac{i c_{k_1} a_{k_2}}{\Delta x} (\sin k_2 \Delta x) e^{i(k_1 + k_2) j \Delta x}. \tag{3.99}$$

The analogous result for the averaging scheme is most easily obtained by noting that

$$\langle(c_j)^x \delta_x \phi_j\rangle^x = \frac{1}{2}\left[\delta_{2x}(c_j\phi_j) + c_j\delta_{2x}\phi_j - \phi_j\delta_{2x}c_j\right]$$

$$= \frac{ic_{k_1}a_{k_2}}{2\Delta x}\left(\sin[(k_1 + k_2)\Delta x] + \sin k_2\Delta x - \sin k_1\Delta x\right)e^{i(k_1+k_2)j\Delta x}.$$

$$(3.100)$$

In the limit of good numerical resolution, $k_1\Delta x$, $k_2\Delta x \to 0$, and each of the above expressions is equivalent, which is to be expected, since (3.99) and (3.100) are both second-order approximations to (3.98). As $(k_1 + k_2)\Delta x$ approaches π, the numerical formulae may become inaccurate, but the most serious problems develop when $\pi < |(k_1 + k_2)\Delta x| \leq 2\pi$ and the product term is aliased into a longer wavelength.

Suppose that a wave of wave number k_2 in the ϕ field is interacting with some disturbance in the velocity field to force $d\phi/dt$ at an aliased wave number $\tilde{k}$. According to (3.91), there is only one resolvable wave number in the velocity field that could alias into $\tilde{k}$ through interaction with k_2 during the approximation of the product $c\partial\psi/\partial x$. The rate at which this aliasing occurs can be computed as follows. Without loss of generality, assume that the unresolvable wave number is positive (i.e., $k_1 + k_2 > \pi/\Delta x$), in which case $\tilde{k}$ is negative and

$$\tilde{k} = k_1 + k_2 - 2\pi/\Delta x. \qquad (3.101)$$

Suppose that at a given instant both interacting waves have unit amplitude, i.e., $|c_{k_1}| = |a_{k_2}| = 1$. Let $C_{k_2 \to \tilde{k}} = d|a_{\tilde{k}}|/dt$ denote the rate at which interactions between the wave numbers k_1 and k_2 force the growth at the aliased wave number. Using (3.101) to eliminate k_1 from (3.100), one obtains a growth rate for the averaging scheme of

$$C^A_{k_2 \to \tilde{k}} = \frac{|\sin \tilde{k}\Delta x + \sin k_2\Delta x - \sin(\tilde{k} - k_2)\Delta x|}{2\Delta x}.$$

The growth rate for the nonaveraging scheme,

$$C^N_{k_2 \to \tilde{k}} = \frac{|\sin k_2\Delta x|}{\Delta x},$$

can be obtained directly from (3.99).

A contour plot of $C^N_{k_2 \to \tilde{k}}$ as a function of k_2 and $\tilde{k}$ appears in Fig. 3.11a. Contours of the wave number k_1 involved in these interactions (computed from (3.101)) also appear plotted as a function of $(k_2, \tilde{k})$ as the dashed diagonal lines in Fig. 3.11. Since $k_1 \leq \pi/\Delta x$ (because every wave must be resolved on the numerical mesh), no aliasing can occur for the $(k_2, \tilde{k})$ combinations above the diagonal in Fig. 3.11a, and this region of the plot is left blank. Equivalent data, showing contours of $C^A_{k_2 \to \tilde{k}}$ for the averaging scheme, appear in Fig. 3.11b.

As indicated in Fig. 3.11, the nonaveraging scheme allows every wave number on the resolvable mesh ($0 \leq k_2 \leq \pi/\Delta x$) to combine with some disturbance

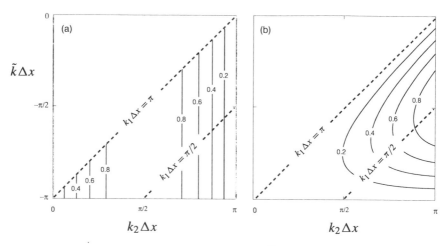

FIGURE 3.11. Spurious growth rate at wave number $\tilde{k}$ due to interactions between wave numbers k_1 and k_2 plotted as a function of k_2 and $\tilde{k}$ for (a) the nonaveraging scheme $C^N_{k_2 \to \tilde{k}}$ and (b) the averaging scheme $C^A_{k_2 \to \tilde{k}}$. Contours of the wave number k_1 involved in these interactions are plotted as diagonal dashed lines.

in the velocity field to produce aliasing into $2\Delta x$ waves ($\tilde{k}\Delta x = -\pi$). On the other hand, the averaging scheme does not allow any aliasing into the $2\Delta x$ wave, although aliasing is permitted into the longer wavelengths. The practical impact of this difference in aliasing is evident in the numerical comparisons shown in Fig. 3.10, in which the aliasing error in the nonaveraging scheme appears primarily at $2\Delta x$, whereas the errors that eventually develop in the averaging scheme appear at at longer wavelengths.

3.5.2   Conservation and Stability

Numerical schemes for the simulation of advective transport by a nondivergent flow are more stable and have better conservation properties if the spatial derivatives of the discretized tracer field ϕ are approximated using the averaging operator $\langle \langle u \rangle^x \, \delta_x \phi \rangle^x$ in preference to the nonaveraging operator $u\delta_{2x}\phi$. The contrast in the conservation properties associated with these operators is not apparent in one-dimensional problems because all nondivergent one-dimensional flows have uniform velocity, and the operators are equivalent when u is spatially uniform. The differences can be revealed by considering the advection of a passive scalar by a two-dimensional nondivergent flow, which is governed by the equation

$$\frac{\partial \psi}{\partial t} + \mathbf{v} \cdot \nabla \psi = 0, \tag{3.102}$$

where ψ is the passive scalar and $\mathbf{v} = (u, v)$ is the vector velocity. Solutions to this equation conserve $\|\psi\|_2$, provided that

$$\nabla \cdot \mathbf{v} = 0 \tag{3.103}$$

and that the spatial domain D is periodic or there is no flow through to the boundary of D. Under these assumptions

$$
\begin{aligned}
\frac{d\,\|\psi\|_2^2}{dt} &= 2\int_D \psi\,\frac{\partial\psi}{\partial t}\,dV \\
&= -2\int_D \psi\,(\nabla\psi\cdot\mathbf{v})\,dV \\
&= -\int_D \nabla\cdot(\psi^2\mathbf{v}) - \psi^2\nabla\cdot\mathbf{v}\,dV \\
&= 0.
\end{aligned}
$$

The first term in the final integrand is zero by the assumed boundary conditions; the second term is zero because the flow is nondivergent.

The conservation of $\|\psi\|_2$ may be used to formulate stability conditions for numerical approximations to (3.102) by demanding that the finite-difference approximation conserve the discrete analogue of $\|\psi\|_2$. Let $\mathbf{u}$ be a column vector containing the approximate solution at each spatial grid point and $\mathbf{A}$ a matrix containing the finite-difference approximation to $\mathbf{v}\cdot\nabla\psi$. Then the approximate solution satisfies a set of linear differential–difference equations of the form

$$
\frac{d\mathbf{u}}{dt} + \mathbf{A}\mathbf{u} = \mathbf{0}. \tag{3.104}
$$

This system will conserve $\|\mathbf{u}\|_2$ if the matrix $\mathbf{A}$ is skew-symmetric, as may be verified by noting that if $\mathbf{A} = -\mathbf{A}^T$, then

$$
\frac{d\,\|\mathbf{u}\|_2^2}{dt} = \frac{d}{dt}(\mathbf{u}^T\mathbf{u}) - (\mathbf{A}\mathbf{u})^T\mathbf{u} + \mathbf{u}^T(\mathbf{A}\mathbf{u}) = \mathbf{u}^T\mathbf{A}^T\mathbf{u} + \mathbf{u}^T\mathbf{A}\mathbf{u} = 0.
$$

Assuming that ϕ, u, and v are colocated on an unstaggered mesh in a spatially periodic domain, the nonaveraging operator yields the following differential–difference approximation to (3.102):

$$
\frac{d\phi}{dt} + u\delta_{2x}\phi + v\delta_{2y}\phi = 0, \tag{3.105}
$$

and the averaging operator gives

$$
\frac{d\phi}{dt} + \left\langle\langle u\rangle^x\delta_x\phi\right\rangle^x + \left\langle\langle v\rangle^y\delta_y\phi\right\rangle^y = 0. \tag{3.106}
$$

Expressing (3.106) in the form (3.104) shows that the matrix $\mathbf{A}$ generated by the averaging scheme is skew-symmetric whenever the velocities satisfy

$$
\delta_{2x}u + \delta_{2y}v = 0, \tag{3.107}
$$

which is the discrete analogue of (3.103). On the other hand, the differential–difference approximation (3.105) generated by the nonaveraging operator does not yield a skew-symmetric matrix even when the velocities satisfy (3.107), and as a consequence, it does not guarantee the conservation of $\|\mathbf{u}\|_2$.

As an alternative to the construction of $\mathbf{A}$ and the evaluation of its symmetry properties, one can examine the stability and conservation properties of approximate solutions to (3.106) and (3.107) using the energy method to compute the time tendency of the ℓ_2-norm of the semi-discrete solution from the relation

$$\frac{d\,\|\phi\|_2^2}{dt} = \frac{d}{dt}\left(\sum_i\sum_j \phi_{i,j}^2\,\Delta x\,\Delta y\right) = 2\sum_i\sum_j \phi_{i,j}\frac{d\phi_{i,j}}{dt}\,\Delta x\,\Delta y.$$

First, consider the averaging scheme (3.106). After some manipulation one obtains

$$\frac{d\,\|\phi\|_2^2}{dt} = -\Delta x\,\Delta y\sum_i\sum_j\Bigg[-\left(\frac{u_{i+1,j}-u_{i-1,j}}{2\Delta x} + \frac{v_{i,j+1}-v_{i,j-1}}{2\Delta y}\right)\phi_{i,j}^2$$

$$+\left(\frac{u_{i+1,j}+u_{i,j}}{2}\right)\frac{\phi_{i+1,j}\phi_{i,j}}{\Delta x} - \left(\frac{u_{i,j}+u_{i-1,j}}{2}\right)\frac{\phi_{i,j}\phi_{i-1,j}}{\Delta x}$$

$$+\left(\frac{v_{i,j+1}+v_{i,j}}{2}\right)\frac{\phi_{i,j+1}\phi_{i,j}}{\Delta y} - \left(\frac{v_{i,j}+v_{i,j-1}}{2}\right)\frac{\phi_{i,j}\phi_{i,j-1}}{\Delta y}\Bigg].$$

The first line on the right-hand side will vanish if the discrete velocity field satisfies (3.107). The terms in the second line sum to zero over the index i due to the periodicity in x, and the terms in the third line sum to zero over the index j due to the periodicity in y. Thus the averaging scheme (3.106) conserves $\|\phi\|_2$ in problems where the flow field satisfies the discrete continuity equation (3.107).

Now consider the nonaveraging scheme (3.107) and again assume that the problem is periodic in x and y. Then

$$\frac{d\,\|\phi\|_2^2}{dt} = -2\Delta x\,\Delta y\sum_i\sum_j\Bigg[u_{i,j}\left(\frac{\phi_{i+1,j}\phi_{i,j}-\phi_{i-1,j}\phi_{i,j}}{2\Delta x}\right)$$

$$+ v_{i,j}\left(\frac{\phi_{i,j+1}\phi_{i,j}-\phi_{i,j-1}\phi_{i,j}}{2\Delta y}\right)\Bigg]$$

$$= \Delta x\,\Delta y\sum_i\sum_j\Bigg[\left(\frac{u_{i,j}-u_{i-1,j}}{\Delta x}\right)\phi_{i,j}\phi_{i-1,j}$$

$$+ \left(\frac{v_{i,j}-v_{i,j-1}}{\Delta y}\right)\phi_{i,j}\phi_{i,j-1}\Bigg],$$

where the last equality is obtained using the periodicity in x and y. It follows that $\|\phi\|_2$ is not conserved by the nonaveraging scheme, regardless of the numerical form of the discrete continuity equation. In summary, (3.106) is preferable to (3.105) because it better preserves the ℓ_2-norm stability and the conservation properties of the continuous problem. The averaging operator is also a natural choice on staggered meshes in which the velocity normal to the interface between each pair of grid cells is located at the center of that interface (as in Fig. 3.6).

Comparison of Flux and Advective Form

The transport of a passive scalar in a nondivergent flow field is also described by the equation

$$\frac{\partial \psi}{\partial t} + \nabla \cdot (\psi \mathbf{v}) = 0, \tag{3.108}$$

which can either be derived from physical principles or obtained by combining (3.102) and (3.103). Equation (3.108) is referred to as being in *flux* form, whereas (3.102) is in *advective* form. The flux form (3.108) is an example of a *conservation law*. Conservation laws can be expressed by equations in the general form

$$\frac{\partial \psi}{\partial t} + \nabla \cdot \mathbf{f} = 0,$$

which states that the local rate of change of ψ is determined by the convergence of a flux $\mathbf{f}$. It follows from the divergence theorem that the integral of ψ over the entire domain, denoted by $\overline{\psi}$, is determined by the net flux through the boundaries, and thus, $\overline{\psi}$ is conserved if $\mathbf{f}$ is periodic over the domain or if the component of $\mathbf{f}$ normal to the boundary vanishes at the boundary. Solutions to arbitrary conservation laws need not, however, conserve $\|\psi\|_2^2$. For example, solutions to (3.108) need not conserve $\|\psi\|_2$ when $\nabla \cdot \mathbf{v} \neq 0$.

Conservation laws may admit solutions that contain shocks or discontinuities, and as discussed in Chapter 5, when simulating solutions with shocks or discontinuities it is essential to use a scheme that conserves the discretized equivalent of $\overline{\psi}$. Even when the solution remains smooth and well-resolved, it is generally advantageous to choose a scheme that conserves $\overline{\psi}$. A natural way to achieve this conservation is to evaluate the flux at each cell interface and then difference those fluxes across each cell. Assuming that the fluxes are originally available on the same mesh points as the scalar field, the fluxes at the cell boundaries can be computed by spatial averaging, in which case the time tendency of ϕ is given by

$$\frac{d\phi}{dt} + \delta_x \langle f_x \rangle^x + \delta_y \langle f_y \rangle^y = 0, \tag{3.109}$$

where f_x and f_y are the x and y components of $\mathbf{f}$. Two possible ways to arrive at a finite-difference approximation to the flux form of the transport equation (3.108) are

$$\frac{d\phi}{dt} + \delta_x \left(\langle u\phi \rangle^x \right) + \delta_y \left(\langle v\phi \rangle^y \right) = 0$$

and

$$\frac{d\phi}{dt} + \delta_x \left(\langle u \rangle^x \langle \phi \rangle^x \right) + \delta_y \left(\langle v \rangle^y \langle \phi \rangle^y \right) = 0. \tag{3.110}$$

Both of the preceding schemes are in the general form (3.109), and therefore both conserve the domain integral of ϕ. The domain integral of ϕ^2 is, however, only conserved by (3.110), and is only conserved when the velocity field satisfies the discretized continuity equation (3.107).

It should be emphasized that desirable conservation properties are not limited to finite-difference approximations to equations in flux form. In order for (3.110) to conserve $\|\phi\|_2$, the flow field must satisfy the discrete continuity equation (3.107), but in that case the finite-difference approximation to the advective form (3.106) is algebraically equivalent to the flux form (3.110), and both schemes conserve $\overline{\phi}$ and $\|\phi\|_2$.

The Effect of Time-Differencing on Conservation

Differential-difference equations that conserve $\|\phi\|_2$, such as (3.106) and (3.110), generally cease to be conservative when the time derivative is approximated by finite differences. Nevertheless, one type of time-differencing that does preserve the conservation properties of linear differential–difference equations is trapezoidal differencing. The conservation properties of trapezoidal time differencing may be demonstrated by writing the differential–difference equation in the general form

$$\frac{d\phi_j}{dt} + L(\phi_j) = 0, \tag{3.111}$$

where L is a linear finite-difference operator including all the spatial differences. As a preliminary step, note that in order for the differential–difference equation (3.111) to conserve $\|\phi\|_2$, the linear operator L must have the algebraic property

$$\sum_j \varphi_j L(\varphi_j) = 0, \tag{3.112}$$

where φ_j is any discrete function defined on the numerical mesh and the summation is taken over all the grid points.

Approximating (3.111) with trapezoidal time differences yields

$$\frac{\phi_j^{n+1} - \phi_j^n}{\Delta t} + \frac{L(\phi_j^{n+1}) + L(\phi_j^n)}{2} = 0.$$

Multiplying the preceding by $(\phi_j^{n+1} + \phi_j^n)$, using the linearity of L, summing over the discrete mesh, and using (3.112), one obtains

$$\sum_j \left[(\phi_j^{n+1})^2 - (\phi_j^n)^2 \right] = \frac{\Delta t}{2} \sum_j \left[\left(\phi_j^{n+1} + \phi_j^n \right) L \left(\phi_j^{n+1} + \phi_j^n \right) \right] = 0,$$

which implies that $\|\phi^{n+1}\|_2 = \|\phi^n\|_2$.

3.6   Nonlinear Instability

As discussed in the preceding section, the stability of finite-difference approximations to linear equations with variable coefficients can be determined by examining the stability of the associated family of frozen-coefficient problems—provided that the solution, and the variable coefficients, are sufficiently smooth and well-resolved on the numerical mesh. One may attempt to analyze the stability of nonlinear equations through a similar procedure. First, the nonlinear equations are linearized; then a frozen-coefficient analysis is performed to determine stability conditions for the linearized problem. This approach gives necessary conditions for stability, but as was the case with variable-coefficient linear equations, it may give misleading results in situations where the solution is dominated by poorly resolved short-wavelength perturbations. Unfortunately, the caveat that the solution must remain smooth and well-resolved is a much more serious impediment to the analysis of nonlinear finite-difference equations because such equations can rapidly generate unresolvable short-wave perturbations from very smooth initial data.

In the following we will examine techniques for stabilizing the finite-difference approximation of two nonlinear equations: Burgers's equation and the barotropic vorticity equation. Solutions to Burgers's equation often develop shocks and discontinuities whose accurate approximation requires the use of methods that will be presented in Chapter 5. The schemes that will be considered in this section provide very simple examples illustrating the stabilization of numerical approximations to a nonlinear problem by a judicious choice of finite-difference formula. These schemes are not, however, recommended for practical applications involving the simulation of problems with shocks or discontinuous solutions. The opportunity for practical application of the ideas illustrated using Burgers's equation arises in attempting to avoid nonlinear instabilities in numerical solutions to the barotropic vorticity equation. Solutions to the barotropic vorticity equation never develop shocks and remain essentially as smooth as the initial data.

3.6.1   Burgers's Equation

The inviscid Burgers's equation,

$$\frac{\partial \psi}{\partial t} + \psi \frac{\partial \psi}{\partial x} = 0, \qquad (3.113)$$

is an example of a nonlinear partial differential equation whose solution rapidly develops unresolvable short-wavelength components. If $\psi(x, 0) = f(x)$ at some initial time $t = 0$, solutions to this problem can be written in the implicit form

$$\psi(x, t) = f(x - \psi(x, t)t),$$

which implies that f is constant along the characteristic curves

$$x - \psi(x, t)t = x_0.$$

Here x_0 is the x-intercept of the curve at $t = 0$. Since $\psi = f$, ψ must also be constant along each characteristic curve, and every characteristic is therefore a straight line. In any region where $\partial\psi/\partial x$ is negative, the characteristics will converge, and for some sufficiently large value of t, these converging characteristics must cross. At those points where two (or more) characteristics intersect, the solution is multivalued and exhibits a discontinuity, or shock. If the initial condition is smooth, the time when the solution first develops a shock t_c can be determined by examining the rate at which gradients of ψ steepen. Define $S(x, t) = \partial\psi/\partial x$ and note that

$$\frac{dS}{dt} = \frac{\partial S}{\partial t} + \psi\frac{\partial S}{\partial x} = \frac{\partial}{\partial x}\left(\frac{\partial\psi}{\partial t} + \psi\frac{\partial\psi}{\partial x}\right) - \left(\frac{\partial\psi}{\partial x}\right)^2 = -S^2.$$

Integration of the preceding yields

$$S = \left(t + S(x_0, 0)^{-1}\right)^{-1}. \tag{3.114}$$

The first discontinuity, or shock, develops when S becomes infinite at a time $t_c = -S(x_0, 0)^{-1}$ determined by the most negative initial value of $\partial\psi/\partial x$. This behavior may be compared with that for the linear problem with variable coefficients shown in Fig. 3.10ab, in which the characteristic curves never cross (but rather approach the lines $x = \frac{1}{8}$ and $x = \frac{5}{8}$ asymptotically) and true discontinuities do not develop over any finite time interval.

Suppose that solutions to Burgers's equation are sought on the periodic domain $0 \le x \le 1$ subject to the initial condition $\psi(x, 0) = \sin(2\pi x)$. When (3.113) is approximated by the advective-form differential–difference equation

$$\frac{d\phi_j}{dt} + \phi_j\left(\frac{\phi_{j+1} - \phi_{j-1}}{2\Delta x}\right) = 0, \tag{3.115}$$

with $\Delta x = 1/50$, the numerical solution appears as shown in Fig. 3.12.[7] The numerical solution provides a good approximation to the true solution at $t = 0.13$, at which time the true solution is still smooth and easy to resolve on the discrete grid. But by $t = 0.22$ the true solution has developed a shock, and the numerical solution misrepresents the shock as a steep gradient bounded by a series of large-amplitude short-wavelength perturbations. These short-wavelength perturbations are amplifying rapidly and, as a consequence, $\|\phi\|_2$ is growing without bound. The growth in $\|\phi\|_2$ is an instability, since the ℓ_2-norm of the true solution does not increase with time. If the solution is smooth, $\|\psi\|_2$ is conserved along with all other moments, i.e.,

$$\int_0^1 [\psi(x)]^p\, dx = 0 \tag{3.116}$$

---

[7] The solution shown in Fig. 3.12 was obtained using a fourth-order Runge–Kutta method and a very small time step to accurately approximate the time derivative in (3.115).

FIGURE 3.12. Differential–difference solution to Burgers's equation obtained using (3.115) at (a) $t = 0.13$ and (b) $t = 0.22$.

for any positive integer p. The time invariance of (3.116) may be derived by multiplying (3.113) by $p\psi^{p-1}$, which yields

$$0 = p\psi^{p-1}\left(\frac{\partial\psi}{\partial t} + \psi\frac{\partial\psi}{\partial x}\right) = \frac{\partial\psi^p}{\partial t} + \frac{p}{p+1}\frac{\partial\psi^{p+1}}{\partial x},$$

and then integrating this equation over the periodic domain. If the solution contains discontinuities, the preceding manipulations are not valid, but one can show that $\partial\|\psi\|_2/\partial t$ is never positive (see Section 5.1.2).

The inability of the advective-form differential–difference scheme to conserve $\|\phi\|_2$ can be demonstrated by multiplying (3.115) by ϕ_j and summing over the domain to obtain

$$\frac{d}{dt}\sum_j \phi_j^2 = -\sum_j\left(\frac{\phi_j^2\phi_{j+1} - \phi_j^2\phi_{j-1}}{\Delta x}\right)$$

$$= -\frac{1}{\Delta x}\left(\sum_j \phi_j^2\phi_{j+1} - \sum_j \phi_{j+1}^2\phi_j\right)$$

$$= \sum_j \phi_j\phi_{j+1}\left(\frac{\phi_{j+1} - \phi_j}{\Delta x}\right), \tag{3.117}$$

where the second equality follows from the periodicity of the solution. One might attempt to obtain a scheme that conserves $\|\phi\|_2$ by rewriting Burgers's equation in the flux form

$$\frac{\partial\psi}{\partial t} + \frac{1}{2}\frac{\partial\psi^2}{\partial x} = 0$$

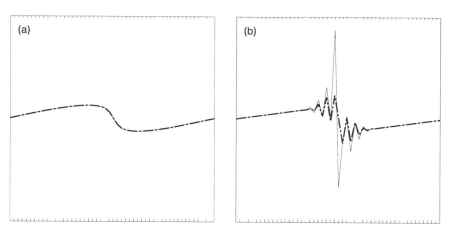

FIGURE 3.13. Differential–difference solution to Burgers's equation at (a) $t = 0.13$ and (b) $t = 0.28$ obtained using the advective form ((3.115), thin solid line) and the conservative form ((3.120) dot-dashed line).

and approximating this with the differential–difference equation

$$\frac{d\phi_j}{dt} + \frac{1}{2}\left(\frac{\phi_{j+1}^2 - \phi_{j-1}^2}{2\Delta x}\right) = 0. \tag{3.118}$$

Multiplying the preceding by ϕ_j and summing over the periodic domain yields

$$\frac{d}{dt}\sum_j \phi_j^2 = -\frac{1}{2}\sum_j \phi_j\phi_{j+1}\left(\frac{\phi_{j+1} - \phi_j}{\Delta x}\right), \tag{3.119}$$

which demonstrates that the flux form also fails to conserve $\|\phi\|_2$. Since the terms representing the nonconservative forcing in (3.117) and (3.119) differ only by a factor of $-\frac{1}{2}$, it is possible to obtain a scheme that does conserve $\|\phi\|_2$ using a weighted average of the advective- and flux-form schemes. The resulting "conservative form" is

$$\frac{d\phi_j}{dt} + \frac{1}{3}\phi_j\left(\frac{\phi_{j+1} - \phi_{j-1}}{2\Delta x}\right) + \frac{1}{3}\left(\frac{\phi_{j+1}^2 - \phi_{j-1}^2}{2\Delta x}\right) = 0. \tag{3.120}$$

Figure 3.13 shows a comparison of the solutions to (3.115) and (3.120). The test problem is the same test considered previously in connection with Fig. 3.12, except that the vertical scale of the plotting domain shown in Fig. 3.13 has been reduced, and the second panel now shows solutions at $t = 0.28$. The unstable growth of the short-wavelength oscillations generated by advective-form differencing can be observed by comparing the solution at $t = 0.22$ (Fig. 3.12b) and $t = 0.28$ (Fig. 3.13b). As illustrated in Fig. 3.13b, short-wavelength oscillations also develop in the conservative-form solution, but these oscillations do not continue to amplify.[8] The flux form (3.118) yields a solution (not shown) to this test problem that looks qualitatively similar to the conservative-form solution shown in Fig. 3.13b, although the spurious oscillations in the flux-form result are actually somewhat weaker. It is perhaps surprising that the short-wavelength oscillations are smaller in the flux-form solution than in the conservative-form solution and that the flux-form solution does not show a tendency toward instability. In fact, practical experience suggests that the flux-form difference (3.118) is not particularly susceptible to nonlinear instability. Fornberg (1973) has, nevertheless, demonstrated that both the advective and flux forms are unstable (and that the conservative form is stable) when the discretized initial condition has the special form $\dots, 0, -1, 1, 0, -1, 1, 0, \dots$.

The instabilities that develop in the preceding solutions to Burgers's equation appear to be associated with the formation of the shock. The development of a shock is not, however, a prerequisite for the onset of nonlinear instability, and such instabilities may occur in numerical simulations of very smooth flow. One example in which nonlinear instability develops in a smooth flow is provided by the viscous Burgers's equation

$$\frac{\partial \psi}{\partial t} + \psi \frac{\partial \psi}{\partial x} = \nu \frac{\partial^2 \psi}{\partial x^2}, \tag{3.121}$$

where ν is a coefficient of viscosity. The true solution to the viscous Burgers's equation never develops a shock, but the advective-form differential difference approximation to (3.121) becomes unstable for sufficiently small values of ν.

3.6.2   The Barotropic Vorticity Equation

A second example involving the development of nonlinear instability in very smooth flow is provided by the equation governing the vorticity in a two-dimensional incompressible homogeneous fluid,

$$\frac{\partial \zeta}{\partial t} + \mathbf{u} \cdot \nabla \zeta = 0. \tag{3.122}$$

Here $\mathbf{u}$ is the two-dimensional velocity vector describing the flow in the x-y plane and ζ is the vorticity component along the z-axis. Since the flow is nondivergent, $\mathbf{u}$ and ζ may be expressed in terms of a stream function ψ such that

$$\mathbf{u} = \mathbf{k} \times \nabla \psi, \qquad \zeta = \mathbf{k} \cdot \nabla \times \mathbf{u} = \nabla^2 \psi,$$

---

[8]Even though they do not lead to instability, the short-wavelength oscillations in the conservative-form solution to Burgers's equation are nonphysical and are not present in the correct generalized solution to Burgers's equation, which satisfies the Rankine–Hugoniot condition (5.10) at the shock and is smooth away from the shock. After the formation of the shock the correct generalized solution ceases to conserve $\|\phi\|^2$, so it can no longer be well-approximated by the numerical solution obtained using the conservative-form difference. In order to obtain good numerical approximations to discontinuous solutions to Burgers's equation it is necessary to use the methods discussed in Chapter 5.

and (3.122) may be written as

$$\frac{\partial \nabla^2 \psi}{\partial t} + J(\psi, \nabla^2 \psi) = 0, \tag{3.123}$$

where J is the Jacobian operator

$$J(p, q) = \frac{\partial p}{\partial x}\frac{\partial q}{\partial y} - \frac{\partial p}{\partial y}\frac{\partial q}{\partial x}.$$

In atmospheric science, (3.123) is known as the barotropic vorticity equation.

Fjørtoft (1953) demonstrated that if the initial conditions are smooth, solutions to (3.123) must remain smooth in the sense that there can be no net transfer of energy from the larger spatial scales into the smaller scales. Fjørtoft's conclusions follow from the properties of the domain integral of the Jacobian operator. Let $\bar{p}$ denote the domain integral of p, and suppose, for simplicity, that the domain is periodic in x and y. Then by the assumed periodicity of the spatial domain[9]

$$\overline{J(p, q)} = \overline{\frac{\partial}{\partial x}\left(p\frac{\partial q}{\partial y}\right)} - \overline{\frac{\partial}{\partial y}\left(p\frac{\partial q}{\partial x}\right)} = 0.$$

As a consequence,

$$\overline{pJ(p, q)} = \overline{J(p^2/2, q)} = 0, \tag{3.124}$$

and

$$\overline{qJ(p, q)} = \overline{J(p, q^2/2)} = 0. \tag{3.125}$$

The preceding relations may be used to demonstrate that the domain-integrated kinetic energy and the domain-integrated enstrophy (one-half the vorticity squared) are both conserved. First consider the enstrophy, $\zeta^2/2 = (\nabla^2\psi)^2/2$. Multiplying (3.123) by $\nabla^2\psi$ and integrating over the spatial domain yields

$$\frac{\partial}{\partial t}\overline{\left(\frac{(\nabla^2\psi)^2}{2}\right)} + \overline{(\nabla^2\psi)J(\psi, \nabla^2\psi)} = 0,$$

which using (3.125) reduces to

$$\frac{\partial}{\partial t}\overline{\left(\frac{(\nabla^2\psi)^2}{2}\right)} = 0.$$

The conservation of the domain-integrated kinetic energy, $\mathbf{u} \cdot \mathbf{u}/2 = \nabla\psi \cdot \nabla\psi/2$, may be demonstrated by first noting that the vector identity

$$\nabla \cdot (\alpha\mathbf{a}) = \nabla\alpha \cdot \mathbf{a} + \alpha(\nabla \cdot \mathbf{a})$$

---

[9]Equivalent conservation properties hold in a rectangular domain in which the normal velocity is zero at all points along the boundary.

implies that

$$\psi \frac{\partial \nabla^2 \psi}{\partial t} = \nabla \cdot \left(\psi \frac{\partial \nabla \psi}{\partial t} \right) - \frac{\partial}{\partial t} \left(\frac{\nabla \psi \cdot \nabla \psi}{2} \right).$$

Then multiplying (3.123) by ψ, using the preceding relation and integrating over the periodic spatial domain one obtains

$$\frac{\partial}{\partial t} \left(\frac{\overline{\nabla \psi \cdot \nabla \psi}}{2} \right) = 0.$$

Now suppose that the stream function is expanded in Fourier series along the x and y coordinates

$$\psi = \sum_k \sum_\ell a_{k,\ell} e^{i(kx + \ell y)} = \sum_{k,\ell} \psi_{k,\ell},$$

and define the total wave number κ such that $\kappa^2 = k^2 + \ell^2$. By the periodicity of the domain and the orthogonality of the Fourier modes,

$$\overline{\mathbf{u} \cdot \mathbf{u}} = \overline{\nabla \psi \cdot \nabla \psi} = \overline{\nabla \cdot (\psi \nabla \psi)} - \overline{\psi \nabla^2 \psi} = -\overline{\psi \nabla^2 \psi} = \sum_{k,\ell} \kappa^2 \overline{\psi_{k,\ell}^2}$$

and

$$\overline{\zeta^2} = \overline{(\nabla^2 \psi)^2} = \sum_{k,\ell} \kappa^4 \overline{\psi_{k,\ell}^2}.$$

The two preceding relations may be used to evaluate an average wave number, κ_{avg}, given by the square root of the ratio of the domain-integrated enstrophy to the domain-integrated kinetic energy,

$$\kappa_{\mathrm{avg}} = \left(\frac{\overline{\zeta^2}}{\overline{\mathbf{u} \cdot \mathbf{u}}} \right)^{1/2}.$$

Since the domain-integrated enstrophy and the domain-integrated kinetic energy are both conserved, κ_{avg} does not change with time. Any energy transfers that take place from larger to smaller scales must be accompanied by a second energy transfer from smaller to larger scales to conserve κ—there can be no systematic energy cascade into the short-wavelength components of the solution.

Suppose that the barotropic vorticity equation (3.123) is approximated using centered second-order differences in space and time such that

$$\delta_{2t}(\tilde{\nabla}^2 \phi) + \tilde{J}(\phi, \tilde{\nabla}^2 \phi) = 0,$$

where the numerical approximation to the horizontal Laplacian operator is

$$\tilde{\nabla}^2 \phi = (\delta_x^2 + \delta_y^2)\phi$$

and the numerical approximation to the Jacobian operator is

$$\tilde{J}(p, q) = (\delta_{2x} p)(\delta_{2y} q) - (\delta_{2y} p)(\delta_{2x} q).$$

Phillips (1959) showed that solutions obtained using the preceding scheme are subject to an instability in which short-wavelength perturbations suddenly amplify without bound. This instability cannot be controlled by reducing the time step, and it occurs using values of Δt that are well below the threshold required to maintain the stability of equivalent numerical approximations to the linearized constant-coefficient problem. Phillips demonstrated that this instability could be controlled by removing all waves with wavelengths shorter than four grid intervals, thereby eliminating the possibility of aliasing error.

A more elegant method of stabilizing the solution was proposed by Arakawa (1966), who suggested reformulating the numerical approximation to the Jacobian to preserve the discrete analogue of the relations (3.124) and (3.125) and thereby obtain a numerical scheme that conserves both the domain-integrated enstrophy and kinetic energy. In particular, Arakawa proposed the following approximation to the Jacobian:

$$\tilde{J}_a(p, q) = \frac{1}{3} \left[(\delta_{2x} p)(\delta_{2y} q) - (\delta_{2y} p)(\delta_{2x} q) \right]$$

$$+ \frac{1}{3} \left[\delta_{2x}(p\, \delta_{2y} q) - \delta_{2y}(p\, \delta_{2x} q) \right] + \frac{1}{3} \left[\delta_{2y}(q\, \delta_{2x} p) - \delta_{2x}(q\, \delta_{2y} p) \right].$$

The Arakawa Jacobian satisfies the numerical analogue of (3.124) and (3.125),

$$\sum_{m,n} p_{m,n}\, \tilde{J}_a(p_{m,n}, q_{m,n}) = \sum_{m,n} q_{m,n}\, \tilde{J}_a(p_{m,n}, q_{m,n}) = 0, \qquad (3.126)$$

where the summation is taken over all grid points in the computational domain. As a consequence of (3.126), solutions to

$$\frac{\partial}{\partial t} \left(\tilde{\nabla}^2 \phi \right) + \tilde{J}_a(\phi, \tilde{\nabla}^2 \phi) = 0 \qquad (3.127)$$

conserve their domain-integrated enstrophy and kinetic energy and must therefore also conserve the discretized equivalent of the average wave number κ_{avg}. Since the average wave number is conserved, there can be no net amplification of the short-wavelength components in the numerical solution. The numerical solution is not only stable, it remains smooth.

Any numerical approximation to the barotropic vorticity equation will be stable if it conserves the domain-integrated kinetic energy, since that is equivalent to the conservation of $\|\mathbf{u}\|_2$. The enstrophy conservation property of the Arakawa Jacobian does more, however, than guarantee stability; it prevents a systematic cascade of energy into the shortest waves resolvable on the discrete mesh. In designing a numerical approximation to the barotropic vorticity equation it is clearly appropriate to chose a finite-difference scheme like the Arakawa Jacobian that inhibits the down-scale cascade of energy. On the other hand, it is not clear that schemes that limit the cascade of energy to small scales are appropriate in those fluid-dynamical applications where there actually is a systematic transfer of kinetic energy from large to small scale. Indeed, any accurate numerical approximation to the equations governing such flows must replicate this down-scale energy transfer.

One natural approach to the elimination of nonlinear instability in systems that support a down-scale energy cascade is through the parametrization of unresolved turbulent dissipation. In high-Reynolds-number (nearly inviscid) flow, kinetic energy is ultimately transferred to very small scales before being converted to internal energy by viscous dissipation, yet the storage limitations of digital computers do not allow most numerical simulations to be conducted with sufficient spatial resolution to resolve all the small-scale eddies involved in this energy cascade. Under such circumstances the kinetic energy transferred down-scale during the numerical simulation will tend to accumulate in the smallest scales resolvable on the numerical mesh, and it is generally necessary to remove this energy by some type of scale-selective dissipation. The scale-selective dissipation constitutes a parametrization of the influence of the unresolved eddies on the resolved-scale flow and should be designed to represent the true behavior of the physical system as closely as possible. Regardless of the exact formulation of the energy removal scheme, it will tend to stabilize the solution and prevent nonlinear instability.

Many fluid flows contain limited regions of active small-scale turbulence and relatively larger patches of dynamically stable laminar flow. Since eddy diffusion will not be active outside the regions of parametrized turbulence, a scale-selective background dissipation, similar to Phillips's (1959) technique of removing all wavelengths shorter than four grid intervals, is often required in order to avoid nonlinear instability. This dissipation may be implicitly included in the time-differencing or in an upwind-biased spatial difference, or it may be explicitly added to an otherwise nondamping method using formulae such as those discussed in Section 2.4.3. Although it is not required for stability, a small amount of background dissipation may also be incorporated in numerical approximations to linear partial differential equations to damp those short-wavelength components of the numerical solution whose phase speed and group velocity are most seriously in error.

Problems

1. Verify that the leapfrog time-differenced shallow-water equations (3.14) and (3.15) support a computational mode, and that the forward-backward-differenced system (3.17) and (3.18) does not, by solving their respective discrete-dispersion relations for ω.

2. Eliminate h from the finite-difference equations for the leapfrog unstaggered scheme (3.14) and (3.15) and compare the resulting higher-order finite-difference approximation to the second-order PDE

$$\frac{\partial^2 u}{\partial t^2} - c^2 \frac{\partial^2 u}{\partial x^2} = 0$$

with the expression that arises when h is eliminated from the forward-backward approximation on the staggered mesh (3.17) and (3.18). What does this comparison suggest about the number of computational modes admitted by each numerical approximation?

3. Suppose that numerical solutions to the two-dimensional Boussinesq system (3.39)–(3.42) are obtained using the staggered grid shown in Fig. 3.6 except that the distribution of the variables is modified so that b is co-located with the w rather than the P points.

   (a) Write down appropriate modifications for the discretized vertical momentum and buoyancy equations (3.44) and (3.45), and derive the discrete dispersion relation for this system.

   (b) Assume that all resolved modes are hydrostatic, so that $\tilde{k}_1^2$ can be neglected with respect to $\tilde{\ell}_1^2$ in the denominator of (3.47) and in the result derived in (a). Compare the horizontal and vertical group velocities for the numerical solutions on each staggered grid with the exact expression from the nondiscretized hydrostatic system.

4. Derive the amplification factor and stability condition given in the text for the CTU method (3.36). Show that including the cross-derivative term in the CTU method always decreases the amplification factor relative to that obtained with the standard two-dimensional upstream scheme (3.31).

5. Show that the false 2-D Lax–Wendroff scheme (3.37) is unstable for all Δt.

6. Determine the range of Δt (if any) over which the backward, forward, and leapfrog schemes give a stable approximation to

$$\frac{d\psi}{dt} = r\psi.$$

Consider both the cases $r > 0$ and $r < 0$. The true solution to this equation preserves the sign of $\psi(t = 0)$. What, if any, additional restrictions must be placed on Δt to ensure that the numerical solution for each method is both stable and sign-preserving.

7. Suppose the Lax–Wendroff method is used to obtain an $O\left[(\Delta t)^2\right]$-accurate approximation to the advection–diffusion equation (3.72). Show that before discretizing the spatial derivatives, the scheme has the form

$$\frac{\phi^{n+1} - \phi^n}{\Delta t} + c\frac{\partial \phi^n}{\partial x} - M\frac{\partial^2 \phi^n}{\partial x^2}$$
$$= \frac{\Delta t}{2}\left(c^2\frac{\partial^2 \phi^n}{\partial x^2} - 2cM\frac{\partial^3 \phi^n}{\partial x^3} + M^2\frac{\partial^3 \phi^n}{\partial x^3}\right).$$

Comment on the probable utility of this approach.

8. Derive expressions for the boundaries for the regions of useful stability for the leapfrog-backward scheme (3.78) and the leapfrog-trapezoidal method (3.79) shown in Fig. 3.7a.

9. The following approximation to the advection–diffusion equation (3.72) is unstable:

$$\delta_{2t}\phi_j^n + c\delta_{2x}\phi_j^n = M\delta_x^2\phi_j^n.$$

Modify the right side of the above equation to stabilize the method, at least for sufficiently small Δt, but *do not make the scheme implicit*. Prove that your modified scheme is indeed stable for sufficiently small values of Δt. It is not necessary to work out the exact range of Δt over which the scheme is stable.

10. The analysis of the frozen-coefficient problem does not always correctly indicate the behavior of solutions to partial differential equations with variable coefficients. Consider the initial value problem

$$\frac{\partial \psi}{\partial t} - i\frac{\partial}{\partial x}\left(\sin x \frac{\partial \psi}{\partial x}\right) = 0, \qquad (3.128)$$

$\psi(x, 0) = f(x)$ on the interval $-\infty < x < \infty$.

(a) Show that the ℓ_2-norm of the solution to this problem does not grow with time.

(b) Freeze the coefficients at $x = 0$ and show that the resulting problem is ill-posed because its solution does not depend continuously on the initial data. (*Hint:* consider

$$\psi_1(x, 0) = e^{ik_1 x} \quad \text{and} \quad \psi_2(x, 0) = e^{ik_2 x}$$

and show that $\|\psi_1(x, 0) - \psi_2(x, 0)\|$ is bounded, while $\|\psi_1(x, t) - \psi_2(x, t)\|$ can be arbitrarily large for any finite t.)

Since stable numerical solutions cannot be obtained for ill-posed problems, the stability of a numerical approximation to (3.128) cannot be determined by examining the stability of the family of all frozen-coefficient problems.

11. Suppose that (3.105) and (3.106) are applied to model tracer advection in a closed rectangular domain with no velocity normal to the boundaries and that the boundaries are located at the edges (as opposed to the centers) of the outermost grid cells. Let the differential–difference equations generated by each scheme be expressed as a linear system of the form (3.104). Write down the coefficient matrix $\mathbf{A}$ for each scheme, and show that the matrix associated with (3.106) is skew-symmetric, whereas that associated with (3.105) is not.

12. *The linearized one-dimensional Rossby adjustment problem for an atmosphere with no mean wind is governed by the equations

$$\frac{\partial u}{\partial t} - fv + g\frac{\partial h}{\partial x} = 0, \qquad \frac{\partial v}{\partial t} + fu = 0, \qquad \frac{\partial h}{\partial t} + H\frac{\partial u}{\partial x} = 0.$$

Compare the approximate solution to these equations obtained using leapfrog differencing on an unstaggered mesh

$$\delta_{2t}u - fv + g\delta_{2x}h = 0,$$
$$\delta_{2t}v + fu = 0,$$
$$\delta_{2t}h + H\delta_{2x}u = 0$$

with those obtained using forward–backward time-differencing on the staggered mesh shown in Fig. 3.1:

$$\delta_t u_j^{n+\frac{1}{2}} - fv_j^n + g\delta_x h_j^n = 0,$$
$$\delta_t v_j^{n+\frac{1}{2}} + fu_j^{n+1} = 0,$$
$$\delta_t h_{j+\frac{1}{2}}^{n+\frac{1}{2}} + H\delta_x u_{j+\frac{1}{2}}^{n+1} = 0.$$

Assume that v, which is not shown in Fig. 3.1, is defined at the same points as u. Let the spatial domain be periodic on the interval $0 \le x \le 2000$ km, but show your solutions only in the domain $600 \le x \le 14000$ km. Let $f = 10^{-4}$ s^{-1} and $c = \sqrt{gH} = 10$ ms^{-1}. For initial conditions choose $u(x,0) = v(x,0) = 0$, and let the height field be given by a slightly smoothed unit-amplitude square wave with nodes at $x = 0$ and 1000 km. Obtain this smoothed square wave by three iterative applications of the filter

$$\phi_j^f = \frac{1}{4}(\phi_{j+1} + 2\phi_j + \phi_{j-1})$$

to a pure square wave. Let $\Delta x = 3\frac{1}{3}$ km.

(a) Show solutions for all three fields at the time step closest to $t = 21000$ s. Use Courant numbers $(c\Delta t/\Delta x)$ of 0.9 and 0.1. Discuss the quality of the two solutions. Explain the source of the difference between the two solutions.

(b) Eliminate the smoothing step from the initialization and discuss the impact on the solution.

(Note that analytic solutions to this problem are given in Gill (1982, Sections 7.2–7.3).)

13. *Compute exact and numerical solutions to the variable-wind-speed advection equation (3.87) in a periodic domain $0 \le x \le 2$. Choose

$$c(x) = \begin{cases} 0.3 - 1.5(x - \frac{1}{3})\sin(3\pi x)\sin(12\pi x), & \text{if } \frac{1}{3} \le x \le \frac{2}{3}, \\ 0.3, & \text{otherwise,} \end{cases}$$

and use the initial condition

$$\psi(x,0) = \begin{cases} \frac{1}{4}(\cos(8\pi(x - \frac{1}{8})) + 1)^2 & \text{if } |x - \frac{1}{8}| \le \frac{1}{8}; \\ 0, & \text{otherwise}. \end{cases}$$

(a) Given that

$$\int_{1/3}^{2/3} \frac{dx}{c(x)} = 1.391,$$

find the correct x-location of the peak of the initial distribution at time $t = 3$. Describe the shape and location of the true solution at $t = 3$.

(b) Compare numerical solutions obtained using the second-order approximations

$$\delta_{2t}\phi + c\delta_{2x}\phi = 0 \tag{3.129}$$

and

$$\delta_{2t}\phi + (\langle c \rangle^x \delta_x \phi)^x = 0$$

with the fourth-order space schemes

$$\delta_{2t}\phi + c\left[\frac{4}{3}\delta_{2x}\phi - \frac{1}{3}\delta_{4x}\phi\right] = 0$$

and

$$\delta_{2t}\phi + \frac{4}{3}(\langle c \rangle^x \delta_x \phi)^x - \frac{1}{3}(\langle c \rangle^{2x} \delta_{2x}\phi)^{2x} = 0.$$

Assume that c and ϕ are located at the same points. Use $\Delta x = 1/32$ and a Courant number of 0.6 based on the maximum wind speed. Take a single forward step to initialize the leapfrog integration. Plot the left $\frac{2}{3}$ of the domain and show the solutions at $t = 0$, 1.5, and 3. Also plot the wind speed.

(c) Retry the above simulations using $\Delta x = 1/64$ and discuss the degree of improvement.

(d) Now try adding a fourth-order spatial filter to each scheme in the $\Delta x = 1/32$ case. Lag the filter in time. For example, (3.129) becomes

$$\delta_{2t}\phi_j^n + c\delta_{2x}\phi_j^n = -\gamma\left(\phi_{j+2}^{n-1} - 4\phi_{j+1}^{n-1} + 6\phi_j^{n-1} - 4\phi_{j-1}^{n-1} + \phi_{j-2}^{n-1}\right).$$

Discuss the dependence of the solution on the parameter $\gamma\Delta t$. As a start, set $\gamma\Delta t = 0.01$.

4
Series-Expansion Methods

Series-expansion methods that are potentially useful in geophysical fluid dynamics include the spectral method, the pseudospectral method, and the finite-element method. The spectral method plays a particularly important role in global atmospheric models, in which the horizontal structure of the numerical solution is often represented as a truncated series of spherical harmonics. Finite-element methods, on the other hand, are not commonly used in multidimensional wave propagation problems because they generally require the solution of implicit algebraic systems and are therefore not as efficient as competing explicit methods. All of these series-expansion methods share a common foundation that will be discussed in the next section.

4.1   Strategies for Minimizing the Residual

Suppose F is an operator involving spatial derivatives of ψ, and that solutions are sought to the partial differential equation

$$\frac{\partial \psi}{\partial t} + F(\psi) = 0, \tag{4.1}$$

subject to the initial condition $\psi(x, t_0) = f(x)$ and to boundary conditions at the edges of some spatial domain S. The basic idea in all series-expansion methods is to approximate the spatial dependence of ψ as a linear combination of a finite number of predetermined expansion functions. Let the general form of the series expansion be written as

$$\phi(x, t) = \sum_{k=1}^{N} a_k(t)\varphi_k(x), \tag{4.2}$$

where $\varphi_1, \ldots, \varphi_N$ are predetermined expansion functions satisfying the required boundary conditions. Then the task of solving (4.1) is transformed into a problem of calculating the unknown coefficients $a_1(t), \ldots, a_N(t)$ in a way that minimizes the error in the approximate solution. One might hope to obtain solvable expressions for the $a_k(t)$ by substituting (4.2) into the governing equation (4.1). For example, if F is a linear function of $\partial^n \psi/\partial x^n$ with constant coefficients and the expansion functions are Fourier series, direct substitution will yield a system of ordinary differential equations for the evolution of the $a_k(t)$. Unfortunately, direct substitution yields a solvable system of equations for the expansion coefficients only when the φ_k are eigenfunctions of the differential operator F—direct substitution works in precisely those special cases for which analytic solutions are available. This, of course, is a highly restrictive limitation.

In the general case where the φ_k are not eigenfunctions of F, it is impossible to specify $a_1(t), \ldots, a_N(t)$ such that an expression of the form (4.2) exactly satisfies (4.1). As an example, suppose $F(\psi) = \psi \partial \psi/\partial x$ and the expansion functions are the Fourier components $\varphi_k = e^{ikx}, -N \leq k \leq N$. If this Fourier series is substituted into (4.1), the nonlinear product in $F(\psi)$ introduces spatial variations at wave numbers that were not present in the initial truncated series, e.g., $F(e^{iNx}) = iN e^{i2Nx}$. A total of $4N + 1$ equations are obtained after substituting the expansion functions into (3.1) and requiring that the coefficients of each Fourier mode sum to zero. It is not possible to choose the $2N + 1$ Fourier coefficients in the original expansion to satisfy these $4N + 1$ equations simultaneously. The best one can do is to select the expansion coefficients to minimize the error.

Since the actual error in the approximate solution $\|\psi - \phi\|$ cannot be determined, the most practical way to try to minimize the error is to minimize the *residual*,

$$R(\phi) = \frac{\partial \phi}{\partial t} + F(\phi), \tag{4.3}$$

which is the amount by which the approximate solution fails to satisfy the governing equation. Three different strategies are available for constraining the size of the residual. Each strategy leads to a system of N coupled *ordinary* differential equations for the time-dependent coefficients $a_1(t), \ldots, a_N(t)$. This transformation of the partial differential equation into a system of ordinary differential equations is similar to that which occurs in grid-point methods when the spatial derivatives are replaced with finite differences.

One strategy for constraining the size of the residual is to pick the $a_k(t)$ to minimize the square of the ℓ_2-norm of the residual:

$$(\|R(\phi)\|_2)^2 = \int_S [R(\phi(x))]^2 \, dx.$$

A second approach, referred to as collocation, is to require the residual to be zero at a discrete set of grid points:

$$R(\phi(j\Delta x)) = 0 \quad \text{for all } j = 1, \ldots, N.$$

The third strategy, known as the Galerkin approximation, requires the residual to be *orthogonal* to each of the expansion functions, i.e.,

$$\int_S R(\phi(x))\varphi_k(x)\,dx = 0 \quad \text{for all } k = 1, \ldots, N. \tag{4.4}$$

Different series-expansion methods rely on one or more of the preceding approaches. The collocation strategy is used in the pseudospectral method and in some finite-element formulations, but not in the spectral method. The ℓ_2-minimization and Galerkin criteria are equivalent when applied to a problem of the form (4.1), and are the basis of the spectral method. The Galerkin approximation is also used extensively in finite-element schemes.

The equivalence of the ℓ_2-minimization criterion and the Galerkin approximation can be demonstrated as follows. According to (4.3), the residual depends on both the instantaneous values of the expansion coefficients and their time tendencies. The expansion coefficients are determined at the outset from the initial conditions and are known at the beginning of any subsequent integration step. The criteria for constraining the residual are not used to obtain the instantaneous values of the expansion coefficients, but rather to determine their time evolution. If the rate of change of the kth expansion function is calculated to minimize $(\|R(\phi)\|_2)^2$, a necessary criterion for a minimum may be obtained by differentiation with respect to the quantity $da_k/dt \equiv \dot{a}_k$

$$
\begin{aligned}
0 &= \frac{d}{d(\dot{a}_k)}\left\{\int_S (R(\phi))^2\,dx\right\} \\
&= \frac{d}{d(\dot{a}_k)}\left\{\int_S \left[\sum_{n=1}^N \dot{a}_n\varphi_n + F\left(\sum_{n=1}^N a_n\varphi_n\right)\right]^2 dx\right\} \\
&= 2\int_S \left[\sum_{n=1}^N \dot{a}_n\varphi_n + F\left(\sum_{n=1}^N a_n\varphi_n\right)\right]\varphi_k\,dx \tag{4.5} \\
&= 2\int_S R(\phi)\varphi_k\,dx. \tag{4.6}
\end{aligned}
$$

The second derivative of $(\|R(\phi)\|_2)^2$ with respect to $\dot{a}_k$ is $2(\|\varphi_k\|_2)^2$, which is positive. Thus, the extremum condition (4.6) is associated with a true minimum of $(\|R(\phi)\|_2)^2$, and the Galerkin requirement is identical to the condition obtained by minimizing the ℓ_2-norm of the residual.

As derived in (4.5), the Galerkin approximation and the ℓ_2-minimization of the residual both lead to a system of ordinary differential equations for the expansion coefficients of the form

$$\sum_{n=1}^{N} I_{nk} \frac{da_n}{dt} = - \int_S \left[F \left(\sum_{n=1}^{N} a_n \varphi_n \right) \varphi_k \right] dx \quad \text{for all } k = 1, \ldots, N, \quad (4.7)$$

where

$$I_{nk} = \int_S \varphi_n \varphi_k \, dx.$$

The initial conditions for the preceding system of differential equations are obtained by choosing $a_1(t_0), \ldots, a_N(t_0)$ such that $\phi(x, t_0)$ provides the "best" approximation to $f(x)$. The possible strategies for constraining the initial error are identical to those used to ensure that the residual is small. As before, the choice that minimizes the ℓ_2-norm of the initial error also satisfies the Galerkin requirement that the initial error be orthogonal to each of the expansion functions,

$$\int_S \left(\sum_{n=1}^{N} a_n(t_0) \varphi_n(x) - f(x) \right) \varphi_k(x) \, dx = 0 \quad \text{for all } k = 1, \ldots, N,$$

or, equivalently,

$$\sum_{n=1}^{N} I_{nk} a_n = \int_S f(x) \varphi_k(x) \, dx \quad \text{for all } k = 1, \ldots, N. \quad (4.8)$$

4.2   The Spectral Method

The characteristic that distinguishes the spectral method from other series-expansion methods is that the expansion functions form an orthogonal set. Since the expansion functions are orthogonal, I_{nk} is zero unless $n = k$, and the system of differential equations for the coefficients (4.7) reduces to

$$\frac{da_k}{dt} = -\frac{1}{I_{kk}} \int_S \left[F \left(\sum_{n=1}^{N} a_n \varphi_n \right) \varphi_k \right] dx \quad \text{for all } k = 1, \ldots, N. \quad (4.9)$$

This is a particularly useful simplification, since explicit algebraic equations for each $a_k(t + \Delta t)$ are obtained when the time derivatives in (4.9) are replaced with finite differences. In contrast, the finite-difference approximation of the time derivatives in the more general form (4.7) introduces a coupling between all the expansion coefficients at the new time level, and the solution of the resulting implicit system of algebraic equations may require considerable computation. The orthogonality of the expansion functions also reduces the expression for the initial value of each expansion coefficient (4.8) to

$$a_k(t_0) = \frac{1}{I_{kk}} \int_S f(x) \varphi_k(x) \, dx. \quad (4.10)$$

The choice of some particular family of orthogonal expansion functions is largely dictated by the geometry of the problem and by the boundary conditions. Fourier series are well suited to rectangular domains with periodic boundary conditions. Chebyshev polynomials are a possibility for nonperiodic domains. Associated Legendre functions are useful for representing the latitudinal dependence of a function on the spherical Earth. Since Fourier series lead to the simplest formulae, they will be used to illustrate the elementary properties of the spectral method. The special problems associated with spherical geometry will be discussed in Section 4.4.

4.2.1   Comparison with Finite-Difference Methods

In Chapter 2, a variety of finite-difference methods were tested on the one-dimensional advection equation

$$\frac{\partial \psi}{\partial t} + c \frac{\partial \psi}{\partial x} = 0. \tag{4.11}$$

Particular emphasis was placed on the simplest case, in which c was constant. When c is constant, it is easy to find expansion functions that are eigenfunctions of the spatial derivative term in (4.11). As a consequence, the problem of advection by a constant wind is almost too simple for the spectral method. Nevertheless, the constant-wind case reveals some of the fundamental strengths and weaknesses of the spectral method and allows a close comparison between the spectral method and finite-difference schemes.

Suppose, therefore, that c is constant and solutions are sought to (4.11) on the periodic domain $-\pi \le x \le \pi$, subject to the initial condition $\psi(x, 0) = f(x)$. A Fourier series expansion

$$\phi(x, t) = \sum_{k=-N}^{N} a_k(t) e^{ikx} \tag{4.12}$$

is the natural choice for this problem. Since individual Fourier modes are eigenfunctions of the differential operator in (4.11), evolution equations for the Fourier coefficients of the form

$$\frac{da_k}{dt} + icka_k = 0 \tag{4.13}$$

may be obtained by directly substituting (4.12) into the advection equation. In this atypically simple case, the residual is zero, and it is not necessary to adopt any particular procedure to minimize its norm. Nevertheless, (4.13) can also be obtained through the Galerkin requirement that the residual be orthogonal to each of the expansion functions. In order to apply the Galerkin formulation it is necessary to generalize the definition of orthogonality to include complex-valued functions. Two complex-valued functions $g(x)$ and $h(x)$ are *orthogonal* over the domain S if

$$\int_S g(x) h^*(x) dx = 0,$$

where $h^*(x)$ denotes the complex conjugate of $h(x)$.[1] As an example, note that for integer values of n and m,

$$\int_{-\pi}^{\pi} e^{inx} e^{-imx}\, dx = \begin{cases} \dfrac{e^{i(n-m)x}}{n-m} \Big|_{-\pi}^{\pi} = 0, & \text{if } m \neq n; \\[2ex] 2\pi, & \text{if } m = n, \end{cases}$$

which is just the well-known orthogonality condition for two Fourier modes. Using this orthogonality relation and setting $f(\psi) = c\,\partial\psi/\partial x$, with c constant, reduces (4.9) to (4.13).

If the ordinary differential equation (4.13) is solved analytically (in practical applications it must be computed numerically), solutions have the form $a_k(t) = \exp(-ickt)$. Thus, in the absence of time-differencing errors, the frequency of the kth Fourier mode is identical to the correct value for the continuous problem $\omega = ck$. The spectral approximation does not introduce phase speed or amplitude errors—even in the shortest wavelengths! The ability of the spectral method to correctly capture the amplitude and phase speed of the shortest resolvable waves is a significant advantage over conventional grid-point methods, in which the spatial derivative is approximated by finite differences, yet surprisingly, the spectral method is not necessarily a good technique for modeling short-wavelength disturbances. The problem lies in the fact that it is only those waves retained in the truncated series expansion that are correctly represented in the spectral solution. If the true solution has a great deal of spatial structure on the scale of the shortest wavelength in the truncated series expansion, the spectral representation will not accurately approximate the true solution.

The problems with the representation of short-wavelength features in the spectral method are illustrated in Fig. 4.1, which shows ten grid-point values forming a $2\Delta x$-wide spike against a zero background on a periodic domain with a uniformly spaced grid. Also shown is the curve defined by the truncated Fourier series passing through those ten grid-point values. The Fourier series approximation to the $2\Delta x$ spike exhibits large oscillations about the zero background state on both sides of the spike. Now suppose that the data in Fig. 4.1 represent the initial condition for a constant-wind-speed advection problem. The over- and under-shoots associated with the Fourier approximation will not be apparent at the initial time if the data are sampled only at the points on the discrete mesh. If time-differencing errors are neglected, the grid-point values will also be exact at those subsequent times at which the initial distribution has translated an integral number of grid intervals. The grid-point values will, however, reveal the oscillatory error at in-

---

[1] Multiplication by the complex conjugate ensures that if $g(x) = a(x) + ib(x)$ with a and b real, then

$$\int_S g(x) g^*(x)\, dx = \int_S (a^2 + b^2)\, dx.$$

FIGURE 4.1. (a) Ten periodic grid-point values exhibiting a piecewise linear $2\Delta x$ spike, and the truncated Fourier series approximation passing through those ten points. (b) Values of the truncated Fourier series sampled at the same grid-point location after translating the curve one-half grid point to the right.

termediate times. The worst errors in the grid-point values will occur at times when the solution has traveled $n + \frac{1}{2}$ grid intervals, where n is any integer. In this particular example, the error on the discrete grid does not accumulate with time; it oscillates instead, achieving a minimum when the solution has translated an integral number of grid intervals. The maximum error is limited by the error generated when the initial condition is projected onto the truncated Fourier series.

The Finite Fourier Transform

There is a simple relationship between the nine[2] independent grid-point values in Fig. 4.1 and the nine coefficients determining the truncated Fourier series passing through those points. If the Fourier-series expansion of a real-valued function is truncated at wave number N, the set of Fourier coefficients contains $2N + 1$ pieces of data. Assuming that the Fourier expansion functions are periodic on the domain $0 \le x \le 2\pi$, an equivalent amount of information is contained by the $2N + 1$ function values $\phi(x_j, t)$, where

$$x_j = j\left(\frac{2\pi}{2N+1}\right) \quad \text{and} \quad j = 1, \ldots, 2N + 1.$$

It is obvious that the set of Fourier coefficients $a_{-N}(t), \ldots, a_N(t)$ defines the grid-point values $\phi(x_j, t)$ through the relation

$$\phi(x_j, t) = \sum_{k=-N}^{N} a_k(t)e^{ikx_j}. \tag{4.14}$$

Although it is not as self-evident, the $2N + 1$ Fourier coefficients may also be determined from the $2N + 1$ grid-point values. An exact algebraic expression for

---

[2]The tenth grid-point value is redundant information because the solution is periodic.

$a_n(t)$ can be obtained by noting that

$$\sum_{j=1}^{2N+1} \phi(x_j, t)e^{-inx_j} = \sum_{j=1}^{2N+1}\left(\sum_{m=-N}^{N} a_m(t)e^{imx_j} \right)e^{-inx_j}$$

$$= \sum_{m=-N}^{N} a_m(t)\left(\sum_{j=1}^{2N+1} e^{imx_j}e^{-inx_j} \right). \qquad (4.15)$$

Further simplification of the preceding equation is possible because the final summation in (4.15) obeys an orthogonality condition on the discrete mesh. Using the definition of x_j,

$$\sum_{j=1}^{2N+1} e^{imx_j}e^{-inx_j} = \sum_{j=1}^{2N+1} \left(e^{\frac{i2\pi(m-n)}{2N+1}} \right)^j. \qquad (4.16)$$

If $m = n$, then (4.16) sums to $2N + 1$; for $m \neq n$ the formula for the sum of a finite geometric series,

$$1 + r + r^2 + \cdots + r^n = \frac{1 - r^{n+1}}{1 - r}, \qquad (4.17)$$

may be used to reduce (4.16) to

$$\sum_{j=1}^{2N+1} e^{imx_j}e^{-inx_j} = \frac{e^{\frac{i2\pi(m-n)}{2N+1}}\left(1 - e^{i2\pi(m-n)}\right)}{1 - e^{\frac{i2\pi(m-n)}{2N+1}}} = 0. \qquad (4.18)$$

Using these orthogonality properties, (4.15) becomes

$$a_n(t) = \frac{1}{2N + 1} \sum_{j=1}^{2N+1} \phi(x_j, t)e^{-inx_j}. \qquad (4.19)$$

The relations (4.19) and (4.14), known as *finite Fourier transforms*, are discretized analogues to the standard Fourier transform and its inverse. The integrals in the continuous transforms are replaced by finite sums in the discrete expressions. These formulae, or more specifically the mathematically equivalent Fast Fourier Transform (FFT) algorithms, are essential for obtaining efficient spectral solutions in many practical applications where it is advantageous to transform the solution back and forth between wave number space and physical space once during the execution of every time step.

The Equivalent Grid-Point Method.

If c is constant, the spectral solution to the advection equation (4.11) can be recast in the form of an equivalent finite-difference method. Observe that

$$\frac{d\phi(x_j, t)}{dt} = \sum_{n=-N}^{N} \frac{da_n}{dt} e^{inx_j}$$

$$= -\sum_{n=-N}^{N} inca_n(t)e^{inx_j}$$

$$= -c \sum_{n=-N}^{N} in \left(\frac{1}{2N+1} \sum_{k=1}^{2N+1} \phi(x_k, t)e^{-inx_k} \right) e^{inx_j}$$

$$= -c \sum_{k=1}^{2N+1} C_{j,k} \phi(x_k, t),$$

where

$$C_{j,k} = \frac{1}{2N+1} \sum_{n=-N}^{N} in e^{in(x_j - x_k)}.$$

The finite-difference coefficient $C_{j,k}$ depends only on the difference between j and k, and is zero if $j = k$. If $j \neq k$, a simpler expression for $C_{j,j+\ell}$ can be obtained by defining

$$s = x_j - x_{j+\ell} = -\ell \left(\frac{2\pi}{2N+1} \right),$$

in which case

$$C_{j,j+\ell} = \frac{1}{2N+1} \frac{d}{ds} \left(\sum_{n=-N}^{N} e^{ins} \right) = \frac{1}{2N+1} \frac{d}{ds} \left(e^{-iNs} \sum_{n=0}^{2N} (e^{is})^n \right).$$

Using (4.17) to sum the finite geometric series, differentiating, and noting that $e^{i(2N+1)s} = 1$, the preceding becomes

$$C_{j,j+\ell} = \frac{(-1)^{\ell+1}}{2 \sin \left(\frac{\ell\pi}{2N+1} \right)},$$

which implies that since $C_{j,j+\ell} = -C_{j,j-\ell}$, the equivalent finite-difference formula is centered in space.

Two grid-point values are used in the centered second-order finite-difference approximation to $\partial\psi/\partial x$. A fourth-order centered difference utilizes four points; the sixth-order difference requires six grid points. Every grid-point on the numerical mesh (except the central point) is involved in the spectral approximation of $\partial\psi/\partial x$. As will be shown in the next section, the use of all these grid points allows the spectral method to compute derivatives of smooth functions with very high accuracy. Merilees and Orszag (1979) have compared the weighting coefficients for the spectral method on a seventeen-point periodic grid with the weighting coefficients for second- through sixteenth-order centered finite differences. Their calculations appear in Table 4.1, which shows that the influence of remote grid points on the spectral calculation is much greater than the remote influence in any of the

Δx away from central point	2nd order	4th order	6th order	16th order	spectral
1	0.500	0.667	0.750	0.889	1.006
2		-0.083	-0.150	-0.311	-0.512
3			0.017	0.113	0.351
4				-0.035	-0.274
5				0.009	0.232
6				-0.001	-0.207
7				0.000	0.192
8				-0.000	-0.186

TABLE 4.1. Comparison of weight accorded each grid point as a function of its distance to the central grid point in centered finite differences and in a spectral method employing 17 expansion coefficients.

finite-difference formulas. The large degree of remote influence in the spectral method has been a source of concern, since the true domain of dependence for the constant-wind-speed advection equation is a straight line. Practical evidence suggests that this remote influence is not a problem provided that enough terms are retained in the truncated Fourier series to adequately resolve the spatial variations in the solution.

Order of Accuracy

The accuracy of a finite difference is characterized by the truncation error, which is computed by estimating a smooth function's values at a series of grid points through the use of Taylor series, and by substituting those Taylor-series expansions into the finite-difference formula. The discrepancy between the finite-difference calculation and the true derivative is the truncation error and is usually proportional to some power of the grid interval. A conceptually similar characterization of accuracy is possible for the computation of spatial derivatives via the spectral method.

The basic idea is to examine the difference between the actual derivative of a smooth function and the approximate derivative computed from the spectral representation of the same function. Suppose that a function $\psi(x)$ is periodic on the domain $-\pi \leq x \leq \pi$ and that the first few derivatives of ψ are continuous. Then ψ and its first derivative can be represented by the convergent Fourier series

$$\psi(x) = \sum_{k=-\infty}^{\infty} a_k e^{ikx}, \qquad (4.20)$$

and

$$\frac{\partial \psi}{\partial x} = \sum_{k=-\infty}^{\infty} ik a_k e^{ikx}.$$

If ψ is represented by a spectral approximation, the series will be truncated at some wave number N, but the Fourier coefficients for all $|k| \le N$ will be identical to those in the infinite series (4.20).[3] Thus, the error in the spectral representation of $\partial\psi/\partial x$ is

$$E = \sum_{|k|>N} ik a_k e^{ikx}.$$

If the pth derivative of ψ is piecewise continuous, and all lower-order derivatives are continuous, the Fourier coefficients satisfy the inequality

$$|a_k| \le \frac{C}{|k|^p}, \tag{4.21}$$

where C is a positive constant (see Problem 10). Thus,

$$|E| \le 2 \sum_{k=N+1}^{\infty} \frac{C}{|k|^{p-1}} \le 2C \int_N^{\infty} \frac{ds}{s^{p-1}} = \frac{2C}{p-2}\left(\frac{1}{N^{p-2}}\right).$$

As demonstrated in the preceding section, a $2N + 1$ mode spectral representation of the derivative is equivalent to some finite-difference formula involving $2N+1$ grid points equally distributed throughout the domain. The spectral computation is therefore equivalent to a finite-difference computation with grid spacing $\Delta x_e = 2\pi/(2N + 1)$. Thus $\Delta x_e \propto N^{-1}$ and

$$|E| \le \tilde{C}(\Delta x_e)^{p-2}, \tag{4.22}$$

where $\tilde{C}$ is another constant. It follows that the effective order of accuracy of the spectral method is determined by the smoothness of ψ. If ψ is infinitely differentiable, the truncation error in the spectral approximation goes to zero faster than any finite power of Δx_e. In this sense, spatial derivatives are represented with infinite-order accuracy by the spectral method.

The preceding error analysis suggests that if a Fourier series approximation to $\psi(x)$ (as opposed to $d\psi/dx$) is truncated at wave number N, the error will be $O(1/N^{p-1})$. This error estimate is actually too pessimistic. As noted by Gottlieb and Orszag (1977, p. 26), (4.21) can be tightened, because if the pth derivative of ψ is piecewise continuous and all lower-order derivatives are continuous,

$$|a_k| \ll \frac{1}{|k|^p} \qquad \text{as} \qquad k \to \pm\infty. \tag{4.23}$$

---

[3] In order to ensure that the a_k are identical in both the infinite and truncated Fourier series, it is necessary to compute the integral in the Fourier transform,

$$a_k = \frac{1}{2\pi}\int_{-\pi}^{\pi} \psi(x)e^{-ikx}\, dx,$$

with sufficient accuracy to avoid aliasing error.

Away from the points where $d^p\psi/dx^p$ is discontinuous, the maximum-norm error in the truncated Fourier series decays at a rate similar to the magnitude of the first few neglected Fourier coefficients, and according to (4.23) this rate is faster than $O(1/N^p)$. In practice, the error is $O(1/N^{p+1})$ away from the points where $d^p\psi/dx^p$ is discontinuous and $O(1/N^p)$ near the discontinuities (Fornberg 1996, p. 13).

Time-Differencing

The spectral representation of the spatial derivatives reduces the original partial differential equation to the system of ordinary differential equations (4.9). In most practical applications, this system must be solved numerically. The time-differencing schemes discussed in Section 2.3 provide a number of possibilities, among which the leapfrog and Adams–Bashforth methods are the most common choices. Integrations performed using the spectral method typically require smaller time steps than those used when spatial derivatives are computed with low-order finite differences. This decrease in the maximum allowable time step is a natural consequence of the spectral method's ability to correctly resolve the spatial gradient in a $2\Delta x$ wave.

In order to better examine the source of this time-step restriction, consider the case of advection by a constant wind speed c. When c is constant, the time dependence of the kth Fourier component is governed by (4.13), which is just the oscillation equation (2.30) with $\kappa = ck$. The maximum value of $|k|$ is $N = \pi/\Delta x_e - \frac{1}{2} \approx \pi/\Delta x_e$, where Δx_e is the equivalent mesh size introduced in connection with (4.22) and $\pi/\Delta x_e \gg \frac{1}{2}$ if the total domain $[-\pi, \pi]$ is divided into at least ten grid intervals. If the oscillation equation is integrated using the leapfrog scheme, the stability requirement is $|\kappa \Delta t| \le 1$. Thus, the time step in a stable leapfrog spectral solution must satisfy $|c\Delta t/\Delta x_e| \le 1/\pi$.

Now suppose the advection equation (4.11) is approximated using a centered second-order spatial difference. The time evolution of the approximate solution at the jth grid point is, once again, governed by the oscillation equation. In this case, however, $\kappa = -c\sin(k\Delta x)/\Delta x$. The misrepresentation of the shorter wavelengths by the finite difference reduces the maximum value of $|\kappa|$ to $c/\Delta x$, and the leapfrog stability criterion relaxes to $|c\Delta t/\Delta x| \le 1$. If higher-order finite differences are used, the error in the shorter wavelengths is reduced, and the maximum allowable value of $|c\Delta t/\Delta x|$ decreases to 0.73 for a fourth-order difference, and to 0.63 for a sixth-order difference. Machenhauer (1979) notes that as the order of a centered finite-difference approximation approaches infinity, the maximum value of $|c\Delta t/\Delta x|$ for which the scheme is stable approaches $1/\pi$. The maximum stable time step for the leapfrog spectral method is therefore consistent with the interpretation of the spectral method as an infinite-order finite-difference scheme.

4.2.2   Improving Efficiency Using the Transform Method

The computational effort required to obtain spectral solutions to the advection equation ceases to be trivial if there are spatial variations in the wind speed. In such circumstances, the Galerkin requirement (4.9) becomes

$$\frac{da_k}{dt} = -\frac{i}{2\pi} \sum_{n=-N}^{N} na_n \int_{-\pi}^{\pi} c(x,t)e^{i(n-k)x}\,dx. \tag{4.24}$$

Although it may be possible to evaluate the integrals in (4.24) exactly for certain special flows, in most instances the computation must be done by numerical quadrature. In a typical practical application, $c(x,t)$ would be available at the same spatial resolution as $\psi(x,t)$; indeed, many models might include equations that simultaneously predict c and ψ. Suppose, therefore, that c is given by the Fourier series

$$c(x,t) = \sum_{m=-N}^{N} c_m(t)e^{imx}. \tag{4.25}$$

Substitution of this series expansion into (4.24) gives

$$\frac{da_k}{dt} = -\frac{i}{2\pi} \sum_{n=-N}^{N} \sum_{m=-N}^{N} nc_m a_n \int_{-\pi}^{\pi} e^{i(n+m-k)x}\,dx,$$

which reduces, by the orthogonality of the Fourier modes, to

$$\frac{da_k}{dt} = -\sum_{\substack{m+n=k \\ |m|,\,|n|\le N}} inc_m a_n. \tag{4.26}$$

The notation below the summation indicates that the sum should be performed for all indices n and m such that $|n| \le N$, $|m| \le N$, and $n+m=k$.

Although the expression (4.26) is relatively simple, it is not suitable for implementation in large, high-resolution numerical models. The number of arithmetic operations required to evaluate the time derivative of the kth Fourier coefficient via (4.26) is proportional to the total number of Fourier coefficients, $M \equiv 2N+1$. The total number of operations required to advance the solution one time step is therefore $O(M^2)$. On the other hand, the number of calculations required to evaluate a finite-difference formula at an individual grid point is independent of the total number of grid points. Thus, assuming that there are M points on the numerical grid, a finite-difference solution may be advanced one time step with just $O(M)$ arithmetic operations. Spectral models are therefore less efficient than finite-difference models when the approximate solution is represented by a large number of grid points—or, equivalently, a large number of Fourier modes. Moreover, the relative difference in computational effort increases rapidly with increases in M. As a consequence, spectral models were limited to just a few Fourier modes until the development of the transform method by Orszag (1970) and Eliasen et al. (1970).

The key to the transform method is the efficiency with which fast Fourier transforms can be used to swap the solution between wave-number space and physical space. Only $O(M \log M)$ operations are needed to convert a set of M Fourier coefficients, representing the Fourier transform of $\phi(x)$, into the M grid-point values

$\phi(x_j)$.[4] The basic idea behind the transform method is to compute product terms like $c\,\partial\psi/\partial x$ by transforming c and $\partial\psi/\partial x$ from wave number space to physical space (which takes $O(M \log M)$ operations), then multiplying c and $\partial\psi/\partial x$ at each grid point (requiring $O(M)$ operations), and finally transforming the product back to wave-number space (which again uses $O(M \log M)$ operations). The total number of operations required to evaluate $c\,\partial\psi/\partial x$ via the transform technique is therefore $O(M \log M)$, and when the number of Fourier components is large, it is far more efficient to perform these $O(M \log M)$ operations than the $O(M^2)$ operations necessary for the direct computation of (4.26) in wave-number space. In order to appreciate the degree to which the transform method can improve efficiency, suppose the spectral method is used in a two-dimensional problem in which the spatial dependence along each coordinate is represented by 128 Fourier modes; then an order-of-magnitude estimate of the increase in speed allowed by the transform method is

$$O\left(\frac{128 \times 128}{\log_2(128 \times 128)}\right) = O(1000).$$

The transform method is implemented as follows. Suppose that one wishes to determine the Fourier coefficients of the product of $\phi(x)$ and $\chi(x)$ such that

$$\phi(x)\chi(x) = \sum_{k=-K}^{K} p_k e^{ikx},$$

where ϕ and χ are periodic on the interval $0 \le x \le 2\pi$ and

$$\phi(x) = \sum_{m=-K}^{K} a_m e^{imx}, \quad \chi(x) = \sum_{n=-K}^{K} b_n e^{inx} \tag{4.27}$$

As just discussed, it is more efficient to transform ϕ and χ to physical space, compute their product in physical space, and to transform the result back to wave-number space than to compute p_k from the "convolution sum"

$$p_k = \sum_{\substack{m+n=k \\ |m|, |n| \le K}} a_m b_n.$$

The values of p_k obtained with the transform technique will be identical to those computed by the preceding summation formula, provided that there is sufficient spatial resolution to avoid aliasing error[5] during the computation of the product

---

[4]To be specific, if M is a power of two, a transform can be computed in $2M \log_2 M$ operations using the FFT algorithm.

[5]Aliasing error occurs when a short-wavelength fluctuation is sampled at discrete intervals and misinterpreted as a longer-wavelength oscillation. As discussed in Section 3.5.1, aliasing error can be generated in attempting to evaluate the product of two poorly resolved waves on a numerical mesh.

FIGURE 4.2. Aliasing of $k_1 + k_2$ into $\tilde{k}$ such that $|\tilde{k}|$ appears as the symmetric reflection of $k_1 + k_2$ about the cutoff wave number on the high-resolution physical mesh.

terms on the physical-space mesh. Suppose that the physical-space mesh is defined such that

$$x_j = \frac{2\pi j}{2N + 1}, \quad \text{where} \quad j = 1, \ldots, 2N + 1. \tag{4.28}$$

It might appear natural to chose $N = K$, thereby equating the number of grid-points on the physical mesh with the number of Fourier modes. It is, however, necessary to choose $N > K$ in order to avoid aliasing error.

The amount by which N must exceed K may be most easily determined using the graphical diagram shown in Fig. 4.2, which is similar to Fig. 3.9 of Section 3.5.1. The wave number is plotted along the horizontal axis; without loss of generality, only positive wave numbers will be considered. The cutoff wave number in the original expansion is K, and the cutoff wave number on the high-resolution physical mesh is $\pi/\Delta x_e$. Any aliasing error that results from the multiplication of waves with wave numbers k_1 and k_2 will appear at wave number $\tilde{k} = k_1 + k_2 - 2\pi/\Delta x_e$. The goal is to choose a sufficiently large value for $\pi/\Delta x_e$ to guarantee that no finite-amplitude signal is aliased into those wave numbers retained in the original Fourier expansion, which lie in the interval $-K \leq k \leq K$. The highest wave number that will have nonzero amplitude after computing the binary product on the physical mesh is $2K$. Thus, there will be no aliasing error if

$$K < \left| 2K - \frac{2\pi}{\Delta x_e} \right| = \frac{2\pi}{\Delta x_e} - 2K.$$

Using the definition of Δx_e implied by (4.28), the criteria for the elimination of aliasing error reduces to $N > (3K - 1)/2$.

The preceding result may be verified algebraically by considering the formula for $\tilde{p}_k$, the kth component of the finite Fourier transform computed from the grid-point values of $\phi\chi$ on the physical mesh,

$$\tilde{p}_k \equiv \frac{1}{M} \sum_{j=1}^{M} \phi(x_j)\chi(x_j)e^{-ikx_j}. \tag{4.29}$$

Here $M = 2N + 1$ is the total number of grid points on the physical mesh. Let the values of $\phi(x_j)$ and $\chi(x_j)$ appearing in the preceding formula be expressed in the form

$$\phi(x_j) = \sum_{m=-N}^{N} a_m e^{imx_j}, \qquad \chi(x_j) = \sum_{n=-N}^{N} b_n e^{inx_j},$$

where those values of a_n and b_m that were not included in the original series expansions (4.27) are zero, i.e.,

$$a_\ell = b_\ell = 0 \quad \text{for} \quad K < |\ell| \le N. \tag{4.30}$$

Substituting these expressions for $\phi(x_j)$ and $\chi(x_j)$ into (4.29), one obtains

$$\tilde{p}_k = \sum_{m=-N}^{N} \sum_{n=-N}^{N} a_m b_n \left(\frac{1}{M} \sum_{j=1}^{M} e^{i(m+n-k)x_j} \right). \tag{4.31}$$

Since $x_j = 2\pi j/M$, each term in the inner summation in (4.31) is unity when $m + n - k$ is 0, M, or $-M$. The inner summation is zero for all other values of m and n by the discrete orthogonality condition (4.18). Thus, (4.31) may be written

$$\tilde{p}_k = \sum_{\substack{m+n=k \\ |m|, |n| \le N}} a_m b_n + \sum_{\substack{m+n=k+M \\ |m|, |n| \le N}} a_m b_n + \sum_{\substack{m+n=k-M \\ |m|, |n| \le N}} a_m b_n, \tag{4.32}$$

where the last two terms represent aliasing error, only one of which can be nonzero for a given value of k. The goal is to determine the minimum resolution required on the physical mesh (or, equivalently, the smallest M) that will prevent aliasing errors from influencing the value of p_k^* associated with any wave number retained in the original Fourier expansion. Any aliasing into a negative wave number will arise through the summation

$$\sum_{\substack{m+n=k+M \\ |m|, |n| \le N}} a_m b_n.$$

If follows from (4.30) that $a_m b_n = 0$ if $m + n > 2K$, so for a given wave number k all the terms in the preceding summation will be zero if $m + n = k + M > 2K$. Thus, there will be no aliasing error in p_k^* for those wave numbers retained in the original expansion if M satisfies $-K + M > 2K$. An equivalent condition expressed in terms of the wave number N is $N > (3K - 1)/2$, which is the same result obtained using Fig. 4.2. A similar argument may be used to show that this same condition also prevents the third term in (4.32) from generating aliasing error in the retained wave numbers. The choice $N = 3K/2$ is therefore sufficient to ensure that $p_k = \tilde{p}_k$ for all $|k| \le K$ and guarantee that the transform method yields the same algebraic result as the convolution sum in wave-number space. In order to maximize the efficiency of the fast Fourier transforms used in practical applications, N is often chosen to be the smallest integer exceeding $(3k - 1)/2$ that contains no prime factor larger than five.

The procedure used to implement the transform method may be summarized as follows. In order to be concrete, suppose that a solution to the variable-wind-speed advection equation is sought on the periodic domain $0 \le x \le 2\pi$ and that $c(x, t)$ is being simultaneously predicted by integrating a second unspecified equation. Let both ϕ and c be approximated by Fourier series expansions of the form (4.14) and (4.25) with cutoff wave numbers $N = K$.

1. Pad the coefficients in the Fourier expansions of c and ϕ with zeros by defining $a_k = c_k = 0$, for $K < |k| \le 3K/2$.

2. Multiply each a_k by ik to compute the derivative of ϕ in wave-number space.

3. Perform two inverse FFTs to obtain $c(x_j)$ and $\partial\phi(x_j)/\partial x$ on the physical-space grid, whose nodal points are located at $x_j = 2\pi j/(3K + 1)$.

4. Compute the product $c(x_j)\partial\phi(x_j)/\partial x$ on the physical-space grid.

5. (If terms representing additional forcing are present in the governing equation, and if those terms are more easily evaluated on the physical mesh than in wave-number space, evaluate those terms now and add the result to to $c(x_j)\partial\phi(x_j)/\partial x$.)

6. Fast-Fourier transform $c(x_j)\partial\phi(x_j)/\partial x$ to obtain the total forcing at each wave number, i.e., to get the right-hand-side of (4.26). Discard the forcing at wave numbers for which $|k| > K$.

7. Step the Fourier coefficients forward to the next time level using an appropriate time-differencing scheme.

Note that the transform method allows processes that are difficult to describe mathematically in wave-number space to be conveniently evaluated during the portion of the integration cycle when the solution is available on the physical mesh. For example, if ϕ represents the concentration of water vapor, any change in ϕ produced by the condensation or evaporation of water depends on the degree to which the vapor pressure at a given grid point exceeds the saturation vapor pressure. The degree of supersaturation is easy to determine in physical space but very difficult to assess in wave-number space.

4.2.3 Conservation and the Galerkin Approximation

The mathematical equations describing non-dissipative physical systems often conserve domain averages of quantities like energy or momentum. When spectral methods are used to approximate such systems, the numerical solution replicates some of the important conservation properties of the true solution. In order to examine the conservation properties of the spectral method for a relatively general class of problems consider those partial differential equations of the form (4.1) for which the forcing has the property that $\overline{\upsilon F(\upsilon)} = 0$, where the overbar denotes the integral over the spatial domain and υ is any sufficiently smooth function that satisfies the boundary conditions.

An example of this type of problem is the simulation of passive tracer transport by nondivergent flow in a periodic spatial domain, which is governed by the equation

$$\frac{\partial \psi}{\partial t} + \mathbf{v} \cdot \nabla \psi = 0.$$

In this case $F(\upsilon) = \mathbf{v} \cdot \nabla \upsilon$. One can verify that $\overline{\upsilon F(\upsilon)} = 0$ if υ is any periodic function with continuous first derivatives by noting that

$$\int_D \upsilon(\mathbf{v} \cdot \nabla \upsilon) \, dV = \frac{1}{2} \int_D \nabla \cdot (\upsilon^2 \mathbf{v}) - \upsilon^2 (\nabla \cdot \mathbf{v}) \, dV = 0,$$

where the second equality follows from periodicity and the nondivergence of the velocity field.

If ϕ is an approximate spectral solution to (4.1) in which the time dependence is not discretized, then

$$\frac{\partial \phi}{\partial t} + F(\phi) = R(\phi), \tag{4.33}$$

where $R(\phi)$ denotes the residual. Suppose that the partial differential equation being approximated is a conservative system for which $\overline{\phi F(\phi)} = 0$, then multiplying (4.33) by ϕ and integrating over the spatial domain yields

$$\frac{1}{2} \overline{\frac{\partial \phi^2}{\partial t}} = \overline{\phi R(\phi)}.$$

The right side of the preceding is zero because ϕ is a linear combination of the expansion functions and $R(\phi)$ is orthogonal to each individual expansion function. As a consequence,

$$\frac{d}{dt} \|\phi\|_2 = 0, \tag{4.34}$$

implying that spectral approximations to conservative systems are not subject to nonlinear instability because (4.34) holds independent of the linear or nonlinear structure of $F(\psi)$. The only potential source of numerical instability is in the discretization of the time derivative.

Neglecting time-differencing errors, spectral methods will also conserve $\overline{\phi}$ provided that $\overline{F(\upsilon)} = 0$, where once again υ is any sufficiently smooth function that satisfies the boundary conditions. The conservation of $\overline{\phi}$ can be demonstrated by integrating (4.33) over the domain to obtain

$$\frac{\partial \overline{\phi}}{\partial t} = \overline{R(\phi)} \propto \overline{R(\phi)\varphi_0} = 0,$$

where φ_0 is the lowest-wave-number orthogonal expansion function, which is a constant.

4.3   The Pseudospectral Method

The spectral method uses orthogonal expansion functions to represent the numerical solution and constrains the residual error to be orthogonal to each of the expansion functions. As discussed in Section 3.1, there are alternative strategies for constraining the size of the residual. The pseudospectral method utilizes one of these alternative strategies: the collocation approximation, which requires the residual to be zero at every point on some fixed mesh. Spectral and pseudospectral methods might both represent the solution with the same orthogonal expansion functions; however, as a consequence of the collocation approximation, the pseudospectral method is basically a grid-point scheme—series expansion functions are used only to compute derivatives.

In order to illustrate the pseudospectral procedure, suppose that solutions are sought to the advection equation (4.11) on the periodic domain $0 \le x \le 2\pi$ and that the approximate solution ϕ and the spatially varying wind speed $c(x)$ are represented by Fourier series truncated at wave number K:

$$\phi(x, t) = \sum_{n=-K}^{K} a_n e^{inx}, \quad c(x, t) = \sum_{m=-K}^{K} c_m e^{imx}.$$

The collocation requirement at grid point j is

$$\sum_{n=-K}^{K} \frac{da_n}{dt} e^{inx_j} + \sum_{m=-K}^{K} c_m e^{imx_j} \sum_{n=-K}^{K} in a_n e^{inx_j} = 0. \tag{4.35}$$

Enforcing $R(\phi(x_j)) = 0$ at $2K + 1$ points on the physical-space grid leads to a solvable linear system for the time derivatives of the $2K + 1$ Fourier coefficients. In the case of the Fourier spectral method, the most efficient choice for the location of these points is the equally spaced mesh

$$x_j = j \left(\frac{2\pi}{2K + 1} \right), \quad j = 1, 2, \dots, 2K + 1. \tag{4.36}$$

There is no need actually to solve the linear system for the da_k/dt. It is more efficient to write (4.35) in the equivalent form

$$\frac{d\phi}{dt}(x_j) + c(x_j) \frac{\partial \phi}{\partial x}(x_j) = 0, \tag{4.37}$$

where

$$\frac{\partial \phi}{\partial x}(x_j) = \sum_{n=-K}^{K} in a_n e^{inx_j}. \tag{4.38}$$

The grid-point nature of the pseudospectral method is apparent in (4.37), which is similar to the time-tendency equations[6] that arise in differential–difference approximations to the advection equation, except that the derivative is computed in a special way. Instead of using finite differences, the spatial derivative is calculated at each time step by first computing the Fourier coefficients through the discrete Fourier transform

$$a_k(t) = \frac{1}{2K+1} \sum_{j=1}^{2K+1} \phi(x_j, t) e^{-ikx_j},$$

then differentiating each Fourier mode analytically and inverse transforming according to (4.38). This procedure requires two fast Fourier transforms (FFT) per time step.

The advantage of the pseudospectral method relative to conventional finite-difference schemes is that provided the solution is smooth, the pseudospectral method is more accurate. As discussed in Section 3.2.1, the error in the Fourier approximation to the derivative of an infinitely differentiable function will decrease more rapidly than any finite power of Δx as the grid resolution is increased. Thus, like the spectral method, the pseudospectral method is essentially an infinite-order finite-difference scheme. The disadvantage of the pseudospectral method is that it requires more computation than conventional finite-difference schemes when both methods are used with the same spatial resolution. If M is the total number of grid points, the FFTs in the pseudospectral computation require $O(M \log(M))$ operations per time step, whereas conventional finite-difference methods need only $O(M)$ operations. The extra work per time step may, however, be easily offset if the increased accuracy of the pseudospectral representation allows the computations to be performed on a coarser mesh.

The advantage of the pseudospectral method relative to the spectral method is that the pseudospectral method requires less computation. The increase in efficiency of the pseudospectral method is achieved by allowing aliasing error in the computation of the products of spatially varying functions. As a consequence of this aliasing error, the residual need not be orthogonal to the individual expansion functions, and the pseudospectral method does not possess the conservation properties discussed in Section 4.2.3. In particular, the pseudospectral method is subject to nonlinear instability.

The difference in aliasing between the pseudospectral and spectral methods can be evaluated by multiplying (4.35) by e^{ikx_j} and summing over all j to obtain

$$\sum_{n=-K}^{K} \frac{da_n}{dt} \sum_{j=1}^{2K+1} e^{i(n-k)x_j} + \sum_{n=-K}^{K} \sum_{m=-K}^{K} inc_m a_n \sum_{j=1}^{2K+1} e^{i(n+m-k)x_j} = 0.$$

---

[6]As in conventional spectral and finite-difference techniques, the time derivative would be discretized using leapfrog, Adams–Bashforth, or some other appropriate scheme.

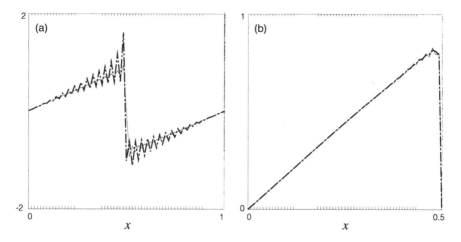

FIGURE 4.3. Spectral (solid) and pseudospectral (dot-dashed) solutions to the viscous Burgers's equation at $t = 0.4$: (a) truncation at wave number 64π, (b) truncation at wave number 128π. Only the left half of the domain is shown in (b).

Using the definition of x_j (4.36) and the discrete-mesh orthogonality condition (4.18), the preceding reduces to

$$\frac{da_k}{dt} + \sum_{\substack{m+n=k \\ |m|, |n| \le K}} inc_m a_n + \sum_{\substack{m+n=k+M \\ |m|, |n| \le K}} inc_m a_n + \sum_{\substack{m+n=k-M \\ |m|, |n| \le K}} inc_m a_n = 0,$$

where $M = 2K + 1$. As when spectral computations are performed using the transform technique, the last two terms represent aliasing error, only one of which can be nonzero for a given value of k. In contrast to the spectral method, however, these aliasing terms do not disappear, because in the pseudospectral method the number of grid points on the physical mesh is identical to the number of Fourier wave numbers, and therefore none of the c_m and a_n need be zero.

One might suppose that aliasing error always decreases the accuracy of the solution, but the impact of aliasing error on accuracy depends on the problem. As an example, suppose that spectral and pseudospectral approximations are computed to the solution of the viscous Burgers's equation

$$\frac{\partial \psi}{\partial t} + \psi \frac{\partial \psi}{\partial x} = \nu \frac{\partial^2 \psi}{\partial x^2} \tag{4.39}$$

on the periodic domain $0 \le x \le 1$ subject to the initial condition $\psi(x, 0) = \sin(2\pi x)$. Spectral and pseudospectral solutions to this problem are shown in Fig. 4.3 at $t = 0.4$ for the case $\nu = 0.002$. In these computations the time-differencing for the nonlinear advection term was leapfrog, and the diffusion term was integrated using a forward difference over an interval of $2\Delta t$. The time step was selected such that

$$\frac{\Delta t}{\Delta x} \max_x [\psi(x,0)] = 0.1 .$$

The true solution to the inviscid problem gradually steepens around the point $x = \frac{1}{2}$ and becomes discontinuous at $t = (2\pi)^{-1}$ (see Section 3.6.1), but the viscous dissipation in (4.39) prevents the gradient at $x = \frac{1}{2}$ from collapsing to a true discontinuity and gradually erodes the amplitude of the solution so that $\psi(x,t) \to 0$ as $t \to \infty$. Fig. 4.3a shows the approximate solutions to (4.39) obtained using a spectral truncation at wave number 64π, which is equivalent to a grid spacing of $\Delta x = 1/64$. Both the spectral and the pseudospectral solutions have trouble resolving the steep gradient at $x = \frac{1}{2}$ and develop significant $2\Delta x$ noise. The amplitude of the $2\Delta x$ ripples remains bounded in the spectral solution, but aliasing error generates a rapidly growing instability in the $2\Delta x$ component of the pseudospectral solution.

Rather different results are, however, obtained if the same problem is repeated with twice the spatial resolution. When the cutoff wave number is 128π, the pseudospectral method remains stable and actually generates a more accurate solution than that obtained with the spectral method. A close-up comparison of the two solutions over the subdomain $0 \le x \le \frac{1}{2}$ appears in Fig. 4.3b. The $2\Delta x$ ripples in the spectral solution are of distinctly larger amplitude than those appearing in the pseudospectral solution. The pseudospectral solution remains stable because the rate at which viscous damping erodes a $2\Delta x$ wave increases by a factor of four as the spatial resolution is doubled from $\Delta x = 1/64$ to $1/128$, and when $\Delta x = 1/128$, the rate of energy removal from the $2\Delta x$ wave by viscous damping exceeds the rate at which $2\Delta x$ waves are amplified by aliasing error. The superiority of the pseudospectral solution over the spectral solution in Fig. 4.3b highlights the fact that although conservation of $\|\phi\|_2$ implies stability, it does not imply better accuracy.

The influence of aliasing error on accuracy is largely a matter of chance. Although it is certainly an error when the pseudospectral method misrepresents interactions between $2\Delta x$ and $3\Delta x$ waves as an aliased contribution to a $6\Delta x$ wave, it is also an error when the spectral method simply neglects the interactions between these same short waves, since the product of $2\Delta x$ and $3\Delta x$ disturbances should properly appear in a $6\Delta x/5$ wave. In Burgers's equation, and in many other fluid-flow problems, there is a cascade of energy to smaller scales. An accurate conservative scheme, such as the spectral method, replicates this down-scale energy transfer except that the cascade is terminated at the shortest scales resolved in the numerical simulation. In the absence of viscous dissipation, the spectral approximation to Burgers's equation conserves energy, and the energy that cascades down scale simply accumulates in the shortest resolvable modes. In order to simulate the continued cascade of energy into the unresolvable scales of motion, it is necessary to remove energy from the shortest resolvable waves. The energy-removal algorithm constitutes a parametrization of the influence of unresolved short-wavelengths on the resolved modes and should be designed to represent the true behavior of the physical system as closely as possible.

Whatever the exact details of the energy-removal scheme, if it prevents the unphysical accumulation of energy at the short wavelengths in the spectral solution, the same energy-removal scheme will often stabilize a pseudospectral solution to the same problem. In the case shown in Fig. 4.3b, for example, the amount of viscous dissipation required to stabilize the pseudospectral solution is less than that required to remove the spurious ripples from the spectral solution. Although it is not generally necessary to filter the solution this heavily, aliasing error can be completely eliminated by removing all energy at wavelengths shorter than or equal to $3\Delta x$ after each time step, or equivalently, by removing the highest onethird of the resolved wave numbers. If such a filter is used in combination with a pseudospectral method truncated at wave number M, the resulting algorithm is identical to that for a Galerkin spectral method truncated at wave number $2M/3$ in which the nonlinear terms are computed via the transform method.

4.4 Spherical Harmonics

The two-dimensional distribution of a scalar variable on the surface of a sphere can be efficiently approximated by a truncated series of spherical harmonic functions. Spherical harmonics can also be used to represent three-dimensional fields defined within a volume bounded by two concentric spheres if grid points or finite-elements are used to approximate the spatial structure along the radial coordinate and thereby divide the computational domain into a series of nested spheres. Let λ be the longitude, θ the latitude, and define $\mu = \sin\theta$. If ψ is a smooth function of λ and μ, it can be represented by a convergent expansion of spherical harmonic functions of the form

$$\psi(\lambda, \mu) = \sum_{m=-\infty}^{\infty} \sum_{n=|m|}^{\infty} a_{m,n} Y_{m,n}(\lambda, \mu), \qquad (4.40)$$

where each spherical harmonic function $Y_{m,n}(\lambda, \mu) = P_{m,n}(\mu)e^{im\lambda}$ is the product of a Fourier mode in λ and an associated Legendre function in μ.

The associated Legendre functions are generated from the Legendre polynomials using the relation

$$P_{m,n}(\mu) = \left[\frac{(2n+1)}{2} \frac{(n-m)!}{(n+m)!}\right]^{1/2} (1-\mu^2)^{m/2} \frac{d^m}{d\mu^m} P_n(\mu), \qquad (4.41)$$

where P_n is the nth-order Legendre polynomial defined such that

$$P_n(\mu) = \frac{1}{2^n n!} \frac{d^n}{d\mu^n}\left[(\mu^2 - 1)^n\right], \qquad (4.42)$$

and the formula that results after substituting (4.42) into (4.41) is valid for $|m| \leq n$. Note that when m is odd[7] the associated Legendre functions are not polynomials in μ.

The leading factor in (4.41) normalizes $P_{m,n}$ so that

$$\int_{-1}^{1} P_{m,n}(\mu) P_{m,s}(\mu) \, d\mu = \delta_{n,s}, \qquad (4.43)$$

where $\delta_{n,s} = 1$ if $n = s$ and is zero otherwise. As a consequence, the orthogonality relation for the spherical harmonics becomes

$$\frac{1}{2\pi} \int_{-1}^{1} \int_{-\pi}^{\pi} Y_{m,n}(\lambda, \mu) Y_{r,s}^{*}(\lambda, \mu) \, d\lambda \, d\mu = \delta_{m,r} \delta_{n,s}, \qquad (4.44)$$

where $Y_{r,s}^{*}$ is the complex conjugate of $Y_{r,s}$. The associated Legendre functions have the property that $P_{-m,n}(\mu) = (-1)^m P_{m,n}(\mu)$, which implies that $Y_{-m,n} = (-1)^m Y_{m,n}^{*}$, and thus the expansion coefficients for any approximation to a real-valued function satisfy

$$a_{-m,n} = (-1)^m a_{m,n}^{*}. \qquad (4.45)$$

Two recurrence relations satisfied by the associated Legendre functions that will be used in the subsequent analysis are

$$\mu P_{m,n} = \epsilon_{m,n+1} P_{m,n+1} + \epsilon_{m,n} P_{m,n-1} \qquad (4.46)$$

and

$$(1 - \mu^2) \frac{dP_{m,n}}{d\mu} = -n \epsilon_{m,n+1} P_{m,n+1} + (n+1) \epsilon_{m,n} P_{m,n-1}, \qquad (4.47)$$

where

$$\epsilon_{m,n} = \left(\frac{n^2 - m^2}{4n^2 - 1} \right)^{1/2}.$$

The spherical harmonics are eigenfunctions of the Laplacian operator on the sphere such that

$$\nabla^2 Y_{m,n} = \frac{-n(n+1)}{a^2} Y_{m,n}, \qquad (4.48)$$

where a is the radius of the sphere and the horizontal Laplacian operator in spherical coordinates is

$$\nabla^2 = \frac{1}{a^2 \cos^2 \theta} \left[\frac{\partial^2}{\partial \lambda^2} + \cos \theta \frac{\partial}{\partial \theta} \left(\cos \theta \frac{\partial}{\partial \theta} \right) \right]$$

$$= \frac{1}{a^2 (1 - \mu^2)} \frac{\partial^2}{\partial \lambda^2} + \frac{1}{a^2} \frac{\partial}{\partial \mu} \left[(1 - \mu^2) \frac{\partial}{\partial \mu} \right].$$

---

[7]Specialized terminology is sometimes used to differentiate between the indices of the associated Legendre function $P_{m,n}$. The m index indicates the "order," whereas the n index indicates the "degree."

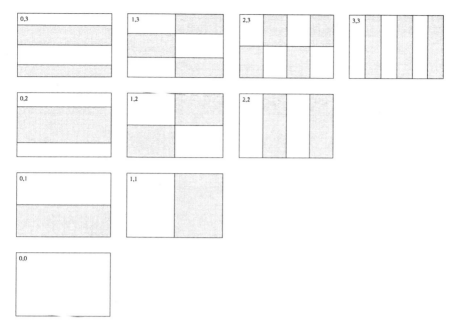

FIGURE 4.4. Schematic indication of the distribution of the nodal lines for the spherical harmonics $Y_{m,n}$ in the set $0 \le m \le n \le 3$. The horizontal axis in each map is linear in λ and includes the domain $-\pi \le \lambda \le \pi$. The vertical axis is linear in $\sin \theta$ and includes the domain $-\pi/2 \le \theta \le \pi/2$. The m, n index of each mode is indicated in the upper left corner of each map.

The eigenvalue associated with each $Y_{m,n}$ can be used to define a total wave number by analogy to the situation on a flat plane, where

$$\nabla^2 e^{i(mx+ny)} = -(m^2 + n^2)e^{i(mx+ny)}$$

and the total wave number is $(m^2 + n^2)^{1/2}$. The total wave number associated with $Y_{m,n}$ is $(n^2 + n)^{1/2}/a$, which is independent of the zonal wave number m. The nonintuitive absence of m in the formula for the total wave number arises, in part, because the effective meridional wave number of a spherical harmonic depends on both m and n.

4.4.1 Truncating the Expansion

The zero-order structure of $Y_{m,n}(\lambda, \mu)$ is described by the distribution of its nodal lines on the surface of the sphere. The nodal lines for the those modes for which $0 \le m \le n \le 3$ are schematically diagrammed in Fig. 4.4. The zonal structure of $Y_{m,n}$ is that of a simple Fourier mode $e^{im\lambda}$, so there are $2m$ nodal lines intersecting a circle of constant latitude. The distribution of the nodal lines along a line of constant longitude is more complex. The formula for the meridional structure of

n	$P_{0,n}$	$P_{1,n}$	$P_{2,n}$	$P_{3,n}$
3	$\sqrt{\frac{7}{8}}\left(5\mu^3 - 3\mu\right)$	$\sqrt{\frac{21}{32}}\left(5\mu^2 - 1\right)\sqrt{1-\mu^2}$	$\frac{\sqrt{105}}{4}\left(\mu - \mu^3\right)$	$\frac{\sqrt{70}}{8}\left(1-\mu^2\right)^{\frac{3}{2}}$
2	$\sqrt{\frac{5}{8}}\left(3\mu^2 - 1\right)$	$\frac{\sqrt{15}}{2}\mu\sqrt{1-\mu^2}$	$\frac{\sqrt{15}}{4}\left(1 - \mu^2\right)$	N/A
1	$\sqrt{\frac{3}{2}}\,\mu$	$\frac{\sqrt{3}}{2}\sqrt{1-\mu^2}$	N/A	N/A
0	$1/\sqrt{2}$	N/A	N/A	N/A

TABLE 4.2. Meridional structure of the low-order spherical harmonics appearing in Fig. 4.4.

each of the modes shown in Fig. 4.4 is given in Table 4.2. $P_{m,n}(\mu)$ is proportional to

$$(1 - \mu^2)^{m/2}\frac{d^{n+m}}{d\mu^{n+m}}(\mu^2 - 1)^n. \tag{4.49}$$

The first factor has no zeros between the north and south poles; the second factor is a polynomial of order $n-m$ that has $n-m$ zeros between the two poles. Thus the modes with zero meridional wave number are $Y_{s,s}$, whereas the modes with zero zonal wave number are $Y_{0,s}$. Figure 4.4 also provides a graphical illustration of the reason why expansions in spherical harmonics are constructed without attempting to define and include modes with $m > n$.

In all practical applications the infinite series (4.40) must be truncated to create a numerical approximation of the form

$$\psi(\lambda, \mu) = \sum_{m=-M}^{M} \sum_{n=|m|}^{N(m)} a_{m,n} Y_{m,n}(\lambda, \mu). \tag{4.50}$$

The *triangular* truncation, in which $N(m) = M$, is unique among the various possible truncations because it is the only one that provides uniform spatial resolution over the entire surface of the sphere. The approximation to $\psi(\lambda, \mu)$ obtained using a triangular truncation is invariant to an arbitrary rotation of the latitude and longitude coordinates about the center of the sphere. This invariance follows from the fact that any spherical harmonic of degree less than or equal to M (i.e., for which $n \leq M$) can be exactly expressed as a linear combination of the spherical harmonics in an Mth-order triangular truncation defined with respect to the arbitrarily rotated coordinates. To be specific, if $Y_{m,n}$ is a spherical harmonic with $n \leq M$, and λ', μ', and $Y'_{r,s}$ are coordinates and spherical harmonics defined with respect to an arbitrarily rotated polar axis, then there exist a set of expansion

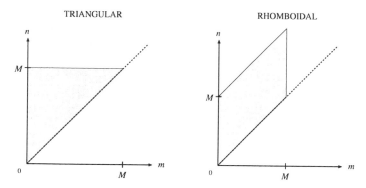

FIGURE 4.5. Shading indicates the portion of the $m \geq 0$ half-plane in which the m and n indices are retained in Mth-order triangular and rhomboidal truncations. Note that the set of indices retained in the rhomboidal truncation define a parallelogram rather than a rhombus.

coefficients $b_{r,s}$ such that

$$Y_{m,n}(\lambda, \mu) = \sum_{r=-M}^{M} \sum_{s=|r|}^{M} b_{r,s} Y'_{r,s}(\lambda', \mu')$$

(Courant and Hilbert 1953, p. 535).

In spite of its elegance, the triangular truncation may not be optimal in situations where the characteristic scale of the approximated field exhibits a systematic variation over the surface of the sphere. In the Earth's atmosphere, for example, the perturbations in the geopotential height field in the tropics are much weaker than those in the middle latitudes. A variety of alternative truncations have therefore been used in low-resolution ($M < 30$) global atmospheric models. The most common alternative is the *rhomboidal* truncation, in which $N(m) = |m| + M$ in (4.50). The set of indices (m, n) retained in triangular and rhomboidal truncations with approximately the same number of degrees of freedom are compared in Fig. 4.5. Only the right half-plane is shown in Fig. 4.5, since whenever $Y_{m,n}$ is included in the truncation, $Y_{-m,n}$ is also retained. In comparison with the triangular truncation, the rhomboidal truncation neglects two families of modes with large n, those for which $n - m \approx 0$ and those for which $n - m \approx n$. The first of these families is composed of high-zonal-wave-number modes that are equatorially trapped, since the first factor in (4.49) has an mth-order zero at each pole. The second family of neglected modes have small zonal wave number but fine meridional structure near the poles. As a consequence, the spatial resolution in a rhomboidal truncation is somewhat concentrated in the middle latitudes, which may be suitable for low-resolution models of the Earth's atmosphere but less appropriate for more general applications. Other truncations in which $N(m)$ is a more complex function of m have also been proposed in order to improve the efficiency of low-resolution climate models (Kiehl et al. 1996). At present there does not seem to be a clear consensus about which truncation is most suitable for use in low-resolution atmospheric models. The triangular truncation is, however, the universal choice in high-resolution global weather forecasting.

4.4.2   Elimination of the Pole Problem

Explicit finite-difference approximations to the equations governing fluid motion on a sphere can require very small time steps to maintain stability if the grid points are distributed over the sphere on a uniform latitude–longitude mesh. This time-step restriction arises because the convergence of the meridians near the poles greatly reduces the physical distance between adjacent nodes on the same latitude circle, and as a consequence, the CFL condition is far more restrictive near the poles than in the tropics. Several approaches have been used to circumvent this problem (Williamson 1979), but they all have at least one cumbersome aspect. One of the most elegant solutions to the pole problem is obtained using spherical harmonic expansion functions in a spectral or pseudospectral approximation.[8]

A simple context in which to compare the stability criteria obtained using spherical harmonics and finite differences is provided by the shallow-water equations linearized about a resting basic state of depth H on a nonrotating sphere:

$$\frac{\partial \delta}{\partial t} + g\nabla^2 h = 0,$$

$$\frac{\partial h}{\partial t} + H\delta = 0,$$

where

$$\delta = \frac{1}{a\cos\theta}\left[\frac{\partial u}{\partial \lambda} + \frac{\partial v\cos\theta}{\partial \theta}\right]$$

is the horizontal divergence of the velocity field and h is the free-surface displacement. Let $\delta^n = \delta(\lambda, \theta, n\Delta t)$, $h^n = h(\lambda, \theta, n\Delta t)$, and $c = \sqrt{gH}$. Then if the time derivatives in these equations are approximated using forward–backward differencing, one obtains the semidiscrete system

$$\frac{\delta^{n+1} - \delta^n}{\Delta t} + g\nabla^2 h^n = 0, \tag{4.51}$$

$$\frac{h^{n+1} - h^n}{\Delta t} + H\delta^{n+1} = 0, \tag{4.52}$$

or, after eliminating δ^{n+1} and δ^n,

$$\frac{h^{n+1} - 2h^n + h^{n-1}}{(\Delta t)^2} - c^2\nabla^2 h^n = 0. \tag{4.53}$$

---

[8]Another attractive approach for minimizing the pole problem in global weather forecasting is provided by the semi-Lagrangian semi-implicit scheme discussed in Section 6.3.2.

Suppose that the spatial structure of h^n is represented by spherical harmonics in the triangular truncation

$$h(\lambda, \mu, n\Delta t) = \sum_{r=-M}^{M} \sum_{s=|r|}^{M} b_{r,s}^n Y_{r,s}(\lambda, \mu),$$

where as before, $\mu = \sin\theta$. Since the spherical harmonics are eigenfunctions of the Laplacian operator on the sphere, a solvable system of equations for the expansion coefficients $b_{r,s}^n$ can be obtained by substituting this expansion in (4.53) and using (4.48) to arrive at

$$\frac{b_{r,s}^{n+1} - 2b_{r,s}^n + b_{r,s}^{n-1}}{(\Delta t)^2} = -\frac{c^2 s(s+1)}{a^2} b_{r,s}^n.$$

Assuming that the expansion coefficients have a time dependence proportional to $e^{-i\omega n \Delta t}$, the dispersion relation for each mode is

$$\sin^2\left(\frac{\omega \Delta t}{2}\right) = s(s+1)\left(\frac{c\Delta t}{2a}\right)^2.$$

The scheme will be stable when the frequencies of the highest meridional wave numbers are real, or

$$\frac{c\Delta t \sqrt{M(M+1)}}{2a} < 1. \tag{4.54}$$

Now suppose that the Laplacian operator in (4.53) is evaluated using finite differences on a latitude–longitude grid in which $\Delta\theta$ and $\Delta\lambda$ are uniform over the globe. Then near the poles, the highest-frequency components in the numerical solution will be forced by short-wavelength spatial variations around a latitude circle. Approximating the second derivative with respect to longitude in (4.53) as $(a\cos\theta)^{-2}\delta_\lambda^2$ and substituting a Fourier mode in time and longitude of the form

$$\hat{h}(\theta) e^{i(rm\Delta\lambda - \omega n \Delta t)}$$

into the resulting semidiscrete equation yields

$$\frac{4\hat{h}}{(\Delta t)^2}\sin^2\left(\frac{\omega \Delta t}{2}\right) = \frac{4c^2\hat{h}}{(a\Delta\lambda\cos\theta)^2}\sin^2\left(\frac{r\Delta\lambda}{2}\right) - \frac{c^2}{a^2\cos\theta}\frac{\partial}{\partial\theta}\left(\cos\theta\frac{\partial\hat{h}}{\partial\theta}\right).$$

For those modes with zero meridional wave number, a necessary condition for the reality of ω and the stability of this semidiscrete approximation is that

$$\frac{c\Delta t}{a\Delta\lambda\cos\theta} \leq 1.$$

Let M be the highest zonal wave number (in radians) resolved on the numerical mesh; then $M \Delta\lambda = \pi$, and the preceding stability condition may be expressed as

$$\frac{cM \Delta t}{a\pi \cos\theta} \leq 1.$$

A comparison of this condition with (4.54) shows that the maximum stable time step that can be used with the finite-difference method on the portion of the mesh where $\theta \rightarrow \pm\pi/2$ is far smaller than that which can be used in a spectral model employing spherical harmonic expansion functions with the same cutoff wave number.

The restrictions on the maximum stable time step can be removed altogether by using trapezoidal time-differencing instead of the forward-backward scheme in (4.51) and (4.52). This is not a particularly efficient approach when the Laplacian is approximated using finite differences, since the trapezoidal approximation generates a large system of implicit algebraic equations that must be solved at every time step. Trapezoidal time-differencing can, however, be implemented very efficiently in spectral approximations that use spherical harmonic expansion functions, because the spherical harmonics are eigenfunctions of the horizontal Laplacian operator on the sphere. As a consequence, the expansion coefficient for each $Y_{r,s}$ can be computed independently of the other modes, and the implicit coupling introduced by trapezoidal time-differencing only generates a trivial two-variable system involving the amplitudes of the divergence and the free-surface elevation of each mode. The ease with which trapezoidal approximations to (4.51) and (4.52) can be integrated using spherical harmonics can be used to great advantage in formulating semi-implicit time-differencing approximations to the nonlinear equations governing fluid flow on a sphere (see Sections 7.2.3 and 7.6.5).

4.4.3   Gaussian Quadrature and the Transform Method

In most practical applications some of the forcing terms in the governing equations contain products of two or more spatially varying functions. Unless the total number of modes retained in the series expansion is very small, a variant of the transform method described in Section 4.2.2 must be used in order to efficiently apply spectral methods to such problems. The transform between grid-point values and the spectral coefficients of the spherical harmonic functions is, however, more cumbersome and computationally less efficient than the fast Fourier transform. The lack of highly efficient transforms is one of the few drawbacks associated with the use of spherical harmonic expansion functions in global spectral models. Even so, it is far more efficient to use the transform method than the alternative "interaction coefficient" method, in which the forcing is computed from a summation of products of pairs of the spectral coefficients (Orszag 1970; Eliasen et al. 1970).

If $\psi(\lambda, \mu)$ is approximated by a truncated series of spherical harmonics of the form (4.50), the transformation from the set of spectral coefficients to points on a latitude–longitude grid can be computed using the relation

$$\psi(\lambda, \mu) = \sum_{m=-M}^{M} \hat{a}_m(\mu)e^{im\lambda}, \tag{4.55}$$

where

$$\hat{a}_m(\mu) = \sum_{n=|m|}^{N(m)} a_{m,n} P_{m,n}(\mu). \tag{4.56}$$

The first summation (4.55) is a discrete Fourier transform with respect to the longitudinal coordinate λ that can be efficiently evaluated using fast Fourier transforms to obtain $2M + 1$ grid-point values around each latitude circle in $O[M \log M]$ operations. The second summation (4.56) is essentially an inner product requiring $O[N^2]$ operations to evaluate $\hat{a}_m$ at N different latitudes. The lack of a fast transform for the latitude coordinate makes the spherical-harmonic spectral model less efficient than spectral models that use two-dimensional Fourier series.

The inverse transform, from physical space to spectral coordinates, is accomplished as follows. The orthogonality properties of the spherical harmonics (4.44) imply that

$$a_{m,n} = \frac{1}{2\pi} \int_{-1}^{1} \int_{-\pi}^{\pi} \psi(\lambda, \mu)Y_{m,n}^*(\lambda, \mu) \, d\lambda \, d\mu,$$

or, equivalently,

$$a_{m,n} = \int_{-1}^{1} \hat{a}_m(\mu) P_{m,n}(\mu) \, d\mu, \tag{4.57}$$

where

$$\hat{a}_m(\mu) = \frac{1}{2\pi} \int_{-\pi}^{\pi} \psi(\lambda, \mu)e^{-im\lambda} \, d\lambda. \tag{4.58}$$

The last of these integrals is a Fourier transform that can be evaluated from data on a discrete mesh using fast Fourier transforms. After computing the Fourier transform of ψ, the integral (4.57) can be evaluated using Gaussian quadrature. Provided that one avoids aliasing error, it is possible to numerically evaluate both (4.57) and (4.58) without introducing errors beyond those associated with the original truncation of the spherical harmonic expansion at some finite wave number. Before discussing how to avoid aliasing error, it may be helpful to review Gaussian quadrature.

Gaussian Quadrature

As will soon be demonstrated, the integrand in (4.57) often turns out to be a polynomial in μ, which is fortuitous, because simple formulae exist for computing the exact definite integral of a polynomial. For example, suppose that $f(x)$ is a polynomial in x of order $m - 1$ and that m arbitrarily spaced grid points x_j are distributed over the domain $a \le x \le b$. Then $f(x)$ can be expressed in the form of a Lagrange interpolating polynomial

$$f(x) = \sum_{j=1}^{m} f(x_j) p_j(x),$$

where

$$p_j(x) = \prod_{\substack{k=1 \\ k \ne j}}^{m} \frac{(x - x_k)}{(x_j - x_k)}.$$

If

$$A_j = \int_a^b p_j(x)\, dx, \tag{4.59}$$

it follows immediately that

$$\int_a^b f(x)\, dx = A_1 f(x_1) + A_2 f(x_2) + \cdots + A_m f(x_m), \tag{4.60}$$

and that (4.59) and (4.60) give the exact integral of all polynomials $f(x)$ of order less than or equal to $m - 1$.

The preceding formula achieves exact results for polynomials up to order $m - 1$ without imposing any constraint on the location of the x_j within the interval $[a, b]$. Gaussian quadrature, on the other hand, obtains exact results for polynomials up to order $2m - 1$ without adding more terms to the quadrature formula by choosing the x_j to be the zeros of the Legendre polynomial of order m. The role played by Legendre polynomials in Gaussian quadrature is essentially independent of their relation to the associated Legendre functions and spherical harmonics. The property of the Legendre polynomials that is important for Gaussian quadrature is that these polynomials satisfy the orthogonality condition[9]

$$\int_{-1}^{1} P_m(x) P_n(x)\, dx = \frac{2\delta_{m,n}}{2n + 1}.$$

In order to appreciate how this judicious choice for the x_j increases the accuracy of (4.59) and (4.60), suppose that $f(x)$ is a polynomial of order $2m - 1$ and let $P_m(x)$ be the Legendre polynomial on $[a, b]$ of order m. If $q(x)$ and $r(x)$ are, respectively, the quotient and the remainder obtained when dividing f by P_m, then $f = q P_m + r$, where both q and r are polynomials of order less than or

---

[9]Other commonly used sets of orthogonal polynomials are orthogonal with respect to a nonconstant weight function. In the case of Chebyshev polynomials, for example,

$$\int_{-1}^{1} T_m(x) T_n(x)(1 - x^2)^{-1/2}\, dx = 0,$$

unless $m = n$.

equal to $m - 1$. Since the polynomial q can be expressed as a linear combination of the Legendre polynomials of order less than or equal to $m - 1$, all of which are orthogonal to P_m,

$$\int_a^b f(x)\, dx = \int_a^b q(x) P_m(x)\, dx + \int_a^b r(x)\, dx = \int_a^b r(x)\, dx. \qquad (4.61)$$

Also,

$$\sum_{j=1}^m A_j f(x_j) = \sum_{j=1}^m A_j q(x_j) P_m(x_j) + \sum_{j=1}^m A_j r(x_j)$$

$$= \sum_{j=1}^m A_j r(x_j)$$

$$= \int_a^b r(x)\, dx, \qquad (4.62)$$

where the second equality holds because the x_j are the zeros of P_m, and the third equality is obtained because r is a polynomial of order less than or equal to $m - 1$. It follows from (4.61) and (4.62) that

$$\int_a^b f(x)\, dx = \sum_{j=1}^m A_j f(x_j),$$

and that m-point Gaussian quadrature is exact for polynomials up to order $2m - 1$. For m-point Gaussian quadrature over the domain $[-1, 1]$,

$$A_j = \frac{2}{(1 - x_j^2)\left[P_m'(x_j)\right]^2} = \frac{2(1 - x_j^2)}{\left[m P_{m-1}(x_j)\right]^2}.$$

Formulae for the x_j are not known in closed form and must be computed numerically. This can be done using Newton's method (Dahlquist and Björck 1974) with first guesses for the m zeros of $P_m(x)$ given by the set of points

$$\tilde{x}_j = -\cos\left[\left(\frac{4j - 1}{2m + 1}\right)\frac{\pi}{2}\right] \quad \text{for} \quad 1 \le j \le m.$$

Avoiding Aliasing Error

The product of two or more truncated spectral harmonic expansions contains high-order Fourier modes in λ and high-order functions in μ that are not present in the original truncation. When using the transform method it is important to retain enough zonal wave numbers in the Fourier transforms and enough meridional grid points in the Gaussian quadrature to ensure that the transform procedure does not generate errors in any of the modes retained in the original truncated expansions.

Suppose that one wishes to compute the spectral coefficients of the binary product $\psi\chi$, where ψ and χ are given by the truncated spherical harmonic expansions

$$\psi(\lambda, \mu) = \sum_{p=-M}^{M} \sum_{q=|p|}^{N(p)} a_{p,q} Y_{p,q}(\lambda, \mu), \qquad (4.63)$$

$$\chi(\lambda, \mu) = \sum_{r=-M}^{M} \sum_{s=|r|}^{N(r)} b_{r,s} Y_{r,s}(\lambda, \mu). \qquad (4.64)$$

Let $c_{m,n}$ be the coefficient of $Y_{m,n}$ in the spherical harmonic expansion of $\psi\chi$. Without loss of generality consider the case $m \geq 0$, since the coefficients for which $m < 0$ can be obtained using (4.45). Then

$$c_{m,n} = \int_{-1}^{1} \hat{c}_m(\mu) P_{m,n}(\mu) \, d\mu, \qquad (4.65)$$

where

$$\hat{c}_m(\mu) = \frac{1}{2\pi} \int_{-\pi}^{\pi} \psi(\lambda, \mu) \chi(\lambda, \mu) e^{-im\lambda} \, d\lambda. \qquad (4.66)$$

The last of the preceding integrals is a Fourier transform, and as discussed in Section 4.2.2, the discrete Fourier transform of binary products of Fourier series truncated at wave number M can be evaluated without aliasing error if the transforms are computed using a minimum of $(3M - 1)/2$ wave numbers. In order to maximize the efficiency of the fast Fourier transforms in practical applications, the actual cutoff wave number may be chosen as the smallest product of prime factors no larger than five that exceeds $(3M - 1)/2$. This criterion for the cutoff wave number can be alternatively expressed as a requirement that the physical mesh include a minimum of $3M + 1$ grid points around each latitude circle.

Now consider the evaluation of (4.65). The associated Legendre functions have the form

$$P_{m,n}(\mu) = (1 - \mu^2)^{m/2} Q_{m,n}(\mu),$$

where $Q_{m,n}$ is a polynomial in μ of degree $n - m$. Since $P_{m,n}$ is not a polynomial when m is odd, it is not obvious that Gaussian quadrature can be used to integrate (4.65) without error. Nevertheless, it turns out that the complete integrand $\hat{c}_m(\mu) P_{m,n}(\mu)$ is a polynomial in μ whose maximum degree can be determined as follows. Substituting the finite series expansions for ψ and χ into (4.66) and using the orthogonality of the Fourier modes,

$$\hat{c}_m(\mu) = \sum_{\substack{p+r=m \\ |p|,|r| \leq M}} \left(\sum_{q=|p|}^{N(p)} a_{p,q} P_{p,q}(\mu) \right) \left(\sum_{s=|r|}^{N(r)} b_{r,s} P_{r,s}(\mu) \right),$$

where the notation below the first summation indicates that the sum should be performed for all indices p and r such that $|p| \leq M$, $|r| \leq M$, and $p + r = m$.

Each term in $\hat{c}_m(\mu) P_{m,n}(\mu)$ is therefore a function of the form

$$(1 - \mu^2)^{(p+r+m)/2} Q_{p,q}(\mu) Q_{r,s}(\mu) Q_{m,n}(\mu).$$

Since the indices in the preceding satisfy $p + r + m = 2m$, each term is a polynomial in μ of degree

$$2m + (q - p) + (s - r) + (n - m) = q + s + n.$$

The degree of the highest-order polynomial in the integrand of (4.65) is the maximum value of $q + s + n$, which is dependent on the type of truncation used in the expansions (4.63) and (4.64). In the case of a triangular truncation, this maximum is simply $3M$, and the exact evaluation of (4.65) by Gaussian quadrature requires a minimum of $(3M + 1)/2$ meridional grid points. In the case of a rhomboidal truncation, the maximum value of $q + s + n$ is $3M + p + r + m = 5M$, and $(5M + 1)/2$ meridional grid points are required for an exact quadrature.

4.4.4 Nonlinear Shallow-Water Equations

Two additional considerations that arise in using spherical-harmonic expansions functions in practical applications are the evaluation of derivatives with respect to the meridional coordinate and the representation of the vector velocity field. The treatment of these matters can be illustrated by considering the algorithm proposed by Bourke (1972) for integrating the nonlinear shallow-water equations on a rotating sphere,

$$\frac{\partial \mathbf{u}}{\partial t} + \mathbf{u} \cdot \nabla \mathbf{u} + f \mathbf{k} \times \mathbf{u} + \nabla \Phi = 0, \tag{4.67}$$

$$\frac{\partial \Phi}{\partial t} + \nabla \cdot \Phi \mathbf{u} = 0. \tag{4.68}$$

Here $\mathbf{u} = u\mathbf{i} + v\mathbf{j}$, where u and v are the eastward and northward velocity components, $f = 2\Omega \sin\theta$ is the Coriolis parameter, k is the vertical unit vector, Φ is the gravitational constant times the free-surface displacement, and ∇ is the horizontal gradient operator.

Prognostic Equations for Vorticity and Divergence

The velocity components u and v are not conveniently approximated by a series of spherical harmonic functions because artificial discontinuities in u and v are present at the poles unless the wind speed at the pole is zero. This problem arises because the direction defined as "east" switches by 180 degrees as an observer traveling northward along a meridian steps across the pole. The same vector velocity that is recorded as "westerly" at a point on the Greenwich meridian is recorded as "easterly" at a point on the international dateline. In a similar way, a southerly velocity becomes a northerly velocity as the observer crosses the pole. This difficulty can be commonly avoided by replacing the prognostic equations for u and v by equations for the vorticity and divergence and by rewriting all remaining expressions involving **v** in terms of the transformed velocities

$$U = u \cos \theta, \qquad V = v \cos \theta.$$

Both U and V are zero at the poles and free of discontinuities. It is also convenient to separate Φ into a constant mean $\overline{\Phi}$ and a perturbation $\Phi'(\lambda, \mu, t)$.

Equations for the divergence δ and the vertical component of vorticity ζ can be derived by substituting for $\mathbf{u} \cdot \nabla \mathbf{u}$ in (4.67) using the identity

$$\mathbf{u} \cdot \nabla \mathbf{u} = (\nabla \times \mathbf{u}) \times \mathbf{u} + \frac{1}{2} \nabla (\mathbf{u} \cdot \mathbf{u}) \tag{4.69}$$

to yield

$$\frac{\partial \mathbf{u}}{\partial t} + (\nabla \times \mathbf{u}) \times \mathbf{u} + \frac{1}{2} \nabla (\mathbf{u} \cdot \mathbf{u}) + f \mathbf{k} \times \mathbf{u} + \nabla \Phi = \mathbf{0}, \tag{4.70}$$

Taking the divergence of the preceding gives

$$\frac{\partial \delta}{\partial t} = \mathbf{k} \cdot \nabla \times (\zeta + f) \mathbf{u} - \nabla^2 \left(\Phi' + \frac{\mathbf{u} \cdot \mathbf{u}}{2} \right), \tag{4.71}$$

and taking the vertical component of the curl of (4.70) yields

$$\frac{\partial \zeta}{\partial t} = -\nabla \cdot (\zeta + f) \mathbf{u}. \tag{4.72}$$

The horizontal velocity may be expressed in terms of a stream function ψ and a velocity potential χ as

$$\mathbf{u} = \mathbf{k} \times \nabla \psi + \nabla \chi,$$

in which case the vertical component of the vorticity is

$$\zeta = \mathbf{k} \cdot \nabla \times \mathbf{u} = \nabla^2 \psi, \tag{4.73}$$

and the divergence is

$$\delta = \nabla \cdot \mathbf{u} = \nabla^2 \chi. \tag{4.74}$$

The governing equations (4.68), (4.71), and (4.72) can be concisely expressed in spherical coordinates by defining the operator

$$\mathcal{H}(A, B) = \frac{1}{a} \left(\frac{1}{1 - \mu^2} \frac{\partial A}{\partial \lambda} + \frac{\partial B}{\partial \mu} \right).$$

Using the relations

$$\nabla \cdot \alpha \mathbf{u} = \mathcal{H}(\alpha U, \alpha V) \tag{4.75}$$

and

$$\mathbf{k} \cdot \nabla \times \alpha \mathbf{u} = \mathcal{H}(\alpha V, -\alpha U), \tag{4.76}$$

(4.68)–(4.72) become

$$\frac{\partial \nabla^2 \chi}{\partial t} = \mathcal{H}(V \nabla^2 \psi, -U \nabla^2 \psi) - 2\Omega \left(\frac{U}{a} - \mu \nabla^2 \psi \right)$$

$$- \nabla^2 \left(\Phi' + \frac{U^2 + V^2}{2(1 - \mu^2)} \right), \tag{4.77}$$

$$\frac{\partial \nabla^2 \psi}{\partial t} = -\mathcal{H}(U \nabla^2 \psi, V \nabla^2 \psi) - 2\Omega \left(\frac{V}{a} + \mu \nabla^2 \chi \right), \tag{4.78}$$

$$\frac{\partial \Phi'}{\partial t} = -\mathcal{H}(U \Phi', V \Phi') - \overline{\Phi} \nabla^2 \chi. \tag{4.79}$$

The preceding system of equations for ψ, χ, and Φ' can be closed using the diagnostic relations

$$U = (1 - \mu^2)\mathcal{H}(\chi, -\psi) \quad \text{and} \quad V = (1 - \mu^2)\mathcal{H}(\psi, \chi). \tag{4.80}$$

Implementation of the Transform Method

The basic strategy used to implement the transform method for spherical-harmonic expansion functions is the same as that used with simpler Fourier series, which is to compute binary products in physical space and then transform the result back to spectral space. During this procedure, those terms involving derivatives are evaluated to within the truncation error of the spectral approximation using the known properties of the expansion functions. The zonal derivative of each spherical harmonic is simply $imY_{m,n}$. The horizontal Laplacian is easily evaluated using (4.48), and the meridional derivative can be determined using the recurrence relation (4.47)

Prognostic equations for the spectral coefficients associated with the vorticity, the divergence, and the free-surface displacement can be derived as follows. Let

$$\psi(\lambda, \mu) = a^2 \sum_{m=-M}^{M} \sum_{n=|m|}^{N(m)} \psi_{m,n} Y_{m,n}(\lambda, \mu), \tag{4.81}$$

$$\chi(\lambda, \mu) = a^2 \sum_{m=-M}^{M} \sum_{n=|m|}^{N(m)} \chi_{m,n} Y_{m,n}(\lambda, \mu), \tag{4.82}$$

$$\Phi'(\lambda, \mu) = \sum_{m=-M}^{M} \sum_{n=|m|}^{N(m)} \Phi_{m,n} Y_{m,n}(\lambda, \mu).$$

Since the nonlinear products in the governing equations (4.77)–(4.79) are $U \nabla^2 \psi$, $V \nabla^2 \psi$, $U \Phi'$, $V \Phi'$, and $U^2 + V^2$, it is also convenient to define expansion coefficients for U and V. These coefficients can be diagnostically computed from the expansion coefficients for ψ and χ as follows. Let $U_{m,n}$ and $V_{m,n}$ be the spectral expansion coefficients for U/a and V/a; then

$$U_{m,n} = \frac{1}{2\pi} \int_{-1}^{1} \int_{-\pi}^{\pi} \left(\frac{\partial \chi}{\partial \lambda} - (1 - \mu^2) \frac{\partial \psi}{\partial \mu} \right) Y_{m,n}^* \, d\lambda \, d\mu$$

$$= im \chi_{m,n} + (n-1) \epsilon_{m,n} \psi_{m,n-1} - (n+2) \epsilon_{m,n+1} \psi_{m,n+1},$$

where the second equality follows from (4.47) and the orthogonality of the spherical harmonics. Similarly,

$$V_{m,n} = im \psi_{m,n} - (n-1) \epsilon_{m,n} \chi_{m,n-1} + (n+2) \epsilon_{m,n+1} \chi_{m,n+1}.$$

Note that nonzero values of $\psi_{m,n}$ and $\chi_{m,n}$ imply nonzero values of $U_{m,n+1}$ and $V_{m,n+1}$, so the expansions for U and V must be truncated at one higher degree than those for ψ and χ, i.e.,

$$U(\lambda, \mu) = a \sum_{m=-M}^{M} \sum_{n=|m|}^{N(m)+1} U_{m,n} Y_{m,n}(\lambda, \mu),$$

$$V(\lambda, \mu) = a \sum_{m=-M}^{M} \sum_{n=|m|}^{N(m)+1} V_{m,n} Y_{m,n}(\lambda, \mu).$$

After computing products such as $U\nabla^2 \psi$ on the physical mesh, the right sides of (4.77)–(4.79) are transformed back to wave-number space. The first step of this transformation is performed using fast Fourier transforms. Suppose, for notational convenience, that $\hat{A}_m, \hat{B}_m, \ldots, \hat{E}_m$ are the Fourier transforms of the preceding binary products such that

$$U\nabla^2 \psi = \sum_{m=-M}^{M} \hat{A}_m e^{im\lambda}, \quad V\nabla^2 \psi = \sum_{m=-M}^{M} \hat{B}_m e^{im\lambda}$$

$$U\Phi' = \sum_{m=-M}^{M} \hat{C}_m e^{im\lambda}, \quad V\Phi' = \sum_{m=-M}^{M} \hat{D}_m e^{im\lambda}$$

$$\tfrac{1}{2}(U^2 + V^2) = \sum_{m=-M}^{M} \hat{E}_m e^{im\lambda}.$$

The remaining step in the transformation back to wave-number space is computed by Gaussian quadrature. The only nontrivial quadratures are those related to the transform of $(U^2 + V^2)/(1 - \mu^2)$ and of functions of the form μR and $\mathcal{H}(R, S)$, where R and S are functions of μ and λ. Functions of the form μR can be transformed analytically using the recurrence relation (4.46). As an example consider the term in (4.78) proportional to $\mu \nabla^2 \chi$, whose (m, n)th spectral coefficient is

$$\frac{1}{2\pi} \int_{-1}^{1} \int_{-\pi}^{\pi} \mu \nabla^2 \chi Y_{m,n}^* \, d\lambda \, d\mu = \int_{-1}^{1} \left[\sum_{s=|m|}^{N(m)} s(s+1) \chi_{m,s} \mu P_{m,s} \right] P_{m,n} \, d\mu$$

$$= (n-1) n \epsilon_{m,n} \chi_{m,n-1} + (n+1)(n+2) \epsilon_{m,n+1} \chi_{m,n+1}.$$

Now consider the transform of $\mathcal{H}(R, S)$, where R and S are binary products of $\nabla^2 \psi$ or ϕ' and U or V. Let the Fourier transforms of $R(\lambda, \mu)$ and $S(\lambda, \mu)$ be denoted by $\hat{R}_m(\mu)$ and $\hat{S}_m(\mu)$, and define the coefficient of the (m, n)th component of the spherical harmonic expansion for $\mathcal{H}(R, S)$ to be $\mathcal{G}_{m,n}(\hat{R}_m, \hat{S}_m)$. Then

$$\mathcal{G}_{m,n}(\hat{R}_m, \hat{S}_m) = \frac{1}{a} \int_{-1}^{1} \left(\frac{im}{1-\mu^2} \hat{R}_m + \frac{\partial \hat{S}_m}{\partial \mu} \right) P_{m,n} \, d\mu. \qquad (4.83)$$

S contains a factor of either U or V, and since U and V are zero at $\mu = \pm 1$, (4.83) may be integrated by parts to obtain

$$\mathcal{G}_{m,n}(\hat{R}_m, \hat{S}_m) = \frac{1}{a} \int_{-1}^{1} \left(\frac{im}{1-\mu^2} \hat{R}_m P_{m,n} - \hat{S}_m \frac{\partial P_{m,n}}{\partial \mu} \right) d\mu.$$

The derivative in the preceding can be evaluated exactly using (4.47), and the result can be integrated exactly wherever it appears in (4.77)–(4.79) using Gaussian quadrature over the same number of nodes required for the transformation of simple binary products. The exactness of this integral follows from the same type of argument used in connection with the ordinary binary product (4.65); the integrand will consist of a sum of terms of the form

$$(1 - \mu^2)^{m-1} Q_{p,q}(\mu) Q_{m-p,s}(\mu) Q_{m,n}(\mu), \qquad (4.84)$$

where $Q_{m,n}$ is a polynomial in μ of order $n - m$. Except for the case $m = 0$, (4.84) is a polynomial of sufficiently low order that it can be computed exactly. When $m = 0$, it is easier to consider the equivalent integral (4.83). Because $m = 0$, the first term in the integrand is zero, and the second is the sum of terms of the form

$$Q_{p,q}(\mu) Q_{-p,s}(\mu) \frac{d}{d\mu} P_{0,n}(\mu).$$

The last factor is the derivative of the nth-order Legendre polynomial, and the entire expression is once again a polynomial of sufficiently low order to be integrated exactly by Gaussian quadrature.

Finally, consider the transform of $\frac{1}{2}(U^2 + V^2)/(1 - \mu^2)$, whose (m, n)th component will be denoted by $E_{m,n}$. From the definition of $\hat{E}_m$,

$$E_{m,n} = \int_{-1}^{1} \frac{\hat{E}_m}{1-\mu^2} P_{m,n} \, d\mu. \qquad (4.85)$$

The numerator in the preceding integrand is an ordinary binary product, and as argued in connection with (4.65), it must be a polynomial of sufficiently low order that it can be integrated exactly by Gaussian quadrature over the same nodes used for the other transforms. Since both U and V have zeros at $\mu = \pm 1$, the polynomial in the numerator has roots at $\mu = \pm 1$ and must be exactly divisible by $(1 - \mu^2)$. As a consequence the entire integrand in (4.85) is a polynomial that can be integrated without error using Gaussian quadrature.

Using the preceding relations, the time tendencies of the spectral coefficients of the divergence, vorticity, and free-surface displacement become

$$-n(n+1)\frac{d\chi_{m,n}}{dt} = \mathcal{G}_{m,n}\left(\hat{B}_m, -\hat{A}_m\right) + \frac{n(n+1)}{a^2}\left(\Phi_{m,n} + E_{m,n}\right)$$
$$- 2\Omega\left[U_{m,n} + (n-1)n\epsilon_{m,n}\psi_{m,n-1} + (n+1)(n+2)\epsilon_{m,n+1}\psi_{m,n+1}\right],$$

$$-n(n+1)\frac{d\psi_{m,n}}{dt} = -\mathcal{G}_{m,n}\left(\hat{A}_m, \hat{B}_m\right) - 2\Omega V_{m,n}$$
$$+ 2\Omega\left[(n-1)n\epsilon_{m,n}\chi_{m,n-1} - (n+1)(n+2)\epsilon_{m,n+1}\chi_{m,n+1}\right], \quad (4.86)$$

and

$$\frac{d\Phi_{m,n}}{dt} = -\mathcal{G}_{m,n}\left(\hat{C}_m, \hat{D}_m\right) + n(n+1)\overline{\Phi}\chi_{m,n}.$$

The extension of this algorithm to three-dimensional models for the simulation of global atmospheric flow is discussed in Section 7.6.2 and in (Machenhauer 1979).

4.5   The Finite-Element Method

The finite-element method has not been widely used to obtain numerical solutions to hyperbolic partial differential equations because it generates implicit equations for the unknown variables at each new time level. The most efficient methods for the solution of wave-propagation problems are generally schemes that update the unknowns at each subsequent time level through the solution of explicit algebraic equations. Nevertheless, in some atmospheric applications computational efficiency can be improved by using semi-implicit differencing to integrate a subset of the complete equations via the implicit trapezoidal method while the remaining terms in the governing equations are integrated explicitly (see Section 7.2), and the finite-element method can be used to efficiently approximate the vertical structure of the flow in such models (Staniforth and Daley 1977). In addition, the finite-element method is easily adapted to problems in irregularly shaped domains, and as a consequence, it has been used in several oceanic applications to model tides and currents in bays and coastal regions (Foreman and Thomson 1997).

In contrast to the situation with hyperbolic partial differential equations, the finite-element method is very widely used to solve time-independent problems.

The tendency of finite-element approximations to produce implicit algebraic equations is no disadvantage in steady-state problems since the finite-difference approximations to such problems also generate implicit algebraic equations. Moreover, in most steady-state systems the fundamental physical problem can be stated in a variational form naturally suited for solution via the finite-element technique (Strang and Fix 1973). Our interest lies in the application of the finite-element method to time-dependent wave-like flows for which variational forms do not naturally arise. The most useful variational criteria for the equations governing most wave-like flows are simply obtained by minimizing the residual as defined by (4.3). The possible strategies for minimizing the residual are those discussed in Section 4.1. The collocation strategy will not be examined here since, at least for piecewise-linear expansion functions, it leads to methods that are identical to simple finite differences. More interesting algorithms with better conservation properties can be achieved using the Galerkin requirement that the error be orthogonal to the residual, or equivalently, by minimizing $(\|R(\phi)\|_2)^2$.

As discussed in Section 4.1, enforcement of the Galerkin requirement leads to the system of ordinary differential equations

$$\sum_{n=1}^{N} I_{nk} \frac{da_n}{dt} = - \int_S \left[F\left(\sum_{n=1}^{N} a_n \varphi_n \right) \varphi_k \right] dx \quad \text{for } k = 1, \dots, N, \qquad (4.87)$$

where

$$I_{nk} = \int_S \varphi_n \varphi_k \, dx.$$

The difference between the spectral method and the Galerkin form of the finite-element method lies in the choice of expansion functions. In the spectral method, the expansion functions form an orthogonal set, and each φ_k is nonzero over most of the spatial domain. The orthogonality of the spectral expansion functions ensures that I_{nk} is zero unless $n = k$, greatly simplifying the left side of (4.87). However, since the spectral expansion functions are nonzero over most of the domain, the evaluation of the right side of (4.87) involves considerable computation.

In the finite-element method, the expansion functions are not usually orthogonal, but each φ_n is nonzero only over a small, localized portion of the total domain. An example of a finite-element expansion function is given by the *chapeau* (or "hat") function shown in Fig. 4.6. In the case of the chapeau function, the total domain is partitioned into N nodes, and φ_k is defined as a piecewise-linear function equal to unity at the kth node and zero at every other node. If the series expansion (4.2) utilizes chapeau functions, the resulting sum will be a piecewise-linear approximation to the true function $\psi(x)$. Because the finite-element expansion functions are not orthogonal, the left side of (4.87) constitutes an implicit relationship between the da_k/dt at a small number of adjacent nodes, and a sparse linear system must be solved every time step. The coefficient matrix multiplying the da_k/dt is often referred to as the "mass matrix."

Because it is necessary to solve a system of linear algebraic equations on every time step, the computational effort required by the Galerkin finite-element method

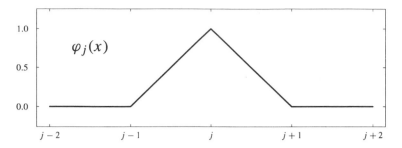

FIGURE 4.6. The chapeau expansion function ϕ_j. The x-axis is labeled in units of Δx.

typically exceeds that associated with finite-difference and spectral methods. Nevertheless, in comparison to the spectral method, the finite-element approach does reduce the computation required to evaluate the right side of (4.87). Since the finite-element expansion functions are nonzero only over a small portion of the total domain, the number of arithmetic operations required to evaluate the right side of (4.87) is $O(N)$, which is comparable to that involved in the calculation of conventional finite differences and can be considerably less than the $O(N \log N)$ operations required to evaluate the same expression using the spectral transform method.

4.5.1 Galerkin Approximation with Chapeau Functions

If the wind speed is constant, the Galerkin approximation to the one-dimensional advection equation (4.11) requires that

$$\sum_{n=1}^{N} I_{nk} \frac{da_n}{dt} + c \sum_{n=1}^{N} a_n \int_S \frac{d\varphi_n}{dx} \varphi_k \, dx = 0 \quad \text{for} \quad k = 1, \dots, N. \tag{4.88}$$

Assuming that the φ_n are chapeau functions and shifting the x-origin to coincide with the left edge of each interval of integration, the integrals involving products of the expansion functions become

$$I_{j-1,j} = I_{j+1,j} = \int_S \varphi_{j+1} \varphi_j \, dx = \int_0^{\Delta x} \left(\frac{x}{\Delta x} \right) \left(\frac{\Delta x - x}{\Delta x} \right) dx = \frac{\Delta x}{6},$$

$$I_{j,j} = 2 \int_0^{\Delta x} \left(\frac{\Delta x - x}{\Delta x} \right)^2 dx = \frac{2\Delta x}{3},$$

$$-\int_S \frac{\partial \varphi_{j-1}}{\partial x} \varphi_j \, dx = \int_S \frac{\partial \varphi_{j+1}}{\partial x} \varphi_j \, dx = \int_0^{\Delta x} \left(\frac{1}{\Delta x} \right) \left(\frac{\Delta x - x}{\Delta x} \right) dx = \frac{1}{2}.$$

Since all other integrals involving products of the expansion functions or their derivatives are zero, (4.88) reduces to

$$\frac{d}{dt}\left(\frac{a_{j+1} + 4a_j + a_{j-1}}{6}\right) + c\left(\frac{a_{j+1} - a_{j-1}}{2\Delta x}\right) = 0. \qquad (4.89)$$

This scheme may be analyzed as if it were a standard differential-difference equation because a_j, the coefficient of the jth chapeau function, is also the nodal value of the approximate solution $\phi(x_j)$. In fact, (4.89) is identical to the the the fourth-order compact differential-difference approximation to the advection equation (2.83), whose properties have been previously discussed in Section 2.4.4. In particular, the scheme's spatial truncation error is $O\left[(\Delta x)^4\right]$ and its phase speed in the limit of good resolution is

$$c^* \approx c\left(1 - (k\Delta x)^4/180\right). \qquad (4.90)$$

This scheme also performs very well at moderately poor spatial resolution. As was shown in Fig. 2.17, (4.89) generates less phase-speed error in moderately short waves than does explicit fourth- or sixth-order spatial differencing.

Now suppose that the time derivatives in (4.89) are approximated using leapfrog time-differencing. The discrete-dispersion relation becomes

$$\sin(\omega\Delta t) = \frac{3\mu\sin(k\Delta x)}{\cos(k\Delta x) + 2}.$$

When $|\mu| < 1/\sqrt{3}$, the right side of the preceding is bounded by unity and the scheme is stable. As was the case with the spectral method, the finite-element method better approximates the spatial derivative of coarsely resolved waves (such as the $3\Delta x$ wave) and thus, the finite-element approximation to the advection equation captures higher-frequency oscillations and the maximum stable time step is reduced relative to that allowed by a centered second-order finite-difference approximation to the spatial derivative.

One way to circumvent this time–step restriction is to use trapezoidal time-differencing. The trapezoidal method is unconditionally stable, more accurate than leapfrog differencing and it does not support a computational mode. Despite these advantages, the trapezoidal scheme is not used in most finite-difference approximations to wave-propagation problems because it leads to implicit equations. The implicit nature of trapezoidal differencing is not, however, a problem in this application, because (4.89) is already a linear system of implicit equations for the da_j/dt and trapezoidal differencing does not increase the bandwidth of the coefficient matrix. The trapezoidal method is, however, less attractive in more general applications where systems of equations must be solved. For example, the implicit coupling among the various nodal values remains tridiagonal when the linearized shallow-water system (3.1)–(3.2) is approximated using explicit time-differencing and the chapeau-function finite-element method, but if the leapfrog difference is replaced by the trapezoidal method, the u^{n+1} and h^{n+1} become implicit functions of each other, and the resulting linear system has a larger bandwidth.

4.5.2 Petrov–Galerkin and Taylor–Galerkin Methods

Another way to increase the maximum stable time step of finite-element approximations to wave-propagation problems is to generalize the orthogonality condition satisfied by the residual. As an alternative to the standard Galerkin requirement that the residual be orthogonal to each of the expansion functions, one may define a different set of "test" functions and require the residual to be orthogonal to each of these test functions. This approach, known as the *Petrov–Galerkin* method, can yield schemes that are stable for Courant numbers as large as unity, and can greatly increase the accuracy of computations performed at Courant numbers near the stability limit. The Petrov–Galerkin method does not, however, share all of the desirable conservation properties of the standard Galerkin method.

Let ϑ_k be an arbitrary member of the set of test functions. If the time derivative is approximated by a forward difference, the Petrov–Galerkin formula for the differential-difference approximation to the general partial differential equation (4.1) is

$$\int_S \left[\sum_{j=1}^N \left(\frac{a_j^{n+1} - a_j^n}{\Delta t} \right) \varphi_j + F \left(\sum_{j=1}^N a_j^n \varphi_j \right) \right] \vartheta_k \, dx = 0 \quad \text{for all } k.$$

As a specific example, suppose that a Petrov–Galerkin approximation is sought to the advection equation (4.11) with c constant and nonnegative. Let the expansion functions $\varphi_j(x)$ be the chapeau functions defined previously, and as suggested by Morton and Parrott (1980), define a family of test functions of the form $\vartheta_k = (1 - v)\varphi_k + v\chi_k$, where v is a tunable parameter and χ_j is the localized sawtooth function

$$\chi_j(x) = \begin{cases} 6(x - x_{j-1})/\Delta x - 2, & \text{if } x \in [x_{j-1}, x_j]; \\ 0, & \text{otherwise.} \end{cases}$$

(χ_j is normalized so that its integral over the domain is unity.) Using these test functions, the Petrov–Galerkin approximation to the constant-wind-speed advection equation may be expressed in terms of the nodal values as

$$\left[1 + \tfrac{1}{6}(1 - v)\tilde{\delta}_x^2 \right] \left(a^{n+1} - a^n \right) + \mu \tilde{\delta}_{2x} a^n = \tfrac{1}{2} \mu v \tilde{\delta}_x^2 a^n, \tag{4.91}$$

where $\tilde{\delta}_x = \Delta x \, \delta_x$ is a nondimensional finite-difference operator and $\mu = c\Delta t / \Delta x$ is the Courant number. Morton and Parrott (1980) used the energy method to show this scheme is stable for $0 \le \mu \le v \le 1$. The amplification factor for this scheme is

$$A = 1 - \frac{i\mu \sin(k\Delta x) - \mu v[\cos(k\Delta x) - 1]}{1 + (1 - v)[\cos(k\Delta x) - 1]/3}.$$

In the limit $k\Delta x \to 0$

$$A = 1 - i\mu k\Delta x - \tfrac{1}{2}\mu\nu(k\Delta x)^2 + \tfrac{1}{6}\mu\nu(k\Delta x)^3 + O\left[(k\Delta x)^4\right],$$

which matches the correct amplification factor, $e^{-i\omega\Delta t} = e^{-i\mu k\Delta x}$, through first order except that the scheme is second-order accurate when $\nu = \mu$. Clearly one should choose $\nu = \mu$ since this allows the largest stable time step and gives the best accuracy. When $\nu = \mu$, (4.91) is closely related to the standard Lax–Wendroff approximation (2.102); the only difference appears in the linear operator (i.e., the mass matrix) acting on the forward-time difference. As noted by Morton and Parrott (1980), if ν is set equal to μ the truncation error in (4.91) is always less than that for the standard Lax–Wendroff scheme; the improvement is particularly pronounced for small values of μ. On the other hand, the standard Lax–Wendroff method is stable for $|\mu| \le 1$, whereas the Galerkin–Petrov method (4.91) is a upstream method that requires $\mu \ge 0$ for stability. A formula analogous to (4.91) can be derived for negative flow velocities using test functions in which the sawtooth component has a negative slope.

A better scheme than that just derived via the Petrov–Galerkin approach can be obtained using the *Taylor–Galerkin* method. The Taylor–Galerkin method does not require the specification of a second set of test functions and yields centered-in-space methods. In the Taylor–Galerkin approach, the time derivative is discretized before invoking the finite-element formalism to approximate the spatial derivatives. Donea et al. (1987) present Taylor–Galerkin approximations to several hyperbolic problems in which the time discretization is Lax–Wendroff, leapfrog or trapezoidal. In the following we will focus on Lax–Wendroff-type approximations to the constant-wind-speed advection equation.

If the spatial dependence of the solution is not discretized, an $O\left[(\Delta t)^2\right]$-accurate Lax–Wendroff approximation to the constant-wind-speed advection equation has the form

$$\frac{\phi^{n+1} - \phi^n}{\Delta t} + c\frac{\partial\phi^n}{\partial x} = \frac{c^2\Delta t}{2}\frac{\partial^2\phi^n}{\partial x^2},$$

(see Section 2.5.3). Suppose that the spatial structure in the preceding differential-difference equation is approximated using the Galerkin finite-element method with chapeau expansion functions, then the function value at each node satisfies

$$\left[1 + \tfrac{1}{6}\tilde{\delta}_x^2\right]\left(a^{n+1} - a^n\right) + \mu\tilde{\delta}_{2x}a^n = \tfrac{1}{2}\mu^2\tilde{\delta}_x^2 a^n, \tag{4.92}$$

where once again $\tilde{\delta}_x = \Delta x\,\delta_x$ and $\mu = c\Delta t/\Delta x$. The stability condition for this scheme is $|\mu| \le 1/\sqrt{3}$, which is identical to that for the leapfrog approximation to (4.89). The leading-order errors in the modified equations[10] for this method and the standard finite-difference Lax–Wendroff method (2.102) are shown in Table 4.3. The leading-order errors in both schemes are second order and gen-

---

[10]The "modified equation" is discussed in Section 2.5.2.

Finite difference   (2.102)

$$\psi_t + c\psi_x = -(c/6)(\Delta x)^2(1 - \mu^2)\psi_{xxx} - (c/8)(\Delta x)^3\mu(1 - \mu^2)\psi_{xxxx} + \dots$$

Finite element   (4.92)

$$\psi_t + c\psi_x = (c/6)(\Delta x)^2\mu^2\psi_{xxx} - (c/24)(\Delta x)^3\mu(1 - 3\mu^2)\psi_{xxxx} + \dots$$

Taylor–Galerkin finite-element   (4.93)

$$\psi_t + c\psi_x = -(c/24)(\Delta x)^3\mu(1 - \mu^2)\psi_{xxxx} + \dots$$

TABLE 4.3. Modified equations for Lax–Wendroff type finite-difference and finite-element approximations and the Taylor–Galerkin method. Subscripts denote partial derivatives. After Donea et al. (1987).

erate numerical dispersion. The dispersion, or phase-speed error, in each method is the net result of accelerative time-differencing error and decelerative spatial differencing error. These errors partially cancel in the standard finite-difference Lax–Wendroff method, and are eliminated entirely when $|\mu| = 1$. On the other hand, the leading-order error in the Lax–Wendroff finite-element method is due entirely to accelerative time-differencing error. The decelerative phase error generated by the finite-element approximation to the spatial derivatives is $O\left[(\Delta x)^4\right]$ (cf. (4.90)), and as a consequence there is no beneficial cancellation between time-differencing error and space-differencing error in the leading-order error for the Lax–Wendroff finite-element method.

Donea (1984) observed that much better results can be obtained using a third-order Lax–Wendroff approximation. Expanding the true solution to the constant-wind-speed advection equation at time $(n + 1)\Delta t$ in a Taylor series about its value at time $n\Delta t$, and using the governing equation to replace the first- and second-order time derivatives by expressions involving derivatives with respect to x, gives

$$\psi^{n+1} - \psi^n = -c\Delta t\frac{\partial\psi^n}{\partial x} + \frac{(c\Delta t)^2}{2}\frac{\partial^2\psi^n}{\partial x^2} + \frac{c^2(\Delta t)^3}{6}\left(\frac{\partial^3\psi}{\partial t\partial x^2}\right)^n + O\left[(\Delta t)^4\right],$$

where ψ^n is the value of the true solution at $t = n\Delta t$. The mixed third-order derivative in the preceding is not replaced by an expression proportional to $\partial^3\psi/\partial x^3$ because the finite-element approximation to such a term would require smoother expansion functions than the piecewise-linear chapeau functions. Instead, the derivative with respect to time in $\partial^3\psi/(\partial t\,\partial x^2)$ can be conveniently approximated by a forward difference to obtain the following $O\left[(\Delta t)^3\right]$-accurate approximation to the advection equation

$$\left(1 - \frac{(c\Delta t)^2}{6}\frac{\partial^2}{\partial x^2}\right)\left(\phi^{n+1} - \phi^n\right) + c\Delta t\frac{\partial\phi^n}{\partial x} = \frac{(c\Delta t)^2}{2}\frac{\partial^2\phi^n}{\partial x^2},$$

in which $\phi^n(x)$ is a semidiscrete approximation to $\psi(n\Delta t, x)$ Using chapeau functions to approximate the spatial dependence of ϕ^n and demanding that the residual be orthogonal to each expansion function, one obtains the Taylor–Galerkin formula for the function value at each node

$$\left[1 + \tfrac{1}{6}(1 - \mu^2)\tilde{\delta}_x^2\right]\left(a^{n+1} - a^n\right) + \mu\tilde{\delta}_{2x}a^n = \tfrac{1}{2}\mu^2\tilde{\delta}_x^2 a^n. \qquad (4.93)$$

This scheme is stable for $|\mu| \leq 1$. Examination of the modified equation for (4.93), which appears in Table 4.3, shows that, in contrast to the Lax–Wendroff finite-difference and finite-element methods, the Taylor–Galerkin method is free from second-order dispersive errors. The leading-order error in the Taylor–Galerkin method is third-order and weakly dissipative. The difference between the Taylor–Galerkin scheme (4.93), the Petrov–Galerkin method (4.91) and the Lax–Wendroff finite-element method (4.92) involves only minor perturbations to the coefficients in the tridiagonal mass matrix. All three schemes require essentially the same computation per time step, but the Taylor–Galerkin method is the most accurate and is stable over the widest range of Courant numbers.

4.5.3 Quadratic Expansion Functions

Higher-order expansion functions are widely used in finite-element approximations to elliptic partial differential equations. Higher-order expansion functions are not, however, commonly used in finite-element simulations of wave-like flow. One serious disadvantage of higher-order expansion functions is that they increase the implicit coupling in the equations for the time-evolution of the expansion coefficients. Another disadvantage is that the accuracy obtained using higher-order expansion functions in hyperbolic problems is generally lower than that which can be achieved using the same expansion functions in finite-element approximations to elliptic and parabolic partial differential equations (Strang and Fix 1973). In fact, the order of accuracy of the function values at the nodes given by quadratic and Hermite-cubic finite-element approximations to hyperbolic equations is lower than that obtained with piecewise-linear chapeau functions. High orders of accuracy can be obtained using cubic splines (Thomée and Wendroff 1974), but splines introduce a non-local coupling between the coefficients of the finite-element expansion functions that makes them too inefficient for most applications involving wave-like flows. In this section we will consider the behavior of quadratic finite-element approximations to the constant-wind-speed advection equation. Hermite-cubic expansion functions will be considered in Section 4.5.4.

Suppose the piecewise-linear approximation generated by the superposition of chapeau expansion functions is replaced by piecewise quadratics of the form

$$q(x) = C_1 + C_2 x + C_3 x^2. \qquad (4.94)$$

In contrast to linear interpolation, the three coefficients C_1, C_2 and C_3 cannot be uniquely determined by the two nodal values at the ends of each element. The most straightforward way to proceed is extend the piecewise quadratic across an interval of $2\Delta x$ and to choose C_1, C_2 and C_3 so that (4.94) matches the function values at the "midpoint" node and at both "endpoint" nodes. Suppose that a

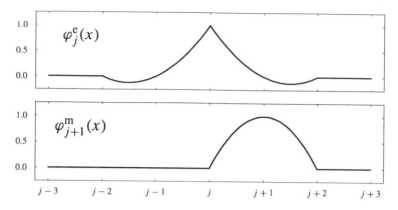

FIGURE 4.7. Quadratic expansion functions for φ_j^e, an endpoint node centered at grid point j, and φ_{j+1}^m, a midpoint node centered at $j+1$. The x-axis is labeled in units of Δx.

function assumes the values a_1 and a_3 at the endpoint nodes x_1 and x_3, and assumes the value b_2 at the midpoint node x_2. The quadratic Lagrange interpolating polynomial that assumes these values at the nodes is

$$\frac{(x-x_2)(x-x_3)}{(x_1-x_2)(x_1-x_3)}a_1 + \frac{(x-x_1)(x-x_3)}{(x_2-x_1)(x_2-x_3)}b_2 + \frac{(x-x_1)(x-x_2)}{(x_3-x_1)(x_3-x_2)}a_3.$$

On the interval $x_1 \le x \le x_3$, the preceding is algebraically identical to

$$a_1\varphi_1^e(x) + b_2\varphi_2^m(x) + a_3\varphi_3^e(x),$$

where φ_j^e is the endpoint quadratic expansion function

$$\varphi_j^e(x) = \begin{cases} 1 - 3\dfrac{|x-x_j|}{2\Delta x} + \dfrac{1}{2}\left(\dfrac{x-x_j}{\Delta x}\right)^2, & \text{if } |x-x_j| \le 2\Delta x, \\ 0, & \text{otherwise,} \end{cases} \tag{4.95}$$

and φ_j^m is the midpoint quadratic expansion function

$$\varphi_j^m(x) = \begin{cases} \left(\dfrac{x-x_{j-1}}{\Delta x}\right)\left(2 - \dfrac{x-x_{j-1}}{\Delta x}\right), & \text{if } |x-x_j| \le \Delta x, \\ 0, & \text{otherwise.} \end{cases} \tag{4.96}$$

These expansion functions are plotted in Fig. 4.7. The jth endpoint expansion function is zero outside an interval of length $4\Delta x$ centered at the node x_j; it is unity at x_j and is zero at every other node. The jth midpoint expansion function is zero outside an interval of length $2\Delta x$ centered at x_j; it is equal to unity at x_j and zero at the other nodes. As was the case with chapeau expansion functions, the coefficient of the jth quadratic expansion function is also the value of the approximate solution at the jth node.

Let a finite-element approximation to the solution to the constant-wind-speed advection equation be constructed from the preceding quadratic expansion functions such that

$$\phi(x, t) = \sum_{j \text{ odd}} a_j(t)\varphi_j^e(x) + \sum_{\ell \text{ even}} b_\ell(t)\varphi_\ell^m(x).$$

Enforcing the Galerkin requirement that the residual be orthogonal to each expansion function yields two families of equations for the evolution of the expansion coefficients in the constant-wind-speed advection problem. The equations centered at the endpoint nodes are

$$\frac{d}{dt}\left(\frac{-a_{j-2} + 2b_{j-1} + 8a_j + 2b_{j+1} - a_{j+2}}{10}\right)$$
$$+ c\left(\frac{b_{j+1} - b_{j-1}}{\Delta x} - \frac{a_{j+2} - a_{j-2}}{4\Delta x}\right) = 0, \quad (4.97)$$

whereas those centered on the midpoints are

$$\frac{d}{dt}\left(\frac{a_{\ell-1} + 8b_\ell + a_{\ell+1}}{10}\right) + c\left(\frac{a_{\ell+1} - a_{\ell-1}}{2\Delta x}\right) = 0. \quad (4.98)$$

Although (4.97) and (4.98) are expressions for the coefficients of the quadratic finite-element expansion functions, they can be alternatively interpreted as finite-difference approximations for the function values at each node. The truncation error in the function values at the nodes can therefore be assessed by a conventional Taylor series analysis which shows that both (4.97) and (4.98) are $O\left[(\Delta x)^2\right]$-accurate finite-difference approximations to the one-dimensional advection equation. The truncation error at the nodes is considerably worse than the $O\left[(\Delta x)^4\right]$ error obtained using chapeau expansion functions!

The quadratic finite-element method requires more work per time step per element than the linear finite-element scheme because a pentadiagonal matrix must be inverted to evaluate the time derivatives in (4.97) and (4.98), whereas the mass matrix associated with (4.89) is only tridiagonal. It is therefore tempting to conclude that quadratic finite elements are decidedly inferior to linear elements, at least for the constant-wind-speed advection problem. In fact, the error in quadratic finite-element solutions to some constant-wind-speed advection problems can be significantly smaller than that obtained using chapeau functions over the same number of nodes. There are two reasons why the preceding comparison of truncation errors can be misleading. The first reason is that, unlike finite-difference approximations, finite-element methods involve an explicit assumption about the functional dependence of the solution between the nodal points, and the error at all nonnodal points is $O\left[(\Delta x)^2\right]$ for both the linear and quadratic finite-element methods. In the case of chapeau expansion functions, the function values between the nodal points are obtained by linear interpolation and are therefore only $O\left[(\Delta x)^2\right]$ accurate. In general, a smooth function can be interpolated to

$O\left[(\Delta x)^{n+1}\right]$ by piecewise polynomials of order n. Thus, if the nodal values could be specified with negligible error, quadratic expansion functions could provide $O\left[(\Delta x)^3\right]$ accuracy between the nodes. But since the quadratic finite-element method only predicts the nodal values to $O\left[(\Delta x)^2\right]$, the accuracy between the nodes is also limited to $O\left[(\Delta x)^2\right]$. As discussed in detail by Cullen and Morton (1980), it is necessary to specify how the error will be measured (e.g., pointwise errors at the nodes or the square integral of the error over the entire spatial domain) before attempting to determine the truncation error.

The second, and perhaps more important, reason why the preceding comparison of truncation error can be misleading is that it does not provide reliable information about the errors in the poorly resolved waves, and in many fluid dynamical applications, the total error can be dominated by the errors in the shortest waves. The error in quadratic finite-element solutions to the constant-wind-speed advection equation can be evaluated as a function of the spatial resolution by examining the phase-speed and amplitude errors in semi-discrete wave solutions to (4.97) and (4.98). There is, however, no single wave of the form $\phi_j(t) = e^{i(kj\Delta x - \omega t)}$ that will simultaneously satisfy (4.97) and (4.98) (Hedstrom 1979a; Cullen 1982). If the initial disturbance consists of a single wave, at subsequent times the approximate numerical solution will split into two traveling disturbances. Each disturbance resembles an ordinary traveling wave except that the wave amplitude at the endpoint nodes is different from that at the midpoint nodes. The nodal values in each traveling disturbance have the form

$$a_j(t) = e^{ik(j\Delta x - c^* t)} \qquad b_\ell(t) = r_a e^{ik(\ell\Delta x - c^* t)}, \tag{4.99}$$

where r_a is the ratio of the amplitude at a midpoint node to the amplitude at an endpoint node, and c^* is the phase-speed. Substitution of (4.99) into (4.97) and (4.98) yields

$$\frac{1}{5}(8 - 2\cos 2\theta + 4r_a \cos\theta)\,c^* - \frac{c}{\theta}(4r_a \sin\theta - \sin 2\theta) = 0$$

and

$$\frac{1}{5}(\cos\theta + 4r_a)\,c^* - \frac{c}{\theta}\sin\theta = 0,$$

where $\theta = k\Delta x$. Solutions to this system of two equations in the two unknowns c^* and r_a have the form

$$\frac{c^*}{c} = \frac{[(19 - \cos 2\theta)(1 - \cos 2\theta)]^{1/2} \pm 2\sin 2\theta}{\theta(3 - \cos 2\theta)} \tag{4.100}$$

and

$$r_a = \frac{1}{4}\left(5\frac{c}{c^*}\frac{\sin\theta}{\theta} - \cos\theta\right). \tag{4.101}$$

The negative root in (4.100) gives the phase speed of the physical mode; the positive root is associated with a computational mode. The phase-speed and amplitude

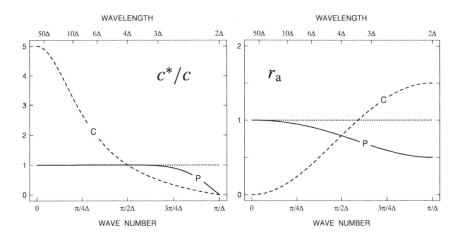

FIGURE 4.8. Normalized phase speed, c^*/c, and amplitude ratio, r_a, for the physical (P) and computational (C) modes in the quadratic finite-element solution to the advection equation.

errors in the physical mode vanish in the limit of good spatial resolution. In particular,

$$\left(\frac{c^*}{c}\right)_{\text{phys}} \approx 1 + \frac{(k\Delta x)^4}{270} \quad \text{and} \quad (r_a)_{\text{phys}} \approx 1 - \frac{(k\Delta x)^2}{12}.$$

Normalized phase speeds and amplitude ratios for the physical and computational modes are plotted as a function of horizontal wave number in Fig. 4.8. The adjectives "physical mode" and "computational mode" have been chosen to describe the behavior of each mode as $k\Delta x \to 0$. However, as indicated in Fig. 4.8 and by (4.100), the $4\Delta x$ physical and computational modes are identical. The $4\Delta x$ computational mode is redundant because the two linearly independent components of the physical mode, $\sin(\pi(x - c^*t)/2\Delta x)$ and $\cos(\pi(x - c^*t)/2\Delta x)$, superimpose to produce any arbitrary relation between the function values at the midpoint and endpoint nodes. Given the close relation between the two modes for wavelengths near $4\Delta x$, the interpretation of one mode as "physical" and the other as "computational" is less meaningful at poor spatial resolution.

Unlike most finite-difference schemes, the phase-speed error in a well-resolved physical mode is much less than the amplitude error, r_a. The nature of the amplitude error in the quadratic finite-element solution is, however, very different from that in conventional finite-difference schemes. The wave amplitude does not grow or decay by a constant factor each time step. Instead, the amplitude decreases as a wave crest travels from an endpoint node to the adjacent midpoint node, and reamplifies as the wave approaches the next endpoint node. The height of the traveling crest oscillates as the wave propagates, but there is no cumulative amplification.

Linear and quadratic finite-element solutions to a constant-wind-speed advection problem are compared in Fig. 4.9. Also shown is the solution generated by the explicit fourth-order finite-difference method (2.65). These solutions were

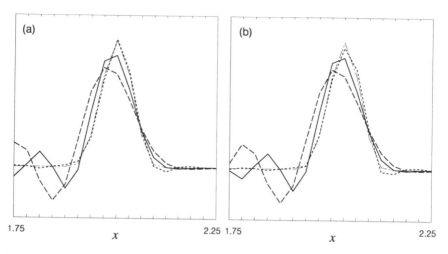

FIGURE 4.9. Comparison of solutions to the constant-wind-speed advection equation at (a) $t = 10$ and (b) $t = 10\frac{5}{16}$: quadratic finite-element (short dashed), linear finite-element (solid), fourth-order explicit finite difference (long dashed) and exact (dot-dashed).

obtained using trapezoidal time-differencing with a very small Courant number ($c\Delta t/\Delta x = 1/16$), so essentially all the error is produced by the spatial discretization. Solutions were computed on the periodic domain $0 \le x \le 3$ subject to the initial condition

$$\psi(x, 0) = \begin{cases} \frac{1}{4}(\cos(8\pi(x - 1)) + 1)^2, & \text{if } |x - 1| \le \frac{1}{8}, \\ 0, & \text{otherwise .} \end{cases}$$

In order to facilitate the comparison with the finite-difference method, the nodal values were initialized by collocation, i.e., $a_j(0) = \psi(j\Delta x, 0)$. The horizontal mesh spacing is $\Delta x = 1/32$, implying that the total width of the initial spike is $8\Delta x$, which is sufficiently narrow to reveal short-wavelength errors without allowing the solution to be completely dominated by $2\Delta x$ disturbances. The wind speed is $c = 0.1$. The solution at $t = 10$ is shown in Fig. 4.9a, at which time the peak in the true solution is centered at $x = 2$. Only the central portion of the total domain is shown in Fig. 4.9. For simplicity, the quadratic finite-element solution is plotted as a piecewise-linear function between the nodes. The superiority of the quadratic finite-element solution over the linear finite-element solution is clearly evident. The linear finite-element solution is, nevertheless, substantially better than the solution obtained with explicit fourth-order finite differences.

The nature of the amplitude error in the quadratic finite-element solution can be seen by comparing Fig. 4.9a with Fig. 4.9b. The exact solution propagates exactly one grid interval between the times shown in panels (a) and (b). There are essentially no changes in the shapes of the linear finite-element and the finite-difference solutions over this short period of time, but the quadratic finite-element solution is damped noticeably. This damping is followed by reamplification as the solu-

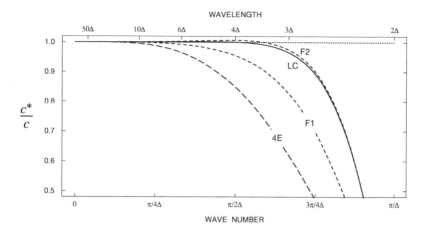

FIGURE 4.10. Phase speed error as a function of spatial resolution for linear finite-elements (F1), quadratic finite-elements: physical mode (F2), explicit 4th-order centered differences (4E), and Lele's 4th-order tridiagonal compact scheme (LC).

tion translates another Δx, and the amplitude of the peak in the quadratic finite-
element solution continues to oscillate as it moves alternatively over the midpoint
and endpoint nodes. Nevertheless, even when the quadratic finite-element solution looks its worst, it is still much better than the solutions generated by the other
schemes. Although the amplitude of the computational mode remains small in
this linear constant-coefficient test problem, the computational mode may be amplified during the computation of spatially varying products in linear equations
with variable coefficients, or by nonlinear wave interactions in nonlinear problems. Cullen (1979, 1982) discusses strategies for minimizing the error in the
approximation of the product of two spatially varying functions via the quadratic
finite-element method. An example of the amplification of computational modes
via nonlinear interaction is shown in Fig. 4.14.

The results shown in Fig 4.9 are consistent with the comparison of the phase
speeds for each scheme plotted in Fig. 4.10. The phase speeds of the $3\Delta x$ or $4\Delta x$
waves are captured much better by the quadratic finite-element method than by
the linear finite-element method or fourth-order finite-differencing. The quadratic
finite-element method is not, however, an optimal choice for this problem. Also
plotted in Fig. 4.10 are the phase speeds produced when the spatial derivative in
the constant-wind-speed advection equation is approximated using Lele's tridiagonal compact finite-difference formula (2.85). The compact scheme exhibits essentially the same accuracy as the quadratic finite-element method for the poorly
resolved waves, but it is distinctly superior because it has no computational mode,
its truncation error is $O\left[(\Delta x)^4\right]$, and it requires less work per time step since it
leads to a tridiagonal implicit system, whereas the mass matrix for the quadratic
finite-element method is pentadiagonal.

4.5.4   *Hermite-Cubic Expansion Functions*

The use of quadratic finite-elements leads to a system of ordinary differential equations for the function values at each node that are unlike those generated by typical finite-difference schemes. The contrast between the finite-element method and conventional finite differences is even more obvious when the expansion functions are Hermite-cubic polynomials. The four coefficients of the Hermite cubic defined on the interval $x_j \leq x \leq x_{j+1}$ are determined by the function values and the first derivatives at each end of the interval. If a_j and a_{j+1} are the function values at x_j and x_{j+1}, and if b_j and b_{j+1} are Δx times the first derivatives at the same nodes, the Hermite-cubic polynomial on the interval $x_j \leq x \leq x_{j+1}$ may be written as

$$a_j \varphi_j^v(x) + b_j \varphi_j^d(x) + a_{j+1} \varphi_{j+1}^v(x) + b_{j+1} \varphi_{j+1}^d(x),$$

where φ_j^v and φ_j^d are the finite-element expansion functions

$$\varphi_j^v(x) = \begin{cases} \left(\dfrac{|x - x_j|}{\Delta x} - 1 \right)^2 \left(2\dfrac{|x - x_j|}{\Delta x} + 1 \right), & \text{if } |x - x_j| \leq \Delta x, \\ 0, & \text{otherwise,} \end{cases} \quad (4.102)$$

and

$$\varphi_j^d(x) = \begin{cases} \left(\dfrac{x - x_j}{\Delta x} \right) \left(\dfrac{|x - x_j|}{\Delta x} - 1 \right)^2, & \text{if } |x - x_j| \leq \Delta x, \\ 0, & \text{otherwise.} \end{cases} \quad (4.103)$$

As illustrated in Fig. 4.11, φ_j^v has unit amplitude and a zero first derivative at the jth node, whereas the amplitude of φ_j^d is zero at x_j, but its first derivative is $(\Delta x)^{-1}$. In contrast to linear or quadratic finite-element approximations, two pieces of information are available at each node. The expansion coefficients a_j are the function values at x_j, and the b_j are the first derivatives normalized by the mesh spacing.

Let a finite-element approximation to the solution of the constant-wind-speed advection equation be constructed using Hermite-cubic expansion functions such that

$$\phi(x, t) = \sum_j \left[a_j(t) \varphi_j^v(x) + b_j(t) \varphi_j^d(x) \right]. \quad (4.104)$$

The Galerkin requirement that the residual be orthogonal to each expansion function yields two types of equations for the evolution of the expansion coefficients. Recalling that $\tilde{\delta}_x = \Delta x\, \delta_x$ and defining

$$H^v(a_j, b_j) = \left(54a_{j+1} + 312a_j + 54a_{j-1} \right) - 13 \left(b_{j+1} - b_{j-1} \right),$$

$$H^d(a_j, b_j) = - \left(3b_{j+1} - 8b_j + 3b_{j-1} \right) + 13 \left(a_{j+1} - a_{j-1} \right),$$

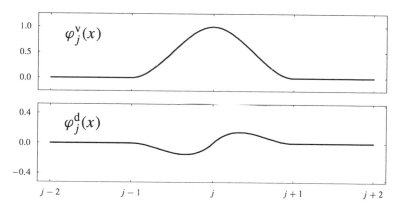

FIGURE 4.11. Hermite-cubic expansion functions φ_j^{v} and φ_j^{d}. The x-axis is labeled in units of Δx.

the Hermite-cubic approximation to the constant-wind-speed advection equation is given by the pair of equations

$$\Delta x \, H^{\mathrm{v}} \left(\frac{da}{dt}, \frac{db}{dt} \right) + 420c \, \tilde{\delta}_{2x} a - 42c \, \tilde{\delta}_x^2 b = 0, \qquad (4.105)$$

$$\Delta x \, H^{\mathrm{d}} \left(\frac{da}{dt}, \frac{db}{dt} \right) - 14c \, \tilde{\delta}_{2x} b + 42c \, \tilde{\delta}_x^2 a = 0. \qquad (4.106)$$

Replacing a_j with $\psi(x_j)$, $b_j \Delta x$ with $(\partial \psi / \partial x)(x_j)$ and performing the usual Taylor series analysis of the truncation error at each node shows that (4.105) is an $O\left[(\Delta x)^4\right]$-accurate approximation to the advection equation, and that (4.106) is an $O\left[(\Delta x)^2\right]$ approximation to its spatial derivative:

$$\frac{\partial^2 \psi}{\partial t \partial x} + c \frac{\partial^2 \psi}{\partial x^2} = 0.$$

As a consequence, the overall accuracy of the Hermite-cubic finite-element method is $O\left[(\Delta x)^3\right]$ (Dupont 1973).

Perhaps the most interesting aspect of the Hermite-cubic approximation to the constant-wind-speed advection problem is that regardless of the numerical resolution, the solution associated with the physical mode is almost free of phase-speed error. In order to evaluate the phase-speed error as a function of the numerical resolution, solutions to (4.105) and (4.106) may be obtained in the form

$$a_j(t) = r_a e^{ik(j\Delta x - c^* t)} \qquad b_j(t) = ik e^{ik(j\Delta x - c^* t)},$$

where r_a represents a factor whose deviation from unity indicates an inconsistency between the wave amplitude in the coefficients of the expansion functions for the displacement field (the a_j) and the coefficients of the expansion functions for the spatial derivative of the displacement field (the $b_j \Delta x$). Substituting the preceding

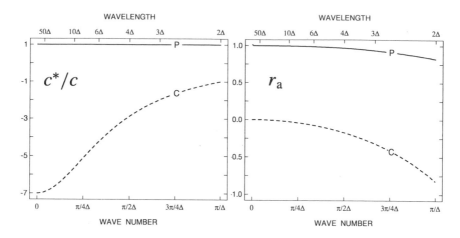

FIGURE 4.12. Normalized phase speed, c^*/c, and amplitude ratio, r_a, for the physical (P) and computational (C) modes in the Hermite-cubic finite-element solution to the advection equation.

expressions into (4.105) and (4.106), and simplifying the result with a symbolic algebra program (Maple),

$$r_a = \frac{13(c^*/c)\theta \sin\theta + 42(\xi - 1)}{(210/\theta) \sin\theta - (c^*/c)(54\xi + 156)}$$

and

$$\frac{c^*}{c} = \frac{6(\xi - 16) \sin\theta \pm [6(\xi^4 - 144\xi^3 + 2026\xi^2 - 8424\xi + 6541)]^{1/2}}{\theta(\xi^2 - 36\xi + 65)}$$

where $\theta = k\Delta x$, and $\xi = \cos\theta$. The physical mode is given by the positive root in the preceding; the other root is associated with a computational mode. In the limit of good spatial resolution, the phase speed and amplitude mismatch for the physical mode are[11]

$$\frac{c^*}{c} \approx 1 + \frac{(k\Delta x)^6}{241920} \quad \text{and} \quad r_a \approx 1 - \frac{(k\Delta x)^2}{96}.$$

Not only is the phase-speed error sixth order, the coefficient multiplying the leading-order error is extremely small. The normalized phase speeds and amplitude factors for both the physical and computation modes are plotted as a function of spatial resolution in Fig. 4.12. As apparent in Fig. 4.12, the phase-speed errors in all the physical modes are essentially zero—even the $2\Delta x$ wave moves at the correct speed. It is not necessarily surprising that the $2\Delta x$ wave propagates. The Hermite-cubic finite-element method can resolve changes in the phase of a

---

[11] See Hedstrom (1979a) for an alternative derivation of these results.

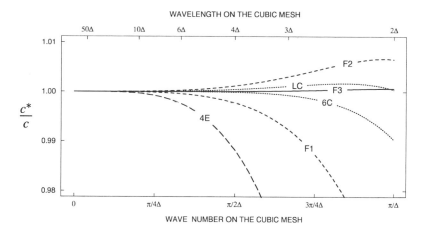

FIGURE 4.13. Phase speed error as a function of spatial scale for the Hermite-cubic finite-element method (F3). Also plotted are curves for the error generated by linear finite-elements (F1), quadratic finite-elements (F2), 6th-order compact differences (6C), Lele's tridiagonal compact scheme (LC) and explicit 4th-order centered differences (4E).

$2\Delta x$ wave that can't be captured in ordinary finite-difference representations because the expansion functions contain information about the function value and its derivative at each point. It is, nevertheless, surprising that the phase speed of $2\Delta x$ waves is approximated with such high accuracy.

The phase-speed error in the Hermite-cubic physical mode is compared with that generated by several other schemes in Fig. 4.13. In constructing Fig. 4.13, it has been assumed that all finite-element approximations use the same number of expansion functions. Thus, since there are two Hermite-cubic expansion functions (φ_j^{v} and φ_j^{d}) at each node, the spacing of the Hermite-cubic nodes is assumed to be twice that of the nodes in the linear and quadratic finite-element approximations (i.e., the behavior of the cubic-finite-element $2\Delta x$ wave is compared with all other schemes' $4\Delta x$ wave). As indicated in Fig. 4.13, the phase-speed errors in the Hermite-cubic finite-element solution are clearly less than those of the other schemes.

As a consequence of its low phase-speed error, the Hermite-cubic finite-element method will give exceptionally good solutions to the constant-wind-speed advection problem. The computational effort required to achieve these results is not, however, insignificant. At every time step, the the implicit coupling in (4.105) and (4.106) necessitates the solution of a block tridiagonal linear system (or alternatively a banded system, whose bandwidth is seven). The effectiveness with which Hermite-cubic expansion functions can be employed in more complex applications depends on the behavior of the computational mode. Even if the initial amplitude of the computational mode associated with every resolvable wave number is insignificant, some of these modes may be amplified by nonlinear interactions in nonlinear problems, or through the computation of the product of two spatially varying functions in linear equations with variable coefficients.

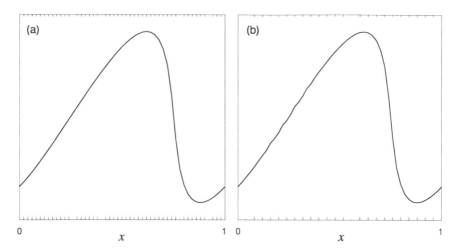

FIGURE 4.14. Comparison of finite-element solutions to the inviscid Burgers's equation obtained using (a) chapeau expansion functions and (b) Hermite-cubic expansion functions.

An example of the difficulty associated with the computational mode is illustrated by the finite-element solutions to the inviscid Burgers's equation (3.113) shown in Fig 4.14. In this example, the domain is $0 \le x \le 1$ and periodic, and the initial condition is $\psi(x, 0) = -\cos(2\pi x)$. The solutions are plotted at $t = 0.12$, at which time the true solution is still continuous and easy to resolve. The solution shown in Fig 4.14a was computed using chapeau expansion functions defined at 50 nodes, trapezoidal time-differencing, and $\Delta t = \Delta x/10$. As might be expected, since the true solution is smooth and well resolved, the linear finite-element approximation is free from any obvious error. Fig 4.14b shows the solution to the same problem computed with 50 Hermite-cubic expansion functions defined at 25 nodes. Using a Hermite-cubic finite-element expansion of the form (4.104), and employing the operator notation defined in connection with (4.105) and (4.106), the evolution of the coefficients a_j and b_j is determined by the block-tridiagonal system of equations

$$\Delta x \, H^{\mathrm{v}}\left(\frac{da}{dt}, \frac{db}{dt}\right) = a\left(50b + 140\tilde{\delta}_{2x}a - 34\,\langle b\rangle^{2x}\right) + 140\tilde{\delta}_{2x}(a^2)$$

$$+ b\left(34\,\langle a\rangle^{2x} - 8\tilde{\delta}_{2x}b\right) - 50\,\langle ab\rangle^{2x} + 5\tilde{\delta}_{2x}(b^2),$$

$$\Delta x \, H^{\mathrm{d}}\left(\frac{da}{dt}, \frac{db}{dt}\right) = -a\left(50a - 16\,\langle a\rangle^{2x} + 3\tilde{\delta}_{2x}b\right) + 34\left\langle a^2\right\rangle^{2x}$$

$$+ b\left(5\tilde{\delta}_{2x}a - \langle b\rangle^{2x}\right) - 11\tilde{\delta}_{2x}(ab) + \left\langle b^2\right\rangle^{2x}.$$

The preceding equations were initialized by setting the function value and its derivative at each node to a_j and $b_j/\Delta x$, respectively. The equations were integrated using trapezoidal time-differencing with the same time step used for the linear finite-element approximation. The result is plotted at the same spatial resolution shown in Fig 4.14a ($\Delta x = 1/50$) by evaluating the piecewise cubics at the midpoint of each element. The solution generated with the Hermite-cubics looks very similar to that obtained with chapeau expansion functions, except in a region centered around $x = 0.25$, where it is degraded by four low-amplitude but nevertheless distinct ripples. These ripples appear to be generated by the nonlinear growth of the computational mode, and although their wavelength can be reduced, their amplitude is not easily diminished by increasing the spatial resolution.

After the formation of the shock at $t = (2\pi)^{-1}$, large-amplitude errors rapidly develop in the piecewise-linear finite-element solution. The error growth is much slower in the Hermite-cubic solution, which continues to resemble a somewhat noisy approximation to the correct generalized solution to the inviscid Burgers's equation until roughly $t = 0.25$. The relative insensitivity of the Hermite-cubic approximation to error growth in the vicinity of the shock may be valuable in some applications (such as simulations of the viscous Burgers's equation), but the method should not be mistaken as a viable candidate for the proper simulation of discontinuous solutions to the inviscid Burgers's equation because the Galerkin finite-element solution incorrectly conserves $\|\phi\|_2$, whereas the ℓ_2-norm of the correct solution begins to decrease after the formation of the shock (see Section 5.1.2). These results are consistent with those obtained by Cullen (1982), who compared linear and quadratic finite-element approximations to the inviscid Burgers's equation, and noted that linear elements performed better than quadratic elements in regions where the solution was smooth and worse elsewhere.

4.5.5 Two-Dimensional Expansion Functions

The construction of finite-element approximations to problems in two or more spatial dimensions is straightforward. In the following we will briefly consider the two-dimensional case. The simplest two-dimensional expansion functions are nonzero only within some rectangular region. One of the simplest types of interpolation that can be performed on a rectangular mesh is bilinear interpolation in which the function is estimated as

$$C_1 + C_2 x + C_3 y + C_4 x y.$$

The four coefficients $C_1, \ldots, C_4$ can be uniquely determined within each rectangle by the function values at the four vertices. Bilinear interpolation reduces to linear interpolation along lines parallel to the x or y coordinate axes. Individual expansion functions for bilinear interpolation, sometimes known as "pagoda" functions, may be expressed as the product of a chapeau function with respect to x times a second chapeau function with respect to y. Each pagoda function is unity at a central node and drops to zero at the eight surrounding nodes.

If the two-dimensional constant-wind-speed advection equation

$$\frac{\partial \psi}{\partial t} + U \frac{\partial \psi}{\partial x} + V \frac{\partial \psi}{\partial y} = 0$$

is approximated using the pagoda-function finite-element method, evaluation of (4.87) yields the following system of ordinary differential equations

$$\mathcal{A}^x \mathcal{A}^y \frac{da_{i,j}}{dt} + U \mathcal{A}^y \delta_{2x} a_{i,j} + V \mathcal{A}^x \delta_{2y} a_{i,j} = 0, \tag{4.107}$$

where $a_{i,j}$ is the expansion coefficient of the (i,j)th pagoda function, or equivalently, the approximate solution at the (i,j)th node, and $\mathcal{A}^x$ and $\mathcal{A}^y$ are the averaging operators

$$\mathcal{A}^x a_{i,j} = \tfrac{1}{6}(a_{i+1,j} + 4a_{i,j} + a_{i-1,j}),$$

$$\mathcal{A}^y a_{i,j} = \tfrac{1}{6}(a_{i,j+1} + 4a_{i,j} + a_{i,j-1}).$$

The product of the averaging operators $\mathcal{A}^x \mathcal{A}^y$ couples the time tendencies at nine different nodes and generates a band matrix with a very wide bandwidth. One technique, known as "mass lumping," that has been occasionally advocated to eliminate this implicit coupling diagonalizes the coefficient matrix via some essentially arbitrary procedure (such as summing the coefficients in each row of the mass matrix and assigning the result to the diagonal). As discussed by Gresho et al. (1978) and Donea et al. (1987), mass lumping degrades the accuracy of finite-element approximations to hyperbolic problems.

An efficient solution to (4.107) can, nevertheless, be obtained by organizing the computations as follows. First evaluate the spatial derivatives at every nodal point by solving the family of tridiagonal systems

$$\mathcal{A}^x \left(\frac{\partial \phi}{\partial x} \right)_{i,j} = \delta_{2x} a_{i,j} \quad \text{and} \quad \mathcal{A}^y \left(\frac{\partial \phi}{\partial y} \right)_{i,j} = \delta_{2y} a_{i,j} \tag{4.108}$$

for $(\partial \phi / \partial x)_{i,j}$ and $(\partial \phi / \partial y)_{i,j}$. Then the time tendency at each nodal point is given by the uncoupled equations

$$\frac{da_{i,j}}{dt} + U \left(\frac{\partial \phi}{\partial x} \right)_{i,j} + V \left(\frac{\partial \phi}{\partial y} \right)_{i,j} = 0. \tag{4.109}$$

Note that the time derivative in (4.109) must be approximated using explicit time-differencing to avoid implicit algebraic equations in the fully discretized approximation. The solution algorithm given by equations (4.108) and (4.109) is exactly that which would be most naturally used to solve the two-dimensional advection equation using the compact finite-difference operator (2.81).

A variety of other two dimensional expansion functions can also be defined, including higher-order piecewise polynomials on a rectangular grid and piecewise-linear functions on a triangular grid. If the computational domain itself is rectangular, it appears that the most efficient schemes are obtained using rectangular elements (Staniforth 1987). On the other hand, if the computational domain is highly irregular it can be advantageous to approximate the solution using a network of triangular elements. This approach has been used when modeling tidal currents in bays (Lynch and Gray 1979).

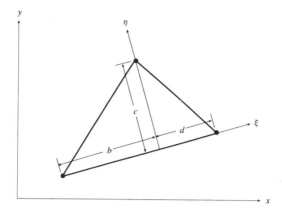

FIGURE 4.15. Local-coordinate system for integrating polynomial expressions over a triangle.

When finite-element expansion functions are defined on triangular grids, several polynomial expressions in x and y must be integrated over triangular domains in order to evaluate the coefficients in the Galerkin approximation (4.87). The calculation of these coefficients is facilitated if each expansion function is defined with respect to the local coordinate system (ξ, η) illustrated in Fig. 4.15. Within each triangular element, the linearly interpolated expansion functions have the general form

$$\alpha\eta + \beta\xi + \gamma = 0,$$

where α, β, and γ are determined by the values at the vertices of the triangle. Polynomial expressions in ξ and η can be integrated over the triangular domain T using the helpful formula

$$\int\int_T \xi^r \eta^s \, d\xi \, d\eta = c^{s+1} \left(d^{r+1} - (-b)^{r+1} \right) \frac{r!s!}{(r+s+2)!},$$

where b, c, and d are the positive dimensions indicated in Fig. 4.15.

Suppose that solutions to the one-dimensional advection equation (4.11) are sought in a domain that has been divided into a uniform grid of equilateral triangles. Then every node not lying along the boundary is surrounded by six triangular elements whose union is a hexagon. These hexagons may be used to define finite-element expansion functions that are unity at the center of the hexagon and zero at each of the surrounding nodes. Let the nodes be numbered as shown in Fig 4.16. Assume that the x-axis is parallel to the line segment connecting nodes 3, 4, and 5, and let Δ denote the distance between any pair of nodes. After considerable algebra, one can show that (4.87) reduces to

$$\frac{1}{2}\left(\frac{da_1}{dt} + \frac{da_2}{dt} + \frac{da_3}{dt} + 6\frac{da_4}{dt} + \frac{da_5}{dt} + \frac{da_6}{dt} + \frac{da_7}{dt} \right)$$
$$+ \frac{c}{6\Delta} [(a_2 - a_1) + 2(a_5 - a_3) + (a_7 - a_6)] = 0.$$

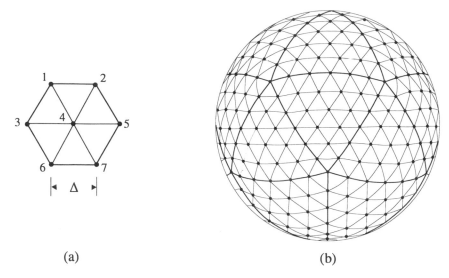

(a)   (b)

FIGURE 4.16. (a) Hexagonal element formed from equilateral triangles. (b) Subdivision of a spherical icosahedron into an almost uniform triangular grid.

One simple nonrectangular domain that can be covered by a quasi-homogeneous lattice of equilateral triangles is the surface of a sphere. A perfectly uniform covering can be obtained using the twelve nodes that are the vertices of a regular icosahedron inscribed within the sphere. If there are more than twelve nodes, the coverage will not be perfectly uniform, but an approximately homogeneous distribution of triangles can be achieved as follows. Beginning with a regular icosahedron inscribed within the sphere, the edges of the icosahedron are projected along great-circle arcs to the surface of the sphere. The resulting spherical triangles are further subdivided into a large number of smaller, almost uniform, triangular elements as illustrated in Fig. 4.16b. All nodes on this mesh, except for the original twelve vertices of the icosahedron, are surrounded by six triangular elements whose union is a hexagon. The original twelve vertices of the icosahedron are surrounded by only five triangular elements, and at these special nodes the elements are pentagonal. The distance between adjacent nodes may vary by as much as 25% over the surface of the sphere, and is smallest in the vicinity of the vertices of the inscribed icosahedron. Williamson (1968), and Sadourny et al. (1968) provide additional details about the properties of geodesic spherical grids. Further discussion of triangular grids in global finite-element models is presented in Cullen (1974), Cullen and Hall (1979), and Priestley (1992).

Problems

1. Show that for $n > 0$, a wavelength of $(n+1)\Delta x/n$ aliases into a wavelength of $-(n + 1)\Delta x$ if it is sampled on a uniform mesh with a grid spacing of

Δx. Sketch an example for $n = 2$ and explain the significance of the change in sign of the wavelength of the aliased wave.

2. Suppose that the spectral method is used to integrate a system including the equation

$$\frac{\partial \phi}{\partial t} \mid \cdots + \phi \chi \psi = 0,$$

and that the term $\phi(x, t)\chi(x, t)\psi(x, t)$ is to be evaluated using the transform technique. If K is the number of modes retained in the Fourier expansions for ϕ, χ, and ψ, derive an expression determining the minimum number of grid points that must be present on the physical-space grid in order to avoid aliasing error in the product $\phi \chi \psi$.

3. In all practical applications, finite Fourier transforms are computed using the fast Fourier transform (FFT) algorithm (Cooley and Tukey, 1965). Suppose that the periodic spatial domain $0 \le x \le 2\pi$ is discretized so that

$$x_j = \frac{2\pi}{M} j, \quad \text{where} \quad j = 1, \ldots, M.$$

In order to be efficient, the FFT algorithm requires that M be the product of small prime numbers. Maximum efficiency is obtained when M is a power of 2. Thus, most FFT codes assume that M is an even number. When the total number of grid points on the physical mesh is even, the finite Fourier transform and inverse transform are given by the relations

$$a_n(t) = \frac{1}{2N} \sum_{j=1}^{2N} \phi(x_j, t)e^{-inx_j}$$

and

$$\phi(x_j, t) = \sum_{k=-N+1}^{N} a_k(t)e^{ikx_j}.$$

The $2N$ data points in physical space uniquely define $2N$ Fourier coefficients. However, in contrast to (4.14), the wave number $k = -N$ does not appear in the expansion. Explain why the $-N$ wave number is retained in finite Fourier transforms when there is an odd number of points on the physical mesh and dropped when the total number of points is even.

4. Solutions to the two-dimensional advection equation

$$\frac{\partial \psi}{\partial t} + u \frac{\partial \psi}{\partial x} + v \frac{\partial \psi}{\partial y} = 0$$

are sought in a domain that is periodic in both x and y. The velocity field is nondivergent.

(a) Show that the domain integral of ψ^3 is conserved by the exact solution to the unapproximated governing equations.

(b) If the effects of time-differencing errors are neglected, is the Galerkin spectral method guaranteed to yield an approximate solution to this problem that conserves the domain integral of ψ^3? Explain your answer.

5. Solutions are sought to the equation

$$\frac{\partial \psi}{\partial t} = \frac{\partial}{\partial x}\left(v(x)\frac{\partial \psi}{\partial x}\right)$$

on the periodic domain $0 \le x \le 2\pi$ using a series-expansion method in which

$$\psi(x, t) = \sum_{|m| \le N} r_m(t)e^{imx}, \qquad v(x) = \sum_{|n| \le N} s_n e^{inx}.$$

(a) If the solution is to be obtained using a Galerkin spectral method, derive the ordinary differential equation for dr_m/dt.

(b) Determine an unconditionally stable $O\left[(\Delta t)^2\right]$-accurate finite-difference method for integrating the ordinary differential equation derived in (a). How would the efficiency of this method change if v did not depend on Δx?

6. Present an algorithm for the solution of the equation described in Problem 5 using a pseudospectral method and second-order Adams–Bashforth time-differencing. Do not assume that v is independent of x.

7. Pseudospectral solutions to the constant-wind-speed advection equation are to be obtained using leapfrog time-differencing such that

$$\frac{\phi_j^{n+1} - \phi_j^{n-1}}{2\Delta t} + c_0\left(\frac{\partial \phi^n}{\partial x}\right)_j = 0.$$

Suppose that the usual formula for calculating the derivative,

$$\left(\frac{\partial \phi^n}{\partial x}\right)_j = \sum_{|k| \le N} ik a_k e^{ikx_j},$$

is replaced by the modified expression

$$\left(\frac{\partial \phi^n}{\partial x}\right)_j = \sum_{|k| \le N} i\left[\frac{\sin(kc_0\Delta t)}{c_0\Delta t}\right] a_k e^{ikx_j}.$$

(a) Determine the phase-speed error and the maximum stable time step for the modified scheme?

(b) What limits the practical utility of this otherwise attractive scheme?

8. Using (4.43), verify the orthogonality relation for the spherical harmonics (4.44). Also use the relation $P_{-m,n}(\mu) = (-1)^m P_{m,n}(\mu)$ to show that $Y_{-m,n}^* = (-1)^m Y_{m,n}$ and that the expansion coefficients for any approximation to a real-valued function satisfy $a_{-m,n} = (-1)^m a_{m,n}^*$.

9. Express the associated Legendre function $P_{4,4}(\mu)$ as an algebraic function of μ (thereby producing an expression similar to those in Table 4.2).

10. Derive (4.21) by repeatedly integrating

$$a_k = \frac{1}{2\pi} \int_{-\pi}^{\pi} \psi(x) e^{-ikx}\, dx$$

by parts.

11. *Consider the family of functions periodic on the interval $[0, 1]$

$$\psi(x) = \begin{cases} \cos^n \left[2\pi \left(x - \frac{1}{2} \right) \right] & \text{if } \left| x - \frac{1}{2} \right| < \frac{1}{4}, \\ 0 & \text{otherwise.} \end{cases}$$

Evaluate the rates of convergence of the Fourier series expansions to this family of functions for the cases $n = 0, 1$, and 2. Use fast Fourier transforms to expand each function as defined on progressively finer meshes for which $\Delta x - 1/(2^m)$, $m - 3, 4, \ldots, 8$. Compute the error in the expansion using both the maximum norm over the entire interval and the maximum norm in the region $|x - \frac{1}{2}| \le \frac{1}{8}$. Evaluate these maximum norms using grid-point values on a mesh with $\Delta x = 1/1024$, and plot the logarithm of the error versus the logarithm of Δx. How do the rates of convergence of the Fourier approximation to these functions compare with the rates suggested in Section 4.2.1?

12. Show that the ℓ_2-norm of the solution to the viscous Burgers's equation (4.39) on the periodic domain $0 \le x \le 1$,

$$\|\psi\|_2 = \left[\int_0^1 \psi^2\, dx \right]^{1/2},$$

is bounded by its value at the initial time.

13. *Use the spectral and pseudospectral methods to compute numerical solutions to the viscous Burgers's equation (4.39) subject to the initial condition $\psi(x, 0) = \sin(2\pi x)$. Set $\nu = 0.002$. Approximate the time derivative using leapfrog time-differencing for the advection term and forward differencing for the diffusion. Initialize the leapfrog scheme with a single forward time step. Use a time step such that

$$\frac{\Delta t}{\Delta x} \max_x(\psi(x, 0)) = 0.16.$$

(a) Use $\Delta x = 1/64$ and 64 Fourier modes (which yields a cutoff wave number of 64π on this spatial domain). Show the solutions at $t = 0.40$ on a scale $-4 \le \psi \le 4$. Which scheme is performing better? How seriously is aliasing error affecting the stability and accuracy of the pseudospectral solution?

(b) Repeat the preceding simulations using $\Delta x = 1/128$ and 128 Fourier modes. Show the solutions at $t = 0.40$ in the subdomain $0 \le x \le \frac{1}{2}$, $0 \le \psi \le 1$. How seriously is aliasing error degrading the stability and accuracy of the pseudospectral solution?

(c) Why is there an improvement in the pseudospectral solution between the simulations in (a) and (b)?

(d) If the spatial resolution is increased to 256 Fourier modes, both the spectral and pseudospectral solutions become unstable. Why? Devise a way around this instability and obtain an approximation to the solution at $t = 0.40$. Again plot this solution on the subdomain $0 \le x \le \frac{1}{2}, 0 \le \psi \le 1$.

14. Solutions to the coupled advection/chemical reaction equations

$$\frac{\partial \phi}{\partial t} + c\frac{\partial \phi}{\partial x} = \phi\psi, \qquad \frac{\partial \psi}{\partial t} + c\frac{\partial \psi}{\partial x} = -\phi\psi,$$

are to be obtained using the Galerkin finite-element approach. Assume that the expansion functions are chapeau functions and that the approximate expressions for ψ and ϕ are

$$\phi \approx \sum_m a_m\varphi_m, \qquad \psi \approx \sum_n b_n\varphi_n.$$

Using (4.89) we know that the first equation will have the form

$$\frac{\Delta x}{6}\left(\frac{da_{j+1}}{dt} + 4\frac{da_j}{dt} + \frac{da_{j-1}}{dt}\right) + c\left(\frac{a_{j+1} - a_{j-1}}{2}\right) = X,$$

where X represents the Galerkin approximation to $\phi\psi$. Evaluate X in terms of the expansion coefficients a_k and b_k.

15. Determine the Galerkin finite-element approximation to $\psi\partial\psi/\partial x$ using chapeau expansion functions. Show that the result is identical to the conservative finite-difference operator appearing in (3.120).

16. *Compute solutions to Problem 13 of Chapter 3 using the spectral and pseudospectral methods. Use the same numerical parameters specified in that problem except choose Δt so that the Courant number based on the maximum wind speed for the shortest wavelength retained in the spectral truncation is 0.3. Do not use any type of smoother. Use leapfrog time-differencing, taking a single forward step to obtain the solution at the first time level.

(a) Obtain solutions using 64 Fourier modes to approximate ψ and $c(x)$. Show your results at $t = 1.5$ and 3.0 as directed in Problem 13 of Chapter 3. Also show the two solutions at some time when the pseudospectral method is clearly showing some aliasing error. (*Hint:* this only happens for a limited period of time during the integration.)

(b) Now retry the solution with 128 Fourier modes and compare your results with those obtained in (a) and, if available, with the finite-difference solutions computed for Problem 13 of Chapter 3.

17. *Compute chapeau-function finite-element method solutions to Problem 16 using the previously specified numerical parameters. Try spatial resolutions of $\Delta x = \frac{1}{32}$ and $\Delta x = \frac{1}{64}$. If a_j is the amplitude at the jth node, the Galerkin chapeau-function approximation to the variable wind speed advection equation is

$$\frac{d}{dt}\left(a_{j-1} + 4a_j + a_{j+1}\right) + (c_{j-1} + 2c_j)\left(\frac{a_j - a_{j-1}}{\Delta x}\right)$$

$$+ (c_{j+1} + 2c_j)\left(\frac{a_{j+1} - a_j}{\Delta x}\right) = 0.$$

(a) Initialize the problem by setting a_j and c_j equal to the exact function values at the nodes.

(b) Initialize the problem by projecting the exact data on to the nodes using the Galerkin (or least-squares) formula (4.8).

5

Finite-Volume Methods

As demonstrated in the preceding chapters, the errors in most numerical solutions increase dramatically as the physical scale of the simulated disturbance approaches the minimum scale resolvable on the numerical mesh. When solving equations for which smooth initial data guarantees a smooth solution at all later times, such as the barotropic vorticity equation (3.123), any difficulties associated with poor numerical resolution can be avoided by using a sufficiently fine computational mesh. But if the governing equations allow an initially smooth field to develop shocks or discontinuities, as is the case with Burgers's equation (3.113), there is no hope of maintaining adequate numerical resolution throughout the simulation, and special numerical techniques must be used to control the development of overshoots and undershoots in the vicinity of the shock. Numerical approximations to equations with discontinuous solutions must also satisfy additional conditions beyond the stability and consistency requirements discussed in Chapter 2 to guarantee that the numerical solution converges to the correct solution as the spatial grid interval and the time step approach zero.

The possibility of erroneous convergence to a function that does not approximate the true discontinuous solution can be demonstrated by comparing numerical solutions to the generalized Burgers's equation in *advective form*

$$\frac{\partial \psi}{\partial t} + \psi^2 \frac{\partial \psi}{\partial x} = 0 \qquad (5.1)$$

with those generated by analogous solutions to the same equation in *flux form*

$$\frac{\partial \psi}{\partial t} + \frac{\partial}{\partial x}\left(\frac{\psi^3}{3}\right) = 0. \qquad (5.2)$$

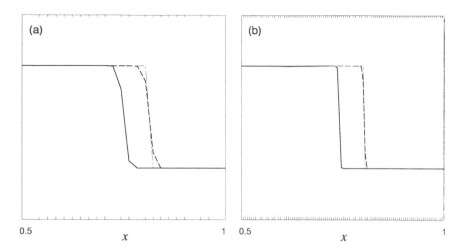

FIGURE 5.1. Exact (dash-dotted), upstream advective-form (solid) and upstream flux-form (dashed) solutions to the generalized Burgers's equation at $t = 2.4$ on the subdomain $0.5 \leq x \leq 1$. (a) $\Delta x = 0.02$, $\Delta t = 0.01$; (b) $\Delta x = 0.005$, $\Delta t = 0.0025$.

As will be explained in Section 5.1, if the initial conditions are specified by the step function

$$\psi(x, 0) = \begin{cases} 1, & \text{if } x \leq 0, \\ 0 & \text{otherwise,} \end{cases} \qquad (5.3)$$

the correct solution consists of a unit-amplitude step propagating to the right at speed $\frac{1}{3}$. An upstream finite-difference approximation to the advective form (5.1) was calculated using

$$\frac{\phi_j^{n+1} - \phi_j^n}{\Delta t} + \left(\frac{\phi_j^n + \phi_{j-1}^n}{2}\right)^2 \left(\frac{\phi_j^n - \phi_{j-1}^n}{\Delta x}\right) = 0, \qquad (5.4)$$

and an upstream approximation to the flux form (5.2) was obtained using

$$\frac{\phi_j^{n+1} - \phi_j^n}{\Delta t} + \frac{(\phi_j^n)^3 - (\phi_{j-1}^n)^3}{3\Delta x} = 0. \qquad (5.5)$$

Figure 5.1a shows a comparison of the exact and the numerical solutions at $t = 2.4$ on the subdomain $0.5 \leq x \leq 1.0$. The computations were performed using $\Delta x = 0.02$ and a time step such that $\max[\psi(x, 0)]\Delta t/\Delta x = 0.5$. Both schemes yield plausible-looking approximations to the correct solution (shown by the thin dot-dashed line), but the numerical solution obtained using advective-form differencing (shown by the solid line) moves at the wrong speed. As illustrated in Fig. 5.1b, in which the numerical solutions are recalculated after reducing Δx and Δt by a factor of four, the speed of the solution generated by the advective-form approximation is not significantly improved by decreasing Δx and Δt. The advective-form approximation simply does not converge to the correct solution in the limit $\Delta x \to 0$ and $\Delta t \to 0$. The difficulties that can be associated with advective-form finite differencing are even more apparent if (5.1) is approximated using the scheme

$$\frac{\phi_j^{n+1} - \phi_j^n}{\Delta t} + \left(\phi_j^n\right)^2 \frac{\phi_j^n - \phi_{j-1}^n}{\Delta x} = 0$$

and the initial data

$$\phi_j^0 = \begin{cases} 1, & \text{if } j \leq j_0, \\ 0, & \text{otherwise.} \end{cases}$$

In this case, the finite-difference approximation to $\psi^2 \partial\psi/\partial x$ is zero at every grid point, and the numerical solution is stationary. In order to understand how advective-form finite-difference approximations can converge to invalid solutions to the generalized Burgers's equation, it is helpful to review the sense in which discontinuous functions constitute solutions to partial differential equations.

5.1   Conservation Laws and Weak Solutions

Many of the partial differential equations arising in fluid dynamics can be expressed as a system of *conservation laws* of the form

$$\frac{\partial \mathbf{u}}{\partial t} + \sum_j \frac{\partial}{\partial x_j} \mathbf{f}_j(\mathbf{u}) = \mathbf{0},$$

which states that the rate of change of $\mathbf{u}$ at each point is determined by the convergence of the fluxes $\mathbf{f}_j$ at that point. An example of this type is provided by the one-dimensional shallow-water equations. Let u denote the velocity and h the fluid depth, and suppose that there is no bottom topography; then conservation of mass requires

$$\frac{\partial h}{\partial t} + \frac{\partial}{\partial x}(hu) = 0,$$

and conservation of momentum implies

$$\frac{\partial}{\partial t}(hu) + \frac{\partial}{\partial x}\left(hu^2 + g\frac{h^2}{2}\right) = 0.$$

If a function contains a discontinuity, it cannot be the solution to a partial differential equation in the conventional sense, because derivatives are not defined at the discontinuity. Instead, the solution is required to satisfy a family of related integral equations. Consider solutions to the scalar conservation law

$$\frac{\partial \psi}{\partial t} + \frac{\partial}{\partial x} f(\psi) = 0 \tag{5.6}$$

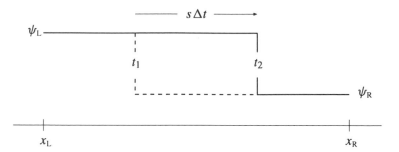

FIGURE 5.2. The displacement of a jump propagating to the right at speed s over time $\Delta t = t_2 - t_1$.

on the unbounded domain $-\infty < x < \infty$. Integrating this conservation law over the intervals $[x_1, x_2]$ and $[t_1, t_2]$, one obtains

$$\int_{x_1}^{x_2} \psi(x, t_2)\,dx = \int_{x_1}^{x_2} \psi(x, t_1)\,dx +$$
$$\int_{t_1}^{t_2} f(\psi(x_1, t))\,dt - \int_{t_1}^{t_2} f(\psi(x_2, t))\,dt, \qquad (5.7)$$

which states that the total change in ψ over the region $x_1 \le x \le x_2$ is determined by the time-integrated fluxes through the boundary of that region. This integral form of the conservation law can usually be derived from first physical principles as easily as the differential form (5.6), and unlike the differential form, the integral form can be satisfied by piecewise-continuous functions. If ψ satisfies the integral equation (5.7) on every subdomain $[x_1, x_2] \times [t_1, t_2]$, then ψ is a *weak solution* of the conservation law. Differentiable weak solutions are also solutions to the partial differential equation (5.6) and are uniquely determined by the initial data. Nondifferentiable weak solutions may, however, be nonunique.

5.1.1   The Riemann Problem

Weak solutions to the conservation law (5.6) are particularly easy to obtain when the initial data are constant on each side of a single discontinuity. This combination of a scalar conservation law and piecewise-constant initial data containing a single discontinuity is known as a *Riemann problem*. Riemann problems have solutions in which the initial discontinuity propagates at a constant speed s, as indicated schematically in Fig. 5.2. Assuming for notational convenience that at $t = 0$ the discontinuity is at $x = 0$, this solution has the form

$$\psi(x, t) = \begin{cases} \psi_L & \text{if } x - st < 0, \\ \psi_R & \text{otherwise,} \end{cases} \qquad (5.8)$$

where $\psi_L = \psi(x_L)$, $\psi_R = \psi(x_R)$, and it has been assumed that x_L and x_R are located sufficiently far upstream and downstream that the jump does not propagate past these points during the time interval of interest. The speed of the shock may be determined as follows. From (5.8),

$$\int_{x_L}^{x_R} \psi(x, t)\, dx = (st - x_L)\psi_L + (x_R - st)\psi_R,$$

and thus

$$\frac{d}{dt} \int_{x_L}^{x_R} \psi(x, t)\, dx = s(\psi_L - \psi_R). \tag{5.9}$$

Integrating (5.6) over the interval $[x_L, x_R]$, one obtains

$$\frac{d}{dt} \int_{x_L}^{x_R} \psi(x, t)\, dx = f(\psi_L) - f(\psi_R),$$

which together with (5.9) implies that

$$s = \frac{f(\psi_L) - f(\psi_R)}{\psi_L - \psi_R}. \tag{5.10}$$

This equation for the speed of the jump is known as the *Rankine–Hugoniot condition*. Note that the Rankine–Hugoniot condition requires the jump in the weak solutions plotted in Fig. 5.1 to propagate at a speed of $\frac{1}{3}$. The Rankine–Hugoniot condition is frequently derived from first principles in various physical applications. For example, Stoker (1957, eqs. 10.6.6 and 10.7.7) derives the Rankine–Hugoniot condition for the one-dimensional shallow-water system by constructing mass and momentum budgets for a control volume containing the shock.

As previously mentioned, nondifferentiable weak solutions need not be uniquely determined by the initial data, and if more than one weak solution exists, it is necessary to select the physically relevant solution. When the solutions to equations representing real physical systems develop discontinuities, one of the physical assumptions used to derive those equations is often violated. Solutions to the inviscid Euler equations may suggest that discontinuities develop in supersonic flow around an airfoil, but the velocity and thermodynamic fields around an airfoil never actually become discontinuous. The discontinuities predicted by the Euler equations actually appear as narrow regions of steep gradients that are stabilized against further scale collapse by viscous dissipation and diffusion. The discontinuous inviscid solution may be considered to be the limit of a series of viscous solutions in which the viscosity is progressively reduced to zero. Thus, one strategy for selecting the physically significant weak solution would be to conduct a series of viscous simulations with progressively smaller viscosities and choose the weak solution toward which the viscous solutions converge. This, of course, is a highly inefficient strategy, and it may be impossible to implement in actual simulations of high-Reynolds-number flow, where any realistic value for the molecular viscosity may be too low to significantly influence the solution on the spatial scales resolvable on the numerical grid. In addition, any attempt to include realistic viscosities in the numerical solution reintroduces precisely those mathematical complications that were eliminated when the full physical system was originally approximated by the simpler inviscid model.

5.1.2   Entropy-Consistent Solutions

It is therefore preferable to obtain alternative criteria for selecting the physically relevant weak solution. These criteria may be derived directly from physical principles. Stoker (1957) eliminated nonphysical shocks in shallow-water flow by requiring that "the water particles do not gain energy upon crossing a shock front." In gas dynamics, thermodynamic principles require that entropy be nondecreasing at the shock. Generalized entropy conditions can also be derived for any system of one or two scalar conservation laws of the form (5.6) by considering the limiting behavior of the corresponding viscous system as the viscosity approaches zero (Lax 1971).

For example, a generalized entropy function for the inviscid Burgers's equation

$$\frac{\partial \psi}{\partial t} + \frac{\partial}{\partial x}\left(\frac{\psi^2}{2}\right) = 0 \tag{5.11}$$

is ψ^2. When ψ is a weak solution to Burgers's equation, ψ^2 is a weak solution to the inequality

$$\frac{\partial \psi^2}{\partial t} + \frac{\partial}{\partial x}\left(\frac{2\psi^3}{3}\right) \le 0. \tag{5.12}$$

If the solutions to Burgers's equation are differentiable, the left side of (5.12) is identically zero and the time rate of change of the integral of ψ^2 over any spatial domain is equal to the divergence of the entropy flux, $2\psi^3/3$, through the edges of the domain. But if the solution of Burgers's equation is discontinuous, (5.12) can no longer be satisfied by an equality. The sense of the inequality demanded by (5.12) is that which matches the limiting behavior of ψ^2 for solutions to the viscous Burgers's equation

$$\frac{\partial \psi}{\partial t} + \frac{\partial}{\partial x}\left(\frac{\psi^2}{2}\right) = \epsilon \frac{\partial^2 \psi}{\partial x^2}$$

as $\epsilon \to 0$ (LeVeque 1992, p. 37).

Consider two possible weak solutions to the inviscid Burgers's equation (5.11), both of which are consistent with the initial condition

$$\psi(x, 0) = \begin{cases} 0, & \text{if } x \le 0, \\ 1, & \text{otherwise.} \end{cases}$$

The first solution, shown in Fig. 5.3a, consists of a unit-amplitude downward[1] jump moving to the right at the speed given by the Rankine–Hugoniot condition, which is a speed of $\frac{1}{2}$. The second solution, shown in Fig. 5.3b, is the *rarefaction*

---

[1] The jump is "downward" in the sense that the fluid level drops during the passage of the discontinuity.

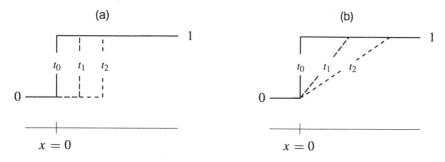

FIGURE 5.3. (a) An entropy-violating shock; (b) the entropy-consistent rarefaction wave.

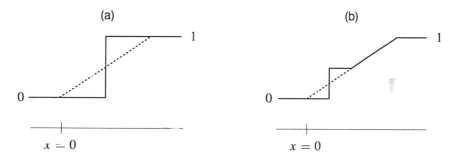

FIGURE 5.4. Spatial distribution of ψ in a rarefaction wave compared to that in: (a) an entropy-violating shock; (b) the combination of a small entropy-violating shock and a rarefaction wave.

wave, or expansion fan, given by

$$\psi(x, t) = \begin{cases} 0, & \text{if } x \leq 0, \\ x/t, & \text{if } 0 < x < t, \\ 1, & \text{otherwise.} \end{cases}$$

Note that the central point in the rarefaction wave moves at the same speed as the shock. The validity of the rarefaction-wave solution on the interval $0 < x < t$ can be confirmed by substituting $\psi = x/t$ into (5.11). The validity of the solution on any larger domain follows from the fact that the shock is a weak solution, since it moves at the speed determined by the Rankine–Hugoniot condition, and as indicated in Fig. 5.4a, the rate of change of $\int \psi dx$ over any domain including the interval $0 \leq x \leq t$ is the same for the shock and the rarefaction wave. An infinite number of other weak solutions also exist, such as the small shock following a rarefaction wave shown in Fig. 5.4b.

Characteristic curves for the shock and rarefaction-wave solutions are plotted in Fig. 5.5. Those characteristics that intersect the trajectory of the shock are directed away from the shock, i.e., they originate at some point along the trajectory of the shock and do not continue back to the line $t = 0$ along which the initial data are specified. As a consequence, the initial data do not determine the value of $\psi(x, t)$ throughout the entire $t > 0$ half-plane, which is clearly a nonphysical situation.

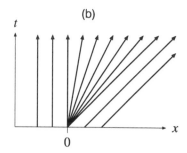

FIGURE 5.5. Characteristic curves for (a) an entropy-violating shock, and (b) the entropy-consistent rarefaction wave. The trajectory of the shock is indicated by the heavy dashed line

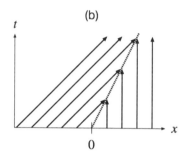

FIGURE 5.6. (a) An entropy-consistent shock, and (b) characteristic curves associated with that shock. The trajectory of the jump is indicated by the heavy dashed line

In contrast, all the characteristics associated with the rarefaction wave originate from the line $t = 0$, and the solution is everywhere determined by the initial data. The rarefaction wave is therefore the physically relevant weak solution.

If the initial data in the preceding example are reflected about the point $x = 0$, the unique and physically relevant solution consists of a unit-amplitude *upward* jump propagating to the right at speed $\frac{1}{2}$. This solution is shown in Fig. 5.6, together with a representative set of its characteristic curves. In this case, the characteristic curves are directed into the jump, so that the initial data do determine the solution. Indeed, the intersecting characteristics indicate the need for a discontinuity in the solution, because otherwise the solution would have to be double valued at the point where two different characteristics meet.

These results are consistent with the entropy condition (5.12), which may be evaluated for jump solutions to Burgers's equation as follows. As in Fig. 5.2, let $\psi_L = \psi(x_L)$, $\psi_R = \psi(x_R)$, and assume that x_L and x_R are located sufficiently far upstream and downstream that the jump does not pass these points during the time interval $[t_1, t_2]$. Following the same derivation that led to (5.9),

$$\frac{d}{dt} \int_{x_L}^{x_R} \psi^2(x, t)\, dx = s(\psi_L^2 - \psi_R^2), \tag{5.13}$$

where s is the speed of the jump $(\psi_L + \psi_R)/2$. Integrating the left side of (5.12) over the domain $[x_L, x_R] \times [t_1, t_2]$, one obtains

$$\int_{t_1}^{t_2} \int_{x_L}^{x_R} \left[\frac{\partial \psi^2}{\partial t} + \frac{\partial}{\partial x} \left(\frac{2\psi^3}{3} \right) \right] dx\, dt =$$

$$\int_{x_L}^{x_R} \psi^2(x, t_2)\, dx - \int_{x_L}^{x_R} \psi^2(x, t_1)\, dx + \int_{t_1}^{t_2} \frac{2}{3} \left(\psi_R^3 - \psi_L^3 \right) dt.$$

Defining $\Delta t = t_2 - t_1$ and expanding the first term on the right side in a Taylor series using (5.13) yields

$$\int_{t_1}^{t_2} \int_{x_L}^{x_R} \left[\frac{\partial \psi^2}{\partial t} + \frac{\partial}{\partial x} \left(\frac{2\psi^3}{3} \right) \right] dx\, dt$$

$$= s\Delta t (\psi_L^2 - \psi_R^2) + O\left[(\Delta t)^2 \right] + \frac{2}{3} \Delta t (\psi_R^3 - \psi_L^3)$$

$$= \frac{\Delta t}{6} (\psi_R - \psi_L)^3 + O\left[(\Delta t)^2 \right]. \tag{5.14}$$

Taking the limit $\Delta t \to 0$, it follows that the only jumps that can satisfy the entropy condition (5.12) are those for which ψ_L exceeds ψ_R.

As suggested by the preceding examples, one criterion for determining the entropy-consistent solution is to demand that all characteristic curves intersecting the shock be directed in toward the trajectory of the shock. For scalar conservation laws, this condition is satisfied, provided that for all ψ between ψ_L and ψ_R,

$$\frac{f(\psi_L) - f(\psi)}{\psi_L - \psi} \geq s \geq \frac{f(\psi) - f(\psi_R)}{\psi - \psi_R} \tag{5.15}$$

(Oleinik 1957). Since s is given by the Rankine–Hugoniot condition (5.10), this inequality reduces to the simple condition that $\psi_L > \psi_R$ when $f(\psi)$ is convex, i.e., when the chord connecting any two points $(\psi_1, f(\psi_1))$ and $(\psi_2, f(\psi_2))$ lies entirely above the graph of f. Because the flux appearing in Burgers's equation is convex, the entropy-consistent shock shown in Fig. 5.6 can be distinguished from the entropy-violating shock shown in Fig. 5.3 by the criterion $\psi_L > \psi_R$, which agrees with (5.14).

5.2   Finite-Volume Methods and Convergence

There are two special difficulties that can arise in attempting to compute discontinuous solutions to partial differential equations. First, as suggested by the spurious solution generated by the advective-form upstream finite-difference approximation (5.1), the numerical scheme might converge to a function that is not a weak solution of the conservation law. Second, the numerical method may converge to a genuine weak solution, but it may fail to converge to the physically relevant entropy-consistent solution.

The possibility of numerical solutions converging to a function that is not a weak solution to the governing equation can be avoided by using finite-difference formulae that can be expressed in *conservation form*. A finite-difference approximation to the scalar conservation law (5.6) is in conservation form if

$$\frac{\phi_j^{n+1} - \phi_j^n}{\Delta t} + \left(\frac{F_{j+\frac{1}{2}} - F_{j-\frac{1}{2}}}{\Delta x}\right) = 0, \tag{5.16}$$

where $F_{j\pm\frac{1}{2}}$ are numerical approximations to $f[\psi(j\pm\frac{1}{2})\Delta x)]$ of the form

$$F_{j+\frac{1}{2}} = F(\phi_{j-p}^n, \phi_{j-p+1}^n, \ldots, \phi_{j+q+1}^n),$$
$$F_{j-\frac{1}{2}} = F(\phi_{j-p-1}^n, \phi_{j-p}^n, \ldots, \phi_{j+q}^n),$$

and p and q are integers. Suppose that the numerical fluxes are smooth functions of the grid-point values (at a minimum, F must be Lipschitz continuous)[2] and that these fluxes are consistent with the conservation law (5.6) in the sense that

$$F(\psi_0, \psi_0, \ldots, \psi_0) = f(\psi_0),$$

i.e., that the numerical fluxes generated by a spatially and temporally uniform ψ_0 are identical to the true flux generated by the same constant value of ψ_0. Then if the numerical solutions converge to some function as $\Delta x \to 0$ and $\Delta t \to 0$, that function must be a weak solution of (5.6) (Lax and Wendroff 1960; LeVeque 1992, p. 130). Note that the results presented in Fig. 5.1 are consistent with this theorem because (5.5) is in conservation form with $F_{j+\frac{1}{2}} = \phi_j^3/3$ but (5.4) is not algebraically equivalent to any scheme in conservation form.

Numerical methods that approximate the integral of a conservation law over the volume of each grid cell are called *finite-volume methods*. The integral of (5.6) over one grid volume and one time step is

$$\int_{x_j-\Delta x/2}^{x_j+\Delta x/2} \psi(x, t^{n+1})\, dx = \int_{x_j-\Delta x/2}^{x_j+\Delta x/2} \psi(x, t^n)\, dx$$
$$+ \int_{t^n}^{t^{n+1}} f(\psi(x_j - \Delta x/2, t))\, dt - \int_{t^n}^{t^{n+1}} f(\psi(x_j + \Delta x/2, t))\, dt.$$

Conservation laws of the form (5.16) may be interpreted as finite-volume approximations to the preceding in which ϕ_j approximates the spatial average of ψ over grid cell j, and $F_{j+\frac{1}{2}}$ approximates the time-averaged flux through the interface between grid cells j and $j + 1$. Finite-volume methods can be used to obtain numerical solutions to any conservation law, but they are particularly appropriate for those problems with discontinuous solutions because they automatically satisfy

$$\sum_{j=j_1}^{j_2} \phi_j^{n+1} = \sum_{j=j_1}^{j_2} \phi_j^n + \Delta t F_{j_1-\frac{1}{2}} - \Delta t F_{j_2+\frac{1}{2}},$$

---

[2] Any differentiable function is Lipschitz continuous.

which is a discrete approximation to an arbitrary member of the family of integral equations (5.7) satisfied by any weak solution to the exact conservation law.

5.2.1 Monotone Schemes

There is no guarantee that a consistent finite-difference method in conservation form will converge to a weak solution. The theorem of Lax and Wendroff assures only that if the numerical solution does converge, it will converge to a weak solution. Convergence to the entropy-consistent weak solution is guaranteed whenever a consistent method in conservation form is *monotone* (Kuznecov and Vološin 1976; Harten et al. 1976; Crandall and Majda 1980b). Recall that a real-valued function is "monotone increasing" if $g(x) \leq g(y)$ whenever $x \leq y$. A finite-difference method is *monotone* if ϕ_j^{n+1} is a monotone increasing function of each grid-point value of ϕ appearing in the finite-difference formula. If the scheme is expressed in the functional form

$$\phi_j^{n+1} = H(\phi_{j-p}^n, \ldots, \phi_{j+q+1}^n),$$

the condition that the method be monotone is

$$\frac{\partial H(\phi_{j-p}, \ldots, \phi_{j+q+1})}{\partial \phi_i} \geq 0 \tag{5.17}$$

for each integer i in the interval $[j - p, \ j + q + 1]$. If the finite-difference method is linear in the ϕ_i^n, the method will be monotone if and only if the coefficients of all the ϕ_i^n are nonnegative.

The upstream approximation to the flux form of the constant-wind-speed advection equation

$$\frac{\partial \psi}{\partial t} + \frac{\partial}{\partial x}(c\psi) = 0 \tag{5.18}$$

is

$$\phi_j^{n+1} = (1 - \mu)\phi_j^n + \mu\phi_{j-1}^n, \tag{5.19}$$

where $\mu = c\Delta t / \Delta x$. According to (5.17), the preceding is monotone for $0 \leq \mu \leq 1$, which is identical to the standard stability condition for the upstream scheme. As suggested by this example, the range of Δt for which a consistent method in conservation form is monotone is a subset of the range of Δt for which the same scheme is stable when used to approximate problems with smooth solutions. The class of monotone methods is, however, far more restrictive than the class of stable finite-volume methods because *any monotone method is at most first-order accurate* (Godunov 1959; Harten et al. 1976). The only exceptions occur in special cases of no practical significance such as when perfect results are obtained using (5.19) with $\mu = 1$. The leading-order truncation error in any monotone first-order approximation to (5.6) is diffusive (e.g., Section 2.5.2), which makes the scheme a higher-order approximation to a viscous problem and ensures that the numerical solution converges to the entropy-consistent solution.

5.2.2   TVD Methods

First-order methods do not provide a particularly efficient way to obtain accurate numerical solutions; better results can often be obtained using higher-order schemes. Although they are not monotone, many of these schemes satisfy the weaker stability condition that they are *total variation nonincreasing*. The total variation of a one-dimensional grid-point function is defined as

$$\text{TV}(\phi) = \sum_{j=1}^{N-1} |\phi_{j+1} - \phi_j|,$$

where N is the total number of grid points in the numerical domain. The total variation of a continuous function on the interval $[a, b]$ may be defined in an analogous manner as the supremum over all possible subdivisions of the domain $a = x_1 < x_2 < \cdots < x_N = b$ of

$$\sum_{j=1}^{N-1} |\psi(x_{j+1}) - \psi(x_j)|,$$

or equivalently as

$$TV(\psi) = \limsup_{\epsilon \to 0} \frac{1}{\epsilon} \int_{-\infty}^{\infty} |\psi(x + \epsilon) - \psi(x)| \, dx = 0.$$

A numerical method is total variation nonincreasing if

$$\text{TV}(\phi^{n+1}) \le \text{TV}(\phi^n). \tag{5.20}$$

Although slightly imprecise, it is common practice and easier on the tongue to refer to a method that is total variation nonincreasing as *total variation diminishing,* or TVD. This convention will be followed in the remainder of this book, so that (5.20) is the working definition of a TVD method.

Solutions to a consistent finite-difference method in conservation form are guaranteed to converge to weak solutions of the exact conservation law whenever the scheme is TVD. The nature of this convergence is, however, complicated by the fact that there may be several nonunique weak solutions to a given conservation law. If the scheme is TVD, the infimum, over the set of all possible weak solutions, of the difference between the numerical solution and each weak solution is guaranteed to go to zero as $\Delta x \to 0$ and $\Delta t \to 0$, but a sequence of numerical solutions computed with successively smaller values of Δx and Δt need not smoothly converge to any particular weak solution (LeVeque 1992, p. 164). Of course, the goal is to obtain an approximation that converges to the entropy-consistent solution, and this is generally accomplished by demanding that every approximate solution also satisfy a discrete form of the entropy condition.

The family of monotone finite-volume schemes is a subset of the family of TVD schemes, which are in turn a subset of an even more general class of monotonicity-preserving methods (Harten 1983). A method is *monotonicity-preserving* if it will

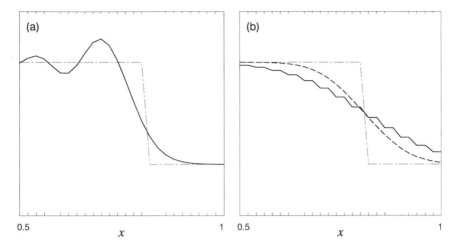

FIGURE 5.7. Numerical and exact solutions to the constant-wind-speed advection equation at $t = 2.4$ on the subdomain $0.5 \leq x \leq 1$: (a) exact (dash-dotted) and filtered leapfrog (solid); (b) exact (dash-dotted), Lax–Fredrichs (solid), and upstream (dashed).

preserve monotone increasing or decreasing initial data. For example, if $\phi_j^0 \geq \phi_{j+1}^0$ for all j, then the numerical solution generated by a monotonicity-preserving method has the property that $\phi_j^n \geq \phi_{j+1}^n$ for all n and j. Monotonicity-preserving methods generate approximate solutions that are free from spurious ripples. In particular, no new local extrema are generated in the numerical solution, and the absolute values of preexisting local extrema are nonincreasing.

One might hope to create a second-order TVD or monotonicity-preserving method by adding enough spatial smoothing to a second-order nondissipative scheme to prevent the development of spurious ripples near the discontinuity. Suppose that numerical solutions to the constant-wind-speed advection equation (5.18) are computed for the case $c = \frac{1}{3}$ using the second-order scheme

$$\delta_{2t}\phi_j^n + \delta_{2x}\left(c\phi_j^n\right) = \delta_x^4\left(\gamma_4\phi_j^{n-1}\right), \tag{5.21}$$

which is a centered leapfrog approximation plus a fourth-derivative filter. Let the strength of the fourth-derivative filter be maximized by setting $\gamma_4\Delta t = 1/32$, which removes all amplitude from the $2\Delta x$ wave in a single leapfrog time step. Solutions computed subject to the initial condition given by the step function (5.3) are shown in Fig. 5.7a at time $t = 2.4$ on the subdomain $0.5 \leq x \leq 1$. This solution was calculated using a Courant number of 0.5 and $\Delta x = 0.02$. Although the strength of the fourth-derivative filter is maximized, spurious ripples still appear behind the leading edge of the jump, implying that (5.21) is neither TVD nor monotonicity preserving. The failure of this attempt to create a second-order monotonicity-preserving method could have been predicted on the basis of the theorem by Godunov (1959), who showed that any linear monotonicity-preserving method is at most first-order accurate.

Godunov's theorem implies that the only way to construct higher-order TVD schemes is through the use of nonlinear finite-difference formulae. Several such nonlinear schemes will be considered in the following sections. In most cases these methods combine some information from a higher-order finite-difference approximation with the smooth solution from a monotone first-order scheme in an attempt to maintain the sharpness of the numerically simulated discontinuity without developing spurious ripples.

In one-dimensional problems the first-order monotone solution is generally computed using upstream differencing because it is superior to most other simple monotone schemes. Figure 5.7b illustrates the superiority of upstream solutions to the advection equation (5.18) over those obtained using the Lax–Fredrichs method

$$\frac{\phi_j^{n+1} - \frac{1}{2}(\phi_{j+1}^n + \phi_{j-1}^n)}{\Delta t} + \left(\frac{c\phi_{j+1}^n - c\phi_{j-1}^n}{2\Delta x}\right) = 0.$$

As in the leapfrog simulation shown in Fig. 5.7a, the initial condition was specified by the step function (5.3), and both solutions were calculated using a Courant number of 0.5 and $\Delta x = 0.02$. Although the Lax–Fredrichs method is monotone, it does generate a spurious $2\Delta x$ stair step in the solution shown in Fig. 5.7b. This $2\Delta x$ perturbation arises from the discontinuity in the initial data and disappears if the initial width of the jump is increased from a single grid interval to $2\Delta x$. Nevertheless, the Lax–Fredrichs scheme diffuses smooth solutions more rapidly than the upstream scheme (see Problem 5), and in spite of the stair step, the long-wavelength components in the Lax–Fredrichs solution are more strongly damped than those in the upstream solution.

5.3   Discontinuities in Geophysical Fluid Dynamics

Although hydraulic jumps can develop from smooth initial conditions in shallow-water flow and fronts can form in association with mid-latitude low-pressure systems, true dynamical discontinuities do not develop from smooth initial data in most other geophysical problems. Geophysically significant motions in a continuously stratified fluid can be well described by filtered sets of equations, such as the Boussinesq system (see Section 1.2). In contrast to the shallow-water system, these filtered equations do not form a hyperbolic system, their linear wave solutions are dispersive, and their nonlinear solutions do not form strong shocks.

Nevertheless, scale contraction does frequently occur in geophysical flows as the result of stretching and shearing deformation by the velocity field. The kinematic effects of flow deformation on an initially circular distribution of a passive tracer are illustrated in Fig. 5.8. As the scale of the tracer distribution shrinks in the direction perpendicular to the axis of dilatation, the concentration field will eventually become difficult to resolve adequately on a given numerical mesh, but a true discontinuity never develops in any finite time. The only discontinuities

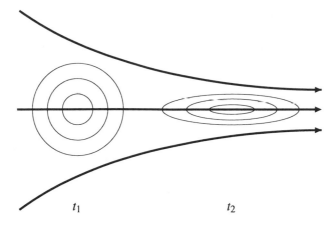

FIGURE 5.8. Deformation of a tracer field in a confluent flow field from a circular pattern at t_1 into a cigar shape at t_2.

supported by the advection equation are a special type of shock known as a contact discontinuity, in which a preexisting discontinuity is simply carried along by the moving fluid. True discontinuities can be generated from initially smooth data at atmospheric fronts (Hoskins and Bretherton 1972), but even in this case the processes producing the scale collapse in the frontal zone are primarily advective.

One might suppose that tracer transport in the scale-contracting flow illustrated in Fig. 5.8 is described by a conservation law of the form

$$\frac{\partial \psi}{\partial t} + \frac{\partial}{\partial x} f(\psi) + \frac{\partial}{\partial y} g(\psi) = 0, \tag{5.22}$$

which is the generalization of (5.6) to two dimensions. In fact, the local rate of change in the mass of a tracer transported by a two-dimensional flow is described by a slightly different conservation law,

$$\frac{\partial \psi}{\partial t} + \frac{\partial}{\partial x} (u\psi) + \frac{\partial}{\partial y} (v\psi) = 0. \tag{5.23}$$

In contrast to (5.22), the fluxes in (5.23) are not determined solely by ψ, but depend on velocity components that are functions of the independent variables x, y, and t.

The conservation laws (5.22) and (5.23) do, nevertheless, have a number of common properties. In particular, pairs of entropy-consistent solutions ψ and ϑ to either (5.22) or (5.23) share the property that if $\psi(x, y, 0) \geq \vartheta(x, y, 0)$ for all x and y at some initial time 0, then $\psi(x, y, t) \geq \vartheta(x, y, t)$ for all x and y, and all $t \geq 0$. If approximate numerical solutions to either (5.22) or (5.23) are computed with a monotone scheme, those solutions have same property, i.e., if $\phi_{i,j}^0 \geq \theta_{i,j}^0$ for all i and j, then $\phi_{i,j}^n \geq \theta_{i,j}^n$ for all i, j, and n. The special case $\theta_{i,j}^0 = 0$ is particularly important, since it implies that monotone schemes will not generate spurious negative values from nonnegative initial data. More generally, monotone schemes yield numerical solutions to either (5.22) or (5.23) that are free from spurious ripples in the vicinity of discontinuities and poorly resolved gradients. This is perhaps the most useful property of monotone approximations to the tracer transport equation, since unlike (5.22), (5.23) is a linear partial differential equation whose weak solutions are uniquely determined by the initial and boundary data. There is therefore no need to employ monotone schemes (or to demand satisfaction of some entropy condition) in order to ensure that consistent, conservation-form approximations to (5.23) converge to the correct solution as Δx, Δy, and Δt approach zero.

As discussed previously, monotone methods are not actually used in most practical applications because they are only first-order accurate and highly diffusive. In regions where the solution is smooth, more accurate approximations to the one-dimensional conservation law (5.6) can be obtained using TVD methods. One might hope to pursue the same strategy in designing approximations to the two-dimensional tracer transport equation, but there are difficulties. The first problem is that except for special cases of no practical importance, all TVD approximations to the two-dimensional nonlinear conservation law (5.22) are at most first-order accurate (Goodman and LeVeque 1985). Thus, in contrast to the one-dimensional case, there are no second-order accurate TVD approximations to (5.22).

The second and more fundamental problem is that although the entropy-consistent solution to the nonlinear conservation law (5.22) is TVD (or, more precisely, total variation nonincreasing), the total variation in the true solution to the tracer transport equation can increase with time—even when the velocity field is non-divergent! The non-TVD nature of the solutions to (5.23) follows from the circumstance that the total variation of $\psi(x, y)$ is not invariant under coordinate rotations.[3] The total variation of a two-dimensional function is conventionally defined as

$$TV(\psi) = \limsup_{\epsilon \to 0} \frac{1}{\epsilon} \int_{-\infty}^{\infty} \int_{-\infty}^{\infty} |\psi(x + \epsilon, y) - \psi(x, y)| \, dx \, dy$$
$$+ \limsup_{\epsilon \to 0} \frac{1}{\epsilon} \int_{-\infty}^{\infty} \int_{-\infty}^{\infty} |\psi(x, y + \epsilon) - \psi(x, y)| \, dx \, dy.$$

Suppose that the initial conditions for (5.23) are

$$\psi(x, y, 0) = \begin{cases} 1 & \text{if } |x| \leq 1 \text{ and } |y| \leq 1, \\ 0 & \text{otherwise,} \end{cases}$$

and that the flow is in solid-body rotation with $u = -y$ and $v = x$. After the distribution of ψ rotates through an angle of $45°$, its total variation will increase by

---

[3]The nonconservation of the total variation under coordinate rotations, which was pointed out to this author by Joe Tenerelli, appears to be a weakness in the mathematical definition of the total variation of a two-dimensional function. A second weakness appears in the physical units that are associated with total variation. If ψ and φ are functions in degrees and x and y are spatial coordinates in meters, then $TV(\psi(x))$ has units of degrees, whereas $TV(\varphi(x, y))$ has units of meters times degrees.

a factor of $\sqrt{2}$. Accurate finite-volume approximations to (5.23) cannot therefore be TVD. Useful schemes for the simulation of tracer transport can nevertheless be obtained by borrowing techniques used to generate TVD approximations to the one-dimensional conservation law (5.6).

Instead of demanding that the scheme be TVD, it is possible to control the development of spurious oscillations by regulating the behavior of the local maxima and minima in the solution. Smooth solutions to the nonlinear conservation law (5.22) also satisfy the advective-form equation

$$\frac{\partial \psi}{\partial t} + \frac{df}{d\psi}\frac{\partial \psi}{\partial x} + \frac{dg}{d\psi}\frac{\partial \psi}{\partial y} = 0. \qquad (5.24)$$

If the velocity field is nondivergent, (5.23) may be written in an analogous advective form,[4]

$$\frac{\partial \psi}{\partial t} + u\frac{\partial \psi}{\partial x} + v\frac{\partial \psi}{\partial y} = 0. \qquad (5.25)$$

Solutions to both (5.24) and (5.25) conserve the amplitude of all local maxima and minima in the initial data. Flux-corrected transport algorithms, which will be the considered in the next section, exploit this property of the true solution to control the development of ripples near a discontinuity.

The remainder of this chapter will be primarily devoted to the examination of methods for the simulation of discontinuities or poorly resolved gradients in nondivergent flow. The first topic considered is one-dimensional nondivergent flow, which can occur only if the velocity is constant. The one-dimensional constant-wind-speed advection equation is also a member of the family of autonomous conservation laws of the form (5.6). As a consequence, the study of the constant-wind-speed advection equation serves as an introduction to both fluid transport problems of the form (5.23) and nonlinear hyperbolic conservation laws of the form (5.22). The extension of these results to nonuniform two-dimensional flow is discussed in Section 5.7. The extension of the one-dimensional constant-wind-speed problem to nonlinear systems of conservation laws, which is beyond the scope of this text, is discussed in LeVeque (1992) and Godlewski and Raviart (1996).

5.4   Flux-Corrected Transport

Flux-corrected transport, or FCT, was proposed by Boris and Book (1973) as a way of approximating a conservation law with a high-order scheme in regions where the solution is smooth while using a low-order monotone scheme where

---

[4]Equation 5.25 states that the tracer concentration (typically expressed as a dimensionless ratio, such as grams per kilogram or parts per billion) is conserved following the motion of each fluid parcel. In contrast, (5.23) states that the local rate of change of the mass of the tracer at a fixed point in space is determined by the divergence of the tracer mass-flux at that point.

the solution is poorly resolved or discontinuous. The concept of FCT and the algorithms for its implementation were further generalized by Zalesak (1979). Zalesak suggested approximating the scalar conservation law (5.6) with a finite-difference formula in the conservation form

$$\frac{\phi_j^{n+1} - \phi_j^n}{\Delta t} + \left(\frac{F_{j+\frac{1}{2}} - F_{j-\frac{1}{2}}}{\Delta x} \right) = 0 \tag{5.26}$$

and then computing the fluxes $F_{j\pm\frac{1}{2}}$ in several steps as follows.

1. Compute a set of low-order fluxes $F_{j+\frac{1}{2}}^l$ using a monotone scheme.

2. Compute a set of high-order fluxes $F_{j+\frac{1}{2}}^h$ using a high-order scheme.

3. Compute the *antidiffusive fluxes*

$$A_{j+\frac{1}{2}} = F_{j+\frac{1}{2}}^h - F_{j+\frac{1}{2}}^l.$$

4. Compute a monotone estimate of the solution at $(n+1)\Delta t$ (also known as the "transported and diffused" solution),

$$\phi_j^{td} = \phi_j^n - \frac{\Delta t}{\Delta x} \left(F_{j+\frac{1}{2}}^l - F_{j-\frac{1}{2}}^l \right).$$

5. Correct the $A_{j+\frac{1}{2}}$ so that the final "antidiffusion" step does not generate new maxima or minima. The correction procedure may be expressed mathematically by defining

$$A_{j+\frac{1}{2}}^c = C_{j+\frac{1}{2}} A_{j+\frac{1}{2}}, \qquad 0 \le C_{j+\frac{1}{2}} \le 1.$$

The procedure for computing $C_{j+\frac{1}{2}}$ will be discussed shortly.

6. Perform the "antidiffusion" step

$$\phi_j^{n+1} = \phi_j^{td} - \frac{\Delta t}{\Delta x} \left(A_{j+\frac{1}{2}}^c - A_{j-\frac{1}{2}}^c \right).$$

If all the $C_{j+\frac{1}{2}}$ were unity, the preceding algorithm would give results identical to the higher-order scheme, and if all the $C_{j+\frac{1}{2}}$ were zero, the solution would be identical to that obtained with the monotone scheme. Criteria for determining $C_{j+\frac{1}{2}}$ are usually designed to prevent the development of new maxima and minima and to prohibit the amplification of existing extrema.

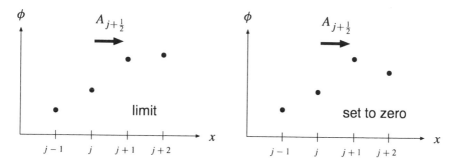

FIGURE 5.9. Two possible configurations in which an antidiffusive flux, indicated by the heavy arrow, may be modified by a flux limiter.

5.4.1   Flux Correction: The Original Proposal

Boris and Book (1973) did not actually give a formula for $C_{j+\frac{1}{2}}$, but offered the following expression for the corrected fluxes:

$$A^c_{j+\frac{1}{2}} = S_{j+\frac{1}{2}} \max \left\{ 0, \min \left[|A_{j+\frac{1}{2}}|, \right. \right.$$

$$\left. \left. S_{j+\frac{1}{2}} \left(\phi^{td}_{j+2} - \phi^{td}_{j+1} \right) \frac{\Delta x}{\Delta t}, \; S_{j+\frac{1}{2}} \left(\phi^{td}_j - \phi^{td}_{j-1} \right) \frac{\Delta x}{\Delta t} \right] \right\},$$

where

$$S_{j+\frac{1}{2}} = \text{sgn} \left(A_{j+\frac{1}{2}} \right).$$

The logic behind this formula can be understood by considering the case where $A_{j+\frac{1}{2}} \geq 0$, so that the preceding may be written as

$$\frac{\Delta t}{\Delta x} A^c_{j+\frac{1}{2}} = \max \left\{ 0, \min \left[A_{j+\frac{1}{2}} \frac{\Delta t}{\Delta x}, \; \left(\phi^{td}_{j+2} - \phi^{td}_{j+1} \right), \; \left(\phi^{td}_j - \phi^{td}_{j-1} \right) \right] \right\}.$$

(5.27)

Two typical configurations for ϕ_j are shown in Fig. 5.9; observe that in both cases the antidiffusive flux is directed up-gradient, which is the most common situation. The increase in ϕ^n_{j+1}, and the decrease in ϕ^n_j, that would be produced by the uncorrected antidiffusive flux $A_{j+\frac{1}{2}}$ is given by the first argument of the "min" function in (5.27). The second argument of the min function ensures that the corrected flux is not large enough to generate a new maximum by rendering $\phi_{j+1} > \phi_{j+2}$. This type of limitation on the antidiffusive flux would apply in the case shown in the left panel of Fig. 5.9. If ϕ_{j+1} is already greater than ϕ_{j+2}, the antidiffusive flux is zeroed to prevent the amplification of the preexisting extrema at ϕ_{j+1}; this situation is illustrated in the right panel in Fig. 5.9. The third argument of the min function ensures that no new minima are created and that no preexisting minima are amplified at ϕ_j.

5.4.2   The Zalesak Corrector

Zalesak (1979) noted that the preceding algorithm limits each antidiffusive flux
without considering the action of the antidiffusive fluxes at neighboring grid points
and that this can lead to an unnecessarily large reduction in the antidiffusive flux.
For example, although the antidiffusive flux shown in the left panel of Fig. 5.9
will tend to decrease ϕ_j, this decrease is likely to be partly compensated by an
up-gradient antidiffusive flux directed from grid point $j - 1$ into grid point j.
Zalesak proposed the following algorithm, which considers the net effect of both
antidiffusive fluxes in order to minimize the correction to those fluxes and thereby
keep the algorithm as close as possible to that which would be obtained using the
higher-order scheme.

1. As an optional preliminary step, set certain down-gradient antidiffusive
   fluxes to zero, such that

$$A_{j+\frac{1}{2}} = 0, \qquad \text{if} \qquad A_{j+\frac{1}{2}}(\phi_{j+1}^{\text{td}} - \phi_j^{\text{td}}) < 0$$

$$\text{and either} \quad A_{j+\frac{1}{2}}(\phi_{j+2}^{\text{td}} - \phi_{j+1}^{\text{td}}) < 0$$

$$\text{or} \quad A_{j+\frac{1}{2}}(\phi_j^{\text{td}} - \phi_{j-1}^{\text{td}}) < 0. \qquad (5.28)$$

Zalesak refers to this as a cosmetic correction, and it is usually omitted. This
cosmetic correction has, nevertheless, been used in the FCT computations
shown in this chapter. It has no effect on the solution shown in Fig. 5.10a,
makes a minor improvement in the solution shown in Fig. 5.10b, and makes
a major improvement in the solution shown in Fig. 5.13b.

2. Evaluate the range of permissible values for ϕ_j^{n+1}:

$$\phi_j^{\text{max}} = \max\left(\phi_{j-1}^n, \phi_j^n, \phi_{j+1}^n, \phi_{j-1}^{\text{td}}, \phi_j^{\text{td}}, \phi_{j+1}^{\text{td}}\right),$$

$$\phi_j^{\text{min}} = \min\left(\phi_{j-1}^n, \phi_j^n, \phi_{j+1}^n, \phi_{j-1}^{\text{td}}, \phi_j^{\text{td}}, \phi_{j+1}^{\text{td}}\right).$$

If the flow is nondivergent, the ϕ^{td} are not needed in the preceding formulae
because the extrema predicted by the monotone scheme will be of lower
amplitude than those at the beginning of the time step. If, however, the flow
is divergent, then the local minima and maxima in the true solution may be
increasing, and the increase predicted by the monotone scheme should be
considered in determining ϕ^{max} and ϕ^{min}.

3. Compute the sum of all antidiffusive fluxes into grid point j,

$$P_j^+ = \max\left(0, A_{j-\frac{1}{2}}\right) - \min\left(0, A_{j+\frac{1}{2}}\right).$$

4. Compute the maximum net antidiffusive flux that will preserve $\phi_j^{n+1} \leq \phi_j^{\text{max}}$,

$$Q_j^+ = \left(\phi_j^{\text{max}} - \phi_j^{\text{td}}\right)\frac{\Delta x}{\Delta t}.$$

5. Compute the required limitation on the net antidiffusive flux into grid point j,

$$R_j^+ = \begin{cases} \min\left(1, Q_j^+/P_j^+\right) & \text{if } P_j^+ > 0, \\ 0 & \text{if } P_j^+ = 0. \end{cases}$$

6. Compute the corresponding quantities involving the net antidiffusive flux out of grid point j,

$$P_j^- = \max\left(0, A_{j+\frac{1}{2}}\right) - \min\left(0, A_{j-\frac{1}{2}}\right).$$

$$Q_j^- = \left(\phi_j^{td} - \phi_j^{min}\right)\frac{\Delta x}{\Delta t}.$$

$$R_j^- = \begin{cases} \min\left(1, Q_j^-/P_j^-\right) & \text{if } P_j^- > 0, \\ 0 & \text{if } P_j^- = 0. \end{cases}$$

7. Limit the antidiffusive flux so that it neither produces an overshoot in the grid cell into which it is directed nor generates an undershoot in the grid cell out of which it flows:

$$C_{j+\frac{1}{2}} = \begin{cases} \min\left(R_{j+1}^+, R_j^-\right) & \text{if } A_{j+\frac{1}{2}} \geq 0, \\ \min\left(R_j^+, R_{j+1}^-\right) & \text{if } A_{j+\frac{1}{2}} < 0. \end{cases}$$

Two examples illustrating the performance of the Zalesak FCT algorithm on the constant-wind-speed one-dimensional advection equation are shown in Fig. 5.10. In these examples, the monotone flux is computed using the upstream method with

$$F_{j+\frac{1}{2}}^l = \frac{c}{2}(\phi_j + \phi_{j+1}) - \frac{|c|}{2}(\phi_{j+1} - \phi_j), \tag{5.29}$$

and the high-order flux is computed using the flux form of the Lax–Wendroff method such that

$$F_{j+\frac{1}{2}}^h = \frac{c}{2}(\phi_j + \phi_{j+1}) - \frac{c^2\Delta t}{2\Delta x}(\phi_{j+1} - \phi_j). \tag{5.30}$$

The calculations were performed in a wide periodic domain, only the center portion of which is shown in each figure. In each case the wind speed is constant, and the Courant number is 0.5.

The curves shown in Fig. 5.10a are solutions to the same traveling-jump problem considered in connection with Fig. 5.7, except that the solutions are plotted at $t = 1.8$. The solution computed using FCT is shown by the solid line. Also shown are the exact solution and the approximate solutions obtained using upstream differencing and using the Lax–Wendroff method without FCT. The FCT scheme is almost identical to the uncorrected Lax–Wendroff method except near the top of the step, where the flux-correction procedure completely eliminates the dispersive

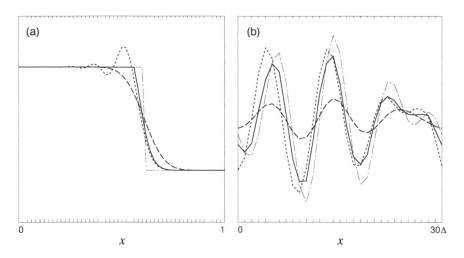

FIGURE 5.10. Results from two constant-wind-speed advection tests: exact solution (thin dash-dotted) and numerical solutions obtained with the Zalesak FCT combination of upstream and Lax–Wendroff differencing (solid), upstream differencing (long dashed), and the Lax–Wendroff scheme (short-dashed).

ripples apparent in the uncorrected Lax–Wendroff solution. The FCT scheme is not only superior to the higher-order scheme, it also captures the steepness of the jump much better than the upstream method.

The curves in Fig. 5.10b show solutions to the test problem considered in Chapter 2 in which the sum of equal-amplitude $7.5\Delta x$ and $10\Delta x$ waves is advected over a distance of twelve grid points (cf. Fig. 2.13). Once again the FCT solution is clearly superior to that obtained using upstream differencing. Although this test case does not involve shocks or discontinuities, the FCT solution remains roughly comparable in quality to that obtained by the uncorrected Lax–Wendroff method. In particular, the FCT solution exhibits more damping but less phase-speed error than that obtained with the Lax–Wendroff method.

As suggested by the preceding tests, the FCT approach allows one to obtain ripple-free solutions that are far superior to those computed by simple upstream differencing. One might attempt to obtain further improvements by computing the high-order flux using an extremely accurate method. Zalesak (1979), for example, gives a formula for an eighth-order-accurate method. Such very high order formulae are seldom used in practical applications. In part, this may be due to the unattractive compromises required to use high-order formulae with forward-in-time differencing. The more fundamental problem, however, is that the scheme reduces to a monotone method near any maxima and minima and is therefore only first-order accurate near extrema. The empirically estimated order of accuracy of the preceding FCT scheme is less than two and is unlikely to be substantially improved by using a more accurate scheme to compute the high-order flux (see Table 5.1 in Section 5.5.2).

5.4.3   Iterative Flux Correction

Substantial improvements in the neighborhood of smooth well-resolved extrema can, nevertheless, be achieved by using a better estimate for the low-order solution. One strategy for obtaining a better low-order solution is to reuse the standard FCT solution in an iterative application of the flux-correction procedure (Schär and Smolarkiewicz 1996). As a consequence of the flux-correction algorithm, the standard FCT solution is free from spurious ripples and can serve as an improved estimate for the "transported and diffused solution" in a second iteration. That portion of the antidiffusive flux that was not applied in the first iteration is the maximum antidiffusive flux available for application in the second iteration. Letting the tilde denote a quantity defined for use in the second iteration, the final step of the first iteration becomes

$$\tilde{\phi}_j^{\mathrm{td}} = \phi_j^{\mathrm{td}} - \frac{\Delta t}{\Delta x}\left(A^{\mathrm{c}}_{j+\frac{1}{2}} - A^{\mathrm{c}}_{j-\frac{1}{2}}\right),$$

and the new antidiffusive flux becomes

$$\tilde{A}_{j+\frac{1}{2}} = A_{j+\frac{1}{2}} - A^{\mathrm{c}}_{j+\frac{1}{2}}.$$

The antidiffusive flux is limited using precisely the same flux-correction algorithm used in the first iteration, and the final estimate for ϕ^{n+1} is obtained using

$$\phi_j^{n+1} = \tilde{\phi}_j^{\mathrm{td}} - \frac{\Delta t}{\Delta x}\left(\tilde{A}^{\mathrm{c}}_{j+\frac{1}{2}} - \tilde{A}^{\mathrm{c}}_{j-\frac{1}{2}}\right).$$

This iteration can be very effective in improving the solution near well-resolved extrema such as the crest of a sine wave, but it does not noticeably improve the solution near a discontinuous step.

5.5   Flux-Limiter Methods

The strategy behind flux-limiter methods is similar to that underlying flux-corrected transport in that the numerical fluxes used in both methods are a weighted sum of the fluxes computed by a monotone first-order scheme and a higher-order method. In flux-limiter methods, however, the limiter that apportions the flux between the high- and low-order schemes is determined without actually computing a low-order solution (ϕ^{td}). This limiter is expressed as a function of the local solution at the previous time level in a manner guaranteeing that the scheme generates TVD approximations to the one-dimensional scalar conservation law (5.6) and that the scheme is second-order accurate except in the vicinity of the extrema of ϕ.

Flux limiter methods approximate (5.6) with a finite-difference scheme in the conservation form (5.26) using the flux

$$F_{j+\frac{1}{2}} = F^{\mathrm{l}}_{j+\frac{1}{2}} + C_{j+\frac{1}{2}}\left(F^{\mathrm{h}}_{j+\frac{1}{2}} - F^{\mathrm{l}}_{j+\frac{1}{2}}\right),$$

where F^l and F^h denote the fluxes obtained using monotone and high-order schemes, and $C_{j+\frac{1}{2}}$ is a multiplicative limiter. As in the FCT algorithm discussed previously, the high-order flux is recovered when $C_{j+\frac{1}{2}} = 1$, and the performance of the scheme is highly dependent on the algorithm for specifying $C_{j+\frac{1}{2}}$. We again demand that $C_{j+\frac{1}{2}} \geq 0$, but as will become evident, it is advantageous to allow $C_{j+\frac{1}{2}}$ to exceed unity. In scalar problems in which the phase speed of the disturbance is greater than zero,[5] $C_{j+\frac{1}{2}}$ is calculated as a nonlinear function of the local solution $C(r_{j+\frac{1}{2}})$, where

$$r_{j+\frac{1}{2}} = \frac{\phi_j - \phi_{j-1}}{\phi_{j+1} - \phi_j}$$

is the ratio of the slope of the solution across the cell interface upstream of $j+\frac{1}{2}$ to the slope of the solution across the interface at $j+\frac{1}{2}$. The parameter $r_{j+\frac{1}{2}}$ is approximately unity where the numerical solution is smooth and is negative when there is a local maximum or minimum immediately upstream of the cell interface at $j+\frac{1}{2}$.

5.5.1 Ensuring That the Scheme Is TVD

Criteria guaranteeing that a flux-limiter method is TVD may be obtained by noting that a finite-difference scheme of the form

$$\phi_j^{n+1} = \phi_j^n - G_{j-\frac{1}{2}}\left(\phi_j^n - \phi_{j-1}^n\right) + H_{j+\frac{1}{2}}\left(\phi_{j+1}^n - \phi_j^n\right) \tag{5.31}$$

will be TVD provided that for all j

$$0 \leq G_{j+\frac{1}{2}}, \quad 0 \leq H_{j+\frac{1}{2}}, \quad \text{and} \quad G_{j+\frac{1}{2}} + H_{j+\frac{1}{2}} \leq 1 \tag{5.32}$$

(Harten 1983). This may be verified by observing that (5.31) and (5.32) imply

$$\left|\phi_{j+1}^{n+1} - \phi_j^{n+1}\right| \leq \left(1 - G_{j+\frac{1}{2}} - H_{j+\frac{1}{2}}\right)\left|\phi_{j+1}^n - \phi_j^n\right|$$
$$+ G_{j-\frac{1}{2}}\left|\phi_j^n - \phi_{j-1}^n\right| + H_{j+\frac{3}{2}}\left|\phi_{j+2}^n - \phi_{j+1}^n\right|.$$

Summing over all j and shifting the dummy index in the last two summations yields

$$\sum_j \left|\phi_{j+1}^{n+1} - \phi_j^{n+1}\right| \leq \sum_j \left(1 - G_{j+\frac{1}{2}} - H_{j+\frac{1}{2}}\right)\left|\phi_{j+1}^n - \phi_j^n\right|$$
$$+ \sum_j G_{j+\frac{1}{2}}\left|\phi_{j+1}^n - \phi_j^n\right| + \sum_j H_{j+\frac{1}{2}}\left|\phi_{j+1}^n - \phi_j^n\right|$$
$$= \sum_j \left|\phi_{j+1}^n - \phi_j^n\right|.$$

---

[5]The general case, in which the phase speed is either positive or negative, is discussed in Section 5.5.3.

Sweby (1984) presented a systematic derivation of the possible functional forms for $C(r)$ that yield TVD flux-limited methods when the monotone scheme is upstream differencing and the high-order scheme is a member of a family of second-order methods that includes the Lax–Wendroff and Warming–Beam methods. Suppose that the constant-wind-speed advection equation (5.18) is approximated using the flux form of the Lax–Wendroff method and that $c > 0$. The Lax–Wendroff flux (5.30) can be expressed as

$$F^{LW}_{j+\frac{1}{2}} = c\phi_j + \frac{c}{2}(1 - \mu)(\phi_{j+1} - \phi_j),$$

where $\mu = c\Delta t/\Delta x$. The first term of the preceding is the numerical flux for upstream differencing (in a flow with $c > 0$). The second term is an increment to the upstream flux that can be multiplied by $C_{j+\frac{1}{2}}$ to obtain the "limited" flux

$$F_{j+\frac{1}{2}} = c\phi_j + \frac{c}{2}(1 - \mu)(\phi_{j+1} - \phi_j)C_{j+\frac{1}{2}}. \tag{5.33}$$

The finite-difference scheme obtained after evaluating the divergence of these limited fluxes may be written

$$\phi^{n+1}_j = \phi^n_j - \left[\mu - \frac{\mu}{2}(1 - \mu)C_{j-\frac{1}{2}}\right]\left(\phi^n_j - \phi^n_{j-1}\right)$$
$$- \frac{\mu}{2}(1 - \mu)C_{j+\frac{1}{2}}\left(\phi^n_{j+1} - \phi^n_j\right). \tag{5.34}$$

In order to arrive at a scheme that is TVD, one natural approach would be to choose

$$G_{j-\frac{1}{2}} = \mu - \frac{\mu}{2}(1 - \mu)C_{j-\frac{1}{2}},$$

$$H_{j+\frac{1}{2}} = -\frac{\mu}{2}(1 - \mu)C_{j+\frac{1}{2}}$$

and attempt to determine a function $C(r_{j+\frac{1}{2}}) \equiv C_{j+\frac{1}{2}}$ that will guarantee satisfaction of (5.32). Unfortunately, this approach fails, since by assumption, $C(r_{j+\frac{1}{2}}) \geq 0$, and thus $H_{j+\frac{1}{2}} < 0$ whenever the Courant number falls in the range $0 \leq \mu \leq 1$.

As an alternative Sweby suggested setting

$$G_{j-\frac{1}{2}} = \mu + \frac{\mu}{2}(1 - \mu)\left[C_{j+\frac{1}{2}}\left(\frac{\phi^n_{j+1} - \phi^n_j}{\phi^n_j - \phi^n_{j-1}}\right) - C_{j-\frac{1}{2}}\right],$$

$$H_{j+\frac{1}{2}} = 0.$$

Then the TVD criteria (5.32) will be satisfied if

$$0 \leq G_{j-\frac{1}{2}} \leq 1$$

for all j, or equivalently, if

$$0 \leq \mu\left[1 + \frac{1}{2}(1 - \mu)\left(\frac{C_{j+\frac{1}{2}}}{r_{j+\frac{1}{2}}} - C_{j-\frac{1}{2}}\right)\right] \leq 1.$$

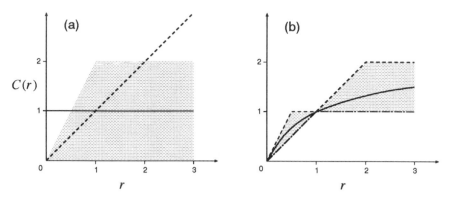

FIGURE 5.11. (a) Shading indicates the region in which $C(r)$ must lie to give a TVD method. Heavy lines indicate $C^{LW}(r)$ (solid) and $C^{WB}(r)$ (dashed). (b) Shading indicates the region in which $C(r)$ must lie to give a TVD scheme that is an internal average of the methods Lax–Wendroff and Warming–Beam. Three possible limiters are also indicated: the "superbee" (dashed), minmod (dot-dashed), and Van Leer (solid).

If the CFL condition ($0 \leq \mu \leq 1$) holds for the upstream scheme, the criteria for the method to be TVD reduce to

$$\frac{-2}{1-\mu} \leq \frac{C_{j+\frac{1}{2}}}{r_{j+\frac{1}{2}}} - C_{j-\frac{1}{2}} \leq \frac{2}{\mu},$$

or

$$\left| \frac{C_{j+\frac{1}{2}}}{r_{j+\frac{1}{2}}} - C_{j-\frac{1}{2}} \right| \leq 2.$$

Suppose that $r_{j+\frac{1}{2}} > 0$. Then since $C(r)$ is assumed to be nonnegative, the preceding inequality is satisfied when

$$0 \leq \frac{C(r)}{r} \leq 2 \quad \text{and} \quad 0 \leq C(r) \leq 2. \tag{5.35}$$

Now consider the case $r_{j+\frac{1}{2}} \leq 0$. Negative values of r occur at the local maxima and minima of ϕ_j, where the flux must be completely determined by the monotone upstream method in order to avoid increasing the total variation; it is therefore necessary[6] to choose $C(r) = 0$ when $r < 0$. Note that the condition $C(r) = 0$ when $r < 0$ is implicitly included in the inequalities (5.35).

The inequalities (5.35) define the shaded region of the (r, C)-plane shown in Fig. 5.11a, which is the locus of all curves $C(r)$ that make the flux-limited method TVD. The range of possible choices for $C(r)$ can be further restricted if it is

---

[6]Although it is necessary to choose $C(r) = 0$ when $r < 0$ to keep the scheme TVD, this is actually a poor choice if the solution is smooth and well-resolved in the vicinity of the extremum. Well-resolved extrema would be captured more accurately using the higher-order scheme.

required that the method be second-order accurate whenever $r > 0$. Noting that $C_{j-\frac{1}{2}}$ depends on the value of ϕ_{j-2}, the flux-limited scheme (5.34) has the form

$$\phi_j^{n+1} = H(\phi_{j-2}^n, \phi_{j-1}^n, \phi_j^n, \phi_{j+2}^n).$$

All second-order approximations to the advection problem that have the preceding form are weighted averages of the methods of Lax–Wendroff and of Warming and Beam. As discussed previously, the flux-limited scheme becomes the Lax–Wendroff method if $C(r) = 1$. In a similar way, specifying $C(r) = r$ converts the scheme to the method of Warming and Beam (2.109). Curves corresponding to

$$C^{LW}(r) = 1 \quad \text{and} \quad C^{WB}(r) = r$$

are also plotted in Fig. 5.11a. Of course, $C^{LW}(r)$ and $C^{WB}(r)$ do not lie entirely within the shaded TVD region because neither the Lax–Wendroff method nor that of Warming and Beam are TVD. Nevertheless, in order to make the flux-limited scheme second-order accurate away from local maxima and minima (i.e., for $r > 0$), $C(r)$ must be a weighted average of $C^{LW}(r)$ and $C^{WB}(r)$. Sweby suggests that the best results are obtained if this weighted average is an internal average such that

$$C(r) = [1 - \theta(r)]C^{LW}(r) + \theta(r)C^{WB}(r), \tag{5.36}$$

where $0 \le \theta(r) \le 1$. This portion of the total TVD region is indicated by the shaded area in Fig. 5.11b, which will be referred to as the "second-order" TVD region, although the true second-order TVD region includes external averages of the Lax–Wendroff and Warming–Beam methods and is larger than the shaded area in Fig. 5.11b.

5.5.2 Possible Flux Limiters

Possible choices for the specific functional form of $C(r)$ that yield a TVD method satisfying (5.36) include the "minmod" limiter

$$C(r) = \max[0, \min(1, r)], \tag{5.37}$$

which is the dot-dashed curve following the lower boundary of the second-order TVD region in Fig. 5.11b; the "superbee" limiter (Roe 1985)

$$C(r) = \max[0, \min(1, 2r), \min(2, r)], \tag{5.38}$$

which lies along the upper boundary of the second-order TVD region; and the van Leer limiter (van Leer 1974)

$$C(r) = \frac{r + |r|}{1 + |r|}, \tag{5.39}$$

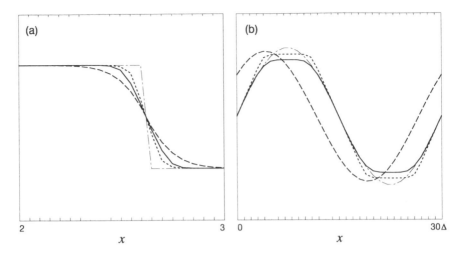

FIGURE 5.12. Comparison of flux-limited approximations using the superbee (short dash) and MC (solid) limiters with (a) the minmod limiter (long dash) in a case with a propagating step, and (b) the Lax–Wendroff solution (dashed) in a case with a well-resolved sinusoidal distribution. The exact solution is shown by the thin dot-dashed line.

which is the smooth curve in Fig. 5.11b. Also of note, but not plotted, is the monotonized centered, or "MC," limiter (van Leer 1977)

$$C(r) = \max\left[0, \min\left(2r, \frac{1+r}{2}, 2\right)\right]. \qquad (5.40)$$

The performance of several different limiters is compared in Fig. 5.12. Figure 5.12a shows results from the same test problem considered in Fig. 5.10a except that the horizontal grid size is reduced from $1/50$ to $1/20$ and the solution is displayed at time 7.8 in order to better reveal small differences between the various solutions. Inspection of Fig. 5.12a shows that the minmod limiter allows the most numerical diffusion, the superbee allows the least, and the MC limiter performs almost as well as the superbee. Although the superbee limiter works best on the example shown in Fig. 5.12a, the MC limiter may be the best choice for general applications. The weakness of the superbee limiter is illustrated in Fig. 5.12b, which shows flux-limited and Lax–Wendroff approximations to a problem whose correct solution is a unit-amplitude sine wave propagating to the right at speed $1/10$ on the periodic domain $0 \le x \le 1$. In this example $\Delta x = 1/30$, the Courant number is $\frac{1}{2}$ and the solution is shown at $t = 200$, at which point the initial distribution has made 20 circuits around the periodic domain. The superbee and MC limiters clearly flatten the crests and troughs in the flux-limited approximation to this well-resolved sine wave. As the superbee limiter flattens the crests and troughs it incorrectly amplifies the solution near the edges of the flattened extrema, but no such spurious amplification is generated by the MC limiter; the MC-limited solution remains within the envelope of the true solution. Although the flux-limited solutions show distortion in the peaks and troughs, they are al-

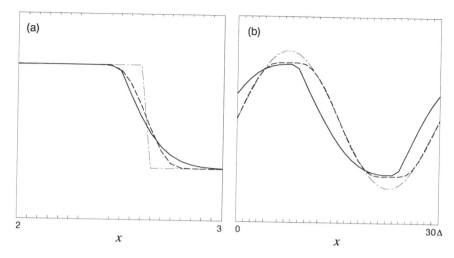

FIGURE 5.13. Comparison of MC flux-limited (long-dashed line) and FCT (solid line) approximate solutions with the exact solution (thin dashed-dotted line) for the two test cases shown in Fig. 5.12.

most completely free from phase-speed error, and as a consequence the overall character of the flux-limited solution is superior to that obtained with the Lax–Wendroff method (i.e., with no limiter), which exhibits a substantial phase lag and very modest damping.

Figure 5.13 shows a comparison of the MC flux-limited scheme with the Zalesak flux-corrected transport algorithm discussed in Section 5.4.3. The two test problems are identical to those just considered in Fig. 5.12. Both methods are implemented using the same methods to evaluate the monotone and high-order fluxes (specifically upstream differencing and the Lax–Wendroff method). The solutions obtained with the MC flux-limited method are clearly superior to those obtained using the FCT scheme. The tendency of the FCT scheme to deform the sine wave into a sawtooth can, however, be eliminated using a second iterative pass of the FCT algorithm as discussed at the end of Section 5.4.2 (Schär and Smolarkiewicz 1996, Fig. 4).

Since flux-corrected transport and flux-limiter methods both revert to first-order schemes in the vicinity of minima and maxima, they do not give fully second-order approximations in problems like the sine-wave-advection test shown in Figs. 5.12b and 5.13b. The effective order of accuracy of these schemes can be empirically determined for the sine-wave-advection test by performing a series of simulations in which both Δx and Δt are repeatedly halved (so that all simulations are performed with the same Courant number of 0.5). Fitting a function of the form $\alpha(\Delta x)^p$ to the error as Δx decreases from 1/40 to 1/320 yields the approximate values for p listed in Table 5.1. As a check on the quality of this calculation, the empirically determined orders of accuracy for the upstream and Lax–Wendroff schemes are also listed in Table 5.1. The result for the upstream scheme is slightly in error, and could be improved by continuing the computa-

Scheme	Estimated Order of Accuracy
Upstream	0.9
Minmod Flux-Limiter	1.6
Superbee Flux-Limiter	1.6
Zalesak FCT	1.7
MC Flux-Limiter	1.9
Lax–Wendroff	2.0

TABLE 5.1. Empirically determined order of accuracy for constant-wind-speed advection of a sine wave

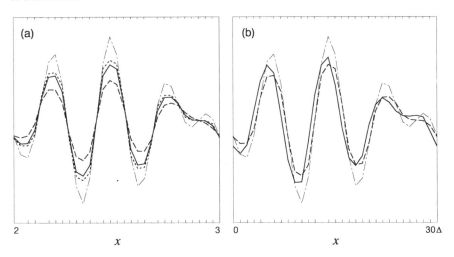

FIGURE 5.14. Comparison of numerical approximations to an advection problem whose exact solution (shown by the thin dot-dashed line) consists of equal-amplitude $7.5\Delta x$ and $10\Delta x$ waves. (a) Flux-limited approximations using the superbee (short dash), MC (solid), and minmod (long dash) limiters. (b) MC flux-limited solution (long-dashed) and the FCT solution (solid).

tion using double precision on still finer grids. The effective order of accuracy of the MC flux-limited scheme is higher than that of the other flux-limited and FCT methods. In fact, at all resolutions between $\Delta x = 1/40$ and $1/320$ the actual error computed with the MC flux-limited scheme is lower than that obtained using any of the other methods (including the Lax–Wendroff scheme).

A final example is provided by the test problem from Chapter 2 in which the initial condition is the superposition of equal-amplitude $7.5\Delta x$ and $10\Delta x$ waves. The numerical parameters for this test are identical to those described in connection with Fig. 5.10b. Figure 5.14a compares the minmod, MC, and superbee flux-limited solutions to these problems. As was the case with the propagating step considered in Fig. 5.12a, the superbee limiter gives the best results, the minmod limiter is too diffusive, and the MC limiter is almost as good as the superbee. Figure 5.14b shows a comparison of the MC flux-limited and FCT solutions to the same problem. The superiority of the MC flux-limited solution over the FCT solution is not as clear-cut as in the examples shown in Fig. 5.13. The flux-limited solution is more heavily damped but exhibits less phase-speed error than that obtained with flux-correct transport.

The performance of all the schemes shown in Figs. 5.12–5.14 is improved by increasing the Courant number towards unity, since the upstream and Lax–Wendroff methods both give perfect results when $\mu = 1$. In practical applications with a temporally and spatially varying wind field there is, however, no hope of stepping the solution forward at each grid point using a local Courant number of unity. The solutions shown in Figs. 5.10–5.14 were obtained using $\mu = 0.5$ and are similar to those obtained at smaller Courant numbers. Note that neither the FCT nor the flux-limited methods approach the accuracy obtained on the test problems in Figs. 5.13b and 5.14 using a simple explicit fourth-order spatial difference and an accurate time difference. Although FCT and flux-limiter methods are highly useful in problems with discontinuities and unresolved gradients, conventional finite-difference schemes may perform much better when the solution is at least moderately resolved. The resolution required to make a high-order finite-difference method attractive need not be particularly high; the $7.5\Delta x$-wavelength component in the solution shown in Fig. 5.14 is certainly not well-resolved.

5.5.3   Flow Velocities of Arbitrary Sign

In order to accommodate velocities of arbitrary sign, the definitions of $F_{j+\frac{1}{2}}$ and $r_{j+\frac{1}{2}}$ must be modified as follows. The Lax–Wendroff flux (5.30) may be expressed in terms of the upstream flux for advection by a velocity of arbitrary sign (5.29) as

$$F^h_{j+\frac{1}{2}} = F^l_{j+\frac{1}{2}} + \frac{|c|}{2}\left(1 - \frac{|c|\Delta t}{\Delta x}\right)\left(\phi^n_{j+1} - \phi^n_j\right). \tag{5.41}$$

The total corrected flux may therefore be expressed as

$$F^n_{j+\frac{1}{2}} = \frac{c}{2}\left(\phi^n_{j+1} + \phi^n_j\right) - \frac{1}{2}\left[(1 - C^n_{j+\frac{1}{2}})|c| + \frac{c^2\Delta t}{\Delta x}C^n_{j+\frac{1}{2}}\right]\left(\phi^n_{j+1} - \phi^n_j\right). \tag{5.42}$$

The value of $r_{j+\frac{1}{2}}$ used in the evaluation of the flux limiter $C^n_{j+\frac{1}{2}}$ should be computed as ratio of the slope of the solution across the cell interface upstream of $j+\frac{1}{2}$ to the slope of the solution across the interface at $j+\frac{1}{2}$. Defining $\gamma = -\text{sgn}(c_j)$, this ratio becomes

$$r^n_{j+\frac{1}{2}} = \frac{\phi^n_{j+\gamma+1} - \phi^n_{j+\gamma}}{\phi^n_{j+1} - \phi^n_j}.$$

If the velocity varies as a function of time, an $O\left[(\Delta t)^2\right]$ approximation to the velocity at $(n+\frac{1}{2})\Delta t$ should be used in (5.42) to preserve second-order accuracy (at least at locations away from the extrema of ϕ). Suitable approximations can be obtained by averaging the velocities between time levels n and $n + 1$, or by extrapolating forward from time levels n and $n-1$ (see Problem 11). A formula for the approximation of advective fluxes in spatially varying nondivergent velocity fields will be presented in Section 5.7.3.

5.6   Approximation with Local Polynomials

Formulae very similar to those obtained with the flux-limiter approach can be derived by approximating the solution as the sum of piecewise-linear functions defined over each grid cell and then computing the evolution of this piecewise-linear approximation over a time interval Δt, e.g., van Leer (1974). Similar methods can be derived using other piecewise-continuous polynomials. The simplest scheme, due to Godunov (1959), is obtained using piecewise-constant functions. Greater accuracy was achieved by Colella and Woodward (1984) using piecewise-parabolic functions. The following sections will discuss piecewise-constant approximations to nonlinear one-dimensional scalar conservation laws of the form (5.6) and piecewise-linear approximations to the constant-wind-speed advection equation.

5.6.1   Godunov's Method

In Godunov's method, the gridpoint values at each individual time step are used to define a piecewise-constant function such that

$$\tilde{\phi}(x, t^n) = \phi_j^n \quad \text{for} \quad x_{j-\frac{1}{2}} \le x \le x_{j+\frac{1}{2}},$$

where $t^n = n\Delta t$ and $x_{j+\frac{1}{2}} = x_j + \Delta x/2$. Using the function $\tilde{\phi}(x, t^n)$ as the initial condition, an approximate solution to the original conservation law at t^{n+1} may be obtained by solving the Riemann problems associated with the discontinuities in $\tilde{\phi}$ at the interface of each grid cell. The exact solution to these Riemann problems can be easily obtained for a scalar conservation law or for linear systems of conservation laws, at least until the signals emanating from each interface begin to interact.[7] The new solution at time ϕ_j^{n+1} is defined to be the average of these individual Riemann solutions over the jth grid cell,

$$\phi_j^{n+1} = \frac{1}{\Delta x} \int_{x_{j-\frac{1}{2}}}^{x_{j+\frac{1}{2}}} \tilde{\phi}(x, t^{n+1}) \, dx.$$

---

[7]See LeVeque (1992) for a discussion of approximate techniques for the solution of Riemann problems involving nonlinear systems of conservation laws.

In fact, it is not necessary to actually compute the solutions to each Riemann problem, since the integral form of the conservation law (5.7) implies that

$$\phi_j^{n+1} = \frac{1}{\Delta x} \int_{x_{j-\frac{1}{2}}}^{x_{j+\frac{1}{2}}} \tilde{\phi}(x, t^n) \, dx - \frac{1}{\Delta x} \int_{t^n}^{t^{n+1}} f\left[\tilde{\phi}(x_{j+\frac{1}{2}}, t)\right] dt$$

$$+ \frac{1}{\Delta x} \int_{t^n}^{t^{n+1}} f\left[\tilde{\phi}(x_{j-\frac{1}{2}}, t)\right] dt. \tag{5.43}$$

The first integral in the preceding is simply ϕ_j^n. The other two integrals may be trivially evaluated, provided that the integrand is constant over the time $t^n < t < t^{n+1}$, which will be the case if the Courant number, $|c\Delta t/\Delta x|$, is less than unity (where c is the speed of the fastest moving wave or shock). Note that the maximum time step permitted by this condition allows the Riemann solutions to interact within each grid cell, but these interactions can be ignored, since they don't change the value of $\tilde{\phi}$ at the cell interfaces and therefore don't complicate the evaluation of the integrals in (5.43).

The solution of the Riemann problem at each cell interface is determined by the initial values of $\tilde{\phi}$ on each side of the interface. In most cases, disturbances in the form of waves or shocks will either propagate rightward or leftward from the cell interface, and the fluxes in (5.43) will be correctly evaluated if $\tilde{\phi}$ is replaced by the value of ϕ^n that is upstream of the interface with respect to the propagation of the wave or shock. For smooth ψ, (5.6) may be expressed in the advective form

$$\frac{\partial \psi}{\partial t} + \frac{df}{d\psi} \frac{\partial \psi}{\partial x} = 0, \tag{5.44}$$

which shows that $df/d\psi$ is the speed at which smooth perturbations in ψ propagate along the x-axis. Thus, one might approximate the solution with a finite-volume method

$$\phi_j^{n+1} = \phi_j^n - \frac{\Delta t}{\Delta x} \left[F(\phi_{j+}^n) - F(\phi_{j-}^n)\right] \tag{5.45}$$

in which the upstream direction is estimated using a numerical approximation to $df/d\psi$ such that

$$F(\phi_{j+}) = \begin{cases} f(\phi_j) & \text{if } [f(\phi_{j+1}) - f(\phi_j)]/[\phi_{j+1} - \phi_j] \geq 0; \\ f(\phi_{j+1}) & \text{otherwise.} \end{cases} \tag{5.46}$$

According to the Rankine–Hugoniot condition (5.10), the upstream flux is also correctly selected when the solution contains a discontinuity in the form of a propagating jump.

An erroneous result can, however, be generated if the entropy-consistent solution to the Riemann problem at a cell interface is a rarefaction wave in which

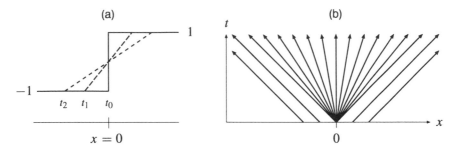

FIGURE 5.15. (a) A transonic rarefaction wave; the position of the left edge of the wave front is indicated at three consecutive time intervals. (b) Characteristic curves associated with this wave.

the disturbance spreads both to the right and left of the interface. For example, if the solution to Burgers's equation (5.11) is approximated using this scheme with initial data

$$\phi_j^0 = \begin{cases} 1 & \text{if } j \geq 0, \\ -1 & \text{if } j < 0, \end{cases}$$

the numerical solution will be a steady entropy-violating shock, since $F(\phi_{j+}^n) = F(\phi_{j-}^n)$ for all j and n. The correct entropy-consistent solution is the rarefaction wave, or expansion fan, illustrated in Fig. 5.15a.

Rarefaction waves in which $df/d\psi$ passes through zero at some point within the wave are known as *transonic rarefaction* waves. As a result of the transonic rarefaction, $\tilde{\phi}(x_{j+\frac{1}{2}}, t)$ assumes the value of ϕ for which the phase speed of the wave is zero (i.e., the value of ϕ for which the characteristics are parallel to the t-axis in the x-t plane—see Fig. 5.15b). Entropy-consistent solutions to the Riemann problem at each interface will be obtained if the upstream fluxes are determined according to the prescription

$$F(\phi_{j+}) = \begin{cases} \displaystyle\min_{\phi_j \leq \phi \leq \phi_{j+1}} f(\phi) & \text{if } \phi_j \leq \phi_{j+1}, \\ \displaystyle\max_{\phi_{j+1} \leq \phi \leq \phi_j} f(\phi) & \text{if } \phi_j > \phi_{j+1} \end{cases} \qquad (5.47)$$

(LeVeque 1992, p. 145). Let ϕ_s be the value of ϕ for which the phase speed of the wave is zero in the transonic rarefaction. In the case shown in Fig. 5.15, the flux obtained from (5.47) will be $f(\phi_s)$ because the minimum value of $f(\phi)$ occurs when the local phase speed, $df/d\phi$, is zero.

5.6.2   Piecewise-Linear Functions

Godunov's method yields a first-order approximation that is essentially identical to that obtained using upstream differencing. A second-order method can be obtained using piecewise-linear functions to approximate the solution over each grid interval, but the resulting method will not be TVD. In order to obtain a TVD method suitable for problems with discontinuous solutions, it is necessary to modify the slope of the piecewise-linear interpolating functions near discontinuities and poorly resolved gradients. This modification of the slope is accomplished using "slope-limiting" algorithms that are closely related to the flux-limiting procedures discussed in Section 5.5.2.

Although the actual computations may be organized in a more efficient manner, the procedure for advancing the numerical solution one time step is equivalent to the following three-step process. In the first step the fully discrete solution is used to define a piecewise-linear function within each grid cell of the form

$$\tilde{\phi}(x, t^n) = \phi_j^n + \sigma_j^n(x - x_j) \quad \text{for} \quad x_{j-\frac{1}{2}} \le x \le x_{j+\frac{1}{2}}.$$

In the second step the conservation law is integrated over a time Δt using $\tilde{\phi}(x, t^n)$ as the initial condition. In the third and final step, ϕ_j^{n+1} is obtained by averaging $\tilde{\phi}(x, t^{n+1})$ over each grid cell. The special considerations required to keep this method TVD are entirely connected with the first step, since if the conservation law is of the form (5.6), the solution obtained in the second step is TVD and the averaging in the third step does not increase the total variation. The first step is kept TVD by imposing limits on the slopes σ_j^n. The most severe limitation would be to set $\sigma_j^n = 0$, in which case the scheme reduces to Godunov's method.

In comparison with Godunov's method, in which exact solutions of the conservation law can be obtained relatively easily at each cell interface by solving a series of Riemann problems, the problems to be solved at each interface in step two of the piecewise-linear method are more difficult, because $\tilde{\phi}$ is not constant on each side of the initial discontinuity. General techniques for obtaining acceptable approximations to the solution required in step two are discussed in LeVeque (1992). In the following we will once again focus on the special case of the constant-wind-speed advection equation (5.18), for which the solution required in step two is simply

$$\tilde{\phi}(x, t^{n+1}) = \tilde{\phi}(x - c\Delta t, t^n).$$

Assuming that $c > 0$ and averaging $\tilde{\phi}(x, t^{n+1})$ over $(x_{j-\frac{1}{2}}, x_{j+\frac{1}{2}})$, steps two and three yield

$$\phi_j^{n+1} = \phi_j^n - \mu \left(\phi_j^n - \phi_{j-1}^n\right) - \frac{\mu}{2}(1 - \mu)\Delta x(\sigma_j - \sigma_{j-1}), \quad (5.48)$$

where $\mu = c\Delta t/\Delta x$. If the slopes of the piecewise-linear functions are defined in step one such that

$$\sigma_j = \frac{\phi_{j+1} - \phi_j}{\Delta x}, \quad (5.49)$$

then (5.48) reduces to the Lax–Wendroff method, which is not TVD.

In order to make the preceding scheme TVD, the slope can be limited by a multiplicative constant $C_{j+\frac{1}{2}}$ such that

$$\sigma_j = \left(\frac{\phi_{j+1} - \phi_j}{\Delta x}\right) C_{j+\frac{1}{2}}, \quad (5.50)$$

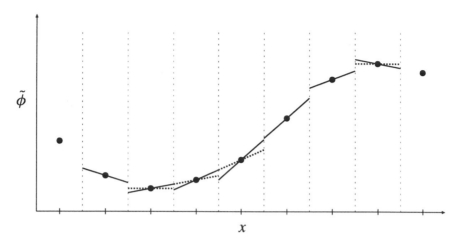

FIGURE 5.16. A piecewise-linear finite-volume approximation. Cell-averaged values are plotted as heavy points. The solid line segments show the $\tilde{\phi}(x)$ obtained using the Lax–Wendroff slopes (5.49). Dashed line segments show the modification to $\tilde{\phi}(x)$ introduced by the minmod slope limiter assuming $c > 0$. Slopes in the rightmost and leftmost grid cells are omitted.

in which case (5.48) is identical to the flux-limited Lax–Wendroff method and may be written in conservation form using the flux (5.33). As with the family of flux-limiter methods, there are a variety of reasonable choices for $C_{j+\frac{1}{2}}$, and every flux limiter defined in Section 5.5.2 can be reinterpreted as a slope limiter via (5.50). Indeed, the behavior of some flux limiters are easier to understand when they are interpreted as slope limiters. For example, letting

$$a = \frac{\phi_{j+1} - \phi_j}{\Delta x}$$

and

$$b = \begin{cases} \dfrac{\phi_j - \phi_{j-1}}{\Delta x} & \text{if } c \geq 0, \\[2mm] \dfrac{\phi_{j+2} - \phi_{j+1}}{\Delta x} & \text{if } c < 0, \end{cases}$$

the minmod limiter defined by (5.37) implies that

$$\sigma_j = \begin{cases} 0 & \text{if } ab \leq 0, \\ \text{sgn}(a) \min\left(|a|, |b|\right) & \text{otherwise,} \end{cases}$$

which guarantees that the magnitude of the slope of ϕ between grid points j and $j + 1$ is no larger than the slope over the next interval upstream, except that the slope is set to zero at an extremum of ϕ. The effect of the minmod limiter is illustrated in Fig. 5.16.

5.7   Two Spatial Dimensions

The preceding discussion has focused almost exclusively on problems in one spatial dimension. The most straightforward way to extend these methods to multiple dimensions is through fractional steps, and fractional steps have been used successfully in problems whose solutions contain discontinuities or poorly resolved gradients. A theoretical basis for the success of these methods was provided by Crandall and Majda (1980a), who showed that convergent approximations to the entropy-consistent solution of two-dimensional scalar conservation laws can be achieved using the method of fractional steps, provided that consistent conservation-form monotone schemes are used in each individual step.

Nevertheless, as discussed in Section 3.3.1, operator splitting generates only an $O\,[\Delta t]$-accurate approximation unless the finite-difference operators associated with each split step commute (and are individually at least second-order accurate). The finite-difference operators cannot be expected to commute unless the corresponding operators in the unapproximated problem commute, and in many practical cases the unapproximated operators do not commute. For example, the advective operators

$$u\frac{\partial}{\partial x} \quad \text{and} \quad v\frac{\partial}{\partial y}$$

do not commute unless

$$u\frac{\partial v}{\partial x} = v\frac{\partial u}{\partial y} = 0.$$

Thus, one drawback to the fractional-step procedure is the likelihood that it will lead to increased time-truncation errors. Strang splitting (3.57) can be used to retain $O\left[(\Delta t)^2\right]$ accuracy when the finite-difference operators don't commute, but if the value of the prognostic variable is required at every time step,[8] Strang splitting requires 50% more work per time step than does conventional splitting.

More accurate results, and a more isotropic finite-difference solution, can be obtained using unsplit algorithms. In the following, we will consider two representative methods: flux-corrected transport (Zalesak 1979) and a flux-limiter algorithm for two-dimensional nondivergent flow proposed by LeVeque (1996). Several other schemes with varying degrees of similarity have also appeared in the literature, including those by Smolarkiewicz (1984), Colella (1990), Saltzman (1994), Leonard et al. (1993), Thuburn (1996), and Stevens and Bretherton (1996).

5.7.1   FCT in Two Dimensions

The extension of the flux-correction algorithm described in Section 5.4.2 to multidimensional problems is straightforward and is discussed in detail by Zalesak

---

[8]The values of the prognostic variables are required at every time step during the integration of systems of equations in which a chemical or physical process (such as cloud condensation and precipitation) is parametrized as a function of the prognostic variables.

(1979). Only the two-dimensional case will be considered here, for which a monotone low-order solution could be computed using the two-dimensional upstream difference (3.31). As discussed in Section 3.2.1, however, the CTU method (3.36) is a better choice. The high-order fluxes could be estimated using an appropriate form of the upstream biased Lax–Wendroff method (3.38). The generalization of (3.38) to problems with spatially varying nondivergent winds will be discussed in Section 5.7.3.

Suppose that i and j are the grid-point indices along the two spatial coordinates. In contrast to the one-dimensional case, there will now be four antidiffusive fluxes into each grid cell, and one must compute four coefficients $C_{i\pm\frac{1}{2},j}$ and $C_{i,j\pm\frac{1}{2}}$ to limit these fluxes. The formulae for $C_{i\pm\frac{1}{2},j}$ and $C_{i,j\pm\frac{1}{2}}$ are identical to those given for $C_{j\pm\frac{1}{2}}$ in Section 5.4.2, except for the inclusion of the second dimension in the subscript notation and the computation of the total antidiffusive fluxes in and out of grid point i, j as

$$P_{i,j}^+ = \max\left(0, A_{i-\frac{1}{2},j}\right) - \min\left(0, A_{i+\frac{1}{2},j}\right)$$
$$+ \max\left(0, A_{i,j-\frac{1}{2}}\right) - \min\left(0, A_{i,j+\frac{1}{2}}\right),$$

$$P_{i,j}^- = \max\left(0, A_{i+\frac{1}{2},j}\right) - \min\left(0, A_{i-\frac{1}{2},j}\right)$$
$$+ \max\left(0, A_{i,j+\frac{1}{2}}\right) - \min\left(0, A_{i,j-\frac{1}{2}}\right).$$

The formula for the permissible range of values for $\phi_{i,j}^{n+1}$ also needs to be generalized to two dimensions; the most natural choice is to define

$$\phi_{i,j}^a = \max\left(\phi_{i,j}^n, \phi_{i,j}^{\text{td}}\right),$$
$$\phi_{i,j}^b = \min\left(\phi_{i,j}^n, \phi_{i,j}^{\text{td}}\right),$$

and then let

$$\phi_{i,j}^{\max} = \max\left(\phi_{i,j}^a, \phi_{i,j-1}^a, \phi_{i,j+1}^a, \phi_{i-1,j}^a, \phi_{i+1,j}^a\right),$$
$$\phi_{i,j}^{\min} = \min\left(\phi_{i,j}^b, \phi_{i,j-1}^b, \phi_{i,j+1}^b, \phi_{i-1,j}^b, \phi_{i+1,j}^b\right).$$

The preceding technique for determining $\phi^{\min}$ and $\phi^{\max}$ does not completely prevent the development of small undershoots and overshoots in situations where ϕ is being transported in a direction almost perpendicular to the gradient of ϕ. Nevertheless, the spurious oscillations are typically very small and can be completely eliminated using additional correction steps discussed by Zalesak.

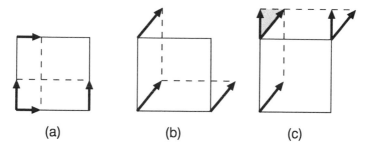

FIGURE 5.17. (a) Transmission of the upstream fluxes parallel to the coordinate axes in grid cell (i, j); heavy arrows denote the vector displacements $u\Delta t$ and $v\Delta t$. (b) Transmission of the fluxes parallel to the wind field; heavy arrows denote displacements over time Δt along the total wind vector. (c) Area in cell $(i, j + 1)$ into which flux is actually transmitted from cell $(i - 1, j)$ (stippled triangle) and the area into which the axis-parallel flux is incorrectly transmitted from cell (i, j) (triangle with diagonal fill).

5.7.2   Flux-Limiter Methods for Uniform 2-D Flow

As was the case for the one-dimensional flux-limiter methods discussed previously, the solution strategy consists of using a monotone method to compute low-order fluxes near poorly resolved gradients and then correcting these low-order fluxes in regions where the solution is well-resolved using fluxes obtained from a higher-order scheme. A finite-difference approximation to the equation governing the advection of a passive scalar in a two-dimensional flow (5.23) can be written in the conservation form

$$\phi_{i,j}^{n+1} = \phi_{i,j}^n - \frac{\Delta t}{\Delta s} \left[F_{i+\frac{1}{2},j}^n - F_{i-\frac{1}{2},j}^n + G_{i,j+\frac{1}{2}}^n - G_{i,j-\frac{1}{2}}^n \right]. \qquad (5.51)$$

The terms $F_{i-\frac{1}{2},j}$ and $G_{i,j-\frac{1}{2}}$ are approximations to the advective fluxes through the left and lower boundaries of the grid cell centered on $\phi_{i,j}$. The horizontal mesh spacing is Δs, and it is assumed to be equal along the x- and y-axes for notational simplicity. In order to present the method in its simplest form, we temporarily assume that both velocity components are positive and spatially uniform. The complete algorithm for an arbitrary nondivergent flow is presented in Section 5.7.3.

A simple monotone approximation to the advective flux is given by the upstream, or donor cell, method, which for positive velocities yields

$$F_{i+\frac{1}{2},j}^{up} = u\phi_{i,j}, \qquad G_{i,j+\frac{1}{2}}^{up} = v\phi_{i,j}. \qquad (5.52)$$

In the standard upwind method, these fluxes are transmitted parallel to the coordinate axes. Each flux induces a change in $\phi_{i,j}$ equal to the upstream value of ϕ times the ratio of the area swept out by the incoming fluid divided by the total area of each grid cell. The situation is schematically illustrated in Fig. 5.17a.

In reality, the fluxes are transmitted parallel to the velocity vector, and an improved monotone scheme can be obtained by accounting for the transmission of the fluxes at their correct angle to the coordinate axes, as illustrated in Fig. 5.17b. The transport of these off-axis fluxes through each grid cell can be accounted for by a modification of the basic upstream fluxes (5.52). Consider the corrections required to the flux through the lower boundary of cell $(i, j + 1)$. As highlighted by the diagonal fill in Fig. 5.17c, the area swept out by the axis-parallel flux $G^{\mathrm{up}}_{i,j+\frac{1}{2}}$ incorrectly includes a triangular region of cell $(i, j + 1)$ that is not penetrated by the true advective flux though the lower cell boundary. The ratio of the area of this triangular region to the area of the full grid cell is $0.5uv(\Delta t)^2/(\Delta s)^2$. There is also a stippled triangular region of the same size that should receive a flux originating from cell $(i - 1, j)$. The axis-parallel upstream flux through the lower boundary of cell $(i, j + 1)$ can be modified to account for these two corrections according to the formula

$$G^{\mathrm{ctu}}_{i,j+\frac{1}{2}} = G^{\mathrm{up}}_{i,j+\frac{1}{2}} - \frac{\Delta t}{2\Delta s} uv(\phi_{i,j} - \phi_{i-1,j}). \tag{5.53}$$

The upstream flux parallel to the x-axis must be similarly modified such that

$$F^{\mathrm{ctu}}_{i+\frac{1}{2},j} = F^{\mathrm{up}}_{i+\frac{1}{2},j} - \frac{\Delta t}{2\Delta s} uv(\phi_{i,j} - \phi_{i,j-1}). \tag{5.54}$$

When the wind speed is constant, the method obtained by accounting for flux propagation along the wind vector is identical to the CTU method (3.36). In Section 3.2.1 the CTU method was derived for a governing equation in advective form using the method of characteristics to compute backward fluid-parcel trajectories. An alternative derivation can be performed using the finite-volume formalism by following trajectories backwards from the corner of each grid cell and then computing the average value of ϕ within the rectangular volume occupied by that cell at the previous time step (Colella 1990). Using back trajectories to define the subareas A_1 through A_4 shown in Fig. 5.18, and assuming that the solution is piecewise constant within each grid cell,

$$\phi^{n+1}_{i,j} = \frac{1}{(\Delta s)^2} \left[A_1 \phi^n_{i,j} + A_2 \phi^n_{i,j-1} + A_3 \phi^n_{i-1,j} + A_4 \phi^n_{i-1,j-1} \right].$$

This finite-difference equation is not in conservation form, but it is equivalent to the conservation form (5.51), with the fluxes given by (5.52), (5.53), and (5.54). The CTU method will be monotone whenever the subareas A_1 through A_4 are positive, which in the general case where u and v can have arbitrary sign requires that

$$\max(|u|, |v|) \frac{\Delta t}{\Delta s} \le 1. \tag{5.55}$$

The CTU method must be first-order accurate because it is a linear monotone scheme (Godunov 1959). As proposed by LeVeque (1996), an essentially

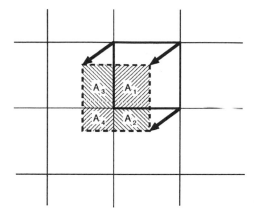

FIGURE 5.18. Backward trajectories from cell (i, j) defining the departure volume in the CTU method.

second-order scheme can be obtained using the same strategy employed for the one-dimensional flux-limiter methods: Corrective fluxes are added to F^{ctu} and G^{ctu} that make the scheme equivalent to the Lax–Wendroff method except in regions where the solution is poorly resolved and the corrective flux is reduced to minimize spurious oscillations. As discussed in Section 2.5.3, the Lax–Wendroff approximation to the constant-wind-speed two-dimensional advection equation has the form

$$\frac{\phi_{i,j}^{n+1} - \phi_{i,j}^{n}}{\Delta t} + \delta_{2x} u \phi_{i,j}^{n} + \delta_{2y} v \phi_{i,j}^{n} = \frac{\Delta t}{2} H(\phi^{n}),$$

where $H(\phi)$ is at least a first-order numerical approximation to

$$\psi_{tt} = u^{2}\psi_{xx} + 2uv\psi_{xy} + v^{2}\psi_{yy},$$

and the subscripts on ψ denote partial derivatives. The divergence of the off-axis fluxes in the CTU method generate a first-order approximation to the mixed partial derivative in the preceding. Thus, the only modifications that need to be made to convert the CTU scheme to the Lax–Wendroff method are to replace the one-sided approximations to $\partial \psi / \partial x$ and $\partial \psi / \partial y$ with centered second-order finite differences and to include an approximation to

$$\frac{\Delta t}{2} \left(u^{2}\psi_{xx} + v^{2}\psi_{yy} \right).$$

This can be accomplished by adding the following terms to the CTU fluxes:

$$F_{i+\frac{1}{2},j} = F_{i+\frac{1}{2},j}^{\text{ctu}} + \frac{|u|}{2} \left(1 - |u|\frac{\Delta t}{\Delta s} \right) (\phi_{i+1,j} - \phi_{i,j}),$$

$$G_{i,j+\frac{1}{2}} = G_{i,j+\frac{1}{2}}^{\text{ctu}} + \frac{|v|}{2} \left(1 - |v|\frac{\Delta t}{\Delta s} \right) (\phi_{i,j+1} - \phi_{i,j}).$$

(Assuming that the CTU fluxes have been computed in the upstream direction, these formulae apply regardless of the sign of the velocity.) The preceding corrections to the CTU flux have exactly the same form as the corrections to the upstream flux in the one-dimensional problem (5.41), which suggests that spurious oscillations in the vicinity of discontinuities or poorly resolved gradients can be controlled if the corrections are limited using one of the flux limiter functions discussed in Section 5.5.2. The resulting flux-limited approximation to the two-dimensional advection problem is neither TVD nor monotone, but the spurious oscillations generated by this scheme are extremely weak.

5.7.3 Nonuniform Nondivergent Flow

The generalization of this method to a nonuniform nondivergent velocity field is most easily presented as the algorithm in Table 5.2, in which the fluxes are initialized to zero at the beginning of each time step and then incrementally built up in the course of two passes through the numerical mesh. The velocities are assumed to be staggered such that $u_{i+\frac{1}{2},j}$ and $v_{i,j+\frac{1}{2}}$ are displaced $(\Delta s/2, 0)$ and $(0, \Delta s/2)$ away from the grid point where $\phi_{i,j}$ is defined.[9]

If the flow is nondivergent, the algorithm in Table 5.2 can easily be recast as an approximation to the transport equation in advective form (5.25). When u and v are positive, the upstream approximation to the spatial derivative operators in (5.25) is

$$\Delta_{i,j} = u_{i-\frac{1}{2},j}\left(\frac{\phi_{i,j}-\phi_{i-1,j}}{\Delta x}\right) + v_{i,j-\frac{1}{2}}\left(\frac{\phi_{i,j}-\phi_{i,j-1}}{\Delta y}\right).$$

Assuming that the discretized velocity field satisfies the natural finite-difference approximation to the nondivergence condition on a staggered mesh,

$$\frac{u_{i+\frac{1}{2},j}-u_{i-\frac{1}{2},j}}{\Delta x} + \frac{v_{i,j+\frac{1}{2}}-v_{i,j-\frac{1}{2}}}{\Delta y} = 0,$$

$\Delta_{i,j}$ may be expressed in the equivalent form

$$\Delta_{i,j} = \frac{F^{up}_{i+\frac{1}{2},j}-F^{up}_{i-\frac{1}{2},j}}{\Delta x} + \frac{G^{up}_{i,j+\frac{1}{2}}-G^{up}_{i,j-\frac{1}{2}}}{\Delta y},$$

where F^{up} and G^{up} are the upstream fluxes defined in (5.52). The algorithm in Table 5.2 may therefore be modified to yield a conservative advective-form approximation by replacing the three lines marked by stars with

---

[9] See Fig. 3.6 for an illustration of the same staggering scheme in a different context.

- *Initialize the fluxes to zero*

  for each i, j do

  $$F_{i-\frac{1}{2},j}^n = 0, \quad G_{i,j-\frac{1}{2}}^n = 0 \qquad (\star)$$

- *Increment F and G due to fluxes through the left cell interface*

  for each i, j do

  $$U = u_{i-\frac{1}{2},j}^{n+\frac{1}{2}}$$
  $$R = \phi_{i,j}^n - \phi_{i-1,j}^n$$
  if $U > 0$, then $I = i - 1$, else $I = i$
  $$F_{i-\frac{1}{2},j}^n = F_{i-\frac{1}{2},j}^n + U\phi_{I,j} \qquad (\star\star)$$
  if $U > 0$, then $I = i$, else $I = i - 1$

  if $v_{i,j+\frac{1}{2}}^{n+\frac{1}{2}} > 0$, then $G_{I,j+\frac{1}{2}}^n = G_{I,j+\frac{1}{2}}^n - \frac{\Delta t}{2\Delta s} R U v_{i,j+\frac{1}{2}}^{n+\frac{1}{2}}$

  if $v_{i,j-\frac{1}{2}}^{n+\frac{1}{2}} < 0$, then $G_{I,j-\frac{1}{2}}^n = G_{I,j-\frac{1}{2}}^n - \frac{\Delta t}{2\Delta s} R U v_{i,j-\frac{1}{2}}^{n+\frac{1}{2}}$

  if $\quad U > 0$, $\quad$ then $\quad r_{i-\frac{1}{2},j}^n (\phi_{i-1,j}^n - \phi_{i-2,j}^n)/R$,

  $\qquad\qquad\qquad$ else $\quad r_{i-\frac{1}{2},j}^n = (\phi_{i+1,j}^n - \phi_{i,j}^n)/R$

  $$F_{i-\frac{1}{2},j}^n = F_{i-\frac{1}{2},j}^n + \tfrac{|U|}{2} \left(1 - \tfrac{\Delta t}{\Delta s}|U|\right) C(r_{i-\frac{1}{2},j}^n)$$

- *Increment F and G due to fluxes through the bottom cell interface*

  (as above, switching the roles of i and j, u and v, and F and G)

- *Update ϕ*

  for each i, j do

  $$\phi_{i,j}^{n+1} = \phi_{i,j}^n - \frac{\Delta t}{\Delta s} \left[F_{i+\frac{1}{2},j}^n - F_{i-\frac{1}{2},j}^n + G_{i,j+\frac{1}{2}}^n - G_{i,j-\frac{1}{2}}^n \right] \qquad (\star\star\star)$$

TABLE 5.2. Algorithm for executing one time step of LeVeque's two-dimensional flux-limited advection scheme.

$$F^n_{i-\frac{1}{2},j} = 0, \quad G^n_{i,j-\frac{1}{2}} = 0, \quad \Delta^n_{i,j} = 0, \qquad (\star)$$

$$\Delta^n_{i,j} = \Delta^n_{i,j} + U(\phi_{I+1,j} - \phi_{I,j}), \qquad (\star\star)$$

$$\phi^{n+1}_{i,j} = \phi^n_{i,j} - \frac{\Delta t}{\Delta s}\left[\Delta^n_{i,j} + F^n_{i+\frac{1}{2},j} - F^n_{i-\frac{1}{2},j} + G^n_{i,j+\frac{1}{2}} - G^n_{i,j-\frac{1}{2}}\right]. \quad (\star\star\star)$$

No additional modification of the second-order correction terms in F and G are required. The equivalence of the second-order corrections in the flux- and advective-form algorithms is a consequence of the nondivergence of the velocity field. Provided that the flow is steady, the advective form of the governing equation (5.25) implies that

$$\psi_{tt} = u(u\psi_x)_x + u(v\psi_y)_x + v(u\psi_x)_y + v(v\psi_y)_y,$$

whereas the flux form (5.23) implies that

$$\psi_{tt} = (u(u\psi)_x)_x + \left(u(v\psi)_y\right)_x + (v(u\psi)_x)_y + \left(v(v\psi)_y\right)_y.$$

If the flow is nondivergent, both of the preceding equations can be expressed as

$$\psi_{tt} = \left(u^2\psi_x\right)_x + \left(uv\psi_y\right)_x + (uv\psi_x)_y + \left(v^2\psi_y\right)_y.$$

This is the form of the second-order Lax–Wendroff correction that is actually approximated by the finite differences in both the advective and flux-form algorithms.

5.7.4 A Numerical Example

In the following, LeVeque's two-dimensional flux-limited scheme will be compared with time-split methods and a linear high-order finite-difference scheme in a test problem in which a passive tracer is advected in a nondivergent deformational flow. The spatial domain is the square $0 \le x \le 1, 0 \le y \le 1$, and the initial concentration of the tracer is given by

$$\phi(x, y, 0) = \frac{1}{2}[1 + \cos(\pi r)],$$

where

$$r(x, y) = \min\left(1, \, 4\sqrt{\left(x - \frac{1}{4}\right)^2 + \left(y - \frac{1}{4}\right)^2}\right).$$

The velocity field is a swirling shear flow defined such that

$$u(x, y) = \sin^2(\pi x)\sin(2\pi y)\cos(\pi t/5),$$
$$v(x, y) = -\sin^2(\pi y)\sin(2\pi x)\cos(\pi t/5).$$

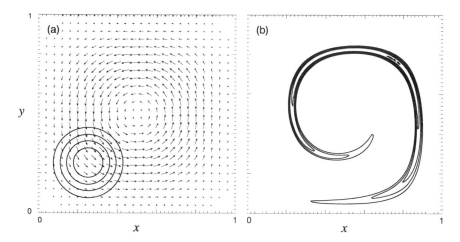

FIGURE 5.19. (a) Velocity vectors and tracer concentration field at $t = 0$. (b) Tracer concentration field at $t = 2.5$. Tracer is contoured at intervals of 0.2 beginning with the 0.2 contour line. The length of each vector is proportional to the speed of the flow.

The initial distribution of the passive tracer and the structure of the flow field are both plotted in Fig. 5.19a. The tracer distribution is most highly deformed at $t = 2.5$, at which time it appears as the long thin arc shown in Fig. 5.19b, which was obtained from a numerical simulation on a high-resolution mesh. Since the velocity periodically reverses direction, every fluid parcel returns to its original position after five time units, and the correct tracer distribution at $t = 5$ is identical to the initial field. The accuracy of the numerical solutions obtained at $t = 5$ can therefore be evaluated by comparing them with the initial tracer distribution. This same problem has been considered by LeVeque (1996).

Pairs of numerical solutions were obtained using horizontal grid intervals of 0.02 and 0.01. The time step was 0.01 for the coarse-mesh simulation and 0.005 for the fine mesh. At the time of maximum deformation, the width of the arc is reduced to approximately 0.05, so the tracer distribution is very poorly resolved on the 0.02 grid. The solutions at $t = 5$ generated by the two-dimensional flux-limited algorithm (Table 5.2) using the superbee limiter (5.38) are shown in Fig. 5.20. Both solutions show a significant loss of amplitude in the region of maximum concentration, which is underestimated by approximately 50% and 20% in the coarse- and fine-mesh simulations, respectively. The shape of the tracer field, which is significantly in error in the coarse mesh simulation, is greatly improved by doubling the numerical resolution. Both simulations exhibit regions where the tracer concentration is slightly negative, but the magnitude of the negative concentrations are very small (on the order of 0.1% of the maximum concentration in the initial distribution). The negative values in these simulations can be completely eliminated by using a different flux limiter (see Fig. 5.23).

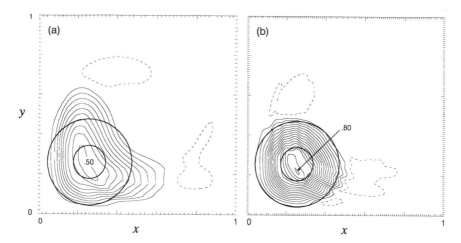

FIGURE 5.20. Comparison of the exact solution at $t = 5$ with the corresponding numerical solution obtained using the two-dimensional flux-limited advection scheme with a super-bee limiter and a horizontal grid interval of (a) 0.02, and (b) 0.01. The heavy circles are the 0.05 and 0.75 contour lines of the exact solution. Thin solid lines are contours of the numerical solution at at intervals of 0.05, beginning with the 0.05 contour. The dashed line is the -0.001 contour.

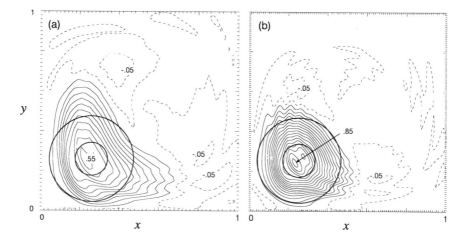

FIGURE 5.21. As in Fig. 5.20 except that the numerical solution is obtained using a linear finite-difference scheme with compact fourth-order spatial differences. Note that a -0.05 contour appears in the solution.

Some idea of the effectiveness of the flux limiter may be obtained by comparing LeVeque's two-dimensional advection scheme with a high-order linear finite-difference scheme in which the spatial derivatives are computed using Lele's compact fourth-order scheme (2.85), time-differencing is third-order Adams–Bashforth, and the shortest wavelengths are smoothed using a sixth-order filter (2.76) with $\gamma_6 \Delta t = 0.001$. Due to the more stringent stability constraint associated with this scheme, the time steps were one-half those used in the corresponding flux-limited simulation. The numerical solution at $t = 5$ obtained with this method is shown in Fig. 5.21. Except for the nontrivial negative concentrations generated by the linear high-order scheme, the overall character of the solution is surprisingly similar to that generated by the two-dimensional flux-limited scheme. In particular, the linear high-order scheme produces a similar tracer distribution and does a slightly better job preserving the maxima in the concentration field. The relative amplitudes of the negative concentrations generated by each method are more clearly indicated in Figs. 5.22a and b, which compare the preceding numerically computed concentrations along the line $y = 0.5$ at $t = 2.5$ with a third numerical solution obtained using very fine spatial resolution. The spurious negative concentrations produced by the high-order linear scheme are clearly evident, but aside from these negative concentrations and slight differences in the amplitude of the peak concentration, the solutions are very similar.

The solution computed using the two-dimensional flux-limited scheme is rather sensitive to the form of the flux limiter. This sensitivity is illustrated in Fig. 5.22c, which compares the results obtained using superbee and Van Leer (5.39) limiters and a 0.01 spatial mesh. The spurious damping of the peak concentration generated by the Van Leer limiter is much more pronounced than that produced by the superbee limiter. The increase in the effective numerical diffusion associated with the Van Leer limiter can also be appreciated by comparing Figs. 5.20 and 5.23, which show the solutions at $t = 5$ computed using the superbee and Van Leer limiters, respectively. Although the Van Leer limiter generates considerably more diffusion, it has the advantage of not producing the spurious negative concentrations obtained using the superbee limiter. Note, however, that the negative concentrations generated by the superbee limiter are too small to be visible in Fig. 5.22.

The performance of a time-split method is shown in Fig. 5.24. The advective transports parallel to the x- and y-coordinates are computed in separate fractional steps, each of which used a superbee limiter in the one-dimensional flux-limited algorithm described in Section 5.5.3. Strang splitting was not used and the operators don't commute, so this method is not fully second-order in time. The one-dimensional fluxes were calculated by replacing c in (5.42) with either $u_{i+\frac{1}{2},j}^{n+\frac{1}{2}}$ or $v_{i,j+\frac{1}{2}}^{n+\frac{1}{2}}$, although this treatment of the spatial variations in the velocity field does not yield a fully second-order Lax–Wendroff approximation to the one-dimensional variable-wind-speed advection equation.

Since the scheme used in each fractional step is the one-dimensional equivalent of LeVeque's two-dimensional flux-limited method, the difference between

FIGURE 5.22. The solution at $y = 0.5$, $t = 2.5$ plotted as a function of x. In each panel a reference solution obtained with a high-resolution simulation is shown as the thin solid line. (a) two-dimensional flux-limited solution using the superbee limiter (heavy solid) and fourth-order compact solution (heavy dot-dashed) when $\Delta x = \Delta y = 0.2$. (b) as in (a) except that the horizontal grid spacing is 0.1. (c) two-dimensional flux-limited solution obtained using the 0.1 mesh spacing and a superbee limiter (solid) or the Van Leer limiter (heavy dot-dashed).

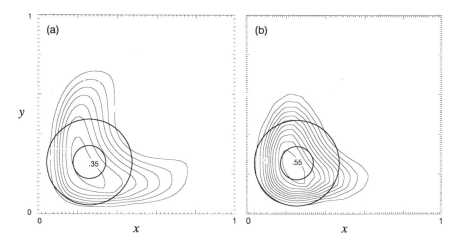

FIGURE 5.23. As in Fig. 5.20 except that the numerical solution is obtained using the Van Leer instead of the superbee limiter. No negative concentrations are generated.

the solutions shown in Figs. 5.20 and 5.24 provides a good indication of the errors generated by a simple fractional-step approach. As evidenced by a comparison of Figs. 5.20 and 5.24, the split solution incorrectly loses more amplitude and is smeared further "northward" than the nonsplit solution. As a consequence, the unsplit scheme is clearly superior, but since the overall character of both solutions is very similar, the degree of superiority is rather small. Note also that the superbee limiter does not generate negative concentrations in the split solution, although it does generate negatives in the nonsplit result. If this comparison is

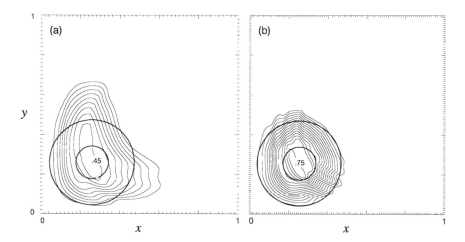

FIGURE 5.24. As in Fig. 5.20 except that the numerical solution is obtained via operator splitting using a one-dimensional flux-limited advection algorithm in each individual step.

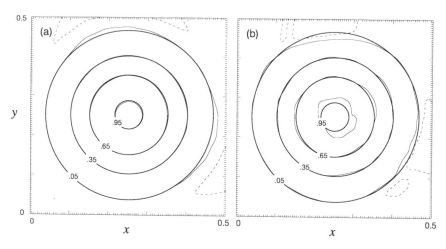

FIGURE 5.25. Comparison of the exact solution (heavy contours) at $t = 5$ with numerical solutions (thin contours) obtained using a grid spacing of 0.005 and (a) the two-dimensional scheme without a limiter, or (b) the two-dimensional scheme with the superbee limiter. The -0.001, 0.05, 0.35, 0.65, and 0.95 contours are plotted. Only the lower left quadrant of the total domain is plotted.

repeated after replacing the superbee limiter with the Van Leer limiter, the fully two-dimensional method is once again superior to the split scheme (not shown), and the degree of superiority is similar to that obtained with the superbee limiter. It is conceivable that the time symmetry in the reversing velocity field cancels some of the splitting error in this test problem. Nevertheless, a comparison of the split and nonsplit solutions at the time of flow reversal ($t = 2.5$, not shown) reveals the same qualitative differences apparent in Fig. 5.24; the split solution is slightly lower in amplitude and shows a slight bias in its spatial distribution.

If the horizontal mesh spacing is decreased to 0.005, the solution always remains well-resolved on the numerical mesh, and a significant increase in the accuracy of the numerical solution is obtained. Under these circumstances, the use of a flux limiter can degrade the solution. Figure 5.25 shows a comparison of the solutions at $t = 5$ produced using the fully two-dimensional method with and without a flux limiter. The solution obtained without a limiter is almost perfect, whereas that obtained with the superbee limiter has broadened and widened the peak. This test provides another example of the tendency of the superbee limiter to square off the peak of a smooth distribution. The comparison of the high-resolution simulations shown in Fig. 5.25 serves as a reminder that the flux limiter decreases the order of accuracy of the scheme, and in situations where the solution is well-resolved the most rapid convergence is obtained using the high-order method.

5.7.5 When Is a Flux Limiter Necessary?

Use of a flux limiter can be essential to ensure the convergence of numerical approximations to problems with shocks or discontinuous solutions. On the other hand, in an advection problem such as that considered in the preceding section, the initially smooth concentration field never develops a discontinuity in a finite time; there are no spurious weak solutions, and the flux limiter is not required to guarantee convergence. The use of a flux limiter in numerical approximations to the advection equation is optional and can be considered as a device for converting one type of error, namely undershoots and overshoots, into a less easily quantifiable but more acceptable form.

Consider the advantages and disadvantages of using flux-limited or flux-corrected methods to obtain approximate solutions of advection problems. As revealed by the example shown in Fig. 5.25, flux limiters are not always helpful and can actually degrade the result if the solution remains well-resolved on the numerical mesh. Nevertheless, in many practical applications the solution is not well-resolved, either due to discontinuities in the initial data or to deformation in the velocity field that stretches a well-resolved initial field until one of the scales characterizing the field contracts to the scale of the numerical grid. In these situations flux limiters can eliminate the spurious overshoots and undershoots that develop as a consequence of poor numerical resolution. Except for the absence of undershoots and overshoots, the overall character of the flux-limited solution may, nevertheless, be rather similar to that obtained using an accurate linear finite-difference scheme (compare Figs. 5.20 and 5.21). The need for the flux limiter is therefore most pronounced in problems where undershoots and overshoots in the tracer concentration field can couple with other physical processes to trigger spurious behaviors. An example of such coupling can occur in simulating the evolution of atmospheric clouds. A spurious cloud can be generated where an error in the advective transport of water vapor produces an overshoot in which the water-vapor mixing ratio exceeds the saturation mixing ratio. Latent heat is released as the water vapor condenses to form the spurious cloud, and this heat generates buoyancy perturbations that feed back on the flow field, thereby altering the subsequent evolution of the system.

A second example of coupling between advectively generated undershoots and other physical processes involves the generation of negative chemical concentrations in simulations of chemically reacting flows. The mixing ratio of a chemical species should never drop below zero, but numerically generated undershoots may produce false negative concentrations that destabilize the integration by triggering nonphysical chemical reactions. As an example, consider the following pair of equations describing the advection and interaction of two chemical species:

$$\frac{\partial \psi_1}{\partial t} + c\frac{\partial \psi_1}{\partial x} = -r\psi_1\psi_2, \tag{5.56}$$

$$\frac{\partial \psi_2}{\partial t} + c\frac{\partial \psi_2}{\partial x} = r\psi_1\psi_2. \tag{5.57}$$

Here ψ_1 and ψ_2 represent the concentration of each chemical species, and r is the rate at which they react, transforming ψ_1 into ψ_2. Suppose there is a nonzero background concentration of ψ_1 throughout the domain and that ψ_2 drops very rapidly to zero outside some localized "plume." If leapfrog-time centered-space differencing is used to simulate the downwind transport of the plume, small dispersive ripples will appear at the edge of the plume, and regions will develop where $\psi_2 < 0$, $\psi_1 > 0$. In the absence of the chemical reactions, these negative regions would remain small and relatively insignificant. However, at any point where $\psi_2 < 0$ and $\psi_1 > 0$, the chemical reaction terms in (5.57) drive ψ_2 more negative while simultaneously increasing ψ_1, thereby amplifying the undershoot and ultimately destabilizing the numerical integration. The difficulties associated with the generation of false negatives in the simulation of physical fields that should never become negative are sufficiently serious that several *positive definite* advection schemes have been specifically proposed to avoid this problem.

5.8   Schemes for Positive Definite Advection

A *positive definite* advection scheme is a method that never generates a negative value from nonnegative initial data.[10] Any monotone scheme is positive definite, but there are no other simple relationships between the sets of methods that are positive definite and those that are monotonicity-preserving or TVD. TVD schemes need not be positive definite, and positive definite schemes need not be TVD.

Early attempts to construct positive definite advection schemes involved "filling algorithms," in which the solution obtained after each integration step was corrected by filling in any negative values. In order to conserve the total mass of the advected species, negatives cannot simply be set to zero; compensating mass must removed from positive regions. There are a variety of filling algorithms designed for this purpose. Some filling algorithms attempt to fill local negative regions from adjacent positive areas (Mahlman and Sinclair 1977). This may be a physically satisfying way to remove dispersive undershoots, but it requires a great deal of logical testing that cannot be performed efficiently on vector computers. In other approaches the compensating mass is removed from the entire field by multiplying the value at every grid point by the ratio of the total original mass to the total nonnegative mass. Multiplicative compensation is computationally efficient, but it preferentially damps the regions of highest tracer concentration. Other filling algorithms are reviewed by Rood (1987). Although empirical testing has shown the value of filling algorithms, the theoretical basis for these schemes is largely undeveloped.

---

[10]*Negative definite* schemes may be similarly defined as any method that never generates positive values from nonpositive initial data. Any positive definite scheme can be trivially converted to a negative definite method.

5.8.1 An FCT Approach

A much better approach can be obtained using flux-corrected transport. Depending on exactly how it is implemented, the standard FCT method gives an essentially monotone scheme. The general FCT algorithm can, however, be greatly simplified if all that is required is a positive definite result. As noted by Smolarkiewicz (1989), any numerical conservation law of the form

$$\frac{\phi_j^{n+1} - \phi_j^n}{\Delta t} + \left(\frac{F_{j+\frac{1}{2}} - F_{j-\frac{1}{2}}}{\Delta x} \right) = 0 \tag{5.58}$$

can be converted to a positive definite method. In order to illustrate the approach in its simplest form, temporarily suppose that the fluxes are always positive (as would be the case if (5.58) were used to approximate an advection problem involving nonnegative flow velocities and tracer concentrations). Then (5.58) will be positive definite if the actual fluxes are replaced by corrected fluxes $C_{j+\frac{1}{2}} F_{j+\frac{1}{2}}$, in which the correction factor is defined by

$$C_{j+\frac{1}{2}} = \max \left(0, \min \left(F_{j+\frac{1}{2}}, \phi_j^n \frac{\Delta x}{\Delta t} \right) \right).$$

This correction ensures that the outgoing flux is not large enough to drive ϕ_j^{n+1} negative.

Now consider the general case, in which the fluxes may have arbitrary sign. The flux-correction coefficient can be determined by omitting steps 1–5 of the Zalesak correction algorithm presented in Section 5.4.2 and modifying steps 6 and 7 as follows. Let P_j^- be the total flux out of grid volume j,

$$P_j^- = \max \left(0, F_{j+\frac{1}{2}} \right) - \min \left(0, F_{j-\frac{1}{2}} \right),$$

and let Q_j^- be the maximum outward flux that can be supported without forcing ϕ_j^{n+1} negative,

$$Q_j^- = \phi_j^n \frac{\Delta x}{\Delta t}.$$

Evaluate a limiter ensuring that negatives will not be created at grid volume j,

$$R_j^- = \begin{cases} \min \left(1, Q_j^- / P_j^- \right) & \text{if } P_j^- > 0, \\ 0 & \text{if } P_j^- = 0. \end{cases}$$

Finally, choose the actual limiter for $F_{j+\frac{1}{2}}$ such that negatives are neither created in grid volume j nor $j + 1$:

$$C_{j+\frac{1}{2}} = \begin{cases} R_j^- & \text{if } A_{j+\frac{1}{2}} \geq 0, \\ R_{j+1}^- & \text{if } A_{j+\frac{1}{2}} < 0. \end{cases}$$

In contrast to the general FCT procedure, there is no initial step involving a low-order monotone scheme because it is not necessary to use a low-order solution to estimate the permissible range of values for ϕ_j^{n+1}. One simply sets $\phi_j^{\min} = 0$, imposes no constraint on $\phi_j^{\max}$, and corrects the fluxes to avoid generating values less than $\phi_j^{\min}$. Clearly, it is possible to further generalize this procedure by setting both $\phi_j^{\min}$ and $\phi_j^{\max}$ to any pair of arbitrarily specified constants.

5.8.2 Antidiffusion via Upstream Differencing

One unique way to obtain a positive definite advection scheme is to use upstream differencing to apply an anti-diffusive correction to a previously computed monotone solution (Smolarkiewicz 1983). The first step of the Smolarkiewicz algorithm is a standard upstream difference in conservation form,

$$\phi_j^* = \phi_j^n - \left[F(\phi_j^n, \phi_{j+1}^n, c_{j+\frac{1}{2}}) - F(\phi_{j-1}^n, \phi_j^n, c_{j-\frac{1}{2}}) \right], \qquad (5.59)$$

where

$$F(\phi_j, \phi_{j+1}, c) = \left[\left(\frac{c + |c|}{2} \right) \phi_j + \left(\frac{c - |c|}{2} \right) \phi_{j+1} \right] \frac{\Delta t}{\Delta x}.$$

The novel aspect of the Smolarkiewicz scheme is that the antidiffusion step is performed using a second upstream difference. Defining an "antidiffusion velocity"

$$\tilde{c}_{j+\frac{1}{2}} = \frac{(|c_{j+\frac{1}{2}}|\Delta x - c_{j+\frac{1}{2}}^2 \Delta t)}{(\phi_j^* + \phi_{j+1}^* + \epsilon)} \left(\frac{\phi_{j+1}^* - \phi_j^*}{\Delta x} \right)$$

(where ϵ is a small positive number whose presence guarantees that the denominator will be nonzero), the antidiffusion step is

$$\phi_j^{n+1} = \phi_j^* - \left[F(\phi_j^*, \phi_{j+1}^*, \tilde{c}_{j+\frac{1}{2}}) - F(\phi_{j-1}^*, \phi_j^*, \tilde{c}_{j-\frac{1}{2}}) \right]. \qquad (5.60)$$

Since the first step (5.59) is the standard upstream method, it is monotone, positive definite, and highly diffusive. If c is constant, the first step provides a second-order approximation to the modified equation

$$\frac{\partial \psi}{\partial t} + \frac{\partial c \psi}{\partial x} = \frac{\partial}{\partial x} \left(K \frac{\partial \psi}{\partial x} \right),$$

in which K is the numerical diffusivity

$$K = \frac{|c|\Delta x - c^2 \Delta t}{2}.$$

The second step compensates for this diffusion by subtracting off a finite-difference approximation to the leading-order truncation error associated with the upstream method. In particular, the second step (5.60) is a numerical approximation to

$$\frac{\partial \psi}{\partial t} = -\frac{\partial \tilde{c}\psi}{\partial x},$$

in which

$$\tilde{c} = \begin{cases} \dfrac{K}{\psi}\dfrac{\partial \psi}{\partial x}, & \text{if } \psi > 0; \\[2mm] 0, & \text{if } \psi = 0. \end{cases}$$

Although the second step utilizes upstream differencing, the ϕ^{n+1} are highly nonlinear functions of the ϕ^*, and the second step is not monotone. The second step will, nevertheless, be positive definite provided that

$$\left| \frac{\tilde{c}_{j+\frac{1}{2}}\Delta t}{\Delta x} \right| \le \frac{1}{2} \quad \text{for all } j,$$

which guarantees that even when both antidiffusive velocities are directed out of a particular grid cell, the antidiffusive fluxes will be too weak to generate a negative value. The preceding condition is satisfied whenever the initial ϕ are nonnegative and the maximum Courant number associated with the physical velocity field satisfies

$$\left| \frac{c_{j+\frac{1}{2}}\Delta t}{\Delta x} \right| \le 1 \quad \text{for all } j.$$

Then

$$\left| \frac{\tilde{c}_{j+\frac{1}{2}}\Delta t}{\Delta x} \right| \le \frac{|c_{j+\frac{1}{2}}|\Delta t}{\Delta x}\left(1 - \frac{|c_{j+\frac{1}{2}}|\Delta t}{\Delta x}\right)\left(\frac{|\phi^*_{j+1} - \phi^*_j|}{|\phi^*_{j+1} + \phi^*_j + \epsilon|}\right)$$
$$\le \frac{1}{4}\left(\frac{|\phi^*_{j+1} - \phi^*_j|}{|\phi^*_{j+1} + \phi^*_j + \epsilon|}\right).$$

Since the first step is monotone, all the ϕ^* are nonnegative and

$$\left| \frac{\tilde{c}_{j+\frac{1}{2}}\Delta t}{\Delta x} \right| \le \frac{1}{4}.$$

The Smolarkiewicz scheme can easily be extended to multidimensional problems (Smolarkiewicz 1984) and can be made monotonicity-preserving by applying limiters in the antidiffusion step (Smolarkiewicz and Grabowski 1990).

One consequence of the nonlinear dependence of the antidiffusive velocities on ϕ^* is that the solution obtained using the Smolarkiewicz scheme will change if a spatially uniform background field is added to the initial tracer concentration. This behavior differs from that of the true solution, in which the time tendency of the tracer concentration is determined only by the velocity field and the derivatives of the tracer-concentration field. Most of the other previously discussed methods for representing discontinuities and steep gradients avoid this dependence on the mean background concentration by using a different formulation of the nonlinear flux corrector. For example, the nonlinear correction used in the flux-limited scheme described in Section 5.5 is computed as a function of the ratio of the slopes of the numerical solution on each side of an individual grid point, and this ratio is independent of the magnitude of any horizontally uniform background concentration.

5.9   Curvilinear Coordinates

If the physical boundary constraining a fluid is nonrectangular, it can be advantageous to solve the governing equations in a curvilinear coordinate system that follows the boundary. In other circumstances, it is possible to simplify the problem by using cylindrical or spherical coordinates to exploit certain symmetries in the fluid system. When the governing equations are expressed in non-Cartesian coordinates, additional "metric" terms arise. These terms should be approximated in a way that preserves the conservation properties of the numerical scheme and the ability of the scheme to represent discontinuities and poorly resolved gradients. One elegant way to treat the metric terms is to begin with the equation formulated for an arbitrary curvilinear coordinate system and to apply one of the preceding methods directly to the transformed system (e.g., Smolarkiewicz and Margolin 1993). As an example, LeVeque's algorithm for two-dimensional tracer transport (described in Section 5.7.3) can be modified for use with curvilinear coordinates as follows.

Suppose that $(x_1, \ldots, x_n)$ is a position vector in Cartesian coordinates, that $(\tilde{x}_1, \ldots, \tilde{x}_n)$ is the corresponding vector in curvilinear coordinates, and that there is a smooth mapping between the two systems for which the Jacobian of the transformation $J = \text{Det}(\partial x_i / \partial \tilde{x}_j)$ is nonsingular. Then the velocities in the curvilinear coordinates are related to the Cartesian velocities such that

$$\tilde{v}_i = \frac{\partial \tilde{x}_i}{\partial x_k} v_k,$$

where repeated subscripts are summed. The divergence of the velocity vector transforms as

$$\frac{\partial v_i}{\partial x_i} = \frac{1}{J} \frac{\partial}{\partial \tilde{x}_k} (J \tilde{v}_k). \tag{5.61}$$

(See Gal-Chen and Somerville 1975.)

The equations governing the transport of a passive tracer in two-dimensional nondivergent flow may be expressed in curvilinear coordinates in either advective or flux form. Let $(x_1, x_2) = (x, y)$ and $(u_1, u_2) = (u, v)$. The advective form

$$\frac{d\psi}{dt} \equiv \frac{\partial\psi}{\partial t} + \tilde{u}\frac{\partial\psi}{\partial\tilde{x}} + \tilde{v}\frac{\partial\psi}{\partial\tilde{y}} = 0$$

can be derived from first principles using the definition of the total derivative in the transformed coordinates. The flux form

$$\frac{\partial\psi}{\partial t} + \frac{1}{J}\frac{\partial}{\partial\tilde{x}}(J\tilde{u}\psi) + \frac{1}{J}\frac{\partial}{\partial\tilde{y}}(J\tilde{v}\psi) = 0,$$

where

$$J = \frac{\partial x}{\partial\tilde{x}}\frac{\partial y}{\partial\tilde{y}} - \frac{\partial x}{\partial\tilde{y}}\frac{\partial y}{\partial\tilde{x}},$$

can also be derived from first principles using the expression for the divergence in transformed coordinates (5.61). The flux form implies conservation of ψ (provided that coordinate transformation is time-independent) and is ready for direct approximation by a numerical conservation law. The numerical fluxes can be limited or corrected as discussed previously to preserve monotonicity.

The proper formulation of a numerical approximation of the advective form is more subtle. As discussed previously, it is important to create a finite-difference approximation to the advective form that is algebraically equivalent to the flux form. This is achieved by the flux-limiter algorithm presented in Table 5.3, provided that the velocities satisfy the incompressible continuity equation on a staggered mesh,

$$\frac{1}{J_{i,j}}\left[(\delta_x(J_{i,j}\tilde{u}_{i,j}) + \delta_y(J_{i,j}\tilde{v}_{i,j})\right] = 0.$$

The velocities are staggered such that $\tilde{u}_{i+\frac{1}{2},j}$ is located $\Delta x/2$ to the "east" and $\tilde{v}_{i,j+\frac{1}{2}}$ is $\Delta y/2$ to the "north" of the grid point where $\phi_{i,j}$ is defined. In the absence of a flux limiter, the last equation in Table 5.3 is a second-order Lax–Wendroff approximation to

$$\frac{\partial\psi}{\partial t} + \frac{1}{J}\left(\tilde{u}J\frac{\partial\psi}{\partial\tilde{x}} + \tilde{v}J\frac{\partial\psi}{\partial\tilde{y}}\right) = 0,$$

where the common factor of J is not canceled out of the numerator and denominator because it is evaluated at different locations on the numerical mesh. The evaluation of J at these slightly different grid points is required to make the finite-difference method in advective form algebraically equivalent to a scheme in flux form.

Problems

1. Consider two sets of equations that might be supposed to govern one-dimensional shallow-water flow:

- *Initialize the fluxes to zero*

  for each i, j do

  $$F^n_{i-\frac{1}{2},j} = 0, \quad G^n_{i,j-\frac{1}{2}} = 0, \quad \Delta^n_{i,j} = 0$$

- *Increment F and G due to fluxes through the left cell interface*

  for each i, j do

  $$U = u^{n+\frac{1}{2}}_{i-\frac{1}{2},j} J_{i-\frac{1}{2},j}$$
  $$R = \phi^n_{i,j} - \phi^n_{i-1,j}$$

  if $U > 0$, then $I = i - 1$, else $I = i$

  $$\Delta^n_{i,j} = \Delta^n_{i,j} + U\left(\phi_{I+1,j} - \phi_{I,j}\right)/\Delta x$$

  if $U > 0$, then $I = i$, else $I = i - 1$

  if $v^{n+\frac{1}{2}}_{i,j+\frac{1}{2}} > 0$, then $G^n_{I,j+\frac{1}{2}} = G^n_{I,j+\frac{1}{2}} - \dfrac{\Delta t}{2\Delta x} R U v^{n+\frac{1}{2}}_{i,j+\frac{1}{2}}$

  if $v^{n+\frac{1}{2}}_{i,j-\frac{1}{2}} < 0$, then $G^n_{I,j-\frac{1}{2}} = G^n_{I,j-\frac{1}{2}} - \dfrac{\Delta t}{2\Delta x} R U v^{n+\frac{1}{2}}_{i,j-\frac{1}{2}}$

  if $U > 0$,   then $r^n_{i-\frac{1}{2},j} = (\phi^n_{i-1,j} - \phi^n_{i-2,j})/R,$

  else $r^n_{i-\frac{1}{2},j} = (\phi^n_{i+1,j} - \phi^n_{i,j})/R$

  $$F^n_{i-\frac{1}{2},j} = F^n_{i-\frac{1}{2},j} + \frac{|U|}{2}\left(1 - \frac{\Delta t}{\Delta x}|u_{i-\frac{1}{2},j}|\right) C(r^n_{i-\frac{1}{2},j})$$

- *Increment F and G due to fluxes through the bottom cell interface*

  (as above, switching the roles of i and j, u and v, and F and G)

- *Update ϕ*

  for each i, j do

  $$\phi^{n+1}_{i,j} = \phi^n_{i,j} - \frac{\Delta t}{J_{i,j}}\left(\Delta^n_{i,j} + \frac{F^n_{i+\frac{1}{2},j} - F^n_{i-\frac{1}{2},j}}{\Delta x} + \frac{G^n_{i,j+\frac{1}{2}} - G^n_{i,j-\frac{1}{2}}}{\Delta y}\right)$$

TABLE 5.3. Algorithm for executing one time step of LeVeque's two-dimensional flux-limited scheme in advective form on a curvilinear grid.

$$\frac{\partial u}{\partial t} + \frac{\partial}{\partial x}\left(\frac{u^2}{2} + gh\right) = 0, \qquad \frac{\partial h}{\partial t} + \frac{\partial hu}{\partial x} = 0$$

and

$$\frac{\partial hu}{\partial t} + \frac{\partial}{\partial x}\left(hu^2 + g\frac{h^2}{2}\right) = 0, \qquad \frac{\partial h}{\partial t} + \frac{\partial hu}{\partial x} = 0.$$

(a) Under what conditions do these systems have identical solutions?

(b) Give an example, including initial conditions and expressions for the time-dependent solutions, for which these systems have different solutions.

(c) In those situations where these systems have different solutions, which one serves as the correct mathematical model for shallow-water flow? (*Hint:* The correct choice must be determined from fundamental physical principles.)

2. Compute the speed at which the unit-amplitude jump (5.3) must propagate to be a weak solution to the conservation law

$$\frac{\partial \psi^2}{\partial t} + \frac{\partial}{\partial x}\left(\frac{2\psi^3}{3}\right) = 0.$$

How does this speed compare to that at which the same jump is propagated by the inviscid Burgers's equation? Explain whether the difference in the speed of these jumps is consistent with the sign of the inequality in the entropy condition for solutions to Burgers's equation (5.12)?

3. Show that if $\psi(x, 0) \geq 0$, the solution to

$$\frac{\partial \psi}{\partial t} + \frac{\partial}{\partial x}[c(x)\psi] = 0$$

remains nonnegative for all $t \geq 0$. Assume that c and ψ have continuous derivatives in order to simplify the argument. (*Hint:* In order to develop negative ψ, there must be a first time t_0 and some point x_0 for which $\psi(x_0, t_0) = 0$ and $\psi_t(x_0, t_0) < 0$. Show that this is impossible.) Does this result generalize to problems in two and three spatial dimensions?

4. Use the results of Problem 3 to show that if

$$\frac{\partial \psi}{\partial t} + \frac{\partial}{\partial x}[c(x)\psi] = 0, \qquad \frac{\partial \varphi}{\partial t} + \frac{\partial}{\partial x}[c(x)\varphi] = 0,$$

and $\psi(x, 0) \geq \varphi(x, 0)$, then $\psi(x, t) \geq \varphi(x, t)$ for all x and $t > 0$.

5. Suppose the constant-wind-speed advection equation (5.18) is approximated using the Lax–Fredrichs scheme

$$\phi_j^{n+1} = \frac{1}{2}(\phi_{j+1}^n + \phi_{j-1}^n) - \frac{\mu}{2}(\phi_{j+1}^n - \phi_{j-1}^n), \qquad (5.62)$$

where $\mu = c\Delta t/\Delta x$. Compare the implicit numerical diffusion generated by this scheme with that produced by upstream differencing. Show that the ratio of the leading-order numerical diffusion in the upstream scheme to that in the Lax–Fredrichs method is $\mu/(1 + \mu)$.

6. Suppose that the constant-wind-speed advection equation (5.18) is approximated using the scheme

$$\phi_j^{n+1} = \left(1 - c\Delta t\,\delta_{2x} + \gamma(\Delta x)^2\delta_x^2\right)\phi_j^n,$$

where γ is a user-specified parameter determining the amount of numerical smoothing.

(a) What is the largest value of γ for which the scheme can be monotone?

(b) Suppose that γ is specified as some value γ_0 for which the scheme can be monotone. For what values of $\mu = c\Delta t/\Delta x$ will the scheme actually be monotone?

(c) For what value of γ is this scheme equivalent to the Lax–Fredrichs scheme (5.62)?

7. Explain why a scheme that is monotonicity-preserving need not be TVD (or more precisely, total variation nonincreasing). Explain why being TVD does not imply that a scheme is monotone.

8. Show that no new maxima or minima can develop in smooth solutions to the conservation law (5.6).

9. Show that Harten's criterion (5.32) insuring that schemes of the form (5.31) are TVD is not sufficient to guarantee that they are monotone.

10. Show that the MC flux limiter (5.40) is equivalent to the following slope limiter

$$\sigma_j = \begin{cases} 0 & \text{if } ab \le 0, \\ \mathrm{sgn}(a)|a + b|/2 & \text{otherwise}, \end{cases}$$

where

$$a = \frac{\phi_{j+1} - \phi_j}{\Delta x}$$

and

$$b = \begin{cases} \dfrac{\phi_j - \phi_{j-1}}{\Delta x} & \text{if } c \ge 0, \\[2mm] \dfrac{\phi_{j+2} - \phi_{j+1}}{\Delta x} & \text{if } c < 0. \end{cases}$$

11. Suppose that Lax–Wendroff solutions are sought to a one-dimensional advection equation (5.18) and that the velocity $c(t)$ depends on time but not on x.

(a) Derive an expression for $\partial^2 \psi / \partial t^2$ in terms of the spatial derivatives of ψ and functions of the velocity field.

(b) Show that a fully second-order Lax–Wendroff approximation to this problem can be obtained using (5.41) with c replaced by $(c^{n+1} + c^n)/2$.

12. Show that the antidiffusion step (5.60) of the Smolarkiewicz positive definite advection scheme is not monotone.

13. Suppose that $f(s)$ is a continuously differentiable function of s and that $\psi(x, t)$ is a solution to the scalar conservation law (5.6). Show that the characteristic curves for this hyperbolic partial differential equation are straight lines.

14. *Compute solutions to the advection equation (5.18) on the periodic domain $0 \le x \le 1$ subject to the initial condition $\psi(x, 0) = \sin^6(2\pi x)$. Let $c = 0.1$.

(a) Compare the exact solution with numerical solutions obtained using forward, Lax–Wendroff, and flux-limited methods. In the flux-limited methods compute the low-order flux using the upstream scheme and the high-order flux using the Lax–Wendroff method, but try three different flux limiters: the MC, the minmod, and the superbee. Perform the simulations using a Courant number $c\Delta t/\Delta x = 0.5$ and $\Delta x = 1/40$. As part of your discussion submit two plots of the solution at time $t = 20$, one comparing the exact solution with that obtained using the three different flux limiters, and one comparing the exact, upstream, Lax–Wendroff, and MC flux-limited solutions. Scale the vertical axis so that $-0.4 \le \psi(x) \le 1.4$.

(b) Repeat the preceding simulations for the initial condition

$$\psi(x, 0) = \begin{cases} 1 & \text{if } |x - \tfrac{1}{2}| \le \tfrac{1}{4}; \\ 0 & \text{otherwise.} \end{cases}$$

Again submit two plots of the solution at time $t = 20$, one comparing the exact solution with that obtained using the three different flux limiters, and one comparing the exact, upstream, Lax–Wendroff, and MC flux-limited solutions. Discuss your results.

15. *Determine the effective order of accuracy of the minmod, MC, and superbee flux-limited approximations to the advection equation considered in Problem 14 except use the very smooth initial data $\psi(x, 0) = \sin(2\pi x)$. In addition, compute results for the Zalesak FCT method using upstream differencing for the low-order solution and the Lax–Wendroff scheme for the higher-order solution. Also try the iterative FCT scheme discussed at the end of Section 5.4.2 using the preceding noniterated FCT solution for the low-order scheme during the second iteration. Keep the Courant number fixed at 0.5, and use $\Delta x = 1/20$, $1/40$, $1/80$, $1/160$, and $1/320$. Compute the ℓ_2-norm of the difference between the exact and approximate solutions and

plot the log of the error versus the log of Δx to estimate the power of Δx that is proportional to the error as $\Delta x \to 0$. Compare these results to the theoretical order of accuracy for the standard upstream and Lax–Wendroff methods.

16. *Compare simulations of the geostrophic adjustment problem described in Problem 12 of Chapter 3 obtained using a flux-limiter scheme with the MC limiter and flux-corrected transport. Consider both the initial conditions: the discontinuous step and the slightly smoothed step. In order to use the constant-wind-speed advection algorithms presented in this chapter, transform the governing equations to an equivalent system for the unknown functions $u + g\eta/c$, $u - g\eta/c$, and v (where $c^2 = gH$). Use upstream differencing and the Lax–Wendroff scheme for the monotone and second-order methods. Treat the Coriolis terms in the transformed system via operator splitting.

6

Semi-Lagrangian Methods

Most of the fundamental equations in fluid dynamics can be derived from first principles in either a *Lagrangian* form or an *Eulerian* form. Lagrangian equations describe the evolution of the flow that would be observed following the motion of an individual parcel of fluid. Eulerian equations describe the evolution that would be observed at a fixed point in space (or at least at a fixed point in a coordinate system such as the rotating Earth whose motion is independent of the fluid). If $S(x, t)$ represents the sources and sinks of a chemical tracer $\psi(x, t)$, the evolution of the tracer in a one-dimensional flow field may be alternatively expressed in Lagrangian form as

$$\frac{d\psi}{dt} = S, \tag{6.1}$$

or in Eulerian form as

$$\frac{\partial \psi}{\partial t} + u\frac{\partial \psi}{\partial x} = S.$$

The mathematical equivalence of these two equations follows from the definition of the total derivative,

$$\frac{d}{dt} = \frac{\partial}{\partial t} + \frac{dx}{dt}\frac{\partial}{\partial x},$$

and the definition of the velocity,

$$\frac{dx}{dt} = u. \tag{6.2}$$

One strategy for the solution of (6.1) as an initial value problem would be to choose a regularly spaced distribution of fluid parcels at the initial time, assign a value for ψ to each of these fluid parcels from the initial condition, and then integrate the ordinary differential equations (6.1) and (6.2) to determine the location and the tracer concentration of each parcel as a function of time. The difficulty with this strategy is that in most practical applications the distribution of the fluid parcels eventually becomes highly nonuniform, and the numerical approximation of $\psi(x, t)$ becomes inaccurate in regions where the fluid parcels are widely separated. In theory, this situation can be improved by adding new parcels to those regions where the initial parcels have become widely separated and removing parcels from regions where the parcels have become too concentrated. It is, however, difficult to create a simple algorithm for adding and removing fluid parcels in response to their evolving distribution within the fluid.

A much better scheme for regulating the number and distribution of the fluid parcels can be obtained by choosing a completely new set of parcels at every time step. The parcels making up this set are those arriving at each node on a regularly spaced grid at the end of each step. As noted by Wiin-Nielsen (1959), this approach, known as the *semi-Lagrangian method*, keeps the fluid parcels evenly distributed throughout the fluid and facilitates the computation of spatial derivatives via finite differences. As an illustration of this approach, let $t^n = n\Delta t$ and $x_j = j\Delta x$; then a semi-Lagrangian approximation to (6.1) can be written using the trapezoidal scheme

$$\frac{\phi\left(x_j, t^{n+1}\right) - \phi\left(\tilde{x}_j^n, t^n\right)}{\Delta t} = \frac{1}{2}\left[S\left(x_j, t^{n+1}\right) + S\left(\tilde{x}_j^n, t^n\right)\right], \qquad (6.3)$$

where ϕ is the numerical approximation to ψ, and $\tilde{x}_j^n$ is the estimated x-coordinate of the departure point of the trajectory originating at time t^n and arriving at (x_j, t^{n+1}). The value of $\tilde{x}_j^n$ is computed by numerically integrating (6.2) backward over a time interval of Δt starting from the initial condition $x(t^{n+1}) = x_j$. Then, since the endpoint of the backward trajectory is unlikely to coincide with a grid point, $\phi(\tilde{x}_j^n, t^n)$ and $S(\tilde{x}_j^n, t^n)$ must be obtained by interpolation.

Semi-Lagrangian methods are of considerable practical interest because in some applications they are more efficient than competing Eulerian schemes. Another advantage of the semi-Lagrangian approach is that it is easy to use in problems with nonuniform grids. In addition, semi-Lagrangian schemes avoid the primary source of nonlinear instability in most geophysical wave-propagation problems because the nonlinear advection terms appearing in the Eulerian form of the momentum equations are eliminated when those equations are expressed in a Lagrangian frame of reference.

As a result of the pioneering work by Robert (1981, 1982), semi-Lagrangian semi-implicit methods have become one of the most popular architectures used in global weather forecast models. An extensive review of the application of semi-Lagrangian methods to atmospheric problems is provided by Staniforth and Côté (1991). Semi-Lagrangian methods are also used in a variety of other fluid-dynamical applications, where they are sometimes referred to as *Eulerian–Lagrangian* methods (e.g., Oliveira and Baptista 1995).

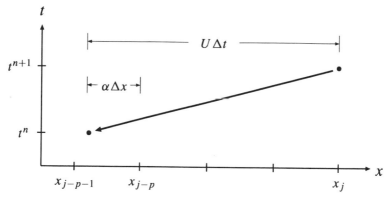

FIGURE 6.1. Backward trajectory from (x_j, t^{n+1}) to $(\tilde{x}_j^n, t^n)$.

6.1   The Scalar Advection Equation

The stability and accuracy of Eulerian finite difference methods were first examined in Chapter 2 by studying the constant-wind-speed advection equation. We will begin the analysis of semi-Lagrangian schemes by considering the same problem, and then investigate the additional considerations that arise when variations in the velocity field make the backward trajectory calculation nontrivial.

6.1.1   Constant Velocity

A semi-Lagrangian approximation to the advection equation for a passive tracer can be written in the form

$$\frac{\phi(x_j, t^{n+1}) - \phi(\tilde{x}_j^n, t^n)}{\Delta t} = 0, \tag{6.4}$$

where $\tilde{x}_j^n$ again denotes the departure point of a trajectory originating at time t^n and arriving at (x_j, t^{n+1}). If the velocity is constant, the backward trajectory computation is trivial, and letting U denote the wind speed,

$$\tilde{x}_j^n = x_j - U \Delta t.$$

Let p be the integer part of $U \Delta t / \Delta x$ and without loss of generality suppose that $U \geq 0$; then $\tilde{x}_j^n$ lies in the interval $x_{j-p} \leq x < x_{j-p-1}$, as shown in Fig. 6.1. Defining

$$\alpha = \frac{x_{j-p} - \tilde{x}_j^n}{\Delta x}$$

and approximating $\phi(\tilde{x}_j^n, t^n)$ by linear interpolation, (6.4) becomes

$$\phi_j^{n+1} = (1 - \alpha)\phi_{j-p}^n + \alpha \phi_{j-p-1}^n, \tag{6.5}$$

where $\phi_j^n = \phi(x_j, t^n)$. Note that if Δt is small enough that

$$0 \le U \frac{\Delta t}{\Delta x} \le 1,$$

(6.5) reduces to the formula for Eulerian upstream differencing.

Stability

Following Bates and McDonald (1982), the stability of the preceding semi-Lagrangian approximation can be analyzed using Von Neumann's method. Substituting a solution of the form $\phi_j^n = A_k^n e^{i(kj\Delta x)}$ into (6.5), one obtains

$$A_k = \left[1 - \alpha(1 - e^{-ik\Delta x})\right] e^{-ikp\Delta x},$$

from which it follows that

$$|A_k|^2 = 1 - 2\alpha(1 - \alpha)(1 - \cos k\Delta x).$$

This is the same expression obtained for the amplification factor associated with upstream differencing, except that α has replaced the Courant number in (2.26). Based on the analysis given in Section 2.2.2, the amplification factor for all waves resolved on the numerical mesh will be less than or equal to unity, provided that

$$0 \le \alpha \le 1,$$

which is always satisfied because the estimated departure point always lies between grid points x_{j-p} and x_{j-p-1}. It is possible to take arbitrarily large time steps without violating the Courant–Fredrichs–Lewy condition because the backward trajectory calculation ensures that the numerical domain of dependence includes the domain of dependence of the true solution.

Errors will be made in the backward trajectory calculation in practical applications where the wind speed is not constant. These errors will affect the accuracy of the solution, and if they do not go to zero as $\Delta x \to 0$ and $\Delta t \to 0$, they may prevent the numerical solution from converging to the correct solution. Nevertheless, as long as the interpolation is performed using data from the two grid points surrounding the estimated departure point, the maximum norm of the solution will not grow with time.

Accuracy

The truncation error of the preceding semi-Lagrangian scheme can be determined by substituting appropriate Taylor series expansions of the continuous solution into the finite difference scheme[1]

$$\frac{\phi_j^{n+1} - \left[(1 - \alpha)\phi_{j-p}^n + \alpha\phi_{j-p-1}^n\right]}{\Delta t} = 0. \tag{6.6}$$

---

[1] According to the discussion in Section 2.3.2, the global truncation error is of same the order as the leading-order errors in the numerical approximation to the differential form of the governing equation (6.6) and is one power of Δt lower than the truncation error in the integrated form (6.5).

In the case of the unforced scalar advection equation, it is helpful to perform the Taylor series expansions about the point $(\tilde{x}_j^n, t^n)$ because this isolates the errors in the trajectory calculations from those generated by the interpolation of the tracer field. Let $\psi_d = \psi(\tilde{x}_j^n, t^n)$; then the error produced by linear interpolation is

$$(1 - \alpha)\psi_{j-p}^n + \alpha\psi_{j-p-1}^n = \psi_d + \alpha(1 - \alpha)\frac{(\Delta x)^2}{2}\frac{\partial^2\psi}{\partial x^2}\bigg|_d + O\left[(\Delta x)^3\right]. \quad (6.7)$$

Since the wind speed is constant, the backward trajectory is exact and $\psi_j^{n+1} = \psi_d$. This can be verified by expanding ψ_j^{n+1} in a Taylor series. Defining $s = x_j - \tilde{x}_j^n$,

$$\psi_j^{n+1} = \psi_d + \Delta t\frac{\partial\psi}{\partial t}\bigg|_d + s\frac{\partial\psi}{\partial x}\bigg|_d$$
$$+ \frac{(\Delta t)^2}{2}\frac{\partial^2\psi}{\partial t^2}\bigg|_d + s\Delta t\frac{\partial^2\psi}{\partial t\partial x}\bigg|_d + \frac{s^2}{2}\frac{\partial^2\psi}{\partial x^2}\bigg|_d + \cdots, \quad (6.8)$$

and since $s = U\Delta t$, the preceding reduces to

$$\psi_j^{n+1} = \psi_d + \Delta t\left(\frac{\partial}{\partial t} + U\frac{\partial}{\partial x}\right)\psi\bigg|_d + \frac{(\Delta t)^2}{2}\left(\frac{\partial}{\partial t} + U\frac{\partial}{\partial x}\right)^2\psi\bigg|_d + \cdots$$
$$= \psi_d. \quad (6.9)$$

Substituting (6.7) and (6.9) into (6.6) yields

$$\frac{\psi_j^{n+1} - \left[(1 - \alpha)\psi_{j-p}^n + \alpha\psi_{j-p-1}^n\right]}{\Delta t} \approx -\frac{\alpha(1 - \alpha)}{2}\frac{(\Delta x)^2}{\Delta t}\frac{\partial^2\psi}{\partial x^2}\bigg|_d. \quad (6.10)$$

If the Courant number is held constant as $\Delta t \to 0$ and $\Delta x \to 0$, the truncation error is clearly $O(\Delta x)$. If $\Delta x/\Delta t \to 0$ as $\Delta x \to 0$ and $\Delta t \to 0$, the error is no larger than $O(\Delta x)$. It may appear that the semi-Lagrangian scheme could be inconsistent in the limit $\Delta t/\Delta x \to 0$, but once the Courant number drops below unity, $\alpha = U\Delta t/\Delta x$, and using

$$\frac{\partial^2\psi}{\partial t^2} = U^2\frac{\partial^2\psi}{\partial x^2},$$

the leading-order truncation error reduces to

$$\frac{\Delta t}{2}\frac{\partial^2\psi}{\partial t^2} - U\frac{\Delta x}{2}\frac{\partial^2\psi}{\partial x^2}.$$

This is identical to the leading-order truncation error in Eulerian upstream differencing (2.12). The global truncation error of the semi-Lagrangian scheme (6.5) is therefore of first order in space and time.

Consistent with (6.10), the preceding semi-Lagrangian scheme is exact whenever the Courant number is an integer, i.e., whenever the departure point exactly coincides with a grid point. This may be compared with Eulerian upstream differencing, which is exact for a Courant number of unity (see Sections 2.5.1 and 2.5.2). In practical applications the wind speed and therefore the Courant number are functions of space and time. Stability constraints require the Eulerian upstream method to be integrated using a time step that ensures that the maximum Courant number will be less than one at every point within the computational domain. As a consequence of this restriction on Δt, the domain-averaged Courant number is often substantially less than the optimal value of unity. In contrast, the unconditional stability of the semi-Lagrangian scheme allows the time step to be chosen such that the average value of the Courant number is unity—or any integer—thereby reducing the average truncation error throughout the computational domain.

Higher-Order Interpolation

Upstream differencing generates too much numerical diffusion to be useful in practical computations involving Eulerian problems with smooth solutions. A similar situation holds in the Lagrangian framework, where linearly interpolating the tracer field also generates too much diffusion. Higher-order interpolation is therefore used in most semi-Lagrangian approximations to equations with smooth solutions. If x_{j-p} is the nearest grid point to the estimated departure point and a quadratic polynomial is fit to the three closest grid-point values of the tracer field,

$$\phi(\tilde{x}_j^n, t^n) = \frac{\alpha}{2}(1 + \alpha)\phi_{j-p-1}^n + (1 - \alpha^2)\phi_{j-p}^n - \frac{\alpha}{2}(1 - \alpha)\phi_{j-p+1}^n, \quad (6.11)$$

where as before, $p + \alpha = U\Delta t$, except that p is now chosen such that $|\alpha| \leq \frac{1}{2}$. Substituting the preceding into (6.4) yields a semi-Lagrangian scheme that approximates the constant wind-speed advection equation to $O\left[(\Delta x)^3/\Delta t\right]$, which gives second-order accuracy. In the limit $\Delta t/\Delta x \to 0$ this scheme is identical to the Lax–Wendroff method (2.102).

Cubic interpolation is widely used in practical applications. If p is the integer part of $U\Delta t/\Delta x$ with $U > 0$ and the cubic is defined to match ϕ at the four closest grid-point values to the departure point, then

$$\phi(\tilde{x}_j^n, t^n) = -\frac{\alpha(1 - \alpha^2)}{6}\phi_{j-p-2}^n + \frac{\alpha(1 + \alpha)(2 - \alpha)}{2}\phi_{j-p-1}^n$$
$$+ \frac{(1 - \alpha^2)(2 - \alpha)}{2}\phi_{j-p}^n - \frac{\alpha(1 - \alpha)(2 - \alpha)}{6}\phi_{j-p+1}^n. \quad (6.12)$$

The preceding is expressed in the form of a Lagrange interpolating polynomial and is an efficient choice if several fields are to be interpolated to the same departure point, since the coefficients of the ϕ_i needn't be recalculated for each field. If only one field is being interpolated, (6.12) can be evaluated more efficiently by writing it as a Newton polynomial (see Dahlquist and Björck 1974 and Problem 2).

The leading-order truncation error in a cubic semi-Lagrangian approximation to the constant-wind-speed advection equation is third order in the perturbations (specifically, $O\left[(\Delta x)^4/\Delta t\right]$). More generally, the local error produced by pth-order polynomial interpolation is $O\left[(\Delta x)^{p+1}\right]$, and the global truncation error in the corresponding semi-Lagrangian approximation to the constant-wind-speed problem is $O\left[(\Delta x)^{p+1}/\Delta t\right]$ (McDonald 1984). In most applications, the motivation for using high-order polynomials to interpolate the tracer field is not to accelerate the convergence of the numerical solution to the correct solution as $\Delta x \to 0$ and $\Delta t \to 0$, but rather to improve the accuracy of the marginally resolved waves. This is the same reason that high-order finite differences are often used to approximate the spatial derivative in Eulerian schemes for the solution of the advection equation.

Two Spatial Dimensions

A semi-Lagrangian approximation to the advection equation in a two-dimensional flow may be expressed as

$$\frac{\phi(x_{m,n}, y_{m,n}, t^{n+1}) - \phi(\tilde{x}^n_{m,n}, \tilde{y}^n_{m,n}, t^n)}{\Delta t} = 0,$$

where $(\tilde{x}^n_{m,n}, \tilde{y}^n_{m,n})$ denotes the departure point of a trajectory originating at time t^n and arriving at $(x_{m,n}, y_{m,n}, t^{n+1})$. Let U and V denote the x and y velocity components, and as before, suppose they are constant and nonnegative. Denote the integer parts of $U\Delta t$ and $V\Delta t$ by p and q respectively and define $\alpha = (U\Delta t - p)/\Delta x$ and $\beta = (V\Delta t - q)/\Delta y$. Then a first-order approximation to $\phi(\tilde{x}^n_{m,n}, \tilde{y}^n_{m,n}, t^n)$ can be obtained using bilinear interpolation, which yields the semi-Lagrangian scheme

$$\phi^{n+1}_{m,n} = (1 - \alpha)\left[(1 - \beta)\phi^n_{m-p,n-q} + \beta\phi^n_{m-p,n-q-1}\right]$$
$$+ \alpha\left[(1 - \beta)\phi^n_{m-p-1,n-q} + \beta\phi^n_{m-p-1,n-q-1}\right].$$

When the Courant numbers along the x- and y-axes are less than unity, $p = q = 0$, and the preceding reduces to the CTU method (3.35).

A second-order approximation can be obtained using biquadratic interpolation. In order to abbreviate the notation, let $\phi_C = \phi_{m-p,n-q}$ and denote the surrounding points using the compass directions (north, northeast, east, ...) such that $\phi_N = \phi^n_{m-p,n-q+1}$, $\phi_{NE} = \phi^n_{m-p+1,n-q+1}$, etc. If p and q are now chosen such that $|\alpha| \leq \frac{1}{2}$ and $|\beta| \leq \frac{1}{2}$, the resulting semi-Lagrangian scheme has the form

$$\phi^{n+1}_C = \frac{\alpha}{2}\left[\frac{\beta}{2}(1 + \beta)\phi^n_{SW} + (1 - \beta^2)\phi^n_W - \frac{\beta}{2}(1 - \beta)\phi^n_{NW}\right]$$
$$+ (1 - \alpha^2)\left[\frac{\beta}{2}(1 + \beta)\phi^n_S + (1 - \beta^2)\phi^n_C - \frac{\beta}{2}(1 - \beta)\phi^n_N\right]$$
$$- \frac{\alpha}{2}(1 - \alpha)\left[\frac{\beta}{2}(1 + \beta)\phi^n_{SE} + (1 - \beta^2)\phi^n_E - \frac{\beta}{2}(1 - \beta)\phi^n_{NE}\right]$$

If the Courant numbers along the x- and y-axes are less than unity, this scheme reduces to the upstream biased Lax–Wendroff method (3.38). Von Neumann stability analysis can be used to show that the preceding bilinear and biquadratic semi-Lagrangian schemes are unconditionally stable (Bates and McDonald 1982).

The preceding interpolation formula generalizes to higher-order polynomials and three dimensions in a straightforward way. Since the evaluation of a three-dimensional high-order interpolating polynomial requires considerable computation, the exact formulae are sometimes approximated. Ritchie et al. (1995), for example, simplify the full expression for three-dimensional cubic interpolation by neglecting the "corner" points.

6.1.2   Variable Velocity

Now consider the case where the velocity is a function of space and time, and the backward trajectory of each fluid parcel must be estimated by a numerical integration. The truncation error in the variable-velocity case can again be determined by expanding $\psi(x, t)$ in a Taylor series about the estimated departure point and evaluating an expression of the form

$$\frac{1}{\Delta t}\left(\psi_j^{n+1} - \psi_d\right) + \frac{1}{\Delta t}\left(\psi_d - \sum_{k=-r}^{s} \beta_k \psi_{j-p+k}^n\right), \tag{6.13}$$

where $\psi_d = \psi(\tilde{x}_j^n, t^n)$ and the summation represents an $(r+s)$-order polynomial interpolation of ψ^n to the departure point. The first term in the preceding is determined by the error in the trajectory calculation, and the second term is determined by the error in the interpolation of ψ^n to the estimated departure point. The error generated in interpolating ψ to the estimated departure point is the same as that for the constant velocity case, but the estimated departure point will not generally coincide with the true departure point, and as a consequence, the ψ_j^{n+1} will no longer be identical to ψ_d. In order to determine the difference between ψ_j^{n+1} and ψ_d, let x^n denote the position of a fluid parcel at time t^n and suppose that backward trajectories are computed subject to the initial condition $x^{n+1} = x_j$.

First suppose the trajectory is computed using Euler's method

$$\tilde{x}_j^n = x^{n+1} - u(x^{n+1}, t^n)\Delta t.$$

As before, define $s = x_j - \tilde{x}_j^n = x^{n+1} - \tilde{x}_j^n$. Then

$$s = u(x^{n+1}, t^n)\Delta t \tag{6.14}$$

$$= \Delta t \left[u(\tilde{x}_j^n, t^n) + s\frac{\partial u}{\partial x}(\tilde{x}_j^n, t^n) + O\left(s^2\right)\right] \tag{6.15}$$

$$= u_d \Delta t + O\left[(\Delta t)^2\right], \tag{6.16}$$

where $u_d = u(\tilde{x}_j^n, t^n)$ and the last equality is obtained by substituting (6.14) into (6.15). The difference between ψ_j^{n+1} and ψ_d is then determined by substituting

(6.16) into (6.8) to obtain

$$\psi_j^{n+1} = \psi_d + \Delta t \left(\frac{\partial}{\partial t} + u \frac{\partial}{\partial x} \right) \psi \bigg|_d + O\left[(\Delta t)^2 \right]$$

$$= \psi_d + O\left[(\Delta t)^2 \right]. \tag{6.17}$$

According to (6.13) the contribution from the trajectory calculation to the global truncation error in the semi-Lagrangian scheme is $\left(\psi_j^{n+1} - \psi_d \right) / \Delta t$, so the error generated by the Euler method is $O(\Delta t)$.

A second-order-accurate result can be obtained using the two-stage trajectory calculation

$$x_* = x^{n+1} - u(x^{n+1}, t^n) \Delta t / 2, \tag{6.18}$$

$$\tilde{x}_j^n = x^{n+1} - u(x_*, t^{n+\frac{1}{2}}) \Delta t. \tag{6.19}$$

As before, the initial condition is $x^{n+1} = x_j$. This differs slightly from the classical Runge–Kutta midpoint method discussed in Section 2.3.3 in that the first stage uses the velocity $u(x^{n+1}, t^n)$ rather than $u(x^{n+1}, t^{n+1})$; the latter is more convenient if u^{n+1} is being predicted at the same time as ϕ^{n+1}. The second-order accuracy of this calculation can be verified as follows. Substitute (6.18) into (6.19) and let the superscript $n + 1$ denote evaluation at $(x(t^{n+1}), t^{n+1})$. Then

$$\tilde{x}_j^n = x^{n+1} - \Delta t\, u \left(x^{n+1} - u(x^{n+1}, t^n) \Delta t / 2,\ t^{n+\frac{1}{2}} \right) \tag{6.20}$$

$$= x^{n+1} - \Delta t \left[u^{n+1} - \frac{\Delta t}{2} \left(u \frac{\partial u}{\partial x} \right)^{n+1} - \frac{\Delta t}{2} \left(\frac{\partial u}{\partial t} \right)^{n+1} + O\left[(\Delta t)^2 \right] \right]$$

$$= x^{n+1} - \Delta t\, u^{n+1} + \frac{(\Delta t)^2}{2} \left(\frac{du}{dt} \right)^{n+1} + O\left[(\Delta t)^3 \right]$$

$$= x^{n+1} - \Delta t \left(\frac{dx}{dt} \right)^{n+1} + \frac{(\Delta t)^2}{2} \left(\frac{d^2 x}{dt^2} \right)^{n+1} + O\left[(\Delta t)^3 \right]. \tag{6.21}$$

Since the right side of (6.21) matches the Taylor series expansion of $\tilde{x}_j^n$ about $x(t^{n+1})$ to within an error of $O\left[(\Delta t)^3 \right]$, the global truncation error in the back-trajectory calculation is $O\left[(\Delta t)^2 \right]$ (Iserles 1996, p. 7).

Now consider the error generated when the Runge–Kutta scheme is used to compute the back trajectory in a semi-Lagrangian scheme. In many practical applications the velocity data are available only at discrete points on a space–time grid, and in order to evaluate (6.19), $u(x_*, t^{n+\frac{1}{2}})$ must be estimated by interpolation or extrapolation. Before examining the errors introduced by such interpolation and extrapolation, consider those cases where the velocity can be evaluated exactly, so the only errors arising in the trajectory calculations are those generated by the Runge–Kutta scheme itself. Using the definition $s = x_j - \tilde{x}_j^n$, (6.20) becomes

$$s = \Delta t\, u\Big(\tilde{x}_j + s - u(\tilde{x}_j + s, t^n)\Delta t/2,\; t^n + \Delta t/2\Big).$$

Thus, $s = u_d\Delta t + O\left[(\Delta t)^2\right]$, which may be substituted into the right side of the preceding to yield

$$s = u_d\Delta t + \frac{(\Delta t)^2}{2}\left(\frac{\partial}{\partial t} + u\frac{\partial}{\partial x}\right)u\bigg|_d + O\left[(\Delta t)^3\right]. \tag{6.22}$$

Substituting (6.22) into (6.8) gives

$$\psi_j^{n+1} = \psi_d + \Delta t\left(\frac{\partial}{\partial t} + u\frac{\partial}{\partial x}\right)\psi\bigg|_d + \frac{(\Delta t)^2}{2}\left(\frac{\partial}{\partial t} + u\frac{\partial}{\partial x}\right)^2\psi\bigg|_d + O\left[(\Delta t)^3\right],$$

which implies that the Runge–Kutta scheme (6.18)–(6.19) generates an $O\left[(\Delta t)^2\right]$ contribution toward the total error in the semi-Lagrangian approximation.

Now suppose that the velocity data are available only at discrete locations on the space–time mesh. Ideally, the velocity at time $t^{n+\frac{1}{2}}$ would be computed by interpolation between times t^n and t^{n+1}. Such interpolation cannot, however, be performed when semi-Lagrangian methods are used to solve prognostic equations for the velocity itself, because the velocity at t^{n+1} will be needed for the trajectory calculations before it has been computed. This problem is generally avoided by extrapolating the velocity field forward in time using data from the two previous time levels such that

$$u(t^{n+\frac{1}{2}}) = \frac{3}{2}u(t^n) - \frac{1}{2}u(t^{n-1}). \tag{6.23}$$

Suppose that the extrapolated velocity field at $u(t^{n+\frac{1}{2}})$ is then linearly interpolated to x_* using data at the nearest spatial nodes, and let u_* denote this interpolated and extrapolated velocity. Since linear interpolation and extrapolation are second-order accurate,

$$u(x_*, t^{n+\frac{1}{2}}) = u_* + O\left[(\Delta x)^2\right] + O\left[(\Delta t)^2\right].$$

Substituting the preceding into (6.20) shows that the use of u_* instead of the exact velocity adds an $O\left[\Delta t(\Delta x)^2\right] + O\left[(\Delta t)^3\right]$ error to the back-trajectory calculation, and thereby contributes a term of $O\left[(\Delta x)^2\right] + O\left[(\Delta t)^2\right]$ to the global truncation error in the semi-Lagrangian solution. A fully second-order semi-Lagrangian scheme can therefore be obtained by (i) using (6.18) to estimate the midpoint of the back trajectory, (ii) computing u_* by linearly interpolating and extrapolating the velocity field, (iii) determining the departure point from

$$\tilde{x}_j^n = x^{n+1} - u_*\Delta t,$$

(iv) evaluating $\phi(\tilde{x}_j^n, t^n)$ using quadratic interpolation, and (v) setting ϕ_j^{n+1} to this value.

A variety of other schemes have been proposed to compute back trajectories. One popular scheme, the second-order implicit midpoint method

$$\tilde{x}_j^n = x_j - u\left((x_j + \tilde{x}_j^n)/2, t^{n+\frac{1}{2}}\right) \Delta t, \tag{6.24}$$

is typically solved by iteration. The Runge–Kutta scheme (6.18)–(6.19) can be considered a two-step iterative approximation to (6.24), but even if the implicit midpoint method is iterated to convergence, its formal order of accuracy is no greater than that obtained using (6.18) and (6.19). A second alternative scheme can be used if the velocities are being calculated as prognostic variables during the integration. Then the value of du/dt can be saved after the evaluation of the forcing terms in the momentum equation and employed in subsequent trajectory calculations using the second-order scheme

$$x_* = x^{n+1} - u(x^{n+1}, t^n)\Delta t,$$
$$\tilde{x}_j^n = x^{n+1} - u(x_*, t^n)\Delta t + \frac{(\Delta t)^2}{2}\frac{du}{dt}(x_*, t^n)$$

(Krishnamurti et al. 1990; Smolarkiewicz and Pudykiewicz 1992).

Higher-order schemes for the computation of back trajectories have also been devised (Temperton and Staniforth 1987). Although a third-order trajectory scheme must be employed as part of any fully third-order semi-Lagrangian method, higher-order trajectory computations are not widely used. In many of the applications where semi-Lagrangian methods are most advantageous it is easier to accurately compute the back trajectory than it is to accurately interpolate all the resolved scales in the tracer field. As a consequence, second-order schemes are often used for the back trajectory calculation even when the tracer field is interpolated using cubic or higher-order polynomials. Moreover, in those problems where there is nonzero forcing in the Lagrangian reference frame, the time integral of the forcing is seldom approximated to more than second-order accuracy, and in such circumstances the use of a higher-order scheme to compute the back trajectory will not reduce the overall time-truncation error of the semi-Lagrangian scheme below $O\left[(\Delta t)^2\right]$.

6.2   Forcing in the Lagrangian Frame

The forced scalar advection equation (6.1) provides a simple example in which to study the treatment of forcing terms in semi-Lagrangian schemes. Defining $\tilde{x}_j^{n-1}$ to be an estimate of the departure point of the fluid parcel at time t^{n-1} that arrives at (x_j, t^{n+1}), second-order approximations to the forcing may be obtained using the trapezoidal method (6.3), the leapfrog scheme

$$\frac{\phi\left(x_j, t^{n+1}\right) - \phi\left(\tilde{x}_j^{n-1}, t^{n-1}\right)}{2\Delta t} = S\left(\tilde{x}_j^n, t^n\right), \tag{6.25}$$

or the second-order Adams–Bashforth method

$$\frac{\phi\left(x_j, t^{n+1}\right) - \phi\left(\tilde{x}_j^n, t^n\right)}{\Delta t} = \frac{1}{2}\left[3S\left(\tilde{x}_j^n, t^n\right) - S\left(\tilde{x}_j^{n-1}, t^{n-1}\right)\right]. \quad (6.26)$$

The most stable and accurate of these schemes is the trapezoidal method, but it may also require more work per time step because it is implicit.

The fundamental stability properties of each scheme can be analyzed by applying them to the prototype problem

$$\frac{d\psi}{dt} = i\omega\psi + \lambda\psi, \quad (6.27)$$

where ω and λ are real, ψ is complex, and the advecting velocity is constant, so that

$$\frac{d}{dt} = \frac{\partial}{\partial t} + U\frac{\partial}{\partial x}.$$

If $\psi(x, 0) = f(x)$, the solution to (6.27) is

$$\psi(x, t) = f(x - Ut)e^{(i\omega+\lambda)t},$$

which is nonamplifying for $\lambda \leq 0$. This prototype problem is similar to that considered in Section 3.4.2 except that (3.75) is a partial differential equation, whereas (6.27) is an ordinary differential equation. In order to simplify the stability analysis, the errors generated during the interpolation of ϕ to $\tilde{x}_j^n$ and $2\tilde{x}_j^n - x_j$ will be neglected, in which case our results describe the limiting behavior of a family of semi-Lagrangian schemes that use increasingly accurate spatial interpolation.

First consider the case where the forcing is approximated with the trapezoidal scheme. Following the standard Von Neumann stability analysis, the Fourier mode $A_k^n e^{ikj\Delta x}$ is substituted for $\phi(x_j, t^n)$ in the trapezoidal approximation to (6.27), which gives

$$A_k e^{ikj\Delta x} - e^{ik(j\Delta x-s)} = \frac{(\tilde{\lambda} + i\tilde{\omega})}{2}\left(A_k e^{ikj\Delta x} + e^{ik(j\Delta x-s)}\right),$$

where $\tilde{\lambda} = \lambda\Delta t$, $\tilde{\omega} = \omega\Delta t$, and as before, $s = x_j - \tilde{x}_j^n$. Solving for the magnitude of the amplification factor and noting that $|e^{iks}| = 1$,

$$|A_k|^2 = \left|A_k e^{iks}\right|^2 = \frac{\left(1 + \tilde{\lambda}/2\right)^2 + \tilde{\omega}^2/4}{\left(1 - \tilde{\lambda}/2\right)^2 + \tilde{\omega}^2/4}.$$

The scheme generates bounded solutions whenever the true solution is bounded, i.e., whenever $\tilde{\lambda} \leq 0$. This stability condition is independent of the Courant number $U\Delta t/\Delta x$, and the magnitude of the amplification factor is identical to that

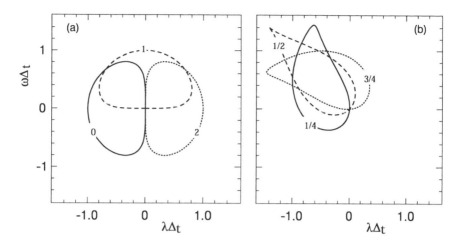

FIGURE 6.2. Region of the $\tilde{\lambda}$-$\tilde{\omega}$ plane in which $2\Delta x$-wavelength solutions to (6.29) are nongrowing when: (a) $U\Delta t/\Delta x$ is 0, 1, or 2, and (b) $U\Delta t/\Delta x$ is $\frac{1}{4}$, $\frac{1}{2}$, or $\frac{3}{4}$. The stable region lies inside of each curve. The stable region for (6.26) is independent of $U\Delta t/\Delta x$ and is defined by the curve labeled "0" in panel (a).

obtained if the ordinary differential equation (3.75) is approximated with a trapezoidal time difference.

A similar Von Neumann analysis of the second-order Adams–Bashforth method yields

$$A_k^2 - A_k e^{-iks} = \frac{(\tilde{\lambda} + i\tilde{\omega})}{2}\left(3A_k e^{-iks} - e^{-2iks}\right).$$

Defining $\hat{A} = A_k e^{iks}$ and $\gamma = \tilde{\lambda} + i\tilde{\omega}$,

$$\hat{A}^2 - \hat{A}\left(\frac{3\gamma}{2} + 1\right) + \frac{\gamma}{2} = 0. \tag{6.28}$$

This quadratic equation is identical to that obtained when the ordinary differential equation (3.75) is approximated using the second-order Adams–Bashforth method. Those values of $\tilde{\lambda}$ and $\tilde{\omega}$ for which the second-order Adams–Bashforth method generates nongrowing solutions lie within the solid curve in Fig. 6.2a. Since $|A_k| = |\hat{A}|$, the amplification factor is independent of the Courant number.

If the leapfrog scheme (6.25) is used to approximate (6.27), the stability condition becomes $\tilde{\lambda} = 0$ and $|\tilde{\omega}| < 1$, which is once again independent of the Courant number and is identical to that for a leapfrog approximation to the ordinary differential equation (3.75). All three of the preceding methods, (6.25), (6.3), and (6.26), yield amplification factors for this prototype problem that are independent of the Courant number because the advecting velocity is a constant, errors in the polynomial interpolation are ignored, and the integration is performed using data lying along the backward trajectory. If the integration does not use data lying along a backward trajectory, the maximum stable time step will depend on the

Courant number, and the stability criteria can become far more restrictive. In the case of Adams–Bashforth-type approximations this consideration has not always been recognized. Alternatives to (6.26) of the form

$$\frac{\phi\left(x_j, t^{n+1}\right) - \phi\left(\tilde{x}_j^n, t^n\right)}{\Delta t} = \frac{3S\left([x_j + \tilde{x}_j^n]/2, t^n\right) - S\left([x_j + \tilde{x}_j^n]/2, t^{n-1}\right)}{2}.$$

(6.29)

and

$$\frac{\phi\left(x_j, t^{n+1}\right) - \phi\left(\tilde{x}_j^n, t^n\right)}{\Delta t}$$

$$= \frac{1}{4}\left[3S\left(x_j, t^n\right) + 3S\left(\tilde{x}_j^n, t^n\right) - S\left(x_j, t^{n-1}\right) - S\left(\tilde{x}_j^n, t^{n-1}\right)\right] \quad (6.30)$$

have been used in atmospheric models. These are both second-order accurate, and (6.30) is potentially more efficient because it requires one spatial interpolation fewer than either (6.26) or (6.29). Both schemes can, however, be substantially less stable than (6.26) when integrations are performed using Courant numbers larger than order unity.

Applying (6.29) to the prototype problem (6.27) and performing a Von Neumann stability analysis yields the following quadratic equation for the amplification factor:

$$A_k^2 - A_k\left(\frac{3\gamma}{2}e^{-iks/2} + e^{-iks}\right) + \frac{\gamma}{2}e^{-iks/2} = 0. \quad (6.31)$$

In contrast to the results obtained previously for schemes that use data lying along a back trajectory, the amplification factor for this scheme does depend on the Courant number. The region of the $\tilde{\omega}$-$\tilde{\lambda}$ plane in which $|A_k| \leq 1$ is plotted in Fig. 6.2 for several values of ks between 0 and 2π. The most severe stability constraints are typically imposed by the $2\Delta x$ wave, for which these values of ks correspond to Courant numbers $U\Delta t/\Delta x$ between 0 and 2. As the Courant number increases, the region of absolute stability rotates clockwise around the origin in the $\tilde{\lambda}$-$\tilde{\omega}$ plane. When $U\Delta t/\Delta x = 2$, all $2\Delta x$ waves that should properly damp are amplified, and the only $2\Delta x$ waves that damp are those that should amplify. The instability that develops in the solutions to (6.29) as the Courant number increases may be qualitatively understood to result from a failure to match the numerical domain of dependence for the data used to compute $S(\psi)$ with the numerical domain of dependence for the true solution. (See Section 2.2.3.)

Polynomial interpolation damps the interpolated field. It is easy to show that this damping further stabilizes the trapezoidal approximation (see Problem 5), but the influence of this damping on the stability of the Adams–Bashforth-type schemes is harder to determine. Numerical simulations are shown in Fig. 6.3 for a series of tests in which (6.26), (6.29), and (6.30) were used to obtain approximate solutions to (6.27) on a periodic spatial domain $0 \leq x \leq 1$. In each test the nu-

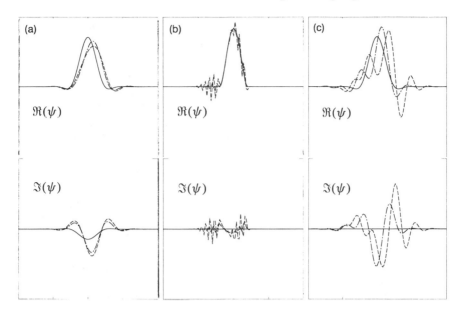

FIGURE 6.3. Real and imaginary parts of the numerical solution to (6.27) at $t = 20$ for (a) $U\Delta t/\Delta x = 0.5$, $\omega\Delta t \approx -0.06$, (b) $U\Delta t/\Delta x = 1.0$, $\omega\Delta t \approx -0.003$, and (c) $U\Delta t/\Delta x = 2.5$, $\omega\Delta t \approx -0.006$. The solid, dashed, and dot-dashed curves show the solutions computed using (6.26), (6.29), and (6.30), respectively. Data are plotted for the interval [0,1] along the horizontal axis; the vertical axis spans the interval $[-1.2, 1.2]$.

merical approximation to ψ at the departure point is obtained by cubic Lagrange interpolation; $\Delta x = 0.01$, $\Delta t = 0.01$, $\lambda = 0$, and the initial condition is

$$\psi(x,0) = \begin{cases} \left\{\left[(x-c)^2 \quad w^2\right]/w^2\right\}^2 + 0i, & \text{if } |x-c| \le w, \\ 0 + 0i, & \text{otherwise}, \end{cases}$$

where $w = 0.1$ and $c = 0.5$. Thus, at $t = 0$ the real part of ψ is a smooth unit-amplitude pulse 20 grid points wide and the imaginary part of ψ is zero.

In the first case, shown in Fig. 6.3a, $U\Delta t/\Delta x = 0.5$, $\omega\Delta t = -\pi/50$, and the solution is plotted at $t = 20$, at which time the energy in the initial pulse has circled the periodic domain ten times and oscillated back and forth between the real and imaginary parts of ψ twenty times. The correct solution is identical to the initial condition: $\Re(\psi)$ is a unit-amplitude pulse centered in the domain and $\Im(\psi)$ is zero everywhere. Although second-order Adams–Bashforth time-differencing generates growing solutions to ordinary differential equations describing purely oscillatory motion, all three numerical solutions shown in Fig. 6.3a have been damped by the diffusion in the cubic interpolation. The effect of the accelerative phase-speed error in the second-order Adams–Bashforth time difference is apparent in the plot of $\Im(\psi)$, which shows that all three solutions develop a negative pulse when the correct solution should be exactly zero. The phase error generated by (6.26) is, however, significantly smaller than that produced by (6.29) and (6.30) and is confined to the time coordinate, whereas the phase errors in the solutions to (6.29) and (6.30) appear in both time and space.

The Courant number is increased to unity and $\omega \Delta t$ is decreased to $-\pi/1000$ in the second test case shown in Fig. 6.3b. At $t = 20$ the correct solution is again identical to the initial condition. Since the wind speed is constant and the Courant number is unity, the semi-Lagrangian advection is exact, and the only source of error is in the integration of the forcing. The solution obtained using (6.26) is stable and almost perfect, whereas the solutions produced by (6.29) and (6.30) are corrupted by growing $2\Delta x$ disturbances. These unstable $2\Delta x$ waves completely dominate the solution computed using (6.29) by $t = 27$ and that obtained using (6.30) by $t = 34$.

In the third case, shown in Fig. 6.3c, the Courant number is 2.5, $\omega \Delta t = -\pi/500$, and the exact solution at $t = 20$ is once again identical to the initial condition. Although it has been somewhat diffused by the cubic interpolation in the semi-Lagrangian advection, the solution produced by (6.26) is free of instability and noticeable phase-speed error. In contrast, the solutions obtained using (6.29) and (6.30) are both contaminated by unstable long-wavelength disturbances.

Instabilities develop in the second and third cases even though the magnitude of the forcing is very small ($|\omega \Delta t| < 0.01$). These results, together with the stability analysis presented in Fig. 6.2, demonstrate the need to compute forcing terms that depend on the solution, i.e., forcing of the form $S(\psi)$, using data along the backward trajectory.

6.3   Systems of Equations

One of the most important applications of semi-Lagrangian methods in atmospheric science is in global weather prediction. This application was pioneered by Robert (1981, 1982), who showed that the equations describing large-scale atmospheric motion could be efficiently integrated using semi-Lagrangian methods in conjunction with a semi-implicit approximation of those terms in the governing equations representing the pressure gradient and velocity divergence. The essential elements of the semi-Lagrangian semi-implicit method will be explored in this section by examining numerical approximations to simple shallow-water equations. The shallow-water system also provides a convenient example in which to illustrate the difference between the semi-Lagrangian approach and the classical method of characteristics.

6.3.1   Comparison with the Method of Characteristics

Semi-Lagrangian approximations to the equations describing the advection and reaction of chemical tracers, such as (6.25), may be regarded as an algorithm for numerically implementing the classical method of characteristics (Courant et al. 1952; see also Gustafsson et al. 1995). The advection equation is, however, a special case because the characteristic curves are identical to the fluid parcel trajectories. The semi-Lagrangian method retains its simplicity and practical utility in more complicated applications precisely because the evolution of the flow continues to be computed following fluid parcel trajectories. The classical method of characteristics, on the other hand, becomes unwieldy or impossible in more general problems where the evolution of the flow along the characteristic curves may be more complicated or characteristic curves may not even be defined. As a simple example consider the nonlinear one-dimensional shallow-water system

$$\frac{\partial}{\partial t}\begin{pmatrix} u \\ h \end{pmatrix} + \begin{pmatrix} u & g \\ h & u \end{pmatrix} \frac{\partial}{\partial x}\begin{pmatrix} u \\ h \end{pmatrix} = 0.$$

A three-time level semi-Lagrangian approximation to the preceding can be written in the form

$$\frac{u^+ - u^-}{2\Delta t} = -g\left(\frac{\partial h}{\partial x}\right)^0, \tag{6.32}$$

$$\frac{h^+ - h^-}{2\Delta t} = -h^0\left(\frac{\partial u}{\partial x}\right)^0, \tag{6.33}$$

where the superscripts "+," "0," and "−" denote evaluation of the function at the points (x_j, t^{n+1}), $(\tilde{x}_j^n, t^n)$, and $(\check{x}_j^{n-1}, t^{n-1})$, respectively. As before, $\tilde{x}_j^n$ is determined by numerically integrating (6.2) backward over a time interval Δt subject to the initial condition $x(t^{n+1}) = x_j$, and $\check{x}_j^{n-1}$ is determined by a similar backward integration over the period $2\Delta t$. The spatial derivatives $\partial u/\partial x$ and $\partial h/\partial x$ are evaluated by centered differences on the regular mesh and then interpolated to $\tilde{x}_j^n$. As long as the solution remains smooth, the numerical evaluation of this system is no more difficult than the integration of a pair of forced advection equations of the form (6.25).

Considerably more computational effort is required to solve this problem using the classical method of characteristics. In order to implement the method of characteristics, the nonlinear shallow-water equations are transformed as described in connection with (1.8) to yield the system

$$\frac{\partial}{\partial t}\begin{pmatrix} d \\ e \end{pmatrix} + \begin{pmatrix} d & 0 \\ 0 & e \end{pmatrix} \frac{\partial}{\partial x}\begin{pmatrix} d \\ e \end{pmatrix} = \mathbf{B}\begin{pmatrix} d \\ e \end{pmatrix}; \tag{6.34}$$

here $d = u - \sqrt{gh}$, $e = u + \sqrt{gh}$, and

$$\mathbf{B} = -\mathbf{T}^{-1}\left[\frac{\partial \mathbf{T}}{\partial t} + \begin{pmatrix} u & g \\ h & u \end{pmatrix}\frac{\partial \mathbf{T}}{\partial x}\right],$$

$$\mathbf{T}^{-1} = \begin{pmatrix} 1 & -\sqrt{g/h} \\ 1 & \sqrt{g/h} \end{pmatrix},$$

$$\mathbf{T} = \frac{1}{2}\begin{pmatrix} 1 & 1 \\ -\sqrt{h/g} & \sqrt{h/g} \end{pmatrix}.$$

The numerical integration of this system requires more computation than that for the semi-Lagrangian scheme because the right side of (6.34) is much more complicated than the right sides of (6.32) and (6.33). Additional effort must also be expended in the trajectory calculations because the evaluation of (6.34) requires the computation of two backward trajectories per grid point (one along each characteristic curve), whereas the semi-Lagrangian method requires only one backward trajectory per grid point.

6.3.2 Semi-implicit Semi-Lagrangian Schemes

If latitudinally varying Coriolis forces are included in the shallow-water equations, the resulting system can support both Rossby and gravity waves. Most physically significant large-scale atmospheric circulations have time scales similar to those of the Rossby waves and much longer than those of the gravity waves. As a consequence, the maximum stable time step dictated by the CFL condition for gravity waves is often much smaller than that required to accurately simulate the physically significant phenomena. A considerable increase in efficiency can be realized by using semi-implicit time-differencing to remove the stability constraint imposed by rapid gravity-wave propagation. Semi-implicit time-differencing is discussed in detail in Section 7.2. In the following we will focus on just one aspect of the semi-implicit method, namely, how it improves the stability of semi-Lagrangian solutions to the one-dimensional shallow-water equations.

First consider the stability properties of the linearized equivalent to (6.32) and (6.33). If the mean wind and fluid depth are constants denoted by U and H respectively, a finite-difference approximation to the linearized system may be written as

$$\frac{u^+ - u^-}{2\Delta t} = -g \left(\frac{\partial h}{\partial x} \right)^0, \tag{6.35}$$

$$\frac{h^+ - h^-}{2\Delta t} = -H \left(\frac{\partial u}{\partial x} \right)^0, \tag{6.36}$$

where u and h now denote the amplitudes of the perturbation velocity and free surface displacement. Defining auxiliary variables a and b such that $a^+ = u^0$ and $b^+ = \eta^0$ and substituting wave solutions of the form

$$\begin{pmatrix} u \\ h \\ a \\ b \end{pmatrix}^n e^{ikx}$$

into the preceding yields the linear system

$$\begin{pmatrix} u \\ h \\ a \\ b \end{pmatrix}^{n+1} = e^{-iks} \begin{pmatrix} 0 & -2i\tilde{g} & 1 & 0 \\ -2i\tilde{H} & 0 & 0 & 1 \\ 1 & 0 & 0 & 0 \\ 0 & 1 & 0 & 0 \end{pmatrix} \begin{pmatrix} u \\ h \\ a \\ b \end{pmatrix}^n ,$$

where $s = U\Delta t$, $\tilde{g} = gk\Delta t$, and $\tilde{H} = Hk\Delta t$. Let λ be an eigenvalue of the amplification matrix for this scheme, and define $\tilde{\lambda} = \lambda e^{-iks}$ and $\tilde{c} = \sqrt{gH}k\Delta t$. Then

$$\tilde{\lambda}^4 + (4\tilde{c}^2 - 2)\tilde{\lambda}^2 + 1 = 0$$

and

$$\tilde{\lambda}^2 = 1 - 2\tilde{c}^2 \pm 2i\tilde{c}\left(1 - \tilde{c}^2\right)^{1/2}.$$

Two of the $\tilde{\lambda}$ are associated with gravity waves and two are associated with computational modes. The magnitude of $\tilde{\lambda}^2$ is unity whenever $|\tilde{c}| \leq 1$, and since $|\lambda| = |\tilde{\lambda}|$, it follows that the eigenvalues of the amplification matrix are bounded by unity whenever $|\tilde{c}| \leq 1$. If the spatial derivatives are evaluated using the centered-difference operator δ_{2x}, this stability condition becomes $\sqrt{gH}\Delta t/\Delta x < 1$, where strict inequality is required to ensure that the norm of the amplification matrix is power bounded (see Section 3.1.1). In contrast to the result obtained for an Eulerian scheme (3.13), the stability of the leapfrog semi-Lagrangian approximation (6.35)–(6.36) depends only on a Courant number defined with respect to the intrinsic gravity-wave phase speed and does not depend on the speed of the mean flow.

An unconditionally stable method can be obtained if the forcing terms in the linearized system are approximated by trapezoidal time-differencing to yield the approximation

$$\frac{u^+ - u^0}{\Delta t} = -\frac{g}{2}\left[\left(\frac{\partial h}{\partial x}\right)^+ + \left(\frac{\partial h}{\partial x}\right)^0\right], \tag{6.37}$$

$$\frac{h^+ - h^0}{\Delta t} = -\frac{H}{2}\left[\left(\frac{\partial u}{\partial x}\right)^+ + \left(\frac{\partial u}{\partial x}\right)^0\right]. \tag{6.38}$$

The eigenvalues for the amplification matrix associated with this scheme are obtained by substituting solutions of the form

$$\begin{pmatrix} u \\ h \end{pmatrix}^n e^{ikx}$$

into the preceding to yield

$$\lambda = e^{iks}\left(\frac{4 - \tilde{c}^2 \pm 4i\tilde{c}}{4 + \tilde{c}^2}\right). \tag{6.39}$$

This scheme is unconditionally stable, since $|\lambda| = 1$ for all Δt, and the norm of the amplification matrix is power bounded because its eigenvectors are linearly independent (and it can therefore be transformed into a diagonal matrix).

The right side of (6.38) becomes nonlinear if the preceding trapezoidal scheme is generalized to approximate the nonlinear shallow-water equations. In order to avoid solving a coupled system of nonlinear algebraic equations at every time step, the total fluid depth is often split into a constant reference value and a perturbation, and the velocity divergence multiplying the perturbation is evaluated using leapfrog time-differencing. Letting $h(x, t) = H + \eta(x, t)$, this approach leads to the following three-time-level scheme:

$$\frac{u^+ - u^-}{2\Delta t} = -\frac{g}{2}\left[\left(\frac{\partial \eta}{\partial x}\right)^+ + \left(\frac{\partial \eta}{\partial x}\right)^-\right], \tag{6.40}$$

$$\frac{\eta^+ - \eta^-}{2\Delta t} = -\frac{H}{2}\left[\left(\frac{\partial u}{\partial x}\right)^+ + \left(\frac{\partial u}{\partial x}\right)^-\right] - \frac{1}{2}\eta^0\left(\frac{\partial u}{\partial x}\right)^0. \tag{6.41}$$

A complete stability analysis of the full nonlinear system is very difficult, but the stability of a linearized system in which

$$u(x, t) = U + u'(x, t), \qquad \eta(x, t) = \bar{\eta} + \eta'(x, t)$$

can be performed following essentially the same steps detailed in Section 7.2.3. This analysis shows that the linearized system is stable for all Δt, provided that $|\bar{\eta}| \leq H$.

One disadvantage of the preceding scheme is that it is potentially half as efficient as a two-time-level method. Both the trajectory calculation and the trapezoidal difference in the preceding are computed over a time interval of $2\Delta t$. In order to evaluate these terms with the same accuracy obtained in a two-level scheme such as (6.37)–(6.38), the time step used in (6.40)–(6.41) must be one-half that used in the two-level scheme. One way to obtain an $O\left[(\Delta t)^2\right]$ approximation to the nonlinear shallow-water equations that preserves the efficiency of the linearized system (6.37)–(6.38) is to use the second-order Adams–Bashforth method to evaluate the portion of the velocity divergence multiplying η, in which case the finite-difference equations become

$$\frac{u^+ - u^0}{\Delta t} = -\frac{g}{2}\left[\left(\frac{\partial \eta}{\partial x}\right)^+ + \left(\frac{\partial \eta}{\partial x}\right)^0\right], \tag{6.42}$$

$$\frac{\eta^+ - \eta^0}{\Delta t} = -\frac{H}{2}\left[\left(\frac{\partial u}{\partial x}\right)^+ + \left(\frac{\partial u}{\partial x}\right)^0\right] - \frac{3}{2}\eta^0\left(\frac{\partial u}{\partial x}\right)^0 + \frac{1}{2}\eta^-\left(\frac{\partial u}{\partial x}\right)^-.$$
$$\tag{6.43}$$

A stability analysis similar to that for the linearized version of (6.40) and (6.41) can be performed by linearizing the preceding about the same basic state:

$$u(x, t) = U + u'(x, t), \qquad \eta(x, t) = \bar{\eta} + \eta'(x, t).$$

Dropping the primes on the perturbation variables, letting $\bar{\eta} = \alpha H$, $s = U \Delta t$, and defining $a^+ = u^0$, the linearized system can be written in the form

$$\begin{pmatrix} 1 & i\tilde{g}/2 & 0 \\ i\tilde{H}/2 & 1 & 3i\alpha\tilde{H}/2 \\ 0 & 0 & 1 \end{pmatrix} \begin{pmatrix} u \\ \eta \\ a \end{pmatrix}^{n+1}$$

$$= e^{-iks} \begin{pmatrix} 1 & -i\tilde{g}/2 & 0 \\ -i\tilde{H}/2 & 1 & i\alpha\tilde{H}/2 \\ 1 & 0 & 0 \end{pmatrix} \begin{pmatrix} u \\ \eta \\ a \end{pmatrix}^n,$$

where $\tilde{g} = gk\Delta t$ and $\tilde{H} = hk\Delta t$. Let λ be an eigenvalue of the amplification matrix for this scheme and define $\tilde{\lambda} = \lambda e^{-iks}$, $\beta^2 = \tilde{g}\tilde{H}/4$, and $\mu = \alpha\beta^2/(1 + \beta^2)$; then $\tilde{\lambda}$ satisfies the cubic equation

$$\tilde{\lambda}^3 + \left(3\mu - \frac{2(1-\beta^2)}{1+\beta^2}\right)\tilde{\lambda}^2 + (2\mu+1)\tilde{\lambda} - \mu = 0. \qquad (6.44)$$

Simmons and Temperton (1997) obtained this cubic equation as part of a more extensive analysis of the stability of similar semi-Lagrangian approximations to the equations governing three-dimensional stratified flow. As noted by Simmons and Temperton, one root of (6.44) is a real number associated with a computational mode, and the other two roots are complex conjugates associated with rightward and leftward propagating gravity waves.

If $\alpha = 0$, then $\mu = 0$, so the linearized equations reduce to (6.37) and (6.38); the computational mode vanishes, and the remaining eigenvalues are given by (6.39). Let $\tilde{\lambda}_0$ denote the value of $\tilde{\lambda}$ when $\mu = 0$:

$$\tilde{\lambda}_0 = \frac{1 - \beta^2 \pm 2i\beta}{1 + \beta^2} = \frac{1 \pm i\beta}{1 \mp i\beta}.$$

For small values of μ the stability of this scheme can be determined by expanding the $\tilde{\lambda}$ in powers of μ. The eigenvalue for the computational mode is $\tilde{\lambda} = \mu + O(\mu^2)$. Since $|\mu|$ is small by assumption, this mode is rapidly damped. Expanding the $\tilde{\lambda}$ for the gravity-wave modes in powers of μ,

$$\tilde{\lambda} = \tilde{\lambda}_0 + \tilde{\lambda}_1 \mu + O(\mu^2), \qquad (6.45)$$

and substituting the preceding into (6.44) yields

$$\tilde{\lambda}_1 = \frac{1 - 3\tilde{\lambda}_0}{\tilde{\lambda}_0 - 1}. \qquad (6.46)$$

Since $|\lambda| = |\tilde{\lambda}|$, the square of the magnitude of the eigenvalues of the amplification matrix is

$$\tilde{\lambda}\tilde{\lambda}^* = \tilde{\lambda}_0\tilde{\lambda}_0^* + 2\mu\Re(\tilde{\lambda}_0\tilde{\lambda}_1^*) + O(\mu^2).$$

Substituting (6.45) and (6.46) into the preceding yields

$$|\lambda|^2 = |\tilde{\lambda}|^2 = 1 + \frac{4\beta^2}{1+\beta^2}\mu + O(\mu^2),$$

which implies that at least for small values of $|\mu|$, the scheme will be stable if $\mu <$ 0 and unstable if $\mu > 0$. Note that μ has the same sign as α and for all Δt, $|\mu| < |\alpha|$. Thus for sufficiently small values of α, the scheme is unconditionally stable when $\alpha < 0$ and unconditionally unstable when $\alpha > 0$. Numerical evaluation of the roots of (6.44) verifies that the scheme is damping independent of Δt for $-1 < \alpha < 0$. In the context of the original nonlinear problem, this analysis shows that a necessary condition for the scheme to be stable independent of Δt is that the reference fluid depth H exceed the maximum height of the actual free-surface displacement.

An alternative to the decomposition of the total fluid depth into $H + \eta$ is to remove the nonlinearity in the velocity divergence by linearizing h about its value at the preceding time step, in which case (6.43) is replaced by

$$\frac{h^+ - h^0}{\Delta t} = -\frac{h^0}{2}\left[\left(\frac{\partial u}{\partial x}\right)^+ + \left(\frac{\partial u}{\partial x}\right)^0\right]. \qquad (6.47)$$

This approach requires the solution of a linear algebraic system with a more complicated coefficient structure than that generated by (6.43), and particularly in two- or three-dimensional problems, the increase in the complexity of the coefficient matrix can be an impediment to numerical efficiency. Promising results have, nevertheless, been obtained using preconditioned conjugate residual solvers (Skamarock et al. 1997), suggesting that this approach can be a viable alternative in those applications where a suitable preconditioning operator can be determined. Yet another possibility has been pursued by Bates et al. (1995), who used a nonlinear multigrid method to solve the nonlinear finite-difference equations generated by a true trapezoidal approximation to the full shallow-water continuity equation.

6.4   Alternative Trajectories

As noted by Smolarkiewicz and Pudykiewicz (1992), the numerical solution can be integrated forward in time along trajectories other than those associated with the standard Eulerian and Lagrangian coordinate frames. Any function $\psi(\mathbf{x}, t)$ with continuous derivatives satisfies the relation

$$\psi(\mathbf{x}_j, t^{n+1}) - \psi(\tilde{\mathbf{x}}, t^n) = \int_C \left(\nabla\psi, \frac{\partial\psi}{\partial t}\right) \cdot (d\mathbf{x}, dt), \qquad (6.48)$$

where $\nabla\psi$ is the gradient of ψ with respect to the spatial coordinates, and C is an arbitrary contour connecting the points $(\tilde{\mathbf{x}}, t^n)$ and $(\mathbf{x}_j, t^{n+1})$. If the time evolution of ψ is governed by the partial differential equation

$$\frac{\partial \psi}{\partial t} + \mathbf{v} \cdot \nabla \psi = S,$$

(6.48) may be expressed in the form

$$\psi(\mathbf{x}_j, t^{n+1}) = \psi(\tilde{\mathbf{x}}, t^n) + \int_C \nabla \psi \cdot (d\mathbf{x} - \mathbf{v}dt) + \int_C S \, dt. \qquad (6.49)$$

Lagrangian and semi-Lagrangian schemes approximate this equation by choosing the integration path to be a fluid parcel trajectory, in which case the first integral in (6.49) is zero, and $\tilde{\mathbf{x}}$ is the departure point of the trajectory arriving at $(\mathbf{x}_j, t^{n+1})$. Eulerian schemes approximate this equation by choosing C to be independent of $\mathbf{x}$, in which case (6.49) becomes

$$\psi(\mathbf{x}_j, t^{n+1}) = \psi(\mathbf{x}_j, t^n) + \int_C (S - \mathbf{v} \cdot \nabla \psi) \, dt.$$

As an alternative to the pure Lagrangian and Eulerian approaches, one may choose $\tilde{\mathbf{x}}$ to coincide with the grid point that is closest to the departure point of the fluid parcel trajectory arriving at $(\mathbf{x}_j, t^{n+1})$. Two such methods will be considered in the following section: In the first method, C is a straight line in $\mathbf{x} - t$ space; in the second approach C is deformed into the union of the true fluid parcel trajectory and a series of straight lines in the hyperplane $t = t^n$. In both of these alternatives the interpolation of ψ to the departure point is accomplished by solving an advection equation rather than by conventional interpolation.

6.4.1 A Noninterpolating Leapfrog Scheme

Some damping is produced in all the previously described semi-Lagrangian schemes when the prognostic fields are interpolated to the departure point. Numerical solutions to the forced one-dimensional advection equation (6.1) can be obtained without interpolation using a modified semi-Lagrangian algorithm due to Ritchie (1986). Let $\check{x}_j^{n-1}$ be the estimated x-coordinate of the departure point of a trajectory originating at time t^{n-1} and arriving at (x_j, t^{n+1}); $\check{x}_j^{n-1}$ is calculated by integrating (6.2) backward over a time interval of $2\Delta t$ using the initial condition $x(t^{n+1}) = x_j$. Define p as the integer for which x_{j-p} is the grid point closest to $\check{x}_j^{n-1}$, and let u' be a residual velocity such that

$$u = \frac{p \Delta x}{2 \Delta t} + u'.$$

Then (6.1) can be expressed as

$$\frac{\partial \psi}{\partial t} + \frac{p \Delta x}{2 \Delta t} \frac{\partial \psi}{\partial x} = -u' \frac{\partial \psi}{\partial x} + S(\psi). \qquad (6.50)$$

The velocities on the left side of the preceding carry a fluid parcel an integral number of grid points over a time interval $2\Delta t$, so the left side of (6.50) can be evaluated as a Lagrangian derivative without numerical error. The right side of (6.50) can be integrated using a leapfrog time difference with the forcing evaluated at $\check{x}_j^{n-1}/2$, which is the midpoint of a back trajectory computed with respect to the velocity $p\Delta x/(2\Delta t)$. Depending on whether p is even or odd, $\check{x}_j^{n-1}/2$ will either coincide with a grid point or lie halfway between two grid points. A second-order finite-difference approximation to (6.50) can therefore be written in the form

$$\frac{\phi_j^{n+1} - \phi_{j-p}^{n-1}}{2\Delta t} = \begin{cases} -u'\delta_{2x}\phi_{j-p/2}^n + S(\phi_{j-p/2}^n) & \text{if } p \text{ is even;} \\ -u'\delta_x\phi_{j-p/2}^n + \left(S(\phi_{j-p/2}^n)\right)^x & \text{if } p \text{ is odd.} \end{cases} \tag{6.51}$$

The stability of the constant-coefficient equivalent of the preceding scheme can be easily investigated. Suppose that the source term is zero and the wind speed is U; substituting the discrete Fourier mode

$$\phi_j^n = e^{i(kj\Delta x - \omega n\Delta t)}$$

into (6.51) and invoking the assumption that $S = 0$ yields the discrete-dispersion relation

$$\sin\left[\omega\Delta t - k(U - u')\Delta t\right] = \begin{cases} u'\Delta t \dfrac{\sin(k\Delta x)}{\Delta x} & \text{if } p \text{ is even;} \\[2mm] 2u'\Delta t \dfrac{\sin(k\Delta x/2)}{\Delta x} & \text{if } p \text{ is odd.} \end{cases} \tag{6.52}$$

By the choice of p,

$$\frac{\Delta x}{2} > |x_j - \check{x}_j^{n-1} - p\Delta x| = |2U\Delta t - p\Delta x|,$$

so

$$\left|\frac{u'\Delta t}{\Delta x}\right| = \left|U - \frac{p\Delta x}{2\Delta t}\right|\frac{\Delta t}{\Delta x} < \frac{1}{4},$$

which implies that the right side of (6.52) is bounded by one-half, ω is real, and the scheme is stable independent of the value of Δt.

If the velocity is a function of space and time, the residual Courant number $|u'\Delta t/\Delta x|$ will vary as a function of x and t. Let the subscript "$*$" indicate that the function is evaluated at the midpoint of the trajectory between x_j and x_{j-p}, in which case

$$u_* = \begin{cases} u_{j-p/2}^n & \text{if } p \text{ is even;} \\ \left(u_{j-p/2}^n\right)^x & \text{if } p \text{ is odd.} \end{cases} \tag{6.53}$$

According to (6.52), a necessary condition for the stability of the variable-velocity algorithm is $|u_*'\Delta t/\Delta x| < \frac{1}{2}$. This condition is not automatically satisfied unless the deviation of u_* from the average velocity along the back trajectory is sufficiently small. Let $\bar{u} = (x_j - \check{x}_j^{n-1})/(2\Delta t)$ be the average velocity required to arrive at the true departure point and define the deviation of the velocity at the midpoint as $v = u_* - \bar{u}$. Then

$$u'_* = u_* - \frac{p\Delta x}{2\Delta t} = \bar{u} - \frac{p\Delta x}{2\Delta t} + v,$$

and by the choice for p,

$$\left| \frac{u'_* \Delta t}{\Delta x} \right| < \frac{1}{4} + \left| \frac{v\Delta t}{\Delta x} \right|.$$

Thus, in order for the variable velocity algorithm to be stable, the deviation of the velocity about the mean along the backward trajectory must be small enough that $|v\Delta t/\Delta x| \le \frac{1}{4}$.

This stability constraint can be removed if the departure point is computed as suggested by Ritchie (1986) by choosing

$$p = \text{nearest integer to } \frac{2\Delta t}{\Delta x} u_*. \tag{6.54}$$

The preceding is an implicit equation for p because u_* is the function of p defined by (6.53). Although this approach stabilizes the scheme in the sense that it keeps the solution bounded, it is still subject to problems if the deviation of u_* from the average velocity along the back trajectory is too large. In particular, the solution to (6.54) need not be unique if

$$\left| \frac{\partial u}{\partial x} \right| \Delta t > \frac{1}{2}.$$

The nonuniqueness of the solution to (6.54) is particularly apparent if the velocity field is defined by the relation $u[(j + n)\Delta x] = -n\Delta x/\Delta t$ at all grid points in a neighborhood surrounding x_j, since (6.54) is then satisfied by any integer p.

6.4.2   Interpolation via Parametrized Advection

Semi-Lagrangian methods will not preserve the nonnegativity of an initially nonnegative tracer concentration field if conventional quadratic or higher-order polynomial interpolation is used to determine the value of ψ at the departure point. Positive definite semi-Lagrangian schemes can be obtained if the interpolating functions are required to satisfy appropriate monotonicity and convex–concave shape-preserving constraints (Williamson and Rasch 1989). As noted by Smolarkiewicz and Rasch (1991), positive definite results can also be obtained if the interpolation step in the standard semi-Lagrangian algorithm is recast as a parametrized advection problem that is approximated using one of the positive definite advection schemes discussed in Section 5.8. If a strictly positive definite result is not required, overshoots and undershoots in the vicinity of poorly

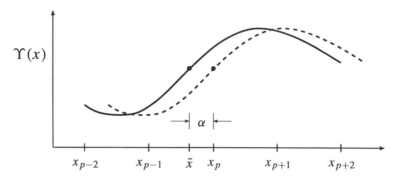

FIGURE 6.4. Interpolation via the solution of a constant-wind-speed advection problem. The initial condition is indicated by the solid line; the solution after translation a distance α is indicated by the dashed line.

resolved gradients can still be minimized by approximating the solution to the parametrized advection equation using any of the various flux-limited or flux-corrected advection schemes discussed in Chapter 5.

If $f(x)$ is a continuously differentiable function, the value of f at some arbitrary point $\tilde{x}$ can be estimated from its value on a regularly spaced mesh by computing a numerical solution to the constant-coefficient advection problem

$$\frac{\partial \Upsilon}{\partial \tau} + \frac{\partial \Upsilon}{\partial x} = 0 \qquad (6.55)$$

subject to the initial condition $\Upsilon(x, 0) = f(x)$. The solution to this advection problem is $\Upsilon(x, \tau) = f(x - \tau)$. Let x_p be the x-coordinate of the grid point nearest to $\tilde{x}$ and define $\alpha = x_p - \tilde{x}$; then

$$\Upsilon(x_p, \alpha) = f(x_p - \alpha) = f(\tilde{x}).$$

Figure 6.4 illustrates how the initial distribution of Υ is shifted along the x-coordinate so that desired value of $f(\tilde{x})$ becomes the value of Υ at grid point x_p when $\tau = \alpha$.

If a single time step is used to integrate forward or backward over the interval $\Delta\tau = \alpha$, the magnitude of the Courant number associated with this integration will be $|\alpha/\Delta x|$. Since $|\alpha/\Delta x| < \frac{1}{2}$ by the definition of α, stable estimates of $f(\tilde{x})$ can be obtained in a single time step using most of the wide variety of schemes available for the numerical approximation of (6.55). Of course, there is no advantage to this approach if (6.55) is solved using an elementary scheme like the Lax–Wendroff method, which will yield exactly the same result that would be obtained if $f(\tilde{x})$ was interpolated from the quadratic polynomial passing through the points $f(x_{p+1})$, $f(x_p)$, and $f(x_{p-1})$. As discussed previously, the advantage of this approach lies in the possibility of using positive definite or flux-limited advection schemes to eliminate spurious negative concentrations or minimize undershoots and overshoots in the semi-Lagrangian solution.

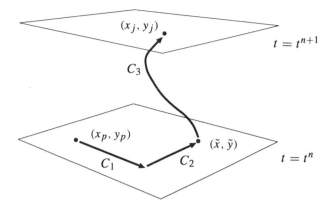

FIGURE 6.5. Contour paths for the integration of (6.49); (x_j, y_j, t^{n+1}) is the arrival point, $(\tilde{x}, \tilde{y}, t^n)$ is the departure point, and (x_p, y_p, t^n) is the nearest grid point to the departure point.

The use of parametrized advection equations to replace the interpolation step in conventional semi-Lagrangian methods can be interpreted as a method for advancing the solution to the new time level by integrating (6.49) along a specially deformed contour between the arrival point and the nearest grid point to the departure point (Smolarkiewicz and Pudykiewicz 1992). In order to easily visualize the geometric structure of this contour, suppose that the spatial domain is two-dimensional and the spatial coordinates are x and y. Let $(\tilde{x}, \tilde{y})$ be the departure point of the trajectory originating at time t^n and arriving at (x_j, y_j, t^{n+1}), and let (x_p, y_p) be the coordinates of the node on the spatial grid that is nearest to $(\tilde{x}, \tilde{y})$. Since the contour integrals in (6.49) are path independent, $\psi(x_j, y_j, t^{n+1})$ can be evaluated by integrating along the path defined by the union of the three contours

$$C_1 = (x_p + (\tilde{x} - x_p)\tau, \ y_p, \ t^n), \quad \tau \in [0, 1],$$
$$C_2 = (\tilde{x}, \ y_p + (\tilde{y} - y_p)\tau, \ t^n), \quad \tau \in [0, 1],$$
$$C_3 = \left(\tilde{x} + \int_{t^n}^{t} u[x(s), y(s), s] \, ds, \ \tilde{y} + \int_{t^n}^{t} v[x(s), y(s), s] \, ds, \ t \right),$$
$$t \in \left[t^n, t^{n+1} \right].$$

A schematic diagram of this integration path is shown in Fig. 6.5. The integral of (6.49) over the contours C_1 and C_2 is independent of the true time variable t and yields the value of ψ at the departure point of the backward trajectory. The final integral along contour C_3 is the standard semi-Lagrangian evaluation of the integral of the forcing along a fluid parcel trajectory.

6.5   Eulerian or Semi-Lagrangian?

The relative efficiency of Eulerian and semi-Lagrangian methods can vary considerably between different physical applications. Semi-Lagrangian methods require more work per time step than their Eulerian counterparts because additional effort is required to compute the backward trajectories. Thus to be more efficient, semi-Lagrangian methods must produce stable and accurate solutions using larger time steps than comparable Eulerian schemes. The feasibility of taking a large semi-Lagrangian time step is primarily determined by two factors: the ease with which an accurate trajectory can be computed that extends several grid intervals upstream and the extent to which the frequency of the forcing in the Lagrangian reference frame is reduced relative to that in the Eulerian frame.

One application where the semi-Lagrangian approach can have a distinct advantage is in the simulation of tracer transport in a smooth, slowly varying flow field. If the tracer is conservative and the flow is inviscid, there is no forcing in the Lagrangian reference frame, and the only factor determining the time step is the need to compute accurate backward trajectories. On the other hand, the highest-frequency forcing in the Eulerian reference frame, ω_E, is determined by the product of the velocity and the largest wave number resolved on the spatial mesh. Stability constraints (for explicit methods) and accuracy considerations require the time step of the Eulerian scheme to be small enough that $|\omega_E \Delta t| < O(1)$. It follows that if the spatial mesh required to adequately resolve the tracer field is much finer than that required to define the flow field, the maximum time step suitable for use with a semi-Lagrangian scheme can be much greater than that suitable for an Eulerian method.

Semi-Lagrangian methods also have an advantage in solving problems in spherical geometry. The most natural coordinate system for such problems is a latitude–longitude grid, but the convergence of the meridians near the poles greatly decreases the east–west distance between grid points in the polar regions. In applications such as global atmospheric modeling, the spatial scale of the disturbances near the poles is similar to that in middle latitudes, and the extra resolution in the polar regions is not required to accurately capture the meteorologically significant phenomena. The maximum stable time step of an Eulerian method must, nevertheless, be small enough to ensure that the CFL condition defined with respect to the wind speed is less than order unity in the polar regions. Semi-Lagrangian methods are free from this time-step restriction, although some care is required in order to accurately compute backward trajectories near the poles (Ritchie 1987; Williamson and Rasch 1989; McDonald and Bates 1989).

In those problems where the frequency of the forcing in the Lagrangian reference frame is similar to that in the Eulerian frame, semi-Lagrangian schemes tend to be at a disadvantage because accuracy considerations often require that both methods use similar time steps. In some cases, such as flow over a topographic barrier, forcing that is stationary in the Eulerian coordinate system is Doppler shifted to a higher frequency in the Lagrangian coordinate frame (Pinty et al. 1995; Héreil and Laprise 1996). One situation in which the frequency of the forcing is similar in both the Lagrangian and Eulerian reference frames occurs in those shallow-water systems where the fluid velocities are much slower than the phase speeds of the gravity waves. In such systems both semi-Lagrangian and Eulerian methods must use essentially the same time step to accurately simulate the most rapidly moving waves.

In some applications the fastest-moving waves are not physically significant, and in these applications the semi-implicit approximation can be used to increase the time step in the numerical integration. When used in conjunction with the semi-implicit method, semi-Lagrangian schemes can be considerably more efficient than Eulerian methods. Semi-implicit semi-Lagrangian schemes have proved particularly useful in global atmospheric modeling. Ritchie et al. (1995) compared Eulerian and semi-Lagrangian versions of the semi-implicit global forecast model developed at the European Center for Medium Range Weather Forecasting and reported that "the semi-Lagrangian version with a 15-minute time step gave an accuracy equivalent to that of an Eulerian version with a 3-min time step, giving an efficiency improvement of about a factor of four after allowing for the 20% ... [overhead for] the semi-Lagrangian computations."

It should be emphasized that the fastest-moving waves in the shallow-water system are artificially decelerated whenever semi-implicit integrations are performed using time steps significantly greater than those permitted by the CFL condition for gravity waves. This loss of accuracy occurs in both Eulerian semi-implicit and semi-Lagrangian semi-implicit models. In contrast, the increase in the time step permitted in semi-Lagrangian approximations to the pure advection equation is achieved without any inherent loss of accuracy because advective forcing generates a zero frequency response in the Lagrangian reference frame.

Problems

1. Show that the phase-speed error associated with the first-order semi-Lagrangian approximation (6.5) is

$$\frac{\tilde{\omega}}{\omega} = \frac{1}{(p+\alpha)k\Delta x}\left(pk\Delta x + \arctan\left[\frac{\alpha \sin k\Delta x}{1-\alpha(1-\cos k\Delta x)}\right]\right),$$

where $\tilde{\omega}$ and ω are the frequencies of the true and numerically approximated waves of wave number k. How does this error depend on the spatial resolution ($k\Delta x$) and the Courant number ($U\Delta t/\Delta x$)?

2. Show that the Lagrange interpolating polynomial in (6.12) is equivalent to the following Newton polynomial:

$$c_0 + (2-\alpha)\left[c_1 + (1-\alpha)\left(\frac{c_2}{2} - \alpha\frac{c_3}{6}\right)\right],$$

where

$$c_0 = \phi^n_{j-p-2},$$
$$c_1 = \phi^n_{j-p-1} - c_0,$$
$$c_2 = \phi^n_{j-p} - \phi^n_{j-p-1} - c_1,$$
$$c_3 = \phi^n_{j-p+1} - 2\phi^n_{j-p} + \phi^n_{j-p-1} - c_2$$

Compare the number of multiplications and additions required to evaluate the preceding with those required to evaluate (6.12).

3. Suppose that a semi-Lagrangian approximation to the constant-wind-speed advection equation uses quadratic polynomial interpolation as specified in (6.11).

(a) Derive the leading-order truncation error for this scheme.

(b) Determine the range of $U \Delta t / \Delta x$ for which the resulting scheme is identical to the Lax–Wendroff method (2.102). Also determine the values of $U \Delta t / \Delta x$ for which the scheme is identical to the method of Warming and Beam (2.109).

4. Determine the values of α for which the semi-Lagrangian approximation to the constant-wind-speed advection equation is stable when quadratic interpolation is used to evaluate $\phi(\tilde{x}^n_j, t^n)$ as in (6.11). Why is this scheme implemented by choosing p such that $|\alpha| \le \frac{1}{2}$?

5. Show that in comparison to a hypothetical scheme that exactly determines the value of ϕ at the departure point, the damping associated with the polynomial interpolation of $\phi(\tilde{x}^n_j, t^n)$ increases the stability of numerical approximations to (6.27) computed using the trapezoidal scheme (6.3).

6. Suppose a noninterpolating three-time-level semi-Lagrangian scheme is used to compute approximate solutions to the variable-wind-speed advection equation

$$\frac{\partial \psi}{\partial t} + u(x)\frac{\partial \psi}{\partial x} = 0.$$

If the approximate solution is defined at the mesh points x_j and the velocity is available at both x_j and $x_{j+\frac{1}{2}}$, determine the strategy for choosing p that minimizes the truncation error in the resulting scheme. Should p be even, odd, or simply the integer such that $p\Delta x$ is closest to the departure point? Does this strategy yield stable solutions? How well does it generalize to two-dimensional problems?

7. Suppose that (6.29) is used to obtain approximate solutions to the prototype equation for forced scalar advection (6.27). How do the stability properties of the $2\Delta x$ waves compare with those of the $4\Delta x$ waves as a function of the Courant number $U \Delta t / \Delta x$?

8. Suppose that the two-level forward extrapolation (6.23) is replaced by the three-level scheme

$$u(t^{n+\frac{1}{2}}) = \frac{1}{8}\left[15u(t^n) - 10u(t^{n-1}) + 3u(t^{n-2})\right].$$

Determine the order of accuracy to which this method estimates $u(t^{n+\frac{1}{2}})$.

9. Show that backward trajectories computed with the Euler method in a spatially varying wind field can cross if $|\partial \mathbf{v}/\partial x|\Delta t$ exceeds unity.

7

Physically Insignificant Fast Waves

One reason that explicit time-differencing is widely used in the simulation of wave-like flows is that accuracy considerations and stability constraints often yield similar criteria for the maximum time step in numerical integrations of systems that support a single type of wave motion. Many fluid systems, however, support more than one type of wave motion, and in such circumstances accuracy considerations and stability constraints can yield very different criteria for the maximum time step. If explicit time-differencing is used to construct a straightforward numerical approximation to the equations governing a system that supports several types of waves, the maximum stable time step will be limited by the Courant number associated with the most rapidly propagating wave, yet that rapidly propagating wave may be of little physical significance.

As an example, consider the Earth's atmosphere, which supports sound waves, gravity waves, and Rossby waves. Rossby waves propagate more slowly than gravity waves which in turn move more slowly than sound waves. The maximum stable time step with which an explicit numerical method can integrate the full equations governing atmospheric motions will be limited by the Courant number associated with sound-wave propagation. If finite differences are used in the vertical, and the vertical grid spacing is 300 m, the maximum stable time step will on the order of one second. Since sound waves have no direct meteorological significance, they need not be accurately simulated in order to obtain a good weather forecast. The quality of the weather forecast depends solely on the ability of the model to accurately simulate atmospheric disturbances that evolve on much slower time scales. Gravity waves can be accurately simulated with time steps on the order of 10 to 100 seconds; Rossby waves require a time step on the order of 500 to 5000 seconds. In order to obtain a reasonably efficient numerical model for the simulation of atmospheric circulations, it is necessary to circumvent the stability constraint associated with sound-wave propagation and bring the maximum stable time step into closer agreement with the time step limitations arising from accuracy considerations.

There are two basic approaches for circumventing the time step constraint imposed by a rapidly moving, physically insignificant wave. The first approach is to approximate the full governing equations with a set of "filtered" equations that does not support the rapidly moving wave. As an example, the full equations for stratified compressible flow might be approximated by the Boussinesq equations for incompressible flow. In this approach fundamental approximations are introduced to the original continuous equations prior to any numerical approximations that may be subsequently employed to generate finite-difference or spectral solutions to the filtered governing equations. The use of the filtered equation set may be motivated entirely by numerical considerations, or it may arise naturally from the standard approximations used in the study of a given physical phenomenon. Gravity waves, for example, are often studied in the context of Boussinesq incompressible flow in order to simplify and streamline the mathematical description of the problem.

The second approach for circumventing the time step constraint imposed by a rapidly moving, physically insignificant wave leaves the continuous governing equations unmodified and relies on numerical techniques to stabilize the fast-moving wave. These numerical techniques achieve efficiency by sacrificing the accuracy with which the fast-moving wave is represented. Note that neither one of these approaches is appropriate in situations where the fast-moving wave needs to be accurately simulated, since small time steps are required to adequately resolve the fast-moving wave.

This chapter begins by examining techniques for the numerical solution of the Boussinesq equations, which is one of the most fundamental systems of filtered equations. Methods for the solution of a second system of filtered equations, the primitive equations, are presented in Sections 7.5–7.6. Numerical methods for stabilizing the solution to problems that simultaneously support fast- and slow-moving waves are considered in Sections 7.2–7.4. One of these techniques, the semi-implicit method, is frequently used to integrate the primitive equations in applications where the phenomena of primary interest are slow-moving Rossby waves. In such applications the numerical integration is stabilized with respect to two different types of physically insignificant, rapidly moving waves. Sound waves are filtered by the primitive-equation approximation, and the most rapidly moving gravity waves (and the Lamb wave) are stabilized by the semi-implicit time integration.

7.1   The Projection Method

The unapproximated mass conservation equation (1.32) is a prognostic equation for the density that can be combined with the equation of state to form a prognostic equation for pressure. The equation of state can also be used to eliminate density from the momentum equations and thereby express the Euler equations as a closed system of five prognostic equations in five unknowns. When sound waves are filtered from the governing equations using the incompressible, Boussinesq, anelastic, or pseudo-incompressible approximations, the approximate mass conservation equations for these filtered systems do not depend on the local time derivative of the true density, and as a consequence, they cannot be used to obtain a prognostic equation for pressure. Since a prognostic equation is not available for the calculation of pressure, the filtered systems are not strictly hyperbolic, and their numerical solution cannot be obtained entirely through the use of explicit finite-difference schemes.

As an example, consider the Boussinesq equations (1.51), (1.56), and (1.60), which may be written in the form

$$\nabla \cdot \mathbf{v} = 0, \tag{7.1}$$

$$\frac{d\rho'}{dt} + w\frac{d\overline{\rho}}{dz} = 0, \tag{7.2}$$

$$\frac{\partial \mathbf{v}}{\partial t} + \frac{1}{\rho_0}\nabla p' = \mathbf{F}(\mathbf{v}, \rho'), \tag{7.3}$$

where

$$\mathbf{F}(\mathbf{v}, \rho') = -\mathbf{v} \cdot \nabla \mathbf{v} - g\frac{\rho'}{\rho_0}\mathbf{k}$$

and p' and ρ' are the deviations of the pressure and density from their values in a hydrostatically balanced reference state, $\overline{p}(z)$ and $\overline{\rho}(z)$. The unknown variables are the three velocity components, the perturbation density, and the perturbation pressure. There is no prognostic equation available to determine the time tendency of p'. The perturbation pressure field at a given instant can, however, be diagnosed from the instantaneous velocity and perturbation density fields by solving the Poisson equation

$$\nabla^2 p' = \rho_0 \nabla \cdot \mathbf{F}, \tag{7.4}$$

which can be derived by taking the divergence of (7.3) and then using (7.1). The perturbation pressure satisfying (7.4) is the instantaneous pressure distribution that will keep the evolving velocity field nondivergent.

7.1.1 Forward-in-Time Implementation

The *projection method* (Chorin 1968) is a classical technique that may be used to obtain numerical solutions to the Boussinesq system. Suppose the momentum equation is integrated over a time interval Δt to yield

$$\int_{t^n}^{t^{n+1}} \frac{\partial \mathbf{v}}{\partial t}\, dt = -\int_{t^n}^{t^{n+1}} \frac{1}{\rho_0}\nabla p'\, dt + \int_{t^n}^{t^{n+1}} \mathbf{F}(\mathbf{v}, \rho')\, dt, \tag{7.5}$$

where $t^n = n\Delta t$. Define the quantity $\tilde{p}^{n+1}$ such that

$$\Delta t \nabla \tilde{p}^{n+1} = \int_{t^n}^{t^{n+1}} \frac{1}{\rho_0} \nabla p' \, dt.$$

Note that $\tilde{p}^{n+1}$ is not necessarily equal to the actual perturbation pressure at any particular time. Using the definition of $\tilde{p}^{n+1}$, (7.5) may be written as

$$\mathbf{v}^{n+1} - \mathbf{v}^n = -\Delta t \nabla \tilde{p}^{n+1} + \int_{t^n}^{t^{n+1}} \mathbf{F}(\mathbf{v}, \rho') \, dt. \tag{7.6}$$

Define $\tilde{\mathbf{v}}$ such that

$$\tilde{\mathbf{v}} = \mathbf{v}^n + \int_{t^n}^{t^{n+1}} \mathbf{F}(\mathbf{v}, \rho') \, dt. \tag{7.7}$$

As noted by Orszag et al. (1986), the preceding integral can be conveniently evaluated using an explicit finite-difference scheme such as the third-order Adams–Bashforth method (2.58). Equations (7.6) and (7.7) imply that

$$\mathbf{v}^{n+1} = \tilde{\mathbf{v}} - \Delta t \nabla \tilde{p}^{n+1}, \tag{7.8}$$

which provides a formula for updating $\tilde{\mathbf{v}}$ to obtain the new velocity field $\mathbf{v}^{n+1}$ once $\tilde{p}^{n+1}$ has been determined.

A Poisson equation for $\tilde{p}^{n+1}$ that is analogous to (7.4) is obtained by taking the divergence of (7.8) and noting that $\nabla \cdot \mathbf{v}^{n+1} = 0$, in which case

$$\nabla^2 \tilde{p}^{n+1} = \frac{\nabla \cdot \tilde{\mathbf{v}}}{\Delta t}. \tag{7.9}$$

Boundary conditions for this equation are obtained by computing the dot product of the unit vector normal to the boundary with each term of (7.8) to yield

$$\frac{\partial \tilde{p}^{n+1}}{\partial n} = -\frac{1}{\Delta t} \mathbf{n} \cdot (\mathbf{v}^{n+1} - \tilde{\mathbf{v}}). \tag{7.10}$$

If there is no flow normal to the boundary, the preceding reduces to

$$\frac{\partial \tilde{p}^{n+1}}{\partial n} = \frac{\mathbf{n} \cdot \tilde{\mathbf{v}}}{\Delta t}, \tag{7.11}$$

which eliminates the implicit coupling between $\tilde{p}^{n+1}$ and $\mathbf{v}^{n+1}$ that is present in the general boundary condition (7.10). In this particularly simple case in which an inviscid fluid is bounded by rigid walls, the projection method is implemented by first updating (7.7), which accounts for the time tendencies produced by advection and buoyancy forces, and then solving (7.9) subject to the boundary conditions (7.11). As the final step of the algorithm, $\mathbf{v}^{n+1}$ is obtained by projecting $\tilde{\mathbf{v}}$ onto the subspace of nondivergent vectors using (7.8).

The preceding algorithm loses some of its simplicity when the computation of $\mathbf{v}^{n+1}$ is coupled with that of $\tilde{p}^{n+1}$, as would be the case if a wave-permeable boundary condition replaced the rigid-wall condition that $\mathbf{n} \cdot \mathbf{v}^{n+1} = 0$. In practice, the coupling between $\mathbf{v}^{n+1}$ and $\tilde{p}^{n+1}$ is eliminated by imposing some approximation to the full, implicitly coupled boundary condition. Coupling between $\mathbf{v}^{n+1}$ and $\tilde{p}^{n+1}$ may also occur when the projection method is applied to viscous flows with a no-slip condition at the boundary. The no-slip condition that $\mathbf{v} = 0$ at the boundary reduces (7.10) to

$$\frac{\partial \tilde{p}^{n+1}}{\partial n} = \frac{1}{\Delta t} \, \mathbf{n} \cdot \int_{t^n}^{t^{n+1}} -g\frac{\rho'}{\rho_0}\mathbf{k} + \nu\nabla^2\mathbf{v} \, dt, \tag{7.12}$$

where viscous forcing is now included in the momentum equations and ν is the kinematic viscosity. High spatial resolution is often required to resolve the boundary layer in no-slip viscous flow. In order to maintain numerical stability in the high-resolution boundary layer without imposing an excessively strict limitation on the time step, the viscous terms are often integrated using implicit differencing[1] (Karniadakis et al. 1991). When the time integral of $\mathbf{F}(\mathbf{v}, \rho')$ includes viscous terms that are approximated using implicit finite differences, (7.12) is an implicit relation between $\tilde{p}^{n+1}$ and $\mathbf{v}^{n+1}$ whose solution is often computed via a fractional-step method. As noted by Orszag et al. (1986), the accuracy with which this boundary condition is approximated can significantly influence the accuracy of the overall solution. The design of optimal approximations to (7.12) has been the subject of considerable research. However, the emphasis in this book is not on viscous flow, and especially not on highly viscous flow in which the diffusion terms need to be treated implicitly for computational efficiency. The reader is referred to Boyd (1989) for further discussion of the use of the projection method in viscous no-slip flow.

7.1.2 Leapfrog Implementation

In atmospheric science the projection method is often implemented using leapfrog time differences, in which case (7.5) becomes

$$\mathbf{v}^{n+1} = \mathbf{v}^{n-1} - \frac{2\Delta t}{\rho_0}\nabla p'^n + 2\Delta t \mathbf{F}\left(\mathbf{v}^n, \rho'^n\right).$$

The solution procedure is very similar to the algorithm described in the preceding section. The velocity field generated by advection and buoyancy forces acting over the time period $2\Delta t$ is defined as

$$\tilde{\mathbf{v}} = \mathbf{v}^{n-1} + 2\Delta t \mathbf{F}\left(\mathbf{v}^n, \rho'^n\right);$$

---

[1] Explicit time-differencing can still be used for the advection terms because the wind speed normal to the boundary decreases as the fluid approaches the boundary.

then the Poisson equation for p'^n is

$$\nabla^2 p'^n = \frac{\rho_0}{2\Delta t} \nabla \cdot \tilde{\mathbf{v}},$$

and the velocity field is updated using the relation

$$\mathbf{v}^{n+1} = \tilde{\mathbf{v}} - 2\Delta t \nabla p'^n.$$

Some technique, such as Asselin time filtering (2.50), must also be used to prevent time-splitting instability in the leapfrog solution to nonlinear problems.

One difference between this approach and the standard projection method is that by virtue of the leapfrog time difference, the pressure field that ensures the nondivergence of $\mathbf{v}^{n+1}$ is the actual pressure at time $n\Delta t$. The pressure must, nevertheless, be updated at the same point in the integration cycle at which $\tilde{p}^{n+1}$ is obtained in the standard projection method, i.e., partway through the calculation of $\mathbf{v}^{n+1}$. Thus, the same problems with implicit coupling between the pressure and $\mathbf{v}^{n+1}$ arise in both the standard and the leapfrog projection methods. If viscosity is included in the momentum equations, stability considerations require that the contribution of viscosity to $\mathbf{F}(\mathbf{v}, \rho')$ be evaluated at time level $n - 1$, so that the viscous terms are treated using forward differencing over a time interval of $2\Delta t$. This is not a particularly accurate way to represent the viscous terms and is not suitable for highly viscous flow, in which the viscous terms are more efficiently integrated using implicit-time differences.

7.1.3   Solving the Poisson Equation for Pressure

Suppose that the Boussinesq equations are to be solved in a two-dimensional x-z domain and that the velocity and pressure variables are staggered as indicated in Fig. 3.6 . Approximating the diagnostic pressure equation (7.9) using the standard five-point finite-difference stencil for the two-dimensional Laplacian, one obtains

$$\delta_x^2 \tilde{p}^{n+1} + \delta_z^2 \tilde{p}^{n+1} = \frac{1}{\Delta t} \left(\delta_x \tilde{u} + \delta_z \tilde{w} \right). \tag{7.13}$$

This is an implicit algebraic relation for the $\tilde{p}_{i,j}^{n+1}$. If pressure is defined at M grid points in x and N points in z, an $M \times N$ system of linear algebraic equations must be solved in order to determine the pressure. Let the unknown grid-point values of the pressure be ordered such that

$$\mathbf{p} = (\tilde{p}_{1,1}^{n+1}, \tilde{p}_{1,2}^{n+1}, \dots, \tilde{p}_{1,N}^{n+1}, \tilde{p}_{2,1}^{n+1}, \dots, \tilde{p}_{M,N}^{n+1}).$$

Then the system may be written as the matrix equation

$$\mathbf{A}\mathbf{p} = \mathbf{f}, \tag{7.14}$$

in which $\mathbf{f}$ is an identically ordered vector containing the numerically evaluated divergence of $\tilde{\mathbf{v}}$. The matrix $\mathbf{A}$ is very sparse, with only five nonzero diagonals.

In practical applications the number of unknown pressures can easily exceed one million, and in order to solve (7.14) efficiently, it is important to take advantage of the sparseness of **A**. Direct methods based on some variant of Gaussian elimination are therefore not appropriate. Direct methods for band matrices are also not suitable because the bandwidth of **A** is not 5, but $2N + 1$, and direct methods for band matrices do not preserve sparseness within the band.

Direct solutions to (7.14) can, nevertheless, be efficiently obtained by exploiting the block structure of **A**. For simplicity, suppose that (7.13) is to be solved subject to Dirichlet boundary conditions. Then the diagonal of **A** contains M copies of the $N \times N$ tridiagonal submatrix

$$\begin{pmatrix} -4 & 1 & & & & \\ 1 & -4 & 1 & & & \\ & \ddots & \ddots & \ddots & & \\ & & \ddots & \ddots & \ddots & \\ & & & 1 & -4 & 1 \\ & & & & 1 & -4 \end{pmatrix},$$

and the superdiagonal and subdiagonal are made up of $M - 1$ copies of the $N \times N$ identity matrix. This system can be efficiently solved using block cyclic reduction (Golub and van Loan 1996, p. 177). Numerical codes for the solution of two- and three-dimensional Poisson equations subject to the most common types of boundary conditions may be accessed through the Internet at several cites, including the National Institute of Standards and Technology's Guide to Available Mathematical Software (NIST/GAMS, http://gams.nist.gov), the National Center for Atmospheric Research's Mathematical and Statistical Libraries (NCAR, http://www.scd.ucar.edu/softlib/mathlib.html), and the Netlib Repository at the Oak Ridge National Laboratory (ORNL, http://www.netlib.org).

Numerical solutions to the anelastic equations (1.33), (1.52), and (1.64) can be obtained using the projection method in essentially the same manner as that for the Boussinesq system, except that the elliptic equation for the pressure in the anelastic system is slightly more complicated than a Poisson equation. After multiplication by $\bar{\rho}(z)$, the anelastic momentum equation (1.64) may be written

$$\frac{\partial \bar{\rho} \mathbf{v}}{\partial t} + c_p \bar{\rho} \nabla (\bar{\theta} \pi') = \bar{\rho} \mathbf{F}, \tag{7.15}$$

where

$$\mathbf{F}(\mathbf{v}, \theta') = -\mathbf{v} \cdot \nabla \mathbf{v} + g \frac{\theta'}{\bar{\theta}} \mathbf{k}. \tag{7.16}$$

An elliptic equation for pressure is obtained by taking the divergence of (7.15) and using the anelastic continuity equation

$$\nabla \cdot (\bar{\rho} \mathbf{v}) = 0,$$

to yield

$$\nabla \cdot \left[\bar{\rho}\nabla(\bar{\theta}\pi')\right] = \nabla \cdot \left(\frac{\bar{\rho}}{c_p}\mathbf{F}\right).$$

A linear system of algebraic equations for $\pi'_{i,j}$ is obtained after approximating the derivatives in the preceding by finite differences, but since $\bar{\rho}$ and $\bar{\theta}$ are functions of z, the structure of the coefficient matrix for this system is less uniform than that for the Boussinesq system. Nevertheless, the resulting linear system can still be efficiently solved by generalizations of the block-cyclic-reduction algorithm, and numerical codes for the solution of this problem appear in the previously noted software libraries.

When the projection method is used to solve the pseudo-incompressible equations (1.33), (1.37), and (1.54), the elliptic pressure equation becomes

$$\nabla \cdot \left[\bar{\rho}\bar{\theta}\theta\nabla\pi'\right] = \nabla \cdot \left(\frac{\bar{\rho}\bar{\theta}}{c_p}\mathbf{F}\right), \qquad (7.17)$$

where $\mathbf{F}$ is once again defined by (7.16). The finite-difference approximation to this equation still produces a very sparse linear algebraic system, with only five nonzero diagonals, but since the coefficient of each second derivative includes the factor θ, which is an arbitrary function of x, y, and z, it is not possible to solve the system by block-cyclic reduction—iterative methods must be used. Iterative methods may also need to be employed to determine the pressure in the Boussinesq and anelastic systems when those equations are solved on a curvilinear grid (such as a terrain-following coordinate system) because the coefficient structure in the elliptic pressure equation is usually complicated by the coordinate transformation.

The two most commonly used techniques for the iterative solution of the sparse linear-algebraic systems that arise in computational fluid dynamics are the preconditioned conjugate gradient method and the multigrid method. The mathematical basis for both of these methods is very nicely reviewed in Chapter 5 of Ferziger and Perić (1997) and will not be covered in this text. Additional information about multigrid methods may be found in Briggs (1987), Hackbusch (1985), and, in the context of geophysical fluid dynamics, in Adams et al. (1992). Conjugate-residual solvers are discussed in more detail in Golub and van Loan (1996) and in the context of atmospheric science in Smolarkiewicz and Margolin (1994) and Skamarock et al. (1997). Both multigrid and preconditioned conjugate residual solvers are available in the previously mentioned software libraries.

7.2   The Semi-implicit Method

As an alternative to filtering the governing equations to eliminate insignificant fast waves, one can retain the unapproximated governing equations and use numerical techniques to stabilize the simulation of the fast-moving waves. One common way to improve numerical stability is through the use of implicit time differences such as the backward and the trapezoidal methods.[2] Implicit methods can, however, produce rather inaccurate solutions when the time step is too large. It is therefore useful to analyze the effect of the time step on the accuracy of fully implicit solutions to wave-propagation problems before discussing the true semi-implicit method.

7.2.1 Large Time Steps and Poor Accuracy

Suppose that a differential–difference approximation to the one-dimensional advection equation

$$\frac{\partial \psi}{\partial t} + c \frac{\partial \psi}{\partial x} = 0 \qquad (7.18)$$

is constructed in which finite differences are used to represent the time derivative, and the spatial derivative is not discretized. If the time derivative is approximated using leapfrog differencing such that

$$\frac{\phi^{n+1} - \phi^{n-1}}{2\Delta t} + c \left(\frac{d\phi}{dx} \right)^n = 0,$$

then wave solutions of the form

$$\phi^n(x) = e^{i(kx - \omega nt)} \qquad (7.19)$$

must satisfy the semidiscrete dispersion relation

$$\omega = \frac{1}{\Delta t} \arcsin(ck\Delta t). \qquad (7.20)$$

The phase speed of the leapfrog-differenced solution is

$$c_{\mathrm{lf}} = \frac{\omega}{k} = \frac{\arcsin(ck\Delta t)}{k\Delta t}. \qquad (7.21)$$

The stability constraint $|ck\Delta t| < 1$ associated with the preceding leapfrog scheme can be avoided by switching to trapezoidal differencing. Many semi-implicit formulations use a combination of leapfrog and trapezoidal differencing, and in those formulations the trapezoidal time difference is computed over an interval of $2\Delta t$. In order to facilitate the application of this analysis to these semi-implicit formulations, and in order to compare the trapezoidal and leapfrog

---

[2]Higher-order implicit schemes are, however, not necessarily more stable than related explicit methods. Backward and trapezoidal differencing are the first- and second-order members of the Adams–Moulton family of implicit time integration schemes. The third- and fourth-order Adams–Moulton schemes generate amplifying solutions to oscillation equation (2.30) for any choice of time step, whereas their explicit cousins, the third- and fourth-order Adams–Bashforth schemes, produce stable nonamplifying solutions whenever the time step is sufficiently small.

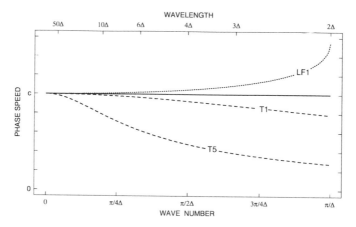

FIGURE 7.1. Phase speed of leapfrog (dotted) and $2\Delta t$-trapezoidal (dashed) approximations to the advection equation when $c\Delta t/\Delta x = 1/\pi$ (LF1 and T1), and for the trapezoidal solution when $c\Delta t/\Delta x = 5/\pi$ (T5).

schemes more directly, (7.18) will be approximated using trapezoidal differencing over a $2\Delta t$-wide stencil such that

$$\frac{\phi^{n+1} - \phi^{n-1}}{2\Delta t} + \frac{c}{2}\left[\left(\frac{d\phi}{dx}\right)^{n+1} + \left(\frac{d\phi}{dx}\right)^{n-1}\right] = 0.$$

Wave solutions to this scheme must satisfy the dispersion relation

$$\omega = \frac{1}{\Delta t}\arctan(ck\Delta t). \qquad (7.22)$$

The phase speed of the trapezoidally differenced solution is

$$c_t = \frac{\arctan(ck\Delta t)}{k\Delta t}.$$

The phase-speed errors generated by the leapfrog and $2\Delta t$ trapezoidal methods are compared in Fig. 7.1. The phase speed at a fixed Courant number is plotted as a function of spatial wave number, with the wave number axis scaled by $1/\Delta x$. These curves may therefore be interpreted as giving the phase speed that would be obtained if the spatial dependence of the numerical solution was represented by a Fourier spectral method with a cutoff wavelength of $2\Delta x$. When $c\Delta t/\Delta x < 1/\pi$ the errors generated by the leapfrog and the $2\Delta t$-trapezoidal methods are similar in magnitude and opposite in sign. The leapfrog scheme is unstable for Courant numbers greater than $1/\pi$, but solutions can still be obtained using the trapezoidal scheme. The phase-speed errors in the $2\Delta t$-trapezoidal solution computed with $c\Delta t/\Delta x = 5/\pi$ are, however, rather large. Even modes with relatively good spatial resolution, such as a $10\Delta x$ wave, are in significant error.

The deceleration generated by $2\Delta t$-trapezoidal differencing may be concisely described by defining a reduced phase speed

$$\hat{c} = c\cos(\omega\Delta t).$$

Then the $2\Delta t$-trapezoidal dispersion relation (7.22) may be expressed as

$$\omega = \frac{1}{\Delta t} \arcsin(\hat{c} k \Delta t),$$

and the phase speed of the $2\Delta t$-trapezoidal solution becomes

$$c_{\mathrm{t}} = \frac{\omega}{k} = \frac{\arcsin(\hat{c} k \Delta t)}{k \Delta t}.$$

The preceding differ from the corresponding expressions for the leapfrog scheme (7.20) and (7.21) in that the true propagation speed, c, has been replaced by the reduced speed $\hat{c}$. As the time step increases, $\hat{c}$ decreases, so that $|\hat{c} k \Delta t|$ remains less than one and the numerical solution remains stable, but the relative error in $\hat{c}$ can become arbitrarily large. As a consequence, it is not possible to take advantage of the unconditional stability of the trapezoidal method by using very large time steps to solve wave-propagation problems unless one is willing to tolerate a considerable decrease in the accuracy of the solution.

7.2.2 A Prototype Problem

The loss of accuracy associated with the poor temporal resolution that can occur using implicit numerical methods is not a problem if the poorly resolved waves are not physically significant. If the fastest-moving waves are not physically significant, the accuracy constraints imposed on the time step by these waves can be ignored, and provided that the scheme is unconditionally stable, a good solution can be obtained using any time step that adequately resolves the slower-moving features of primary physical interest. A simple but computationally inefficient way to ensure the unconditional stability of a numerical scheme is to use trapezoidal time-differencing throughout the approximate equations. It is, however, more efficient to implicitly evaluate only those terms in the governing equations that are crucial to the propagation of the fast wave and to approximate the remaining terms with some explicit time-integration scheme. This is the fundamental strategy in the "semi-implicit" approach, which gains efficiency relative to a "fully implicit" method by reducing the complexity of the implicit algebraic equations that must be solved during each integration step. Semi-implicit differencing is particularly attractive when all the terms that are evaluated implicitly are linear functions of the unknown variables.

In order to investigate the stability of semi-implicit time-differencing schemes, consider a prototype ordinary differential equation of the form

$$\frac{d\psi}{dt} + i\omega_{\mathrm{H}}\psi + i\omega_{\mathrm{L}}\psi = 0. \tag{7.23}$$

This is simply a version of the oscillation equation (2.30) in which the oscillatory forcing is divided into high-frequency (ω_{H}) and low-frequency (ω_{L}) components. The division of the forcing into two terms may appear to be rather artificial, but the dispersion relation associated with wave-like solutions to more complex systems of governing equations (such as the shallow-water system discussed in the next section) often has individual roots of the form

$$\omega = \omega_H + \omega_L,$$

and (7.23) serves as the simplest differential equation describing the time dependence of such waves.

The simplest semi-implicit approximation to (7.23) is

$$\frac{\phi^{n+1} - \phi^n}{\Delta t} + i\omega_H \phi^{n+1} + i\omega_L \phi^n = 0.$$

The stability and the accuracy of this scheme have already been analyzed in connection with (2.33); it is first-order accurate and is stable whenever $|\omega_L| < |\omega_H|$. Since $|\omega_L| < |\omega_H|$ by assumption, the method is stable for all Δt. The weakness of this scheme is its low accuracy. A more accurate second-order method can be obtained using the centered-in-time formula

$$\frac{\phi^{n+1} - \phi^{n-1}}{2\Delta t} + i\omega_H \left(\frac{\phi^{n+1} + \phi^{n-1}}{2} \right) + i\omega_L \phi^n = 0. \tag{7.24}$$

In order to investigate the stability of this method, consider the behavior of oscillatory solutions of the form $\exp(-i\omega n \Delta t)$, which satisfy (7.24) when

$$\sin \tilde{\omega} = \tilde{\omega}_H \cos \tilde{\omega} + \tilde{\omega}_L, \tag{7.25}$$

where

$$\tilde{\omega} = \omega \Delta t, \quad \tilde{\omega}_H = \omega_H \Delta t, \quad \text{and} \quad \tilde{\omega}_L = \omega_L \Delta t.$$

Defining $\tan \beta = \tilde{\omega}_H$, (7.25) becomes

$$\sin \tilde{\omega} = \tan \beta \cos \tilde{\omega} + \tilde{\omega}_L,$$

or equivalently,

$$\sin \tilde{\omega} \cos \beta - \sin \beta \cos \tilde{\omega} = \tilde{\omega}_L \cos \beta.$$

By the Pythagorean theorem, $\cos \beta = (1 + \tilde{\omega}_H^2)^{-1/2}$, and the preceding reduces to

$$\sin(\tilde{\omega} - \beta) = \tilde{\omega}_L (1 + \tilde{\omega}_H^2)^{-1/2},$$

or equivalently,

$$\tilde{\omega} = \arctan(\tilde{\omega}_H) + \arcsin\left(\tilde{\omega}_L (1 + \tilde{\omega}_H^2)^{-1/2} \right).$$

The semi-implicit scheme (7.24) will be stable when the $\tilde{\omega}$ satisfying this equation are real and distinct, which is guaranteed when

$$\tilde{\omega}_L^2 \leq 1 + \tilde{\omega}_H^2. \tag{7.26}$$

Since by assumption $|\omega_L| \leq |\omega_H|$, (7.24) is stable for all Δt. Note that (7.26) will also be satisfied whenever $|\omega_L \Delta t| \leq 1$, implying that semi-implicit differencing permits an increase in the maximum stable time step relative to that for a fully explicit approximation even in those cases where $|\omega_L| > |\omega_H|$, because the terms approximated with the trapezoidal difference do not restrict the maximum stable time step.

As discussed in Section 3.4.2, semi-implicit time-differencing may also be used to stabilize the diffusion operator in some advection–diffusion problems. The gain in the maximum stable time step achieved using a trapezoidal time difference for the diffusion term in conjunction with a leapfrog approximation to the advection is not, however, particularly impressive. A much more stable approximation to the advection–diffusion problem is obtained using the third-order Adams–Bashforth method to integrate the advection terms and trapezoidal differencing to approximate the diffusion term, but the advantages of the semi-implicit Adams–Bashforth–trapezoidal formulation do not carry over to the fast-wave–slow-wave problem in a completely straightforward manner. If, for example, λ is replaced by $i\omega_H$ in (3.81) so that the trapezoidally differenced term represents a fast-moving wave, the stability of the resulting scheme is still quite limited (the scheme is unstable for all $\omega_H \Delta t$ greater than approximately 1.8).

7.2.3 Semi-implicit Solution of the Shallow-Water Equations

The shallow-water equations (1.25)–(1.27) support rapidly moving gravity waves. If there are spatial variations in the potential vorticity of the undisturbed system f/H, the shallow-water equations can also support slowly propagating potential-vorticity (or Rossby) waves. In many large-scale atmospheric and oceanic models, the Rossby waves are of greater physical significance than the faster-moving gravity waves, and the Rossby waves can be efficiently simulated using semi-implicit time-differencing to circumvent the CFL stability condition associated with gravity-wave propagation. The simplest example in which to illustrate the influence of semi-implicit differencing on the CFL condition for gravity waves is provided by (3.1) and (3.2), which are the one-dimensional shallow-water equations linearized about a reference state with a constant fluid velocity U and fluid depth H. If the mean-flow velocity is less than the phase speed of a shallow-water gravity wave, the numerical integration can be stabilized by evaluating those terms responsible for gravity-wave propagation with trapezoidal differencing; leapfrog differencing can be used for the remaining terms (Kwizak and Robert 1971). The terms essential to gravity-wave propagation are the pressure–gradient term in (3.1) and the velocity divergence in (3.2), so the semi-implicit approximation to the linearized shallow-water system is

$$\delta_{2t} u^n + U \frac{du^n}{dx} + g \left\langle \frac{dh^n}{dx} \right\rangle^{2t} = 0, \tag{7.27}$$

$$\delta_{2t} h^n + U \frac{dh^n}{dx} + H \left\langle \frac{du^n}{dx} \right\rangle^{2t} = 0, \tag{7.28}$$

where the finite-difference operator δ_t and the averaging operator $\langle \ \rangle^t$ are defined by (A.1) and (A.2) in the Appendix. Solutions to (7.27) and (7.28) exist of the form $e^{i(kx-\omega j \Delta t)}$, provided that k and ω satisfy the semidiscrete dispersion relation

$$\sin \omega \Delta t = U k \Delta t \pm c k \Delta t \cos \omega \Delta t,$$

in which $c = \sqrt{gH}$. This dispersion relation has the same form as (7.25), and as demonstrated in the preceding section, the method will be stable, provided that $|U| \leq c$, or equivalently, whenever the phase speed of shallow-water gravity waves exceeds the speed of the mean flow.

The Coriolis force has been neglected in the preceding shallow-water system, and as a consequence, there are no Rossby-wave solutions to (7.27) and (7.28). In a more general system that does include the Coriolis force,[3] semi-implicit time-differencing leads to a system that is stable whenever the CFL condition for the Rossby waves is satisfied. This general case is examined in more detail in Problems 1–3 at the end of this chapter.

Instead of considering the complications introduced by the presence of Rossby waves, consider the nonlinear equivalent of the preceding linearized system:

$$\frac{\partial u}{\partial t} + u \frac{\partial u}{\partial x} + g \frac{\partial h}{\partial x} = 0, \tag{7.29}$$

$$\frac{\partial h}{\partial t} + u \frac{\partial h}{\partial x} + h \frac{\partial u}{\partial x} = 0. \tag{7.30}$$

As before, a semi-implicit algorithm can be obtained using trapezoidal time differences to evaluate the pressure gradient in (7.29) and the velocity divergence in (7.30). The term involving the velocity divergence is, however, nonlinear, and an implicit system of nonlinear algebraic equations will be generated if the time integral of $h \, \partial u/\partial x$ is approximated using the trapezoidal method. In order to avoid solving a nonlinear algebraic equation at every time step, the velocity divergence in (7.30) is split into two terms such that

$$\frac{\partial \eta}{\partial t} + u \frac{\partial \eta}{\partial x} + H \frac{\partial u}{\partial x} + \eta \frac{\partial u}{\partial x} = 0,$$

where the total fluid depth has been divided into a constant-mean component H and a perturbation $\eta(x, t)$. The standard semi-implicit approximation to the preceding takes the form

$$\delta_{2t} \eta^n + u^n \frac{d\eta^n}{dx} + H \left(\frac{du^n}{dx} \right)^{2t} + \eta^n \frac{du^n}{dx} = 0; \tag{7.31}$$

only the linear term involving the constant depth H is treated implicitly. The time-differencing of the nonlinear momentum equation is identical to that for the lin-

---

[3] The inclusion of the Coriolis force also requires the inclusion of an additional prognostic equation for the other component of the horizontal velocity.

earized system

$$\delta_{2t} u^n + u^n \frac{du^n}{dx} + g \left(\frac{d\eta^n}{dx} \right)^{2t} = 0. \tag{7.32}$$

Leaving aside possible problems with nonlinear instability, one would intuitively expect that solutions to (7.31) and (7.32) would be unconditionally stable, provided that $\eta \ll H$, which is to say that stability would require the gravity-wave phase speed determined by the mean fluid depth to greatly exceed any local augmentation of the phase speed induced by a local increase in depth.

The impact of a perturbation in the fluid depth on the stability of the semi-implicit scheme is most easily evaluated if (7.31) and (7.32) are linearized about a reference state with no mean flow and a horizontally uniform perturbation in the depth $\bar{\eta}$. The semi-implicit approximation to this linearized system is

$$\delta_{2t} u^n + g \left(\frac{d\eta^n}{dx} \right)^{2t} = 0,$$

$$\delta_{2t} \eta^n + H \left(\frac{du^n}{dx} \right)^{2t} + \bar{\eta} \frac{du^n}{dx} = 0.$$

Letting $v = \sqrt{gH} \, k \Delta t$ and $r = \bar{\eta}/H$, solutions to this system satisfy the dispersion relation

$$\sin^2 \omega \Delta t = v^2 (\cos^2 \omega \Delta t + r \cos \omega \Delta t),$$

which is a quadratic equation in $\cos \omega \Delta t$,

$$(v^2 + 1) \cos^2 \omega \Delta t + r v^2 \cos \omega \Delta t - 1 = 0, \tag{7.33}$$

whose roots are

$$\cos \omega \Delta t = \frac{-r v^2 \pm (r^2 v^4 + 4v^2 + 4)^{1/2}}{2(v^2 + 1)}.$$

The scheme will be stable when ω is real. Since the radicand is always positive, the right side of the preceding expression is always real, and real solutions for ω are obtained when the magnitudes of both roots of (7.33) are less than or equal to unity.

Let $s = \cos \omega \Delta t$ be one of the roots of (7.33). The identity

$$(x - s_1)(x - s_2) = x^2 - (s_1 + s_2)x + s_1 s_2$$

implies that the sum and product of the roots of the quadratic equation (7.33) satisfy

$$s_1 + s_2 = \frac{-r v^2}{v^2 + 1} \quad \text{and} \quad s_1 s_2 = \frac{-1}{v^2 + 1}. \tag{7.34}$$

When $r = 0$,

$$|s_1| = |s_2| = \frac{1}{(v^2 + 1)^{1/2}} \leq 1,$$

and the scheme is stable. As $|r|$ increases, the magnitude of one of the roots eventually exceeds unity, and the scheme becomes unstable. The critical values of r beyond which the scheme becomes unstable may be determined by substituting $s_1 = 1$ and then $s_1 = -1$ into (7.34) to obtain the stability criterion $|r| \le 1$ or $|\bar{\eta}| \le H$. Instability will not occur unless the perturbation fluid depth exceeds the mean depth (in the $U = 0$ case). This may appear to be a very generous criterion; however, the local phase speed of a shallow-water gravity wave is

$$c = \sqrt{g(H + \bar{\eta})},$$

so numerical stability requires that the local wave speed be no faster than a factor of $\sqrt{2}$ times the mean wave speed. As will be discussed in the next section, the horizontal phase speed of hydrostatic internal gravity waves is proportional to the Brunt–Väisälä frequency, and the ratio of the local mean Brunt–Väisälä frequency in the polar regions of the Earth's atmosphere to that in the middle latitudes can easily exceed the square root of two. Thus, it is generally necessary to specify the horizontally uniform reference state using numerical values from that region in the domain where gravity waves propagate at maximum speed. As a consequence, the reference-state stratification in most global atmospheric models is chosen to be isothermal (Simmons et al. 1978). The application of the semi-implicit method to global atmospheric models is discussed further in Section 7.6.5.

7.2.4 Semi-implicit Solution of the Euler Equations

Now consider how semi-implicit differencing can be used to eliminate the stability constraint imposed by sound waves in the numerical solution of the Euler equations for stratified flow. In order to present the numerical approach with a minimum of extraneous detail, it is useful to consider a simplified set of compressible equations that can be obtained from the linearized system (1.41)–(1.44) by the transformation of variables

$$u = \left(\frac{\bar{\rho}}{\rho_0}\right)^{1/2} u', \qquad P = \left(\frac{\bar{\rho}}{\rho_0}\right)^{1/2} c_p \bar{\theta} \pi' \qquad (7.35)$$

$$w = \left(\frac{\bar{\rho}}{\rho_0}\right)^{1/2} w', \qquad b = \left(\frac{\bar{\rho}}{\rho_0}\right)^{1/2} \frac{g}{\bar{\theta}} \theta', \qquad (7.36)$$

which removes the influence of the decrease in the mean density with height on the magnitudes of the dependent variables. This transformation does not symmetrize the system as nicely as (1.45)–(1.46), but P and b have more direct interpretations as normalized pressure and buoyancy than do the thermodynamic variables introduced in (1.45)–(1.46).

The transformed vertical momentum and pressure equations take the form

$$\left(\frac{\partial}{\partial t} + U\frac{\partial}{\partial x}\right) w + \left(\frac{\partial}{\partial z} + \Gamma\right) P = b,$$

and

$$\left(\frac{\partial}{\partial t} + U \frac{\partial}{\partial x}\right) P + c_s^2 \left[\frac{\partial u}{\partial x} + \left(\frac{\partial}{\partial z} - \Gamma\right) w\right] = 0,$$

where

$$\Gamma = \frac{1}{2\overline{\rho}} \frac{\partial \overline{\rho}}{\partial z} + \frac{g}{c_s^2} \qquad \text{and} \qquad c_s^2 = \frac{c_p}{c_v} R\overline{T}. \tag{7.37}$$

In an isothermal atmosphere at temperature T_0,

$$\Gamma = \frac{3}{14} \frac{g}{RT_0}.$$

If $T_0 = 0°C$, Γ is 2.7×10^{-5} ms^{-1}, implying that the vertical derivative in the operators

$$\left(\frac{\partial}{\partial z} \pm \Gamma\right)$$

will exceed the term involving Γ by an order of magnitude in all waves with vertical wavelengths shorter than 100 km. The terms involving Γ can therefore be neglected in most applications,[4] in which case the transformed Euler equations become

$$\left(\frac{\partial}{\partial t} + U \frac{\partial}{\partial x}\right) u + \frac{\partial P}{\partial x} = 0, \tag{7.38}$$

$$\left(\frac{\partial}{\partial t} + U \frac{\partial}{\partial x}\right) w + \frac{\partial P}{\partial z} = b, \tag{7.39}$$

$$\left(\frac{\partial}{\partial t} + U \frac{\partial}{\partial x}\right) b + N^2 w = 0, \tag{7.40}$$

$$\left(\frac{\partial}{\partial t} + U \frac{\partial}{\partial x}\right) P + c_s^2 \left(\frac{\partial u}{\partial x} + \frac{\partial w}{\partial z}\right) = 0. \tag{7.41}$$

The preceding simplified compressible system (7.38)–(7.41) can also be regarded as the linearization of a "compressible Boussinesq" system in which

$$b = -g \frac{\rho - \overline{\rho}(z)}{\rho_0}, \qquad P = \frac{p - \overline{p}(z)}{\rho_0}, \qquad N^2 = -\frac{g}{\rho_0} \frac{\partial \overline{\rho}}{\partial z}. \tag{7.42}$$

As in the standard Boussinesq approximation, the compressible Boussinesq system neglects the influence of density variations on inertia while retaining the in-

---

[4]The isothermal atmosphere does support a free wave (known as the Lamb wave, see Section 7.5) that disappears in the limit $\Gamma \to 0$, but it is not necessary to account for the Lamb wave in this discussion of semi-implicit differencing.

fluence of density variations on buoyancy and assumes that buoyancy is conserved following a fluid parcel. In contrast to the standard Boussinesq system, the compressible Boussinesq system does retain the influence of density fluctuations on pressure and thereby allows the formation of the prognostic pressure equation (7.41).

Suppose that the simplified compressible system (7.39)–(7.40) is approximated using leapfrog time-differencing and that the spatial derivatives are computed using a Fourier pseudospectral method. Waves of the form

$$(u, w, b, P) = (u_0, w_0, b_0, P_0)e^{i(kx+\ell z-\omega n \Delta t)}$$

are solutions to this system, provided that ω, k, and ℓ satisfy the dispersion relation

$$\hat{\omega}^4 - c_s^2 \left(k^2 + \ell^2 + N^2/c_s^2 \right) \hat{\omega}^2 + N^2 k^2 c_s^2 = 0,$$

where

$$\hat{\omega} = \frac{\sin \omega \Delta t}{\Delta t} - Uk.$$

This dispersion relation is quadratic in $\hat{\omega}^2$ and has solutions

$$\hat{\omega}^2 = \frac{c_s^2}{2} \left(k^2 + \ell^2 + \frac{N^2}{c_s^2} \pm \left[\left(k^2 + \ell^2 + \frac{N^2}{c_s^2} \right)^2 - \frac{4N^2 k^2}{c_s^2} \right]^{1/2} \right). \quad (7.43)$$

The positive root yields the dispersion relation for sound waves; the negative root yields the dispersion relation for gravity waves. The individual dispersion relations for sound and gravity waves may be greatly simplified whenever the last term inside the square root in (7.43) is much smaller than the first term. One sufficient condition for this simplification is that

$$\frac{N^2}{c_s^2} \ll \ell^2, \quad (7.44)$$

in which case

$$\frac{4N^2 k^2}{c_s^2} \ll \frac{2N^2 k^2}{c_s^2} + 2k^2 \ell^2 \le \left(k^2 + \ell^2 + \frac{N^2}{c_s^2} \right)^2.$$

In most applications (7.44) is easily satisfied, so the sound-wave-dispersion relation simplifies to

$$\hat{\omega}^2 = c_s^2 \left(k^2 + \ell^2 + N^2/c_s^2 \right), \quad (7.45)$$

and the gravity-wave-dispersion relation becomes

$$\hat{\omega}^2 = \frac{N^2 k^2}{k^2 + \ell^2 + N^2/c_s^2}. \quad (7.46)$$

Consider the time-step limitation imposed by sound-wave propagation. Using the definition of $\hat{\omega}$, (7.45) may be expressed as

$$\sin \omega \Delta t = \Delta t \left(Uk \pm c_s \left(k^2 + \ell^2 + N^2/c_s^2 \right)^{1/2} \right).$$

Stable leapfrog solutions are obtained when the right side of this expression is a real number whose absolute value is less than unity. A necessary condition for stability is that

$$\left(|U|K + c_s(K^2 + L^2)^{1/2} \right) \Delta t < 1, \tag{7.47}$$

where K and L are the largest horizontal and vertical wave numbers retained in the truncation. In many applications the vertical resolution is much higher than the horizontal resolution, and the most severe restriction on the time step is associated with vertically propagating sound waves. The preceding is also a good approximation to the sufficient condition for stability, since the term involving N^2/c_s^2 is typically insignificant for the highest-frequency waves.

The dispersion relation for gravity waves (7.46) may be written as

$$\sin \omega \Delta t = \Delta t \left(Uk \pm \frac{Nk}{\left(k^2 + \ell^2 + N^2/c_s^2 \right)^{1/2}} \right).$$

Since

$$\frac{N|k|}{\left(k^2 + \ell^2 + N^2/c_s^2 \right)^{1/2}} \leq c_s|k|,$$

the necessary condition for sound-wave stability (7.47) is sufficient to ensure the stability of the gravity waves. Although (7.47) guarantees the stability of the gravity-wave modes, it is far too restrictive. Since

$$\frac{N|k|}{\left(k^2 + \ell^2 + N^2/c_s^2 \right)^{1/2}} \leq N,$$

the gravity waves will be stable, provided that

$$(|U|K + N)\Delta t < 1.$$

This is also a good approximation to the necessary condition for stability, because the term involving N^2/c_s^2 is usually dominated by K^2.

In most geophysical applications

$$c_s(K^2 + L^2)^{1/2} \gg |U|K + N,$$

and the maximum stable time step with which the gravity waves can be integrated is therefore far larger than the time step required to maintain stability in the sound-wave modes. In such circumstances, the sound waves can be stabilized using a semi-implicit approximation in which the pressure–gradient and velocity–divergence terms are evaluated using trapezoidal differencing (Tapp and White 1976). The resulting semi-implicit system is

$$\delta_{2t} u^n + U \frac{\partial u^n}{\partial x} + \left\langle \frac{\partial P^n}{\partial x} \right\rangle^{2t} = 0, \tag{7.48}$$

$$\delta_{2t} w^n + U \frac{\partial w^n}{\partial x} + \left\langle \frac{\partial P^n}{\partial z} \right\rangle^{2t} = b^n, \tag{7.49}$$

$$\delta_{2t} b^n + U \frac{\partial b^n}{\partial x} + N^2 w^n = 0, \tag{7.50}$$

$$\delta_{2t} P^n + U \frac{\partial P^n}{\partial x} + c_s^2 \left(\left\langle \frac{\partial u^n}{\partial x} \right\rangle^{2t} + \left\langle \frac{\partial w^n}{\partial z} \right\rangle^{2t} \right) = 0. \tag{7.51}$$

Let $\hat{c}_s = c_s \cos(\omega \Delta t)$. Then the dispersion relation for the semi-implicit system is identical to that obtained for leapfrog differencing, except that c_s is replaced by $\hat{c}_s$ throughout (7.43). The dispersion relation for the sound-wave modes is

$$\hat{\omega}^2 = \hat{c}_s^2 \left(k^2 + \ell^2 + N^2/\hat{c}_s^2 \right),$$

or

$$\sin \omega \Delta t = \Delta t \left(Uk \pm \hat{c}_s \left(k^2 + \ell^2 + N^2/\hat{c}_s^2 \right)^{1/2} \right). \tag{7.52}$$

The most severe stability constraints are imposed by the shortest waves for which the term $N^2/\hat{c}_s^2$ can be neglected in comparison with $k^2 + \ell^2$. Neglecting $N^2/\hat{c}_s^2$, (7.52) becomes

$$\sin \omega \Delta t = Uk\Delta t \pm c_s \Delta t \left(k^2 + \ell^2 \right)^{1/2} \cos \omega \Delta t,$$

which has the same form as (7.25), implying that the sound-wave modes are stable whenever

$$|Uk| \le c_s \left(k^2 + \ell^2 \right)^{1/2}.$$

A sufficient condition for the stability of the sound waves is simply that the flow be subsonic ($|U| \le c_s$), or equivalently, that the Mach number be less than unity.

Provided that the flow is subsonic, the only constraint on the time step required to keep the semi-implicit scheme stable is that associated with gravity-wave propagation. The dispersion relation for the gravity waves in the semi-implicit system is

$$\hat{\omega}^2 = \frac{N^2 k^2}{k^2 + \ell^2 + N^2/\hat{c}_s^2}, \tag{7.53}$$

which differs from the result for leapfrog differencing only in the small term $N^2/\hat{c}_s^2$. Stable gravity-wave solutions to the semi-implicit system are obtained whenever

$$(|U|K + N)\Delta t < 1,$$

which is the same condition obtained for the stability of the gravity waves using leapfrog differencing. Thus, as suggested previously, the semi-implicit scheme allows the compressible equations governing low-Mach-number flow to be integrated with a much larger time step than that allowed by fully explicit schemes. This increase in efficiency comes at a price; whenever the time step is much larger than that allowed by the CFL condition for sound waves, the sound waves are artificially decelerated by a factor of $\cos(\omega\Delta t)$. This error is directly analogous to that considered in Section 7.2.1, in which spurious decelerations were produced by fully implicit schemes using very large time steps. Nevertheless, in many practical applications the errors in the sound waves are of no consequence, and the quality of the solution is entirely determined by the accuracy with which the slower-moving waves are approximated.

How does semi-implicit differencing influence the accuracy of the gravity-wave modes? The only influence is exerted through the reduction in the speed of sound in the third term in the denominator of (7.53). This term is generally small and has little effect on the gravity waves unless $\omega\Delta t$ is far from zero and the waves are sufficiently long that $|k| + |\ell| \leq N/c_s$. It is actually rather hard to satisfy both of these requirements simultaneously. First, since $\omega\Delta t \leq 1$ for the stability of the mode in question, $\hat{c}_s$ can never drop below $0.54c_s$. Second, the maximum value of Δt is limited by the frequency of the most rapidly moving wave ω_m. In most applications the frequencies of the long waves are much lower than ω_m, so for all the long waves, $\omega\Delta t \ll \omega_m\Delta t \leq 1$, and thus $\hat{c}_s \approx c_s$. As a example where the deviation of $\hat{c}_s$ from c_s is maximized, consider a basic state with $N = 0.02$ s^{-1}, $c_s = 318$ ms^{-1}, and $U = 0$, together with the mode $(k, \ell) = (N/c_s, 0.1N/c_s)$ and time steps in the range $0 \leq \Delta t \leq 1/N$. The approximate solution obtained using leapfrog time-differencing exhibits an accelerative phase error that reaches 11% when $\Delta t = 1/N$. This accelerative phase-speed error is reduced by the semi-implicit method to a -5.7% decelerative error when $N = \Delta t$. The wave in this example is a rather pathological mode with horizontal and vertical wavelengths of 100 and 1000 km, respectively. The difference between the leapfrog and semi-implicit gravity-wave solutions is much smaller in most realistic examples.

The semi-implicit differencing scheme (7.48)–(7.51) provides a way to circumvent the CFL stability criterion for sound-wave propagation without losing accuracy in simulation of the gravity-wave modes. In global-scale atmospheric models the gravity waves may actually be of minor physical significance, and the features of primary interest may evolve on an even slower time scale.[5] If the fastest-moving gravity-wave modes do not need to be accurately represented, it is possible to generalize the preceding semi-implicit scheme to allow even larger

[5] In particular, the most important features may consist of slow-moving Rossby waves, which appear as additional solutions to the Euler equations when latitudinal variations in the Coriolis force are included in the horizontal-momentum equations.

time steps by replacing (7.49) and (7.50) with

$$\delta_{2t} w^n + U \frac{\partial w^n}{\partial x} + \left\langle \frac{\partial P^n}{\partial z} \right\rangle^{2t} = \langle b^n \rangle^{2t},$$

$$\delta_{2t} b^n + U \frac{\partial b^n}{\partial x} + N^2 \langle w^n \rangle^{2t} = 0$$

(Cullen 1990; Tanguay et al. 1990). Note that the buoyancy forcing in the vertical-momentum equation and the vertical advection of the mean-state buoyancy in the buoyancy equation are now treated by trapezoidal differences. The gravity-wave dispersion relation for this generalized semi-implicit system is

$$\hat{\omega}^2 = \frac{N^2 k^2 \cos(\omega \Delta t)}{k^2 + \ell^2 + N^2/c_s^2},$$

or

$$\sin \omega \Delta t = \Delta t \left(U k \pm \frac{N k \cos(\omega \Delta t)}{\left(k^2 + \ell^2 + N^2/c_s^2\right)^{1/2}} \right).$$

This dispersion relation has the same form as that for the prototype semi-implicit scheme (7.25), and as discussed in connection with (7.26), stable solutions will be obtained, provided that $|Uk|\Delta t \leq 1$.

7.2.5 Numerical Implementation

The semi-implicit approximation to the compressible Boussinesq system discussed in the preceding section generates a system of implicit algebraic equations that must be solved at every time step. The solution procedure will be illustrated in a relatively simple example using the nonlinear compressible Boussinesq equations

$$\frac{db}{dt} + N^2 w = 0, \tag{7.54}$$

$$\frac{d\mathbf{v}}{dt} + \nabla P = b\mathbf{k}, \tag{7.55}$$

$$\frac{dP}{dt} + c_s^2 \nabla \cdot \mathbf{v} = 0. \tag{7.56}$$

The definitions of b, P, and N given in (7.42) may be used to show that (7.54) and (7.55) are identical to the buoyancy and momentum equations in the standard Boussinesq system (7.2) and (7.3). The standard incompressible continuity equation has been replaced by (7.56) and is recovered in the limit $c_s \to \infty$.

First consider the situation where only the sound waves are stabilized by semi-implicit differencing and suppose that the spatial derivatives are not discretized.

Then the resulting semi-implicit system has the form

$$b^{n+1} = b^{n-1} - 2\Delta t \left(\mathbf{v}^n \cdot \nabla b^n + N^2 w^n \right), \qquad (7.57)$$

$$\mathbf{v}^{n+1} + \Delta t \nabla P^{n+1} = \mathbf{G}, \qquad (7.58)$$

$$P^{n+1} + c_s^2 \Delta t \nabla \cdot \mathbf{v}^{n+1} = h. \qquad (7.59)$$

Here

$$\mathbf{G} = \mathbf{v}^{n-1} - \Delta t \left[\nabla P^{n-1} - 2b^n \mathbf{k} + 2\mathbf{v}^n \cdot \nabla \mathbf{v}^n \right]$$

and

$$h = P^{n-1} - c_s^2 \Delta t \left[\nabla \cdot \mathbf{v}^{n-1} + 2\mathbf{v}^n \cdot \nabla P^n \right].$$

A single Helmholtz equation for P^{n+1} can be obtained by substituting the divergence of (7.58) into (7.59) to yield

$$\nabla^2 P^{n+1} - \frac{P^{n+1}}{(c_s \Delta t)^2} = \frac{\nabla \cdot \mathbf{G}}{\Delta t} - \frac{h}{(c_s \Delta t)^2}. \qquad (7.60)$$

The numerical solution of this Helmholtz equation is trivial if the Fourier spectral method is employed in a rectangular domain or if spherical harmonic expansion functions are used in a global spectral model. If the spatial derivatives are approximated by finite differences, (7.60) yields a sparse linear-algebraic system that can be solved using the techniques described in Section 7.1.3. After solving (7.60) for P^{n+1}, the momentum equations can be stepped forward, and the buoyancy equation (7.57), which is completely explicit, can be updated to complete the integration cycle.

This implementation of the semi-implicit method is closely related to the projection method for incompressible Boussinesq flow. Indeed, in the limit $c_s \to \infty$ the preceding approach will be identical to the leapfrog projection method (described in Section 7.1.2) if $(P^{n+1} + P^{n-1})/2$ is replaced by P^n in (7.60). Although the leapfrog projection method and the semi-implicit method yield algorithms involving very similar algebraic equations, these methods are derived via very different approximation strategies. The projection method is an efficient way to solve a set of continuous equations that is obtained by filtering the exact Euler equations to eliminate sound waves. In contrast, the semi-implicit scheme is obtained by directly approximating the full compressible equations and using implicit time-differencing to stabilize the sound waves. Neither approach allows one to correctly simulate sound waves, but both approaches allow the accurate and efficient simulation of the slower-moving gravity waves.

Now consider the version of the semi-implicit approximation in which those terms responsible for gravity-wave propagation are also approximated by trapezoidal differences; in this case (7.57) and (7.58) become

$$b^{n+1} + \Delta t N^2 w^{n+1} = f_b, \tag{7.61}$$

$$\mathbf{v}^{n+1} + \Delta t \left(\nabla P^{n+1} - \mathbf{k} b^{n+1} \right) = \tilde{\mathbf{G}},$$

where

$$f_b = b^{n-1} - \Delta t \left[N^2 w^{n-1} + 2\mathbf{v}^n \cdot \nabla b^n \right],$$

$$\tilde{\mathbf{G}} = \mathbf{v}^{n-1} - \Delta t \left[\nabla P^{n-1} - \mathbf{k} b^{n-1} + 2\mathbf{v}^n \cdot \nabla \mathbf{v}^n \right].$$

The implicit coupling in the resulting semi-implicit system can be reduced to a single Helmholtz equation for P^{n+1} as follows. Let $\tilde{\mathbf{G}} = (g_u, g_v, g_w)$; then using (7.61) to substitute for b^{n+1} in the vertical-momentum equation, one obtains

$$\left(1 + (N\Delta t)^2 \right) w^{n+1} + \Delta t \frac{\partial P^{n+1}}{\partial z} = g_w + f_b \Delta t. \tag{7.62}$$

Using the horizontal-momentum equations to eliminate u and v from (7.59) yields

$$\left(\frac{\partial^2}{\partial x^2} + \frac{\partial^2}{\partial y^2} - \frac{1}{(c_s \Delta t)^2} \right) P^{n+1} - \frac{1}{\Delta t} \frac{\partial w^{n+1}}{\partial z} = \frac{1}{\Delta t} \left(\frac{\partial g_u}{\partial x} + \frac{\partial g_v}{\partial y} \right) - \frac{h}{(c_s \Delta t)^2}.$$

As the final step, w^{n+1} is eliminated between the two preceding equations to obtain

$$\left[\left(1 + (N\Delta t)^2 \right) \left(\frac{\partial^2}{\partial x^2} + \frac{\partial^2}{\partial y^2} - \frac{1}{(c_s \Delta t)^2} \right) + \frac{\partial^2}{\partial z^2} \right] P^{n+1}$$

$$= \left(1 + (N\Delta t)^2 \right) \left[\frac{1}{\Delta t} \left(\frac{\partial g_u}{\partial x} + \frac{\partial g_v}{\partial y} \right) - \frac{h}{(c_s \Delta t)^2} \right] + \frac{\partial}{\partial z} \left(\frac{g_w}{\Delta t} + f_b \right).$$

After this elliptic equation is solved for P^{n+1}, then u and v are updated using the horizontal momentum equations, w is updated using (7.62), and finally, b is updated using (7.61).

Notice that the vertical advection of density in (7.61) is split between a term involving the mean vertical density gradient (N^2), which is treated implicitly, and a term involving the gradient of the perturbation density field ($\partial b/\partial z$), which is treated explicitly. As discussed in Section 7.2.3, when terms are split between a reference state that is treated implicitly and a perturbation that is treated explicitly, stability considerations demand that the term treated implicitly dominate the term treated explicitly. Thus, in most atmospheric applications the reference stability is chosen to be isothermal, thereby ensuring that $N^2 \geq \partial b/\partial z$. When semi-implicit differencing is used to integrate the complete Euler equations, the terms involving the pressure gradient and velocity divergence must also be partitioned into implic-itly differenced terms involving a reference state and the remaining explicitly differenced perturbations. Since the speed of sound is relatively uniform throughout the atmosphere, it is easy to ensure that the terms evaluated implicitly dominate those computed explicitly and thereby guarantee that the scheme is stable.

7.3   Fractional-Step Methods

The semi-implicit method requires the solution of an elliptic equation for the pressure during each step of the integration. This can be avoided by splitting the complete problem into fractional steps and using a very small time step to integrate the subproblem containing the terms responsible for the propagation of the fast-moving wave. Consider a general partial differential equation of the form

$$\frac{\partial \psi}{\partial t} + \mathcal{L}(\psi) = 0, \tag{7.63}$$

and suppose that

$$\mathcal{L}(\psi) = \mathcal{L}_1(\psi) + \mathcal{L}_2(\psi).$$

As discussed in Section 3.3, if $\mathcal{L}$ does not depend on time, (7.63) can be formally integrated over an interval Δt to obtain

$$\psi(t + \Delta t) = \exp(\Delta t \mathcal{L})\psi(t).$$

7.3.1   Complete Operator Splitting

Let $\mathcal{F}_1(\Delta t)$ and $\mathcal{F}_2(\Delta t)$ be numerical approximations to the exact operators $\exp(\Delta t \mathcal{L}_1)$ and $\exp(\Delta t \mathcal{L}_2)$. In the standard fractional-step approach, the approximate solution is stepped forward over a time interval Δt using

$$\phi^s = \mathcal{F}_1(\Delta t)\phi^n, \tag{7.64}$$
$$\phi^{n+1} = \mathcal{F}_2(\Delta t)\phi^s, \tag{7.65}$$

but it is not necessary to use the same time step in each subproblem. If $\mathcal{L}_2$ contains those terms responsible for the propagation of fast-moving waves and the maximum stable time step with which (7.64) can be integrated is M times that with which (7.65) can be integrated, the numerical solution could be evaluated using the formula

$$\phi^{n+1} = [\mathcal{F}_2(\Delta t/M)]^M \mathcal{F}_1(\Delta t)\phi^n. \tag{7.66}$$

This approach can be applied to the linearized one-dimensional shallow-water system by writing (3.1) and (3.2) in the form

$$\frac{\partial \mathbf{r}}{\partial t} + \mathcal{L}_1(\mathbf{r}) + \mathcal{L}_2(\mathbf{r}) = \mathbf{0}, \tag{7.67}$$

where

$$\mathbf{r} = \begin{pmatrix} u \\ h \end{pmatrix}, \quad \mathcal{L}_1 = \begin{pmatrix} U\partial_x & 0 \\ 0 & U\partial_x \end{pmatrix}, \quad \mathcal{L}_2 = \begin{pmatrix} 0 & g\partial_x \\ H\partial_x & 0 \end{pmatrix},$$

and ∂_x denotes the partial derivative with respect to x. The first fractional step, which is an approximation to

$$\frac{\partial \mathbf{r}}{\partial t} + \mathcal{L}_1(\mathbf{r}) = \mathbf{0},$$

involves the solution of two decoupled advection equations. Since this is a fractional-step method, it is generally preferable to approximate the preceding with a two-time-level method. In order to avoid using implicit, unstable, or Lax–Wendroff methods, the first step can be integrated using the third-order Runge–Kutta scheme

$$\mathbf{r}^* = \mathbf{r}^n + \frac{\Delta t}{3} \mathcal{L}_1(\mathbf{r}^n), \tag{7.68}$$

$$\mathbf{r}^{**} = \mathbf{r}^n + \frac{\Delta t}{2} \mathcal{L}_1(\mathbf{r}^*), \tag{7.69}$$

$$\mathbf{r}^{n+1} = \mathbf{r}^n + \Delta t\, \mathcal{L}_1(\mathbf{r}^{**}). \tag{7.70}$$

This Runge–Kutta method is stable and damping for $|U|K\Delta t < 1.73$, where K is the maximum retained wave number.

The second fractional step, which approximates

$$\frac{\partial \mathbf{r}}{\partial t} + \mathcal{L}_2(\mathbf{r}) = \mathbf{0},$$

can be efficiently integrated using forward-backward differencing. Defining $\Delta\tau = \Delta t/M$ as the length of a small time step, the forward-backward scheme is

$$\frac{u^{m+1} - u^m}{\Delta\tau} + g\frac{dh^m}{dx} = 0, \tag{7.71}$$

$$\frac{h^{m+1} - h^m}{\Delta\tau} + H\frac{du^{m+1}}{dx} = 0. \tag{7.72}$$

This scheme is stable for $cK\Delta\tau < 2$ and is second-order accurate in time. Since the operators used in each fractional step commute, the complete method will be $O\left[(\Delta t)^2\right]$ accurate and stable whenever each of the individual steps is stable.[6]

Although the preceding fractional-step scheme works fine for the linearized one-dimensional shallow water system, it does not generalize as nicely to problems in which the operators do not commute. As an example, consider the com-

---

[6] See Section 3.3 for a discussion of the impact of operator commutativity on the performance of fractional-step schemes.

pressible two-dimensional Boussinesq equations, which could be split into the form (7.67) by defining

$$\mathbf{r} = \begin{pmatrix} u & w & b & P \end{pmatrix}^T,$$

$$\mathcal{L}_1 = \begin{pmatrix} \mathbf{v} \cdot \nabla & 0 & 0 & 0 \\ 0 & \mathbf{v} \cdot \nabla & 0 & 0 \\ 0 & 0 & \mathbf{v} \cdot \nabla & 0 \\ 0 & 0 & 0 & \mathbf{v} \cdot \nabla \end{pmatrix},$$

$$\mathcal{L}_2 = \begin{pmatrix} 0 & 0 & 0 & \partial_x \\ 0 & 0 & -1 & \partial_z \\ 0 & N^2 & 0 & 0 \\ c_s^2 \partial_x & c_s^2 \partial_z & 0 & 0 \end{pmatrix}. \tag{7.73}$$

Suppose that N and c_s are constant and that the full nonlinear system is linearized about a reference state with a mean horizontal wind $U(z)$. The operators associated with this linearized system will not commute unless dU/dz is zero.

As in the one-dimensional shallow-water system, the advection operator $\mathcal{L}_1$ can be approximated using the third-order Runge–Kutta method (7.68)–(7.70). The second fractional step may be integrated using trapezoidal differencing for the terms governing the vertical propagation of sound waves and forward backward differencing for the terms governing horizontal sound-wave propagation and buoyancy oscillations. The resulting scheme is

$$\frac{u^{m+1} - u^m}{\Delta\tau} + \frac{\partial P^m}{\partial x} = 0, \tag{7.74}$$

$$\frac{w^{m+1} - w^m}{\Delta\tau} + \frac{\partial}{\partial z}\left(\frac{P^{m+1} + P^m}{2}\right) - b^m = 0, \tag{7.75}$$

$$\frac{b^{m+1} - b^m}{\Delta\tau} + N^2 w^{m+1} = 0, \tag{7.76}$$

$$\frac{P^{m+1} - P^m}{\Delta\tau} + c_s^2 \frac{\partial u^{m+1}}{\partial x} + c_s^2 \frac{\partial}{\partial z}\left(\frac{w^{m+1} + w^m}{2}\right) = 0, \tag{7.77}$$

This approximation to $\exp(\Delta\tau\mathcal{L}_2)$ is stable and nondamping, provided that the number $\max(c_s K, N)\Delta\tau$ is less than 2. The trapezoidal approximation of the terms involving vertical derivatives does not significantly increase the computations required on each small time step because it generates a simple tridiagonal system of algebraic equations for the w^{m+1} throughout each vertical column within the domain. If the horizontal resolution is very coarse, so that $K \ll N/c_s$, further efficiency can be also obtained by treating the terms involving buoyancy oscillations with trapezoidal differencing. Since these terms do not involve derivatives, the resulting implicit algebraic system remains tridiagonal.

As an alternative to the trapezoidal method, the terms involving the vertical pressure gradient and the divergence of the vertical velocity could be integrated using forward-backward differencing, in which case the stability criterion for the small time step would include an additional term proportional to $c_s L \Delta \tau$, where L is the maximum resolvable vertical wave number. It may be appropriate to use forward-backward differencing instead of the trapezoidal scheme in applications with identical vertical and horizontal grid spacing, but if the vertical resolution is much finer than the horizontal resolution, the additional stability constraint imposed by vertical sound-wave propagation will reduce efficiency by requiring an excessive number of small time steps.

The performance of the preceding scheme is evaluated in a problem involving flow past a compact gravity-wave generator. The wave generator is modeled by including forcing terms in the momentum equations of the form

$$\frac{du}{dt} + \frac{\partial P}{\partial x} = -\frac{\partial \Psi}{\partial z}, \tag{7.78}$$

$$\frac{dw}{dt} + \frac{\partial P}{\partial z} - b = \frac{\partial \Psi}{\partial x}, \tag{7.79}$$

where
$$\Psi(x, z, t) = E(x, z) \sin \omega t \sin k_1 x \cos \ell_1 z$$

and

$$E(x, z) = \begin{cases} \alpha \, (1 + \cos k_2 x) \, (1 + \cos \ell_2 z) & \text{if } |x| \leq \pi/k_2 \text{ and } |z| \leq \pi/\ell_2, \\ 0 & \text{otherwise.} \end{cases}$$

This forcing has no influence on the time tendency of the divergence, and as a consequence it does not excite sound waves. The spatial domain is periodic at $x = \pm 50$ km and bounded by rigid horizontal walls at $z = \pm 5$ km. In the following tests $\Delta x = 250$ m, $\Delta z = 50$ m, $N = 0.01$ s^{-1}, $c_s = 350$ ms^{-1}, and the parameters defining the wave generator are $\alpha = 0.2$, $2\pi/k_1 = 10$ km, $2\pi/\ell_1 = 2.5$ km, $2\pi/k_2 = 11$ km, $2\pi/\ell_2 = 1.5$ km, and $\omega = 0.002$ s^{-1}. The forcing is evaluated every $\Delta \tau$ and applied to the solution on the small time step. The computational domain is -50 km $\leq x \leq 50$ km, -5 km $\leq z \leq 5$ km, $\Delta x = 250$ m, $\Delta z = 50$ m, $N = 0.01$ s^{-1}, and $c_s = 350$ ms^{-1}.

The spatial derivatives are approximated using centered differencing on a staggered grid identical to that shown in Fig. 3.6, except that b is colocated with the w points rather than the P points. As a consequence of the mesh staggering, the horizontal wave number obtained from the finite-difference approximations to the pressure gradient and velocity divergence is $(2/\Delta x)(\sin k\Delta x/2)$, and the small-step stability criterion is max $(2c_s/\Delta x + N)\Delta \tau < 2$. The horizontal wave number generated by the finite-difference approximation to the advection operator is $(\sin k\Delta x)/\Delta x$, so the large time step is stable when $|U|\Delta t/\Delta x < 1.73$. Strang splitting,

$$\phi^{n+1} = [\mathcal{F}_2(2\Delta t/M)]^{(M/2)} \, \mathcal{F}_1(\Delta t) \, [\mathcal{F}_2(2\Delta t/M)]^{(M/2)} \, \phi^n,$$

is used in preference to (7.66) in order to preserve $O\left[(\Delta t)^2\right]$ accuracy in those cases where $\mathcal{F}_1$ and $\mathcal{F}_2$ do not commute.

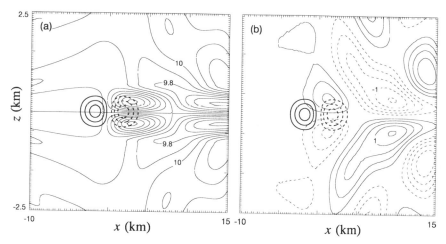

FIGURE 7.2. (a) contours of $U + u$ at intervals of 0.1 ms^{-1} and Ψ at intervals of 0.1 s^{-1} at $t = 8000$ s. (b) as in (a) except that P is contoured at intervals of 0.25 m^{2}s^{-2}. No zero contour is shown for the P and Ψ fields. Minor tick marks indicate the location of the P points on the numerical grid. Only the central portion of the total computational domain is shown.

In the first simulation $\Delta t = 12.5$ s, there are twenty small time steps per large time step, and $U = 10$ ms^{-1} throughout the domain. In this case $(2c_s \Delta x + N)\Delta \tau = 1.76$, so the small time step is being integrated using time steps near the stability limit. The horizontal velocity field and the pressure field obtained from this simulation are plotted in Fig. 7.2. The velocity field is essentially identical to that obtained using the full compressible equations. Very small errors are detectable in the pressure field, but the overall accuracy of the solution is excellent.

Now consider a second simulation that is identical to the first in every respect except that the mean wind U increases linearly from 5 to 15 ms^{-1} between the bottom and the top of the domain. The pressure perturbations that develop in this simulation are shown in Fig. 7.3a, along with streamlines for the forcing function Ψ. Spurious pressure perturbations appear throughout the domain. The correct pressure field is shown in Fig. 7.3d, which was computed using a scheme that will be described in the next subsection. Although the pressure field in Fig. 7.3a is clearly in error, most of the spurious signal in the pressure field relates to sound waves whose velocity perturbations are very weak. The velocity fields associated with all the solutions shown in Fig. 7.3 are essentially identical. The extrema in the pressure perturbations shown in Fig. 7.3a are approximately twice those in the other panels and are growing very slowly, suggesting that the solution is subject to a weak instability. Since the operators for each fractional step do not commute, the stability of each individual operator no longer guarantees the stability of the overall scheme.

FIGURE 7.3. Contours of P at intervals of 0.25 m^{2}s^{-2} and Ψ at intervals of 0.1 s^{-1} at $t = 3000$ s for the case with vertical shear in the mean wind and (a) $\Delta t = 12.5$ s, $M = 20$, (b) $\Delta t = 6.25$ s, $M = 20$, (c) $\Delta t = 6.25$ s, $M = 10$, (d) the solution is computed using the partial splitting method described in the next section with $\Delta t = 12.5$ s, $M = 20$. No zero contours are plotted. Major tick marks appear every 20 grid intervals.

As will be discussed in Section 7.3.2, this scheme can be stabilized by damping the velocity divergence on the small time step. Divergence damping yields only a modest improvement in the solution, however, because the completely split method also has accuracy problems due to inadequate temporal resolution. Cutting Δt by a factor of 2 while leaving $M = 20$ so that $\Delta \tau$ is also reduced by a factor of 2 gives the pressure distribution shown in Fig. 7.3b, which is clearly a significant improvement over that obtained using the original time step. Similar results are obtained if both Δt and M are cut in half, as shown in Fig. 7.3c, which demonstrates that it is the decrease in Δt, rather than $\Delta \tau$, that is responsible for the improvement. Further discussion of the source of the error in the completely split method is provided in Section 7.4.

7.3.2 Partially Split Operators

The first task involved in implementing the fractional-step methods discussed in the previous section is to identify those terms in the governing equations that need to be updated on a shorter time step. Having made this identification, it is possible to leave all the terms in the governing equations coupled together and to update those terms governing the slowly evolving processes less frequently than those terms responsible for the propagation of high-frequency physically insignificant waves. This technique will be referred to as a partial splitting, since the individual fractional steps are never completely decoupled in the conventional manner given by (7.64) and (7.65).

Once again the linearized one-dimensional shallow-water system provides a simple context in which to illustrate partial splitting. As before, it is assumed that the gravity-wave phase speed is much larger than the velocity of the mean flow U. Klemp and Wilhelmson (1978) and Tatsumi (1983) have suggested a partial splitting in which the terms on the right sides of

$$\frac{\partial u}{\partial t} + g\frac{\partial h}{\partial x} = -U\frac{\partial u}{\partial x}, \tag{7.80}$$

$$\frac{\partial h}{\partial t} + H\frac{\partial u}{\partial x} = -U\frac{\partial h}{\partial x} \tag{7.81}$$

are updated as if the time derivative were being approximated using a leapfrog difference, but rather than advancing the solution from time level $t - \Delta t$ to $t + \Delta t$ in a single step of length $2\Delta t$, the solution is advanced through a series of $2M$ "small time steps." During each small time step the terms on the right sides of (7.80) and (7.81) are held constant at their value at time level t and the remaining terms are updated using forward-backward differencing. Let m and n be time indices for the small and large time steps, respectively, and define $\Delta \tau = \Delta t/M$ as the length of a small time step. The solution is advanced from time level $n - 1$ to $n + 1$ in $2M$ small time steps of the form

$$\frac{u^{m+1} - u^m}{\Delta \tau} + g\frac{dh^m}{dx} = -U\frac{du^n}{dx},$$

$$\frac{h^{m+1} - h^m}{\Delta \tau} + H\frac{du^{m+1}}{dx} = -U\frac{dh^n}{dx}.$$

Note that the left sides of the preceding equations are identical to those appearing in the completely split scheme (7.71) and (7.72).

The complete small-step–large-step integration cycle for this problem can be written as a four-dimensional linear system as follows. Define $\hat{u}^m = u^n$, $\hat{h}^m = h^n$, and let

$$\mathbf{r} = (u, h, \hat{u}, \hat{h})^T.$$

Then an individual small time step can be expressed in the form

$$\mathbf{r}^{m+1} = \mathbf{A}\mathbf{r}^m,$$

where

$$
\mathbf{A} = \begin{pmatrix} 1 & -\tilde{g}\partial_x & -\tilde{U}\partial_x & 0 \\ -\tilde{H}\partial_x & 1 + \tilde{c}^2\partial_{xx}^2 & \tilde{U}\tilde{H}\partial_{xx}^2 & -\tilde{U}\partial_x \\ 0 & 0 & 1 & 0 \\ 0 & 0 & 0 & 1 \end{pmatrix},
$$

and the tilde denotes multiplication of the parameter by $\Delta\tau$ (e.g., $\tilde{c} = c\Delta\tau$). At the beginning of the first small time step in a complete big-step–small-step integration cycle

$$
\mathbf{r}^{m=1} = (u^{n-1}, h^{n-1}, u^n, h^n)^T.
$$

At the end of the $(2M)$th small step

$$
\mathbf{r}^{m=2M} = (u^{n+1}, h^{n+1}, u^n, h^n)^T.
$$

Thus, if $\mathbf{S}$ is a matrix interchanging the first pair and second pair of elements in $\mathbf{r}$,

$$
\mathbf{S} = \begin{pmatrix} 0 & 0 & 1 & 0 \\ 0 & 0 & 0 & 1 \\ 1 & 0 & 0 & 0 \\ 0 & 1 & 0 & 0 \end{pmatrix},
$$

the complete big-step–small-step integration cycle is given by

$$
\mathbf{r}^{n+1} = \mathbf{S}\mathbf{A}^{2M}\mathbf{r}^n.
$$

Since the individual operators commute, the completely split approximation to this problem is stable whenever both of the individual fractional steps are stable. One might hope that the stability of the partially spilt method could also be guaranteed whenever the large- and small-step subproblems are stable. Unfortunately, there are many combinations of Δt and $\Delta\tau$ for which the partially split method is unstable even though the subproblems obtained by setting either U or c to zero are both stable (Tatsumi 1983; Skamarock and Klemp 1992). Suppose that the partially split scheme is applied to an individual Fourier mode with horizontal wave number k. Then the amplification matrix for an individual small time step is given by a matrix in which the partial-derivative operators in $\mathbf{A}$ are replaced by ik; let this matrix be denoted by $\hat{\mathbf{A}}$.

Consider the case $M = 1$, for which the amplification matrix is $\mathbf{S}\hat{\mathbf{A}}^2$. The magnitude of the maximum eigenvalue, or spectral radius ρ, of $\mathbf{S}\hat{\mathbf{A}}^2$ is plotted in Fig. 7.4a as a function of $\hat{c} = ck\Delta\tau$ and $\hat{u} = Uk\Delta t$. The domain over which ρ is contoured, $0 \leq \hat{c} \leq 2$ and $0 \leq \hat{u} \leq 1$, is that for which the individual small- and large-step problems are stable. When $M = 1$, ρ exceeds unity, and the partially split scheme is unstable throughout two regions of the $\hat{c}$-$\hat{u}$ plane whose boundaries intersect at $(\hat{c}, \hat{u}) = (\sqrt{2}, 0)$. If $U \ll c$, only a limited subset of the $\hat{c}$-$\hat{u}$ plane shown in Fig. 7.4a is actually relevant to the solution of the shallow-water problem. Once the number of small time steps per large time step is fixed,

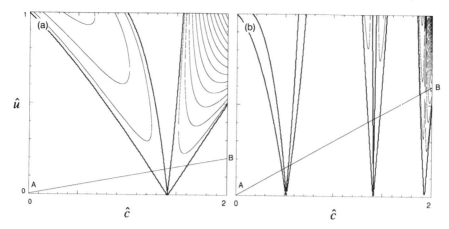

FIGURE 7.4. Spectral radius of the amplification matrix for the partially split method contoured as a function of $\hat{c}$ and $\hat{u}$ for (a) $M = 1$ (b) $M = 3$. Unstable regions are enclosed in the wedged-shaped areas. Contour intervals are 1.0 (heavy line), 1.2, 1.4, ... Line AB indicates the possible combinations of $\hat{c}$ and $\hat{u}$ that can be realized when $U/c = 1/10$ and M is specified as 1 or 3.

the possible combinations of $\hat{u}$ and $\hat{c}$ will lie along a straight line of slope

$$\frac{\hat{u}}{\hat{c}} = \frac{U \Delta t}{c \Delta \tau} = M \frac{U}{c}.$$

Suppose that $U/c = 1/10$. Then if the partial splitting method is used with $M = 1$, the only possible combinations of $\hat{u}$ and $\hat{c}$ are those lying along line AB in Fig. 7.4a. The maximum stable value of $\Delta \tau$ is determined by the intersection of the line AB and the left boundary of the leftmost region of instability. Thus, for $U/c = 1/10$ and $M = 1$, the stability requirement is that $\hat{c}$ be less than approximately 1.25.

As demonstrated in Fig. 7.4b, which shows contours of the spectral radius of $S\tilde{A}^6$, the restriction on the maximum stable time step becomes more severe as M increases to 3. The regions of instability are narrower and the strength of the instability in each unstable region is reduced, but additional regions of instability appear, and the distance from the origin to the nearest region of instability decreases. When $M = 3$ and $U/c = 1/10$, the maximum stable value of $\hat{c}$ is roughly 0.48. Further reductions in the maximum stable value for $\hat{c}$ occur as M is increased, and as a consequence, the gain in computational efficiency that one might expect to achieve by increasing the number of small time steps per large time step is eliminated by a compensating decrease in the maximum stable value for $\Delta \tau$.

The partial splitting method has, nevertheless, been used extensively in many practical applications. The method has proved useful because in most applications it is very easy to remove these instabilities by using a filter. As noted by Tatsumi (1983) and Skamarock and Klemp (1992), the instability is efficiently removed by the Asselin time filter (2.50), which is often used in conjunction with leapfrog time-differencing to prevent the divergence of the solution on the odd and even time steps. Other filtering techniques have also been suggested and will be discussed after considering a partial-splitting approximation to the compressible Boussinesq system.

The equations evaluated at each small time step in a partial-splitting approximation to the two-dimensional compressible Boussinesq equations linearized about a basic-state flow with Brunt–Väisälä frequency N and horizontal velocity U are

$$\frac{u^{m+1} - u^m}{\Delta \tau} + \frac{\partial P^m}{\partial x} = -U\frac{\partial u^n}{\partial x} - w^n\frac{\partial U}{\partial z}, \quad (7.82)$$

$$\frac{w^{m+1} - w^m}{\Delta \tau} + \frac{\partial}{\partial z}\left(\frac{P^{m+1} + P^m}{2}\right) - b^m = -U\frac{\partial w^n}{\partial x}, \quad (7.83)$$

$$\frac{b^{m+1} - b^m}{\Delta \tau} + N^2 w^{m+1} = -U\frac{\partial b^n}{\partial x}, \quad (7.84)$$

$$\frac{P^{m+1} - P^m}{\Delta \tau} + c_s^2\frac{\partial u^{m+1}}{\partial x} + c_s^2\frac{\partial}{\partial z}\left(\frac{w^{m+1} + w^m}{2}\right) = -U\frac{\partial P^n}{\partial x}, \quad (7.85)$$

where as before, m and n are the time indices associated with the small and large time steps. The left sides of these equations are identical to the small-time-step equations in the completely split method (7.74)–(7.77). The right sides are updated at every large time step.

This method is applied to the problem previously considered in connection with Fig. 7.2, in which fluid flows past a compact gravity-wave generator. The forcing from the wave generator appears in the horizontal- and vertical-momentum equations as in (7.78) and (7.79) and is updated on the small time step. In this test U is a constant 10 ms^{-1}, $\Delta t = 12.5$ s, and $\Delta \tau = 0.0625$ s. The horizontal velocity field and the pressure field from this simulation are plotted in Fig. 7.5. The horizontal velocity field is very similar, though slightly noisier than that shown in Fig. 7.2a. The pressure field is, however, complete garbage. Indeed, it is surprising that errors of the magnitude shown in Fig. 7.5b can exist in the pressure field without seriously degrading the velocity field. These pressure perturbations are growing with time (the contour interval in Fig. 7.5b is twice that in Fig. 7.2b); the velocity field eventually becomes very noisy, and the solution eventually blows up.

This instability can be prevented by applying an Asselin time filter (2.50) at the end of each big-step–small-step integration cycle. Skamarock and Klemp (1992) have shown that filtering coefficients on the order of $\gamma = 0.1$ may be required to stabilize the partially split solution to the one-dimensional shallow-water system. A value of $\gamma = 0.1$ is sufficient to completely remove the noise in the pressure field and to eliminate the instability in the preceding test. Nevertheless, as discussed in Section 2.3.5, Asselin filtering reduces the accuracy of the leapfrog scheme to $O(\Delta t)$, so it is best not to rely exclusively on the Asselin filter to stabilize the partially split approximation. Other techniques for stabilizing the preced-

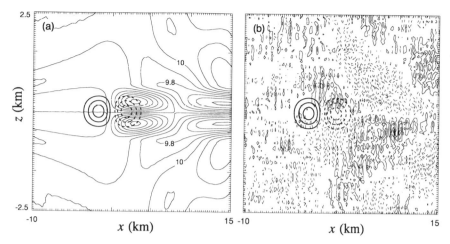

FIGURE 7.5. (a) contours of $U + u$ at intervals of 0.1 ms^{-1} and Ψ at intervals of 0.1 s^{-1} at $t = 8000$ s. (b) as in (a) except that P is contoured at intervals of 0.5 m^{2}s^{-2}.

ing partially split approximation include divergence damping and forward biasing the trapezoidal integral of the vertical derivative terms (7.83) and (7.85). Forward biasing the trapezoidal integration is accomplished without additional computational effort by replacing those terms of the form $(\phi^{m+1} + \phi^m)/2$ with

$$\left(\frac{1+\epsilon}{2}\right)\phi^{m+1} + \left(\frac{1-\epsilon}{2}\right)\phi^m,$$

where $0 \le \epsilon \le 1$. A value of $\epsilon = 0.2$ provides an effective filter that does not noticeably modify the gravity waves (Durran and Klemp 1983).

Since trapezoidal time-differencing is used only to approximate the vertical derivatives, forward biasing those derivatives will not damp horizontally propagating sound waves. Skamarock and Klemp (1992) recommended including a "divergence damper" in the momentum equations such that the system of equations that is integrated on the small time step becomes

$$\frac{\partial u}{\partial t} + \frac{\partial P}{\partial x} - \alpha_x \frac{\partial \delta}{\partial x} = F_u,$$

$$\frac{\partial w}{\partial t} + \frac{\partial P}{\partial z} - b - \alpha_z \frac{\partial \delta}{\partial z} = F_w,$$

$$\frac{\partial b}{\partial t} + N^2 w = F_b,$$

$$\frac{\partial P}{\partial t} + c_s^2 \delta = F_p, \qquad (7.86)$$

where

$$\delta = \frac{\partial u}{\partial x} + \frac{\partial w}{\partial z},$$

and F_u, F_w, F_b, and F_p represent the forcing terms that are updated every Δt. Damping coefficients of $\alpha_x = 0.001(\Delta x)^2/\Delta \tau$ and $\alpha_z = 0.001(\Delta z)^2/\Delta \tau$ removed all trace of noise and instability in the test problem shown in Fig. 7.5 without a supplemental Asselin filter.

The role played by divergence damping in stabilizing the small-time-step integration in the partial-splitting method can be appreciated by noting that if a single damping coefficient α is used in all components of the momentum equation, the divergence satisfies

$$\frac{\partial \delta}{\partial t} + \nabla^2 P - \alpha \nabla^2 \delta = G, \tag{7.87}$$

where $G = -\nabla \cdot (\mathbf{v} \cdot \nabla \mathbf{v}) + \partial b/\partial z$. Eliminating the pressure between (7.86) and (7.87), one obtains

$$\frac{\partial^2 \delta}{\partial t^2} - \alpha \nabla^2 \frac{\partial \delta}{\partial t} - c_s^2 \nabla^2 \delta = \frac{\partial G}{\partial t} - \nabla^2 F_p,$$

The forcing on the right side of this equation will tend to produce divergence in an initially nondivergent flow. Substituting a single Fourier mode into the homogeneous part of this equation, one obtains the classic equation for a damped harmonic oscillator:

$$\frac{d^2 \tilde{\delta}}{dt^2} + \alpha \kappa^2 \frac{d\tilde{\delta}}{dt} + c_s^2 \kappa^2 \tilde{\delta} = 0, \tag{7.88}$$

where $\tilde{\delta}(t)$ is the amplitude and $\kappa = \sqrt{k^2 + \ell^2}$. The damping increases with wave number and is particularly effective in eliminating the high-wave-number modes at which the instability in the partial-splitting method occurs. Gravity waves, on the other hand, are not significantly impacted by the divergence damper because the velocity field in internal gravity waves is almost nondivergent. Skamarock and Klemp (1992) have shown that divergence damping only slightly reduces the amplitude of the gravity waves.

At this point it might appear that the partial-splitting approach is inferior to the complete-splitting method considered previously, since filters are required to stabilize the partially split approximation in situations where the completely split scheme performs quite nicely. Recall, however, that the completely split method does not generate usable solutions to the compressible Boussinesq equations when there is a vertical shear in the basic-state horizontal velocity impinging on the gravity-wave generator. The same filtering strategies that stabilize the partially split method in the no-shear problem remain effective in the presence of vertical wind shear. This is demonstrated in Fig. 7.3d, which shows the pressure perturbations in the test case with vertical shear as computed by the partially split method using a divergence damper with the values of α_x and α_z given previously. Results similar to those in Fig. 7.3d may also be obtained using Asselin time filtering with $\alpha = 0.1$ in lieu of the divergence damper. The advantages of the partial splitting method are not connected with its performance in the simplest test cases, for which it can indeed be inferior to a completely split approximation, but in its adaptability to more complex problems.

One might inquire whether divergence damping can also be used to stabilize the completely split approximation to the test case with vertical shear in the horizontal wind. The norm of the amplification matrix for the large-time-step third-order Runge–Kutta integration (7.68)–(7.70) is strictly less than unity for all sufficiently small Δt. Divergence damping makes the norm of the amplification matrix for the small time step strictly less than unity for all sufficiently small $\Delta \tau$ and thereby stabilizes the completely split scheme by guaranteeing that the norm of the amplification matrix for the overall scheme will be less than unity. Nevertheless, divergence damping only modestly improves the solution obtained with the completely split scheme; the pressure field remains very noisy and completely unacceptable.[7] The fundamental problem with the completely split method appears to be one of inaccuracy, not instability. This will be discussed further in the next section.

7.4   Summary of Schemes for Nonhydrostatic Models

One way to compare the preceding methods for increasing the efficiency of numerical models for the simulation of fluids that support physically insignificant sound waves is to compare the way each approximation treats the velocity divergence. As before, the mathematics of this discussion will be streamlined by using the compressible Boussinesq equations (7.54)–(7.56) as a simple model for the Euler equations. The pressure and the divergence in the compressible Boussinesq system satisfy

$$\frac{\partial P}{\partial t} + c_s^2 \delta = F_p, \tag{7.89}$$

$$\frac{\partial \delta}{\partial t} + \nabla^2 P = G, \tag{7.90}$$

where $\delta = \nabla \cdot \mathbf{v}$, $F_p = -\mathbf{v} \cdot \nabla P$, and $G = -\nabla \cdot (\mathbf{v} \cdot \nabla \mathbf{v}) + \partial b / \partial z$. The semi-implicit method approximates the left sides of the preceding equations with a stable trapezoidal time difference. Sound waves are artificially slowed when large time steps are used in this trapezoidal difference, but the gravity-wave modes are accurately approximated. The implicit coupling in the trapezoidal difference leads to a Helmholtz equation for the pressure that must be solved at every time step.

The prognostic pressure equation (7.89) is discarded in the incompressible Boussinesq approximation, and the local time derivative of the divergence is set to zero in (7.90). This leads to a Poisson equation for pressure that must be solved at every time step. The computational effort required to evaluate the pressure is similar to that required by the semi-implicit method. The Boussinesq system does,

---

[7]One way to appreciate the difference in the effectiveness of divergence damping in the completely and partially split schemes is to note the difference in wavelength at which spurious pressure perturbations appear in each solution. The partially split scheme generates errors at much shorter wavelengths than those produced by the completely split method (compare Figs. 7.3a and 7.5b), and the short-wavelength features are removed more rapidly by the divergence damper.

however, have the advantage of allowing a wider choice of methods for the integration of the remaining oscillatory forcing terms, which are approximated using leapfrog differencing in the conventional semi-implicit method.

The elliptic pressure equations that appear when the semi-implicit or projection methods are used are most efficiently solved by sophisticated algorithms such as block-cyclic reduction, conjugate gradient, or multigrid methods. One may think of the small-time-step procedure used in the fractional-step methods as a sort of specialized iterative solver for the Helmholtz equation obtained using the conventional semi-implicit method. The difference in the character of the solution obtained by the complete- and the partial-splitting methods can be appreciated by considering the behavior of the divergence during the small-time-step integration.

During the small-time-step portion of the completely split method the divergence satisfies

$$\frac{\partial^2 \delta}{\partial t^2} - c_s^2 \nabla^2 \delta = \frac{\partial^2 b}{\partial t \partial z}.$$

The initial conditions for δ are those at the beginning of each small-time-step cycle, and since divergence is typically generated by the operators evaluated on the large time step, the initial δ is nonzero. This divergence is propagated without loss during the small-time-step integration (except for minor modification by the buoyancy forcing) and tends to accumulate over a series of large-step–small-step cycles. The test in which the completely split scheme performs well is the case in which the basic-state horizontal velocity is uniform throughout the fluid. When U is constant, the linearized advection operator merely produces a Galilean translation of the fluid that does not generate any divergence. (Recall that the forcing from the wave generator was computed on the small time step.) Nonlinear advection can, of course, generate divergence, as can the linearized advection operator when there is vertical shear in the basic-state wind, and these are the circumstances in which the complete-splitting method produces spurious sound waves.

In contrast, the divergence is almost zero at the start of the first small time step of the partially split method, and only small changes in the divergence are forced during each individual small step. Moreover, the divergence forcing on each small time step closely approximates that which would appear in an explicit small-time-step integration of the full compressible equations, *provided* that the amplitude of all the sound waves is negligible in comparison to slower modes. The divergence damper ensures that the amplitude of the sound waves remains small and thereby preserves the stability and accuracy of the solution.

7.5   The Hydrostatic Approximation

Large-scale atmospheric and oceanic motions are very nearly in hydrostatic balance, and as a consequence, they are well described by an approximate set of governing equations in which the full vertical momentum equation is replaced by the hydrostatic relation

$$\frac{\partial p}{\partial z} = -\rho g. \tag{7.91}$$

The hydrostatic system is not a hyperbolic system of partial differential equations[8] because there is no prognostic equation for the vertical velocity. The numerical solution of the hydrostatic system requires the evaluation of time-independent equations, such as (7.91), at every time step. These diagnostic equations can be evaluated with much less computational effort than that required to solve the diagnostic Poisson equation for the pressure in the Boussinesq system.

The hydrostatic approximation eliminates sound waves, although as will be discussed below, the hydrostatic approximation does not remove all horizontally propagating acoustic modes. In those large-scale geophysical applications where the numerical resolution along the vertical coordinate is much finer than the horizontal resolution, explicit finite-difference approximations to the hydrostatic system can be integrated much more efficiently than comparable approximations to either the nonhydrostatic Boussinesq equations or the nonhydrostatic compressible equations. The considerable improvement in model efficiency associated with the use of the hydrostatic governing equations does not, however, apply to semi-implicit models because these models can easily be modified to compute semi-implicit approximations to the full nonhydrostatic compressible equations without significantly increasing the computational overhead (Cullen 1990; Tanguay et al. 1990).

The influence of the hydrostatic approximation on wave propagation and the stability criteria for explicit finite-difference approximations to the hydrostatic equations may be determined by examining solutions to the linearized hydrostatic system. Small-amplitude hydrostatically balanced perturbations in the x-z plane about a resting isothermally stratified basic state are governed by the system

$$\frac{\partial u}{\partial t} + \frac{\partial P}{\partial x} = 0, \tag{7.92}$$

$$\left(\frac{\partial}{\partial z} + \Gamma\right) P = b, \tag{7.93}$$

$$\frac{\partial b}{\partial t} + N^2 w = 0, \tag{7.94}$$

$$\frac{\partial P}{\partial t} + c_s^2 \left[\frac{\partial u}{\partial x} + \left(\frac{\partial}{\partial z} - \Gamma\right) w\right] = 0, \tag{7.95}$$

where u, w, b, P, Γ, and c_s are defined by (7.35)–(7.37). Waves of the form

$$(u, w, b, P) = (u_0, w_0, b_0, P_0)e^{i(kx + \ell z - \omega n \Delta t)}$$

---

[8]There has been some concern about the well-posedness of initial-boundary value problems involving the hydrostatic equations (Oliger and Sundström 1978). It is not clear how to reconcile these concerns with the successful forecasts obtained twice daily at several operational centers for at least two decades using limited-area weather prediction models based on the hydrostatic governing equations.

are solutions to this system, provided that

$$\omega^2 = \frac{N^2 k^2}{m^2 + \Gamma^2 + N^2/c_{\mathrm{s}}^2}.$$

This is the standard dispersion relation for two-dimensional gravity waves, except that a term k^2 is missing from the denominator. This term is insignificant when $k \ll m$ and the wave is truly hydrostatic, but the absence of this term can lead to a serious overestimate of the gravity-wave phase speed of modes for which $k \gg m$.

Although there are no conventional sound-wave solutions to the hydrostatic system, a horizontally propagating acoustic mode known as the Lamb wave is supported by both the hydrostatic and the nonhydrostatic equations. The vertical velocity and buoyancy perturbations in a Lamb wave in an isothermal atmosphere are zero, and the pressure and horizontal velocity perturbations have the form

$$(u, P) = (u_0, P_0)e^{ik(x \pm c_{\mathrm{s}}t) - \Gamma z}.$$

This may be verified by noting that when $w = 0$, (7.92)–(7.95) reduce to

$$\frac{\partial^2 P}{\partial t^2} - c_{\mathrm{s}}^2 \frac{\partial^2 P}{\partial x^2} = 0 \qquad \text{and} \qquad \frac{\partial}{\partial t}\left(\frac{\partial}{\partial z} + \Gamma\right) P = 0.$$

If leapfrog time-differencing is used to create a differential–difference approximation to (7.92)–(7.95), a necessary and sufficient condition for the stability of the Lamb–wave mode is

$$c_{\mathrm{s}} \Delta t K < 1, \tag{7.96}$$

where K is the magnitude of the maximum horizontal wave number resolved by the numerical model. This condition is also sufficient to guarantee the stability of the gravity-wave modes, since for these modes

$$\sin^2(\omega \Delta t) = \frac{(N \Delta t k)^2}{m^2 + \Gamma^2 + N^2/c_{\mathrm{s}}^2} \leq (c_{\mathrm{s}} \Delta t K)^2.$$

In many geophysical applications the vertical resolution is much higher than the horizontal resolution, in which case (7.96) allows a much larger time step than that permitted by the stability condition for the leapfrog approximation to the full nonhydrostatic compressible equations (given by (7.47) with $U = 0$).

7.6   Primitive Equation Models

The exact equations governing global and large-scale atmospheric flows are often approximated by the so-called *primitive equations*. The primitive equations differ from the exact governing equations in that the hydrostatic assumption is invoked, small "curvature" and Coriolis terms involving the vertical velocity are neglected in the horizontal-momentum equations, and the radial distance between any point within the atmosphere and the center of the Earth is approximated by the mean radius of the Earth. Taken together, these approximations yield a system that conserves both energy and angular momentum (Lorenz 1967, p. 16).

The primitive equations governing inviscid adiabatic atmospheric motion may be expressed using height as the vertical coordinate as follows. Let x, y, and z be spatial coordinates that increase eastward, northward, and upward, respectively. Let $\mathbf{u} = (dx/dt, dy/dt)$ be the horizontal velocity vector, f the Coriolis parameter, $\mathbf{k}$ an upward-directed unit vector parallel to the z-axis, and ∇_z the gradient with respect to x and y along surfaces of constant z. Then the rate of change of horizontal momentum in the primitive equation system is governed by

$$\frac{d\mathbf{u}}{dt} + f\mathbf{k} \times \mathbf{u} + \frac{1}{\rho}\nabla_z p = \mathbf{0},$$

where

$$\frac{d(\,)}{dt} = \frac{\partial(\,)}{\partial t} + \mathbf{u} \cdot \nabla_z(\,) + w\frac{\partial(\,)}{\partial z}.$$

The continuity equation is

$$\frac{\partial \rho}{\partial t} + \nabla_z(\rho\mathbf{u}) + \frac{\partial \rho w}{\partial z} = 0, \tag{7.97}$$

and the thermodynamic equation may be written

$$\frac{dT}{dt} - \frac{\omega}{\rho} = 0, \tag{7.98}$$

where $\omega = dp/dt$ is the change in pressure following a fluid parcel. The preceding system of equations for the unknown variables $\mathbf{u}$, w, p, ω, ρ, and T may be closed using the hydrostatic relation (7.91) and the equation of state $p = \rho RT$.

7.6.1 Pressure and σ Coordinates

The primitive equations are often solved in a coordinate system in which geometric height is replaced by a new vertical coordinate $\zeta(x, y, z, t)$. Simple functions that have been used to define ζ include the hydrostatic pressure and the potential temperature. The most commonly used vertical coordinates in current operational models are generalized functions of the hydrostatic pressure.

The primitive equations may be expressed with respect to a different vertical coordinate as follows. Suppose that $\zeta(x, y, z, t)$ is the new vertical coordinate and that ζ is a monotone function of z for all fixed x, y, and t with a unique inverse $z(x, y, \zeta, t)$. Defining ∇_ζ as the gradient operator with respect to x and y along surfaces of constant ζ and applying the chain rule to the identity

$$p[x, y, z(x, y, \zeta, t), t] = p(x, y, \zeta, t)$$

yields

$$\nabla_z p + \frac{\partial p}{\partial z}\nabla_\zeta z = \nabla_\zeta p.$$

Using the hydrostatic relation (7.91) and defining the geopotential $\phi = gz$,

$$\nabla_z p = \nabla_\zeta p + \rho \nabla_\zeta \phi,$$

and the horizontal momentum equations in the transformed coordinates become

$$\frac{d\mathbf{u}}{dt} + f\mathbf{k} \times \mathbf{u} + \nabla_\zeta \phi + \frac{RT}{p} \nabla_\zeta p = \mathbf{0}, \tag{7.99}$$

where

$$\frac{d(\ )}{dt} = \frac{\partial(\ )}{\partial t} + \mathbf{u} \cdot \nabla_\zeta(\ ) + \dot{\zeta} \frac{\partial(\ )}{\partial \zeta} \tag{7.100}$$

and $\dot{\zeta} = d\zeta/dt$. The thermodynamic equation in the transformed coordinates is identical to (7.98), except that the total time derivative is computed using (7.100). The hydrostatic equation may be written as

$$\frac{\partial \phi}{\partial \zeta} = -\frac{RT}{p} \frac{\partial p}{\partial \zeta}.$$

The continuity equation in the transformed coordinate system can be determined by transforming the partial derivatives in (7.97) (Kasahara 1974). It is perhaps simpler to derive the continuity equation directly from first principles. Let $\mathcal{V}$ be a fixed volume defined with respect to the time-independent spatial coordinates x, y, and z, and let $\mathbf{n}$ be the outward-directed unit vector normal to the surface S enclosing $\mathcal{V}$. Since the rate of change of mass in the volume $\mathcal{V}$ is equal to the net mass flux through S,

$$\frac{\partial}{\partial t} \int_{\mathcal{V}} \rho \, dV = -\int_{S} \rho \mathbf{v} \cdot \mathbf{n} \, dA$$
$$= -\int_{\mathcal{V}} \nabla \cdot (\rho \mathbf{v}) \, dV, \tag{7.101}$$

where $\mathbf{v}$ is the three-dimensional velocity vector. Equation (5.61), which states the general relationship between the divergence in Cartesian coordinates and curvilinear coordinates, implies that

$$\nabla \cdot (\rho \mathbf{v}) = \frac{1}{J} \nabla_\zeta \cdot (J\rho \mathbf{u}) + \frac{1}{J} \frac{\partial}{\partial \zeta} (J\rho \dot{\zeta}),$$

where J is the Jacobian of the transformation between (x, y, z) and (x, y, ζ), which in this instance is simply $\partial z/\partial \zeta$. In the transformed coordinates

$$dV = \frac{\partial z}{\partial \zeta} \, dx \, dy \, d\zeta,$$

and since the boundaries of $\mathcal{V}$ do not depend on time, (7.101) may be expressed as

$$\int \int \int_{\mathcal{V}} \left[\frac{\partial}{\partial t} \left(\rho \frac{\partial z}{\partial \zeta} \right) + \nabla_\zeta \cdot \left(\rho \mathbf{u} \frac{\partial z}{\partial \zeta} \right) + \frac{\partial}{\partial \zeta} \left(\rho \dot{\zeta} \frac{\partial z}{\partial \zeta} \right) \right] dx \, dy \, d\zeta = 0.$$

Using the hydrostatic equation (7.91) to eliminate ρ from the preceding, and noting that the integrand must be identically zero because the volume $\mathcal{V}$ is arbitrary, the continuity equation becomes

$$\frac{\partial}{\partial t}\left(\frac{\partial p}{\partial \zeta}\right) + \nabla_\zeta \cdot \left(\mathbf{u}\frac{\partial p}{\partial \zeta}\right) + \frac{\partial}{\partial \zeta}\left(\dot{\zeta}\frac{\partial p}{\partial \zeta}\right) = 0.$$

Now consider possible choices for ζ. In most respects, the simplest system is obtained by choosing $\zeta = p$; this eliminates one of the two terms that make up the pressure gradient in (7.99) and reduces the continuity equation to the simple diagnostic relation

$$\nabla_p \cdot \mathbf{u} + \frac{\partial \omega}{\partial p} = 0.$$

The difficulty with pressure coordinates arises at the lower boundary because the pressure at the surface of the Earth is a function of horizontal position and time. As a consequence, constant-pressure surfaces intersect the lower boundary of the domain in an irregular manner that changes as a function of time. In order to simplify the lower-boundary condition, Phillips (1957) suggested choosing $\zeta = \sigma = p/p_s$, where p_s is the surface pressure. The upper and lower boundaries in a σ-coordinate model coincide with the coordinate surfaces $\sigma = 0$ and $\sigma = 1$, and $\dot{\sigma} = 0$ at both the upper and lower boundaries.

The σ-coordinate equations include prognostic equations for $\mathbf{u}$, T, and p_s and diagnostic equations for $\dot{\sigma}$, ϕ, and ω. The prognostic equations for the horizontal velocity and the temperature are

$$\frac{d\mathbf{u}}{dt} + f\mathbf{k} \times \mathbf{u} + \nabla_\sigma \phi + \frac{RT}{p_s}\nabla_\sigma p_s = 0 \tag{7.102}$$

and

$$\frac{dT}{dt} = \frac{\kappa T}{\sigma p_s}\omega, \tag{7.103}$$

where

$$\frac{d()}{dt} = \frac{\partial()}{\partial t} + \mathbf{u}\cdot\nabla_\sigma() + \dot{\sigma}\frac{\partial()}{\partial z}.$$

The continuity equation in σ-coordinates takes the form of a prognostic equation for the surface pressure:

$$\frac{\partial p_s}{\partial t} + \nabla_\sigma \cdot (p_s\mathbf{u}) + \frac{\partial}{\partial \sigma}(p_s\dot{\sigma}) = 0. \tag{7.104}$$

Recalling that $\dot{\sigma}$ is zero at $\sigma = 0$ and $\sigma = 1$, (7.104) can be integrated over the depth of the domain to obtain

$$\frac{\partial p_s}{\partial t} = -\int_0^1 \nabla_\sigma \cdot (p_s\mathbf{u})\, d\sigma. \tag{7.105}$$

A diagnostic equation for the vertical velocity $\dot{\sigma}$ is obtained by integrating (7.104) from the top of the domain to level σ, which yields

$$\dot{\sigma}(\sigma) = -\frac{1}{p_s}\left[\sigma\frac{\partial p_s}{\partial t} + \int_0^\sigma \nabla_\sigma \cdot (p_s\mathbf{u})\, d\tilde{\sigma}\right]. \tag{7.106}$$

A diagnostic equation for ω can be derived by noting that

$$\omega = \frac{d}{dt}(\sigma p_s) = \dot{\sigma}\, p_s + \sigma\frac{\partial p_s}{\partial t} + \sigma\mathbf{u}\cdot\nabla_\sigma p_s,$$

and thus

$$\omega(\sigma) = \sigma\mathbf{u}\cdot\nabla_\sigma p_s - \int_0^\sigma \nabla_\sigma \cdot (p_s\mathbf{u})\, d\tilde{\sigma}. \tag{7.107}$$

The geopotential is determined by integrating the hydrostatic equation

$$\frac{\partial\phi}{\partial(\ln\sigma)} = -RT \tag{7.108}$$

from the surface to level σ, which gives

$$\phi(\sigma) = gz_s - R\int_1^\sigma T\, d(\ln\tilde{\sigma}), \tag{7.109}$$

where $z_s(x, y)$ is the elevation of the topography.

The primary disadvantage of the σ-coordinate system is that it makes the accurate computation of horizontal pressure gradients difficult over steep topography. This problem arises because surfaces of constant σ tilt in regions where there are horizontal variations in surface pressure, and such variations are most pronounced over steep topography. When $\nabla_\sigma p_s \neq 0$, some portion of the vertical pressure gradient is projected onto each of the two terms $\nabla_\sigma\phi$ and $(RT/p_s)\nabla_\sigma p_s$. The vertical pressure gradient will not exactly cancel between these terms due to numerical error, and over steep topography the noncanceling residual can be comparable to the true horizontal pressure gradient because the vertical gradient of atmospheric pressure is several orders of magnitude larger than the horizontal gradient. The pressure-gradient error in a σ-coordinate model is not confined to the lower levels near the topography, but it may be reduced at upper levels using a hybrid vertical coordinate that transitions from σ coordinates to p coordinates at some level (or throughout some layer) in the interior of the domain (Sangster 1960; Simmons and Burridge 1981). Although they are widely used in operational weather and climate models (Williamson and Olson 1994; Ritchie et al. 1995; Kiehl et al. 1996), these hybrid coordinates complicate the solution of the governing equations and will not be considered here.

Several other approaches have also been suggested to minimize the errors generated over topography in σ-coordinate models. Phillips (1973) and Gary (1973) suggest performing the computations using a perturbation pressure defined with respect to a hydrostatically balanced reference state. Finite-difference schemes have been proposed that guarantee exact cancellation of the vertical pressure gradient between the last two terms in (7.102) whenever the vertical profiles of temperature and pressure have a specified functional relation, such as $T = a \ln(p) + b$ (Corby et al. 1972; Nakamura 1978; Simmons and Burridge 1981). Mesinger (1984) suggested using "η-coordinates" in which the mountain slopes are discretized as vertical steps at the grid interfaces with flat terrain between each step. More details and additional techniques for the treatment of pressure-gradient errors over mountains in hydrostatic atmospheric models are presented in the review by Mesinger and Janić (1985).

7.6.2 Spectral Representation of the Horizontal Structure

Global primitive-equation models often use spherical harmonics to represent the latitudinal and longitudinal variation of the forecast variables. In the following sections we present the basic numerical procedures for creating a spectral approximation to the σ-coordinate equations in a global atmospheric model. The approach is similar to that in Hoskins and Simmons (1975) and Bourke (1974), which may be consulted for additional details. The latitudinal and longitudinal variations in each field will be approximated using spherical harmonics, and the vertical variations will be represented using grid-point methods.

As was the case for the global shallow-water model described in Section 4.4.4, the spectral representation of the horizontal velocity field is facilitated by expressing the horizontal momentum equations in terms of the vertical vorticity ζ and the divergence δ. In order to integrate this system easily using semi-implicit time-differencing, it is also helpful to divide the temperature into a horizontally uniform reference state and a perturbation such that $T = \overline{T}(\sigma) + T'$. Using the identity (4.69) and taking the divergence of (7.102) yields

$$
\frac{\partial \delta}{\partial t} - \mathbf{k} \cdot \nabla \times (\zeta + f)\mathbf{u} + \nabla \cdot \left(\dot{\sigma}\frac{\partial \mathbf{u}}{\partial \sigma} + RT'\nabla(\ln p_{\mathrm{s}}) \right)
$$
$$
+ \nabla^2 \left(\phi + \frac{\mathbf{u} \cdot \mathbf{u}}{2} + R\overline{T}\ln p_{\mathrm{s}} \right) = 0. \qquad (7.110)
$$

Again using (4.69) and taking the vertical component of the curl of (7.102) one obtains

$$
\frac{\partial \zeta}{\partial t} + \nabla \cdot (\zeta + f)\mathbf{u} + \mathbf{k} \cdot \nabla \times \left(\dot{\sigma}\frac{\partial \mathbf{u}}{\partial \sigma} + RT'\nabla(\ln p_{\mathrm{s}}) \right) = 0. \qquad (7.111)
$$

Following the notation used in Section 4.4.4, let χ be the velocity potential and ψ the stream function for the horizontal velocity. Let λ be the longitude, θ the latitude, and $\mu = \sin\theta$. Define the operator

$$
\mathcal{H}(M, N) = \frac{1}{a} \left(\frac{1}{1 - \mu^2}\frac{\partial M}{\partial \lambda} + \frac{\partial N}{\partial \mu} \right),
$$

where a is the mean radius of the Earth. Then using the formula for the horizontal divergence in spherical coordinates,

$$\nabla M = \frac{1}{a\cos\theta}\frac{\partial M}{\partial\lambda}\mathbf{i} + \frac{1}{a}\frac{\partial M}{\partial\theta}\mathbf{j}$$

$$= \frac{1}{a(1-\mu^2)^{1/2}}\left(\frac{\partial M}{\partial\lambda}\mathbf{i} + (1-\mu^2)\frac{\partial M}{\partial\mu}\mathbf{j}\right),$$

and the relations (4.73)–(4.76), the prognostic equations for the σ-coordinate system may be expressed in the form

$$\frac{\partial\nabla^2\chi}{\partial t} = \mathcal{H}(B, -A) - 2\Omega\left(\frac{U}{a} - \mu\nabla^2\psi\right)$$

$$-\nabla^2\left(\phi + \frac{U^2 + V^2}{2(1-\mu^2)} + R\overline{T}\ln p_s\right), \qquad (7.112)$$

$$\frac{\partial\nabla^2\psi}{\partial t} = -\mathcal{H}(A, B) - 2\Omega\left(\frac{V}{a} + \mu\nabla^2\chi\right), \qquad (7.113)$$

$$\frac{\partial T'}{\partial t} = -\mathcal{H}(UT', VT') + T'\nabla^2\chi - \dot{\sigma}\frac{\partial T}{\partial\sigma} + \frac{\kappa T\omega}{\sigma p_s}, \qquad (7.114)$$

$$\frac{\partial}{\partial t}(\ln p_s) = -\frac{U}{a(1-\mu^2)}\frac{\partial}{\partial\lambda}(\ln p_s) - \frac{V}{a}\frac{\partial}{\partial\mu}(\ln p_s) - \nabla^2\chi - \frac{\partial\dot{\sigma}}{\partial\sigma}, \qquad (7.115)$$

where

$$U = u\cos\theta = (1-\mu^2)\mathcal{H}(\chi, -\psi),$$

$$V = v\cos\theta = (1-\mu^2)\mathcal{H}(\psi, \chi),$$

$$A = U\nabla^2\psi + \dot{\sigma}\frac{\partial V}{\partial\sigma} + \frac{RT'}{a}(1-\mu^2)\frac{\partial}{\partial\mu}(\ln p_s).$$

$$B = V\nabla^2\psi - \dot{\sigma}\frac{\partial U}{\partial\sigma} - \frac{RT'}{a}\frac{\partial}{\partial\lambda}(\ln p_s).$$

The preceding system of equations is formulated using $\ln p_s$ instead of p_s as the prognostic variable to make the term $(RT/p_s)\nabla_\sigma p_s$ into a binary product of the prognostic variables and thereby facilitate the alias-free evaluation of the pressure-gradient force via the spectral transform method.

At each σ level, the unknown functions ψ, χ, T', and ϕ are approximated using a truncated series of spherical harmonics. The unknown function $\ln p_s$ is also approximated by a spherical-harmonic expansion. Expressions for the time tendencies of the expansion coefficients for each spherical-harmonic are obtained using the transform method in a manner analogous to that for the global shallow-water model described in Section 4.4.4. As an example, suppose that the stream function and velocity potential at a given σ level are expanded in spherical harmonics as in (4.81) and (4.82). Then, using the notation defined in Section 4.4.4, the equation for $\partial\psi_{m,n}/\partial t$ is once again given by (4.86) except that $\hat{A}_m$ and $\hat{B}_m$

now satisfy

$$U\nabla^2\psi + \dot\sigma\,\frac{\partial V}{\partial\sigma} + \frac{RT'}{a}(1-\mu^2)\frac{\partial}{\partial\mu}(\ln p_s) = \sum_{m=-M}^{M} \hat{A}_m e^{im\lambda} \qquad (7.116)$$

and

$$V\nabla^2\psi - \dot\sigma\,\frac{\partial U}{\partial\sigma} - \frac{RT'}{a}\frac{\partial}{\partial\lambda}(\ln p_s) = \sum_{m=-M}^{M} \hat{B}_m e^{im\lambda}. \qquad (7.117)$$

The spectral form of the tendency equations for the velocity potential, the perturbation temperature, and the surface pressure may be found in Bourke (1974) and will not be given here. Note that the vertical advection terms in (7.116) and (7.117) involve the product of three spatially varying functions (since $\dot\sigma$ itself depends on the product of two spatially varying functions). The standard transform method cannot be used to transform these triple products between wave-number and physical space without incurring some numerical error. This "aliasing" error is nevertheless very small (Hoskins and Simmons 1975).

7.6.3 Vertical Differencing

The most significant modifications required to extend the shallow-water algorithm to a σ-coordinate model are those associated with the computation of the vertical derivatives. The vertical derivatives are computed using finite differences at that stage of the integration cycle when all the unknown variables are available on the physical mesh. As in (7.116) and (7.117), the results from these finite-difference computations are then combined with the other binary products computed on the physical mesh, and the net forcing is transformed back to wave-number space.

A convenient and widely used vertical discretization for the σ-coordinate equations is illustrated in Fig. 7.6 for a model with N vertical levels. The upper and lower boundaries are located at $\sigma = 0$ and

$$\sigma = 1 = \sum_{k=1}^{N} \Delta\sigma_k,$$

where $\Delta\sigma_k$ is the width of the kth σ layer. The stream function, velocity potential, temperature, and geopotential are defined at the center of each σ layer, and the velocity $\dot\sigma$ is defined at the interface between each layer. The vertical derivatives appearing in (7.112)–(7.114) involve variables, such as the temperature, that are defined at the center of each σ layer. These derivatives are approximated such that

$$\left(\dot\sigma\,\frac{\partial T}{\partial\sigma}\right)_k \approx \langle\dot\sigma_k\delta_\sigma T_k\rangle^\sigma \qquad (7.118)$$

$$= \dot\sigma_{k+\frac{1}{2}}\left(\frac{T_{k+1}-T_k}{\Delta\sigma_{k+1}+\Delta\sigma_k}\right) + \dot\sigma_{k-\frac{1}{2}}\left(\frac{T_k-T_{k-1}}{\Delta\sigma_k+\Delta\sigma_{k-1}}\right).$$

$$
\begin{array}{lll}
\top & \underline{\hspace{3cm}}\ \ \sigma = 0\ \ \underline{\hspace{3cm}} & \dot\sigma_{\frac{1}{2}} \\
\Delta\sigma_1 \quad \sigma_1 & \text{- -} & \psi_1,\ \chi_1,\ T'_1,\ \phi_1 \\
\bot & \underline{\hspace{5cm}} & \dot\sigma_{\frac{3}{2}}
\end{array}
$$

$$
\begin{array}{lll}
\top & \underline{\hspace{5cm}} & \dot\sigma_{k-\frac{1}{2}} \\
\Delta\sigma_k \quad \sigma_k & \text{- -} & \psi_k,\ \chi_k,\ T'_k,\ \phi_k \\
\bot & \underline{\hspace{5cm}} & \dot\sigma_{k+\frac{1}{2}}
\end{array}
$$

$$
\begin{array}{lll}
\top & \underline{\hspace{5cm}} & \dot\sigma_{N-\frac{1}{2}} \\
\Delta\sigma_N \quad \sigma_N & \text{- -} & \psi_N,\ \chi_N,\ T'_N,\ \phi_N \\
\bot & \underline{\hspace{3cm}}\ \ \sigma = 1\ \ \underline{\hspace{3cm}} & \dot\sigma_{N+\frac{1}{2}}
\end{array}
$$

FIGURE 7.6. Vertical distribution of the unknown variables on a σ-coordinate grid. The thickness of the σ layers need not be uniform. The center of each layer is at level $\sigma = \sigma_k$ and is indicated by the dashed lines. Note that the vertical index k increases with σ and decreases with geometric height.

The preceding is the generalization of the "averaging scheme" discussed in Section 3.5 to the nonuniform staggered mesh shown in Fig. 7.6.

The vertical derivative of $\dot\sigma$ in (7.115) is approximated as

$$
\left(\frac{\partial\dot\sigma}{\partial\sigma}\right)_k \approx \delta_\sigma\dot\sigma_k = \frac{\dot\sigma_{k+\frac{1}{2}} - \dot\sigma_{k-\frac{1}{2}}}{\Delta\sigma_k}. \tag{7.119}
$$

Defining $G_k = \nabla_\sigma\cdot\mathbf{u}_k + \mathbf{u}_k\cdot\nabla_\sigma(\ln p_s)$, the preceding implies that the vertically discretized approximation to the surface-pressure-tendency equation (7.105) is

$$
\frac{\partial}{\partial t}(\ln p_s) = -\sum_{k=1}^{N} G_k\Delta\sigma_k, \tag{7.120}
$$

and that (7.106) is approximated as

$$
\dot\sigma_{k+\frac{1}{2}} = \left(\sum_{j=1}^{k}\Delta\sigma_j\right)\sum_{j=1}^{N}G_j\Delta\sigma_j - \sum_{j=1}^{k}G_j\Delta\sigma_j. \tag{7.121}
$$

The hydrostatic equation (7.108) is approximated as

$$
\frac{\phi_{k+1} - \phi_k}{\ln\sigma_{k+1} - \ln\sigma_k} = -\frac{R}{2}(T_{k+1} + T_k),
$$

except in the half-layer between the lowest σ level and the surface, where

$$\frac{\phi_N - \phi_s}{\ln \sigma_N} = -RT_N.$$

Defining $\alpha_N = -\ln \sigma_N$ and $\alpha_k = \frac{1}{2}\ln(\sigma_{k+1}/\sigma_k)$ for $1 \le k < N$, the discrete analogue of (7.109) becomes

$$\phi_k = \phi_s + R \left(\sum_{j=k}^{N} \alpha_j T_j + \sum_{j=k+1}^{N} \alpha_{j-1} T_j \right). \tag{7.122}$$

Finally, as suggested by Corby et al. (1972), the vertical discretization for the ω equation (7.107) is chosen to preserve the energy-conservation properties of the vertically integrated continuous equations. Such conservation is achieved if

$$\frac{\omega_k}{\sigma_k p_s} = \mathbf{u}_k \cdot \nabla_\sigma (\ln p_s) - \frac{\alpha_k}{\Delta \sigma_k} \sum_{j=1}^{k} G_j \Delta \sigma_j - \frac{\alpha_{k-1}}{\Delta \sigma_k} \sum_{j=1}^{k-1} G_j \Delta \sigma_j. \tag{7.123}$$

7.6.4 Energy Conservation

Why does (7.123) give better energy-conservation properties than the simpler formula that would result if both α_k and α_{k-1} were replaced by $\Delta \sigma_k/(2\sigma_k)$? In order to answer this question it is necessary to review the energy-conservation properties of the continuous σ-coordinate primitive equations. Our focus is on the vertical discretization, so it is helpful to obtain a conservation law for the vertically integrated total energy per unit horizontal area. Using the hydrostatic equation (7.91) and the definition $\sigma p_s = p$, the vertical integral of the sum of the kinetic[9] and internal energy per unit volume may be expressed as

$$\int_{z_s}^{\infty} \rho \left(\frac{\mathbf{u} \cdot \mathbf{u}}{2} + c_v T \right) dz = -\frac{1}{g} \int_{p_s}^{0} \left(\frac{\mathbf{u} \cdot \mathbf{u}}{2} + c_v T \right) dp$$

$$= \frac{p_s}{g} \int_{0}^{1} \left(\frac{\mathbf{u} \cdot \mathbf{u}}{2} + c_v T \right) d\sigma.$$

Using the hydrostatic equation twice and integrating by parts, the vertical integral of the potential energy becomes

$$\int_{z_s}^{\infty} \rho g z \, dz = \int_{0}^{p_s} \frac{\phi}{g} \, dp = \frac{1}{g} \int_{0}^{\phi_s p_s} d(\phi p) - \int_{\infty}^{\phi_s} \frac{p}{g} \, d\phi = \frac{\phi_s p_s}{g} + \int_{0}^{p_s} \frac{p}{g\rho} \, dp.$$

---

[9]Note that as a consequence of the primitive-equation approximation, the vertical velocity does not appear as part of the kinetic energy.

Recalling that $p = \rho RT$,

$$\int_0^\infty \rho g z\, dz = \frac{\phi_s p_s}{g} + \frac{p_s}{g} \int_0^1 RT\, d\sigma,$$

and the total vertically integrated energy per unit area is

$$\mathcal{E} = \frac{\phi_s p_s}{g} + \frac{p_s}{g} \int_0^1 \left(\frac{\mathbf{u} \cdot \mathbf{u}}{2} + c_p T \right) d\sigma.$$

The total time derivatives in the momentum and thermodynamic equations must be written in flux form in order to obtain a conservation law governing $\mathcal{E}$. For any scalar γ,

$$p_s \frac{d\gamma}{dt} = p_s \frac{\partial \gamma}{\partial t} + p_s \mathbf{u} \cdot \nabla_\sigma \gamma + p_s \dot\sigma \frac{\partial \gamma}{\partial \sigma} + \gamma \left[\frac{\partial p_s}{\partial t} + \nabla_\sigma \cdot (p_s \mathbf{u}) + \frac{\partial}{\partial \sigma}(p_s \dot\sigma) \right]$$

$$= \frac{\partial}{\partial t}(p_s \gamma) + \nabla_\sigma \cdot (p_s \gamma \mathbf{u}) + \frac{\partial}{\partial \sigma}(p_s \gamma \dot\sigma), \tag{7.124}$$

where the quantity in square brackets is zero by the pressure-tendency equation (7.104). Adding $p_s c_p$ times the thermodynamic equation (7.103) to the dot product of $p_s \mathbf{u}$ and the momentum equation (7.102) and using (7.124), one obtains

$$\frac{\partial E}{\partial t} + \nabla_\sigma \cdot (E\mathbf{u}) + \frac{\partial}{\partial \sigma}(E\dot\sigma) + \mathbf{u} \cdot p_s \nabla_\sigma \phi + \mathbf{u} \cdot RT \nabla_\sigma p_s - \frac{RT\omega}{\sigma} = 0, \tag{7.125}$$

where $E = p_s(\mathbf{u} \cdot \mathbf{u}/2 + c_p T)$. Defining

$$F = -\phi \nabla_\sigma \cdot (p_s \mathbf{u}) + \mathbf{u} \cdot RT \nabla_\sigma p_s - \frac{RT\omega}{\sigma}, \tag{7.126}$$

(7.125) may be expressed as

$$\frac{\partial E}{\partial t} + \nabla_\sigma \cdot [(E + p_s\phi)\mathbf{u}] + \frac{\partial}{\partial \sigma}(E\dot\sigma) + F = 0. \tag{7.127}$$

The forcing F may be written as the vertical divergence of a flux as follows. Substituting for ω using (7.107),

$$F = -\phi \nabla_\sigma \cdot (p_s \mathbf{u}) + \frac{RT}{\sigma} \int_0^\sigma \nabla_\sigma \cdot (p_s \mathbf{u})\, d\tilde\sigma,$$

and then substituting for RT/σ from the hydrostatic equation,

$$F = -\phi \nabla_\sigma \cdot (p_s \mathbf{u}) - \frac{\partial \phi}{\partial \sigma} \int_0^\sigma \nabla_\sigma \cdot (p_s \mathbf{u})\, d\tilde\sigma$$

$$= -\frac{\partial}{\partial \sigma} \left[\phi \int_0^\sigma \nabla_\sigma \cdot (p_s \mathbf{u})\, d\tilde\sigma \right].$$

Thus,

$$
\int_0^1 F \, d\sigma = -\phi_s \int_0^1 \nabla_\sigma \cdot (p_s \mathbf{u}) \, d\sigma
$$

$$
= \phi_s \int_0^1 \left(\frac{\partial p_s}{\partial t} + \frac{\partial}{\partial \sigma}(\dot{\sigma} p_s) \right) d\sigma
$$

$$
= \frac{\partial}{\partial t}(\phi_s p_s).
$$

The preceding may be used to derive a conservation law for $\mathcal{E}$ by integrating (7.127) over the depth of the domain and applying the boundary condition $\dot{\sigma} = 0$ at the upper and lower boundaries to obtain

$$
\frac{\partial \mathcal{E}}{\partial t} + \nabla_\sigma \cdot \int_0^1 (E + p_s \phi) \mathbf{u} \, d\sigma = 0. \tag{7.128}
$$

Of course, (7.128) also implies that if the horizontal domain is periodic, or if there is no flow normal to the lateral boundaries, the σ-coordinate primitive equations conserve the domain-integrated total energy

$$
\int \int \left[\frac{\phi_s p_s}{g} + \frac{p_s}{g} \int_0^1 \left(\frac{\mathbf{u} \cdot \mathbf{u}}{2} + c_p T \right) d\sigma \right] dx \, dy. \tag{7.129}
$$

The domain-integrated total energy is not, however, exactly conserved by global spectral models. As discussed in Section 4.2.3, a Galerkin spectral approximation to a prognostic equation for an unknown function γ will generally conserve the domain integral of γ^2, provided that the domain integral of γ^2 is also conserved by the continuous equations and time-differencing errors are neglected. Unfortunately, the conservation of the squares of the prognostic variables in (7.112)–(7.115) does not imply exact conservation of the total energy. Practical experience has, nevertheless, shown that the deviations from exact energy conservation generated by the spectral approximation of the horizontal derivatives is very small. The nonconservation introduced by the semi-implicit time-differencing used in most global primitive-equation models has also been shown to be very small (Hoskins and Simmons 1975). Nonconservative formulations of the vertical finite-differencing can, however, have a significantly greater impact on the global energy conservation. This appears to be a particularly important issue if long-time integrations are conducted using global climate models with poor vertical resolution.

The energy-conservation properties of the vertical discretization given by (7.118)–(7.123) will therefore be isolated from the nonconservative effects of the spectral approximation and the time-differencing by considering a system of differential–difference equations in which only those terms containing vertical derivatives are discretized. Except for the terms involving vertical derivatives, the total-energy equation for the semidiscrete system must be identical to (7.127) because the time and horizontal derivatives are exact. The semidiscrete system will therefore conserve total energy, provided that it satisfies the discrete analogues of

$$\int_0^1 \frac{\partial}{\partial \sigma}(E\dot\sigma)\,d\sigma = 0 \tag{7.130}$$

and

$$\int_0^1 F\,d\sigma = \frac{\partial}{\partial t}(\phi_s p_s). \tag{7.131}$$

The integrand in (7.130) appears in the total energy equation (7.127) as a mathematical simplification of a linear combination of the vertical derivative terms in the momentum, thermodynamic, and surface-pressure-tendency equations such that

$$\frac{\partial}{\partial \sigma}(E\dot\sigma)\,d\sigma = \mathbf{u}\dot\sigma\frac{\partial \mathbf{u}}{\partial \sigma} + \frac{\mathbf{u}\cdot\mathbf{u}}{2}\frac{\partial \dot\sigma}{\partial \sigma} + c_p\dot\sigma\frac{\partial T}{\partial \sigma} + c_p T\frac{\partial \dot\sigma}{\partial \sigma}.$$

When the vertical derivatives on the right side of the preceding are approximated using (7.118) and (7.119), their summation over the depth of the domain is exactly zero. This may be demonstrated for the pair of terms involving T by noting that since $\dot\sigma_{\frac{1}{2}} = \dot\sigma_{N+\frac{1}{2}} = 0$,

$$\sum_{k=1}^N \left[(\dot\sigma_k\delta_\sigma T_k)^\sigma + T_k\delta_\sigma\dot\sigma_k\right]\Delta\sigma_k = \sum_{k=1}^N \left[\dot\sigma_{k+\frac{1}{2}}\left(\frac{\Delta\sigma_k T_{k+1} + \Delta\sigma_{k+1}T_k}{\Delta\sigma_{k+1} + \Delta\sigma_k}\right)\right.$$
$$\left. -\dot\sigma_{k-\frac{1}{2}}\left(\frac{\Delta\sigma_{k-1}T_k + \Delta\sigma_k T_{k-1}}{\Delta\sigma_k + \Delta\sigma_{k-1}}\right)\right] = 0.$$

A similar relation holds for the two terms involving the horizontal velocity (see Problem 7).

Now consider the discrete analogue of (7.131), or equivalently,

$$\int_0^1 \frac{F}{p_s}\,d\sigma = \phi_s\frac{\partial}{\partial t}(\ln p_s).$$

Defining $G = \nabla_\sigma\cdot\mathbf{u} + \mathbf{u}\cdot\nabla_\sigma(\ln p_s)$ and substituting for F using (7.126) yields

$$\int_0^1 \left(-\phi G + \mathbf{u}\cdot RT\nabla_\sigma(\ln p_s) - \frac{RT\omega}{\sigma p_s}\right)d\sigma = \phi_s\frac{\partial}{\partial t}(\ln p_s).$$

The discrete form of this integral equation may be obtained using (7.120) and (7.123) and is algebraically equivalent to

$$\sum_{k=1}^N (\phi_k - \phi_s)G_k\Delta\sigma_k = \sum_{k=1}^N RT_k\left(\alpha_k\sum_{j=1}^k G_j\Delta\sigma_j + \alpha_{k-1}\sum_{j=1}^{k-1}G_j\Delta\sigma_j\right).$$

It may be verified that the preceding is indeed an algebraic identity by substituting for $\phi_k - \phi_s$ from the discrete form of the hydrostatic equation (7.122) and using the relation

$$\sum_{k=1}^N\sum_{j=1}^k a_k b_j = \sum_{k=1}^N\sum_{j=k}^N a_j b_k.$$

In addition to conserving total energy, the preceding vertical discretization also conserves total mass (see Problem 8). This scheme does not conserve the integrated angular momentum or the integrated potential temperature. Arakawa and Lamb (1977), Simmons and Burridge (1981), and Arakawa and Konor (1996) describe alternative vertical discretizations that conserve angular momentum, potential temperature, or various other vertically integrated functions.

7.6.5 Semi-implicit Time-Differencing

Computational efficiency can be enhanced by using semi-implicit time-differencing to integrate the preceding primitive-equation model. The semi-implicit method can be implemented in σ-coordinate primitive-equation models as follows. Let $\mathbf{d}$ be a column vector whose kth element is the function $\nabla_\sigma^2 \chi$ at level σ_k. Similarly, define $\bar{\mathbf{t}}$, $\mathbf{t}$, and $\mathbf{h}$ to be column vectors containing the σ-level values of the functions $R\bar{T}$, T, and ϕ. Let $\mathbf{h_s}$ be a column vector in which every element is ϕ_s. Then the vertically discretized equations for the divergence, temperature, surface-pressure tendency, and geopotential may be written in the form

$$\frac{\partial \mathbf{d}}{\partial t} = \mathbf{f_d} - \nabla_\sigma^2 \left(\mathbf{h} + \bar{\mathbf{t}} \ln p_s \right), \tag{7.132}$$

$$\frac{\partial \mathbf{t}}{\partial t} = \mathbf{f_t} - \mathbf{H}\mathbf{d}, \tag{7.133}$$

$$\frac{\partial}{\partial t}(\ln p_s) = f_p - \mathbf{p}^T \mathbf{d}, \tag{7.134}$$

$$\mathbf{h} = \mathbf{h_s} + \mathbf{G}\mathbf{t}. \tag{7.135}$$

Here $\mathbf{G}$ and $\mathbf{H}$ are matrices and $\mathbf{p}$ is a column vector, none of which depend on λ, μ, or t. The thermodynamic equation (7.133) is partitioned such that all terms containing the product of $\bar{T}(\sigma)$ and the divergence are collected in $\mathbf{H}\mathbf{d}$.

Equation (7.120) implies that

$$\mathbf{p}^T = (\Delta\sigma_1, \Delta\sigma_2, \ldots, \Delta\sigma_N),$$

and (7.122) requires

$$\frac{\mathbf{G}}{R} = \begin{pmatrix} \alpha_1 & \alpha_1 + \alpha_2 & \alpha_2 + \alpha_3 & \cdots \\ 0 & \alpha_2 & \alpha_2 + \alpha_3 & \cdots \\ 0 & 0 & \alpha_3 & \cdots \\ 0 & 0 & 0 & \cdots \\ \vdots & \vdots & \vdots & \end{pmatrix}.$$

Let $h_{r,s}$ denote the sth element in the rth row of $\mathbf{H}$. Then according to (7.114), $h_{r,s}$ is determined by the contribution of the divergence at level s to $\dot\sigma\,\partial\bar{T}/\partial\sigma - \kappa\bar{T}\omega/(\sigma p_s)$ at level r. Define a step function such that $S(x) = 1$ if $x \geq 0$ and

$\mathcal{S}(x) = 0$ otherwise. Then from (7.118), (7.121), and (7.123),

$$
\begin{aligned}
\frac{h_{r,s}}{\Delta\sigma_s} = &\ \frac{\kappa \overline{T}_r \mathcal{S}(r-s)}{\Delta\sigma_r} \left[\alpha_r + \mathcal{S}(r-s-1)\alpha_{r-1} \right] \\
&- \left(\frac{\overline{T}_{r+1} - \overline{T}_r}{\Delta\sigma_{r+1} + \Delta\sigma_r} \right) \left(\mathcal{S}(r-s) - \sum_{j=1}^{r} \Delta\sigma_j \right) \\
&- \left(\frac{\overline{T}_r - \overline{T}_{r-1}}{\Delta\sigma_r + \Delta\sigma_{r-1}} \right) \left(\mathcal{S}(r-s-1) - \sum_{j=1}^{r-1} \Delta\sigma_j \right).
\end{aligned}
$$

The remaining terms in (7.120) and the vertically discretized versions of (7.112) and (7.114) are gathered into f_p, $\mathbf{f_d}$, and $\mathbf{f_t}$, respectively.

A single equation for the divergence may be obtained by eliminating $\mathbf{t}$, $\mathbf{h}$, and $\ln p_s$ from (7.132)–(7.135) to give

$$
\left(\frac{\partial^2}{\partial t^2} - \mathbf{B}\nabla_\sigma^2 \right) \mathbf{d} = \frac{\partial \mathbf{f_d}}{\partial t} - \nabla_\sigma^2 \left(\mathbf{Gf_t} + f_p \bar{\mathbf{t}} \right), \tag{7.136}
$$

where $\mathbf{B} = \mathbf{GH} + \bar{\mathbf{t}}\mathbf{p}^T$. The solutions to the homogeneous part of this equation comprise the set of gravity waves supported by the vertically discretized model. Hoskins and Simmons (1975) present plots showing the vertical structure of each of the gravity-wave modes in a five-layer model. For typical atmospheric profiles of $\overline{T}(\sigma)$ the fastest mode propagates at a speed on the order of 300 ms^{-1} and thereby imposes a severe constraint on the maximum stable time step with which these equations can be integrated using explicit time-differencing.

Since the fastest-moving gravity waves do not need to be accurately simulated in order to obtain an accurate global weather forecast, (7.132)–(7.134) can be efficiently integrated using a semi-implicit scheme in which those terms that combine to form the left side of (7.136) are integrated using the trapezoidal method over a time interval of $2\Delta t$. The formulae that result from this semi-implicit approximation are

$$
\delta_{2t}\mathbf{d}^n = \mathbf{f_d}^n - \nabla_\sigma^2 \left[\langle \mathbf{h}^n \rangle^{2t} + \bar{\mathbf{t}}\langle (\ln p_s)^n \rangle^{2t} \right], \tag{7.137}
$$

$$
\delta_{2t}\mathbf{t}^n = \mathbf{f_t}^n - \mathbf{H}\langle \mathbf{d}^n \rangle^{2t}, \tag{7.138}
$$

$$
\delta_{2t}(\ln p_s)^n = f_p^n - \mathbf{p}^T \langle \mathbf{d}^n \rangle^{2t}. \tag{7.139}
$$

Using the relation $\delta_{2t}\gamma^n = (\langle \gamma^n \rangle^{2t} - \gamma^{n-1})/\Delta t$ together with (7.135), (7.138), and (7.139) to eliminate $\langle \mathbf{h}^n \rangle^{2t}$ and $\langle (\ln p_s)^n \rangle^{2t}$ from (7.137) gives

$$
\begin{aligned}
\left[\mathbf{I} - (\Delta t)^2 \mathbf{B}\nabla_\sigma^2 \right] &\langle \mathbf{d}^n \rangle^{2t} \\
&= \mathbf{d}^{n-1} + \Delta t \mathbf{f_d}^n - \nabla_\sigma^2 \left\{ \Delta t \left[\mathbf{h}^{n-1} + \bar{\mathbf{t}}(\ln p_s)^{n-1} \right] + (\Delta t)^2 \left[\mathbf{Gf_t}^n + \bar{\mathbf{t}}f_p^n \right] \right\}.
\end{aligned}
\tag{7.140}
$$

Let $\chi_{r,s}$ be a column vector whose kth element is the coefficient of $Y_{r,s}$ in the series expansion for the velocity potential at level k. Since the spherical harmonics are eigenfunctions of the horizontal Laplacian operator on the sphere, (7.140) is equivalent to a linear-algebraic system for $\chi_{r,s}^{n+1}$ of the form

$$\left[\mathbf{1} + (\Delta t)^2 \frac{s(s+1)}{a^2} \mathbf{B} \right] \chi_{r,s}^{n+1} = \mathbf{f},$$

where the right side $\mathbf{f}$ does not involve the values of any unknown functions at time $(n+1)\Delta t$. The N unknown variables in this relatively small linear system can be determined by Gaussian elimination. Additional efficiency can be achieved by exploiting the fact that the coefficient matrix is constant in time, so its "LU" decomposition into upper and lower triangular matrices need only be computed once.

Some of the forcing terms that are responsible for gravity-wave propagation in the σ-coordinate equations are nonlinear. In order to obtain the preceding linear-algebraic equation for $\chi_{r,s}^{n+1}$ these terms have been decomposed into a linear part and a nonlinear perturbation by splitting the total temperature into a constant horizontally uniform reference temperature $\overline{T}(\sigma)$ and a perturbation. As discussed in Section 7.2.3, this decomposition imposes a constraint on the stability of the semi-implicit solution that, roughly speaking, requires the speed of the fastest-moving gravity wave supported by the actual atmospheric structure to be only modestly faster than the speed of the fasting-moving gravity wave in the reference state. This stability constraint is usually satisfied by choosing an isothermal profile for the reference state, i.e., $\overline{T}(\sigma) = T_0$ (Simmons et al. 1978). A typical value for T_0 is 300 K.

Problems

1. Consider small-amplitude shallow-water motions on a "mid-latitude β-plane." In the following, x and y are horizontal coordinates oriented east–west and north–south, respectively; the Coriolis parameter is approximated as $f_0 + \beta y$ where f_0 and β are constant; g is the gravitational acceleration; $U > 0$ is a constant mean flow from west to east; u' and v' are the perturbation west-to-east and south-to-north velocities; h is the perturbation displacement of the free surface. Define the vorticity ζ and the divergence δ as

$$\zeta = \frac{\partial v'}{\partial x} - \frac{\partial u'}{\partial y}, \qquad \delta = \frac{\partial u'}{\partial x} + \frac{\partial v'}{\partial y}.$$

Assume that the mean flow is in geostrophic balance,

$$U = -\frac{g}{f_0} \frac{\partial \overline{h}}{\partial y},$$

and that there is a mean north–south gradient in the bottom topography equal to the mean gradient in the height of the free surface, $\partial \overline{h}/\partial y$, so that the mean fluid depth is a constant H. The linearized shallow-water equations for this system are

$$\left(\frac{\partial}{\partial t} + U\frac{\partial}{\partial x}\right)\zeta + f\delta + \beta v = 0,$$

$$\left(\frac{\partial}{\partial t} + U\frac{\partial}{\partial x}\right)\delta - f\zeta + \beta u + g\left(\frac{\partial^2 h}{\partial x^2} + \frac{\partial^2 h}{\partial y^2}\right) = 0,$$

$$\left(\frac{\partial}{\partial t} + U\frac{\partial}{\partial x}\right)h + H\delta = 0. \quad (7.141)$$

The terms involving β in the preceding vorticity and divergence equations can be approximated[10] as

$$\left(\frac{\partial}{\partial t} + U\frac{\partial}{\partial x}\right)\zeta + f_0\delta + \frac{\beta g}{f_0}\frac{\partial h}{\partial x} = 0, \quad (7.142)$$

$$\left(\frac{\partial}{\partial t} + U\frac{\partial}{\partial x}\right)\delta - f_0\zeta + g\left(\frac{\partial^2 h}{\partial x^2} + \frac{\partial^2 h}{\partial y^2}\right) = 0. \quad (7.143)$$

(a) Show that waves of the form

$$(\zeta, \delta, h) = (\zeta_0, \delta_0, h_0)e^{i(kx+\ell y-\omega t)}$$

are solutions to the preceding system if they satisfy the dispersion relation

$$(\omega - Uk)^2 = c^2(k^2 + \ell^2) + f_0^2 + \frac{k\beta c^2}{\omega - Uk},$$

where $c^2 = gH$.

(b) Show that if $\beta/c \ll k^2$, the individual solutions to this dispersion relation are well-approximated by the solutions to either the inertial-gravity-wave dispersion relation

$$(\omega - Uk)^2 = c^2(k^2 + \ell^2) + f_0^2$$

or the Rossby-wave dispersion relation

$$\omega = Uk - \frac{\beta k}{k^2 + \ell^2 + f_0^2/c^2}.$$

---

[10]The approximations used to obtain (7.142) and (7.143) are motivated by the desire to obtain a clean dispersion relation rather than a straightforward scale analysis.

2. Suppose that the time derivatives in (7.141)–(7.143) are approximated by leapfrog differencing and the spatial dependence is represented by a Fourier spectral or pseudospectral method. Recall that we have assumed $U > 0$, as would be the case in the middle latitudes of the Earth's atmosphere.

(a) Determine the constraints on Δt required to keep the gravity waves stable and show that $(U + c)k\,\Delta t \leq 1$ is a necessary condition for stability.

(b) Determine the constraints on Δt required to keep the Rossby waves stable. Let K be the magnitude of the maximum vector wave number retained in the truncation, i.e.,

$$K = \max_{k,\ell} \sqrt{k^2 + \ell^2}.$$

Show that $U K \Delta t \leq 1$ is a sufficient condition for the stability of the Rossby waves unless

$$K^2 \leq \frac{\beta}{2U} - \frac{f_0^2}{c^2}.$$

3. Suppose that (7.141)–(7.143) are integrated using the semi-implicit scheme

$$\delta_{2t}\zeta^n + U\frac{\partial \zeta^n}{\partial x} + f_0\delta^n + \frac{\beta g}{f_0}\frac{\partial h^n}{\partial x} = 0,$$

$$\delta_{2t}\delta^n + U\frac{\partial \delta^n}{\partial x} - f_0\zeta^n + g\left(\frac{\partial^2 h^n}{\partial x^2} + \frac{\partial^2 h^n}{\partial y^2}\right)^{2t} = 0,$$

$$\delta_{2t}h^n + U\frac{\partial h^n}{\partial x} + H\left(\delta^n\right)^{2t} = 0.$$

(a) Determine the conditions under which the gravity waves are stable.

(b) Determine the conditions under which the Rossby waves are stable.

(c) Discuss the impact of semi-implicit differencing on the accuracy of the Rossby and gravity wave modes.

4. Two-dimensional sound waves in a neutrally stratified atmosphere satisfy the linearized equations

$$\frac{\partial u}{\partial t} + c_s\frac{\partial P}{\partial x} = 0.$$

$$\frac{\partial w}{\partial t} + c_s\frac{\partial P}{\partial z} = 0.$$

$$\frac{\partial P}{\partial t} + c_s\left(\frac{\partial u}{\partial x} + \frac{\partial w}{\partial z}\right) = 0,$$

where $P = p'/(\rho_0 c_s)$. Let this system be approximated using forward-backward differencing for the horizontal gradients and trapezoidal differencing for the vertical gradients such that

$$u^{m+1} = u^m - c_s \Delta \tau \frac{\partial P^m}{\partial x},$$

$$w^{m+1} = w^m - c_s \frac{\Delta \tau}{2} \frac{\partial}{\partial z} \left(P^{m+1} + P^m \right),$$

$$P^{m+1} = P^m - c_s \Delta \tau \left(\frac{\partial u^{m+1}}{\partial x} - \frac{1}{2} \frac{\partial}{\partial z} \left(w^{m+1} + w^m \right) \right).$$

Consider an individual Fourier mode with spatial structure $\exp i (kx + \ell z)$ and show that the eigenvalues of the amplification matrix for this scheme are unity and

$$\frac{4 - \tilde{\ell}^2 - 2\tilde{k}^2 \pm 2\sqrt{(\tilde{k} - 2)(\tilde{k} + 2)(\tilde{k}^2 + \tilde{\ell}^2)}}{\tilde{\ell}^2 + 4},$$

where $\tilde{k} = c_s k \Delta \tau$ and $\tilde{\ell} = c_s \ell \Delta \tau$. What is the stability condition that ensures that this method will not have any eigenvalues with absolute values exceeding unity?

5. Compare the errors generated in gravity waves using semi-implicit differencing to those produced in a compressible Boussinesq system in which the true speed of sound c_s is artificially reduced to $\tilde{c}_s$ in an effort to increase efficiency by increasing the maximum stable value for $\Delta \tau$. *Hint:* Consider waves with wave numbers on the order of $N/\tilde{c}_s$ but larger than N/c_s.

6. The oscillations of the damped-harmonic oscillator (7.88) are "over-damped" when $\alpha^2 \kappa^2 > c_s^2$. Suppose that the mesh is isotropic with grid interval Δ, the Courant number for sound-wave propagation on the small time step is $\frac{1}{2}$, and $\alpha = \gamma \Delta^2 / \Delta \tau$. Estimate the minimum value of α required make the divergence damper over-damp a mode resolved on the numerical mesh. Do the values of α_x and α_z used in the test problem shown in Fig. 7.3d over-damp any of the resolved modes in that test problem?

7. Show that discrete integral of the finite-difference approximation to the vertical divergence of the vertical advective flux of kinetic energy,

$$\sum_{k=1}^{N} \left[(\langle \mathbf{u}_k \rangle^\sigma \dot{\sigma}_k \delta_\sigma \mathbf{u}_k)^\sigma + \frac{\mathbf{u}_k \cdot \mathbf{u}_k}{2} \delta_\sigma \dot{\sigma}_k \right] \Delta \sigma_k,$$

is zero.

8. Examine the mass-conservation properties of σ-coordinate primitive-equation models.

(a) Show that the vertically integrated mass per unit area in a hydrostatically balanced atmosphere is p_s/g.

(b) Show that the vertical finite-difference scheme for the σ-coordinate primitive-equation model described in Section 7.6.3 will conserve total mass if nonconservative effects due to time-differencing and the horizontal spectral representation are neglected.

(c) Suppose that instead of predicting $\ln p_s$ as in (7.115), the actual surface pressure was predicted using (7.104). Show that except for nonconservative effects due to time-differencing, total mass will be exactly conserved in a numerical model in which the Galerkin spectral method is used to evaluate the horizontal derivatives, and the vertical derivative is approximated by (7.119). This approach is not used in practice because it generates a noisier solution that that obtained using $\ln p_s$ as the prognostic variable (Kiehl et al. 1996, p. 15).

8

Nonreflecting Boundary Conditions

If the boundary of a computational domain coincides with a true physical boundary, an appropriate boundary condition can generally be derived from physical principles and can be implemented in a numerical model with relative ease. It is, for example, easy to derive the condition that the fluid velocity normal to a rigid boundary must vanish at that boundary, and if the shape of the boundary is simple, it is easy to impose this condition on the numerical solution. More serious difficulties may be encountered if the computational domain terminates at some arbitrary location within the fluid. When possible, it is a good idea to avoid artificial boundaries by extending the computational domain throughout the entire fluid. Nevertheless, in many problems the phenomena of interest occur in a localized region, and it is impractical to include all of the surrounding fluid in the numerical domain. As a case in point, one would not simulate an isolated thunderstorm with a global atmospheric model just to avoid possible problems at the lateral boundaries of a limited domain. Moreover, in a fluid such as the atmosphere there is no distinct upper boundary, and any numerical representation of the atmosphere's vertical structure will necessarily terminate at some arbitrary level.

When the computational domain is terminated at an arbitrary location within a larger body of fluid, the conditions imposed at the edge of the domain are intended to mimic the presence of the surrounding fluid. The boundary conditions should therefore allow outward-traveling disturbances to pass through the boundary without generating spurious reflections that propagate back toward the interior. Boundary conditions designed to minimize spurious backward reflection are known as *nonreflecting*, *open*, *wave-permeable*, or *radiation* boundary conditions. The terminology "radiation boundary condition" is due to Sommerfeld

(1949, p. 189), who defined it as the condition that "the sources must be *sources*, not *sinks*, of energy. The energy which is radiated from the sources must scatter to infinity; *no energy may be radiated from infinity into . . . the field.*" As formulated by Sommerfeld, the radiation condition applies at infinity; however, in all practical computations a boundary condition must be imposed at some finite distance from the energy source, and this creates two problems. The first problem is that the radiation condition itself may not properly describe the physical behavior occurring at an arbitrarily designated location within the fluid when that location is only a finite distance from the energy source. The second problem is that it is typically more difficult to express the radiation condition mathematically at a boundary that is only a finite distance from the energy source.

The radiation condition is obviously not appropriate in situations where energy must be transmitted inward through the boundary; yet inward radiation may even be required in problems where the surrounding fluid is initially quiescent and all the initial disturbances are contained within the computational domain. Two nonlinear waves that propagate past an artificial boundary within a fluid may interact outside the artificial boundary and generate an inward-propagating disturbance that should reenter the domain. This point was emphasized by Hedstrom (1979b), who provided a simple example from compressible gas dynamics demonstrating that a shock overtaking a contact discontinuity must generate an echo that propagates back toward the wave generator. Numerical simulations of thunderstorms provide another example where the documented sensitivity of numerical simulations to the lateral boundary conditions (Clark 1979; Hedley and Yau 1988) may be not only a consequence of poorly approximating the radiation condition, but also the result of inadequately representing important feedbacks on the convection arising through interactions with the storm's environment that occur outside the numerical domain.

Errors resulting from a failure to incorporate inward-propagating signals generated by real physical processes occurring outside the boundaries of the computational domain cannot be avoided without enlarging the domain. Nevertheless, nonreflecting boundary conditions often become reasonable approximations to the true physical boundary condition as the size of the computational domain increases, provided that the local energy density of a disturbance arriving at the boundary is reduced as the result of wave dispersion or absorption in the large domain. When the disturbances arriving at the boundary are sufficiently weak, the governing equations in the region near the boundary can be approximated by their linearized equivalents, and it can be relatively easy to ensure that the radiation condition correctly describes the boundary conditions for the linearized system. The situation is particularly simple in the case of constant-coefficient linear hyperbolic systems, for which radiation boundary conditions are clearly appropriate because the characteristic curves for such systems are straight lines that cannot exit and subsequently reenter the domain.

Even when it is clear that the radiation condition is appropriate, it is not always easy to translate Sommerfeld's physical description into a mathematical formula. As noted in the monograph by Givoli (1992), it is generally easier to express the radiation condition at infinity mathematically than to formulate the same condition for a boundary that is only a finite distance from the wave source. In fact, in most practical problems it is impossible to express the radiation condition exactly as an algebraic or differential equation involving the prognostic variables in the neighborhood of the boundary. Instead of the exact radiation condition one must use an approximation, and this approximation introduces an error in the mathematical model of the physical system that is distinct from the errors subsequently incurred when constructing a discrete approximation to the mathematical model. As a consequence, the numerical solution will not converge to the correct solution to the underlying physical problem as the spatial and temporal mesh is refined, but rather to the correct solution to the approximate mathematical problem determined by the approximate radiation condition.

The first topic that will be considered in this chapter is therefore the mathematical formulation of well-posed radiation boundary conditions. We begin with examples where this can be done exactly and then consider problems where approximations are required. After formulating exact or approximate radiation boundary conditions for the continuous problem, we will consider their numerical implementation.

8.1   One-Dimensional Flow

Exact open boundary conditions can be obtained for certain simple one-dimensional systems. Two such problems will be considered in this section, the linear advection equation and the linearized shallow-water system. The shallow-water system provides an instructive example illustrating the construction of exact radiation boundary conditions for the continuous equations. In contrast, there is no need to explicitly determine a radiation boundary condition for the nondiscretized linear advection equation because a well-posed mathematical formulation of the advection problem does not require any outflow boundary condition. The linear advection equation is, however, useful for investigating the influence of various numerical approximations to the exact outflow boundary condition on the accuracy and stability of the discretized solution.

8.1.1   Well-Posed Initial–Boundary Value Problems.

Consider the linearized one-dimensional advection equation

$$\frac{\partial \psi}{\partial t} + U \frac{\partial \psi}{\partial x} = 0, \tag{8.1}$$

and suppose ψ is to be determined throughout some limited domain $0 \le x \le L$. For the sake of illustration, assume that $U > 0$. If initial data are given for the

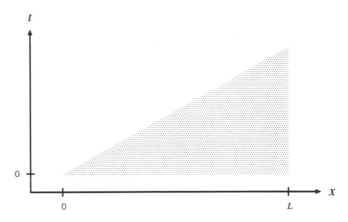

FIGURE 8.1. Portion of the x-t plane (shaded) in which the solution to (8.1) is determined by the initial data $\psi(x, 0), 0 \leq x \leq L$.

domain $0 \leq x \leq L$, how should boundary conditions at $x = 0$ and $x = L$ be specified to yield a well-posed problem?[1]

Since ψ is the solution to a homogeneous hyperbolic equation with constant coefficients, ψ will be constant along the characteristic curves $x - Ut = x_0$. The initial data will therefore uniquely determine ψ in the shaded triangular region in Fig. 8.1. The solution in the remainder of the strip $t > 0, 0 \leq x \leq L$ is determined by the values of $\psi(0, t)$ at the inflow boundary. Thus, in order to uniquely determine the solution (one prerequisite for well-posedness), a boundary condition must be imposed at $x = 0$. On the other hand, no differentiable function will be able to satisfy an arbitrary boundary condition imposed at $x = L$, because $\psi(L, t)$ is already determined by the governing equation (8.1), the initial condition, and the boundary condition at $x = 0$. Since no solution exists to the over-specified problem, it is not well-posed. A consistent solution might be obtained if the correct value of $\psi(x, t)$ is imposed at $x = L$, but even in this case the solution will not depend continuously on the boundary conditions, since it will cease to exist if the downstream boundary values are perturbed, and the problem remains ill-posed.

The preceding example may seem obvious: A boundary condition is required at inflow; no condition should be imposed at outflow. Physical intuition is less likely to yield an obvious answer in the case of the one-dimensional shallow-water equations. If η is gravity times the displacement of the free surface about its equilibrium height H, and u and U are respectively the perturbation and constant basic-state fluid velocities, the linearized shallow-water equations are

$$\frac{\partial}{\partial t} \begin{pmatrix} u \\ \eta \end{pmatrix} + \begin{pmatrix} U & 1 \\ c^2 & U \end{pmatrix} \frac{\partial}{\partial x} \begin{pmatrix} u \\ \eta \end{pmatrix} = 0, \tag{8.2}$$

[1] As discussed in Section 1.3.2, a well-posed problem is one in which a unique solution to a given partial differential equation exists and depends continuously on the initial- and boundary-value data.

where $c = \sqrt{gH}$. The preceding is a homogeneous hyperbolic system of partial differential equations of the form

$$\frac{\partial \mathbf{q}}{\partial t} + \mathbf{A}\frac{\partial \mathbf{q}}{\partial x} = 0.$$

The matrices

$$\mathbf{T}^{-1} = \begin{pmatrix} 1 & -1/c \\ 1 & 1/c \end{pmatrix}, \qquad \mathbf{T} = \begin{pmatrix} 1/2 & 1/2 \\ -c/2 & c/2 \end{pmatrix}$$

may be used to transform (8.2) to

$$\frac{\partial \mathbf{T}^{-1}\mathbf{q}}{\partial t} + \mathbf{T}^{-1}\mathbf{A}\mathbf{T}\frac{\partial \mathbf{T}^{-1}\mathbf{q}}{\partial x} = 0,$$

which is the system of two decoupled scalar equations

$$\frac{\partial}{\partial t}\begin{pmatrix} d \\ e \end{pmatrix} + \begin{pmatrix} U - c & 0 \\ 0 & U + c \end{pmatrix}\frac{\partial}{\partial x}\begin{pmatrix} d \\ e \end{pmatrix} = 0 \qquad (8.3)$$

for the Riemann invariants $d = u - \eta/c$ and $e = u + \eta/c$. Each of these scalar equations is a partial differential equation of the form (8.1) and will be well-posed if a boundary value is specified at inflow and no value is specified at outflow. A well-posed shallow-water problem is therefore obtained by specifying d at the boundary through which $U - c$ is directed inward and e at the boundary where $U + c$ is directed inward. Suppose that $U > 0$ and solutions are sought on the interval $0 \le x \le L$. In the "supercritical" case $c < U$, both d and e should be specified at $x = 0$ in a manner directly analogous to the scalar advection problem.

In many geophysical applications $c > |U|$; the flow is "subcritical," and well-posed boundary conditions have the general form

$$e(0, t) = \alpha_1 d(0, t) + f_1(t), \qquad d(L, t) = \alpha_2 e(L, t) + f_2(t). \qquad (8.4)$$

The terms $f_i(t)$ represent external forcing, whereas the terms involving α_i allow information carried along the outward-directed characteristic to be incorporated in the boundary condition. The boundary condition at $x = 0$ can be rewritten as

$$(1 - \alpha_1)u(0, t) + (1 + \alpha_1)\eta(0, t)/c = f_1(t),$$

thereby demonstrating that the value of α_i determines how the forcing is apportioned between u and η. If $\alpha_i = -1$, the boundary conditions on d and e reduce to conditions on u. If $\alpha_i = 1$, the forcing determines η. When $\alpha_i = 0$, $f(t)$ specifies values for the Riemann invariants d and e.

Well-posed boundary conditions for one-dimensional hyperbolic systems with more unknowns may be determined by the same transformation procedure. Since the system is hyperbolic, the coefficient matrix of the spatial derivative term can be diagonalized by a suitable change of variables. Each component of the diagonalized system will have the form (8.1) and will require a boundary value at inflow and no value at outflow. Each positive eigenvalue of the coefficient matrix will therefore be associated with a Riemann invariant requiring a boundary value at $x = 0$, and each negative eigenvalue will necessitate the specification of a boundary value at $x = L$.

8.1.2   The Radiation Condition

Radiation boundary conditions can be easily imposed in the one-dimensional shallow-water system by transforming the equations to the diagonal form (8.3) and setting the incoming Riemann invariant to zero at each boundary. Unfortunately, this approach does not easily generalize to two-dimensional shallow-water flow. In many practical problems it is simpler to retain the velocities and the height (or pressure) field as the unknown prognostic variables and to develop open boundary conditions involving these variables.

Consider, therefore, the problem of expressing the radiation boundary condition at $x = L$ in terms of u and η instead of the Riemann invariants in a case where $c > |U|$. Since the incoming characteristic is zero at $x = L$ for all $t > 0$, $d(x, t)$ must be zero throughout the wedge-shaped region of the x-t plane defined by the inequalities $(U - c)t + L < x < L$. Thus for all $t > 0$,

$$u(x, t) = \frac{\eta(x, t)}{c} \tag{8.5}$$

in a small neighborhood of the $x = L$ boundary. In the same neighborhood of $x = L$, the equation for the outward-directed characteristic is

$$\frac{\partial}{\partial t}\left(u + \frac{\eta}{c}\right) + (U + c)\frac{\partial}{\partial x}\left(u + \frac{\eta}{c}\right) = 0. \tag{8.6}$$

Using (8.5) to eliminate η/c from the preceding equation yields the radiation boundary condition for u at $x = L$,

$$\frac{\partial u}{\partial t} + (U + c)\frac{\partial u}{\partial x} = 0. \tag{8.7}$$

An identical radiation boundary condition for η,

$$\frac{\partial \eta}{\partial t} + (U + c)\frac{\partial \eta}{\partial x} = 0, \tag{8.8}$$

can be derived using (8.5) to eliminate u from (8.6). Similar conditions may also be obtained $x = 0$; they are

$$\frac{\partial u}{\partial t} + (U - c)\frac{\partial u}{\partial x} = 0 \tag{8.9}$$

and

$$\frac{\partial \eta}{\partial t} + (U - c)\frac{\partial \eta}{\partial x} = 0. \tag{8.10}$$

The radiation boundary conditions (8.7) and (8.8) have the following alternative derivation and interpretation as one-way wave equations. The general solution for the perturbation velocity in the linearized shallow-water system may be expressed as

$$u = F_r[x - (U + c)t] + F_l[x - (U - c)t].$$

The first component of the general solution, F_r, represents a wave traveling to the right, and the second component represents a wave traveling to the left. The partial differential equation (8.7) imposed at the boundary is satisfied by solutions of the form F_r but does not admit solutions of the form F_l. All reflection at the right boundary may be eliminated by ensuring that no leftward-propagating waves are present at the boundary, and this may be achieved by imposing (8.7) and (8.8) at the right boundary. Because all the solutions to (8.7) propagate in the same direction, that equation is sometimes known as a *one-way* wave equation. One-way wave equations are particularly useful in problems involving several spatial dimensions.

8.1.3   Time-Dependent Boundary Data

In some applications it is important to allow changes in the surrounding fluid to influence the interior solution while simultaneously radiating outward-propagating waves through the boundary without reflection. Suppose that $u_e(x, t)$ and $\eta_e(x, t)$ are the velocity and the scaled free-surface displacement in a large domain containing the subdomain $0 \le x \le L$, and that information about the solution in the large domain is to be used to generate boundary conditions for a simulation of the linearized shallow-water equations inside the subdomain. It is not generally possible simply to set $u(L, t) = u_e(L, t)$ and $\eta(L, t) = \eta_e(L, t)$ without generating spurious reflections in those waves attempting to pass outward through the boundary at $x = L$. Well-posed boundary conditions are obtained by specifying the value of the incoming Riemann invariant. At the right boundary, $d(L, t)$ is set to $d_e(L, t)$, where

$$d_e(x, t) = u_e(x, t) - \eta_e(x, t)/c,$$

and at the left boundary, $e(0, t)$ is set to $e_e(0, t)$,where

$$e_e(x, t) = u_e(x, t) + \eta_e(x, t)/c.$$

As an alternative to the direct specification of the incoming Riemann invariants, the data from the large domain can be incorporated in the one-way wave equations for u and η as follows. For all $t > 0$, the value of the incoming Riemann invariant in a small neighborhood of $x = L$ will equal $d_e(x, t)$. Thus, in this neighborhood

$$d(x, t) \equiv u(x, t) - \frac{\eta(x, t)}{c} = d_e(x, t).$$

Solving the preceding for $\eta(x, t)$ and substituting in the equation for the outward-directed characteristic (8.6) yields

$$\frac{\partial u}{\partial t} + (U + c)\frac{\partial u}{\partial x} = \frac{1}{2}\left(\frac{\partial d_e}{\partial t} + (U + c)\frac{\partial d_e}{\partial x}\right),\tag{8.11}$$

or alternatively,

$$\frac{\partial u}{\partial t} + (U + c)\frac{\partial u}{\partial x} = \frac{\partial}{\partial t}\left(u_e - \frac{e_e}{2}\right) + (U + c)\frac{\partial}{\partial x}\left(u_e - \frac{e_e}{2}\right).$$

This last form of the boundary condition shows that $u(L, t)$ is equal to $u_e(L, t)$ only in those special cases where e_e is zero in the neighborhood of $x = L$, or equivalently, where there are no waves in the large-domain solution propagating away from the small domain in the neighborhood of the boundary.

If u is eliminated from (8.5), a comparable relation for the free surface displacement is obtained:

$$\frac{\partial \eta}{\partial t} + (U + c)\frac{\partial \eta}{\partial x} = -\frac{c}{2}\left(\frac{\partial d_e}{\partial t} + (U + c)\frac{\partial d_e}{\partial x}\right).\tag{8.12}$$

As in (8.11), this relation simplifies to $\eta(L, t) = \eta_e(L, t)$ only in the special case where the solution on the large domain does not contain any waves that propagate away from the small domain near the boundary at $x = L$.

8.1.4   Reflections at an Artificial Boundary— The Continuous Case

Although exact nonreflecting boundary conditions were derived in Section 8.1.2 for the one-dimensional shallow-water problem, exact formulas for nonreflecting boundary conditions are not generally available in more complicated problems. Even in the relatively simple case of two-dimensional shallow-water flow some approximation to the exact radiation boundary condition is required in order to obtain a useful relationship involving the prognostic variables and their derivatives at the lateral boundary (see Section 8.2). As a consequence of these approximations, errors are introduced in the mathematical model of the physical system before the governing equations are discretized. The errors generated by such approximations are considered in this section.

Consider the one-dimensional shallow-water equations on the domain $0 \le x \le L$ and suppose that a zero gradient condition is used to approximate the correct radiation boundary condition at the lateral boundaries. The strength of the reflections generated at the $x = 0$ boundary may be analyzed by examining the behavior of a unit-amplitude incident wave as it reflects off the boundary. Suppose that the perturbation velocity has the form

$$u(x, t) = \sin\left[k_i(x - (U - c)t)\right] + r\sin\left[k_r(x - (U + c)t) + \phi\right].$$

The first term in the preceding represents the unit-amplitude wave propagating toward the boundary at $x = 0$, and the second term represents an arbitrary reflected wave propagating away from that boundary. The amplitude, wave number, and phase of this reflected wave are determined by the boundary condition at $x = 0$. If the zero gradient condition $\partial u/\partial x = 0$ is enforced at $x = 0$, then

$$k_i \cos\left[k_i(U - c)t\right] + rk_r \cos\left[k_r(U + c)t + \phi\right] = 0. \qquad (8.13)$$

Since this equation must hold for all t, the two terms must be linearly dependent functions of time, which implies that the frequencies of the incident and reflected waves must be identical and that $\phi = 0$ (or $\phi = \pi$, which will only flip the sign of r). Equating the frequencies yields a formula for the wave number of the reflected wave

$$k_r = k_i \left(\frac{U - c}{U + c}\right).$$

This, together with (8.13), implies that the amplitude of the reflected wave is

$$r = -\frac{k_i}{k_r} = -\frac{U + c}{U - c}.$$

If $U > 0$, the wavelength and the amplitude of the reflected wave exceed those of the incident wave. After the initial reflection at $x = 0$, the reflected wave will travel across the domain and experience a second reflection at $x = L$. If a zero gradient condition is also specified at the right boundary, a similar analysis will show that the reflection coefficient at $x = L$ is

$$\hat{r} = -\frac{U - c}{U + c}.$$

After reflecting off both the left and right boundaries, the amplitude of the spurious wave will be $r\hat{r} = 1$. No energy has escaped through the lateral boundaries! Indeed, the net reflection generated by specifying $\partial u/\partial x = 0$ at both boundaries is just as strong as that obtained from a pair of rigid lateral boundaries at which $u = 0$. The same type of reflection is also obtained by specifying $\partial \eta/\partial x = 0$. Evidently, $\partial u/\partial x = 0$ and $\partial \eta/\partial x = 0$ are not acceptable substitutes for the correct nonreflecting boundary conditions (8.9) and (8.10).

8.1.5 Reflections at an Artificial Boundary— The Discretized Case

Once physically appropriate well-posed boundary conditions have been formulated, the problem is ready for numerical solution. Unfortunately, additional difficulties develop when the spatial derivatives in the continuous problem are replaced with finite differences. The first difficulty is that finite-difference formulae often require boundary data where none are specified in the well-posed continuous problem. The second difficulty is that most finite-difference formulae support nonphysical computational modes that propagate in a direction opposite to the correct physical solution and may thereby blur the distinction between inflow and outflow boundaries.

In order to examine the effects of spatial discretization, suppose that the advection equation (8.1) is approximated by the differential–difference equation

$$\frac{d\phi_j}{dt} + U\left(\frac{\phi_{j+1} - \phi_{j-1}}{2\Delta x}\right) = 0 \tag{8.14}$$

at $j = 0, \ldots, N$ discrete grid points in the finite domain $0 \le x \le L$. The preceding differential–difference equation cannot be used to calculate ϕ_0 and ϕ_N because the centered difference cannot be evaluated at a boundary. Suppose that $U < 0$, so $x = L$ is an inflow boundary. Then a boundary condition $\psi(L, t) = f(t)$ must be imposed to render the continuous problem complete and well-posed. This same inflow boundary condition may be used to specify ϕ_N, but the numerical calculation of the solution at the other boundary will be a problem. There is no boundary condition in the formulation of the nondiscretized initial-boundary-value problem that can be used to specify ϕ_0. Since $\psi(0, t)$ is determined by the interior solution, one might try to estimate $\psi(0, t)$ by extrapolating the interior solution outward to the boundary. The simplest extrapolation is

$$\phi_0 = \phi_1. \tag{8.15}$$

This, of course, is a zero-gradient condition, and based on the earlier discussion of zero-gradient conditions in the continuous shallow-water problem, one might expect it to produce reflection.

The numerical boundary condition (8.15) does indeed produce some reflection, but the situation is fundamentally different from that in the shallow-water system. When $c > |U|$, the shallow-water equations support physical modes that propagate both to the right and to the left. Specifying $\partial u/\partial x = 0$ or $\partial \eta/\partial x = 0$ generates reflection from the outgoing physical mode into the incoming physical mode because these conditions are inadequate approximations to the correct open boundary conditions in the shallow-water system. In contrast to the shallow-water system, all physical solutions to the scalar advection equation (8.1) travel in the same direction. Any reflections that occur when (8.15) is used as a boundary condition for the differential–difference equation (8.14) involve nonphysical numerical modes.

In the remainder of this section we will examine the interactions that may occur between physical and nonphysical modes as a result of the numerical approximations that are made at an open boundary. This investigation will focus on the simplest dynamical system in which such interactions occur—the problem of scalar advection. A similar, though more tedious, analysis may be performed for shallow-water flow after allowing for the fact that centered second-order finite-difference approximations to the spatial derivatives in the one-dimensional shallow-water system support two physical and two nonphysical modes. Under such circumstances, a shallow-water wave arriving at a boundary may simultaneously reflect into two inward-moving waves. If $c > |u|$, one reflected wave will be a physical mode and the other a nonphysical mode.

As discussed in Section 2.4.1, solutions of the form

$$\phi(x, t) = e^{i(kj\Delta x - \omega t)}$$

satisfy the differential–difference equation (8.14), provided that

$$\omega = U \frac{\sin k\Delta x}{\Delta x}. \tag{8.16}$$

The group velocities of these waves are given by

$$\frac{\partial \omega}{\partial k} = U \cos k\Delta x.$$

For each ω there are two different wave numbers that satisfy the dispersion relation (8.16): k_1 and $k_2 = \pi/\Delta x - k_1$. Since the group velocities of these two waves are equal in magnitude but opposite in sign, it is useful to divide the waves resolvable on the discrete grid into two categories: physical modes, for which $0 \le k\Delta x < \pi/2$, and computational modes, for which $\pi/2 \le k\Delta x \le \pi$. According to this division of the modes, a group (or wave packet) of physical modes propagates in the same direction as the true physical solution, whereas a group of computational modes propagates backwards ($4\Delta x$ waves have zero group velocity). For each ω except $U/\Delta x$ one of the roots of (8.16) is a physical mode and one is a computational mode.

In order to determine the strength of the reflection introduced by the extrapolation boundary condition (8.15), suppose that a disturbance oscillating at the frequency ω is present at the left boundary. This disturbance, being a solution of the interior differential–difference equation, must have the form

$$\left(\alpha e^{ikj\Delta x} + \beta e^{i(\pi - k\Delta x)j} \right) e^{-i\omega t}.$$

Let the first term in the preceding equation represent the incident wave (with $\alpha = 1$) and the second term the reflected wave (with $\beta = r$). Then the disturbance has the form

$$\left(e^{ikj\Delta x} + r(-1)^j e^{-ikj\Delta x} \right) e^{-i\omega t}. \tag{8.17}$$

Although the interior differential–difference equation will support a single-mode solution with $r = 0$, the boundary condition (8.15) requires nonzero amplitude in both modes. The amplitude of the reflected wave may be evaluated by substituting (8.17) into (8.15), which yields

$$1 + r = e^{ik\Delta x} - r e^{-ik\Delta x}.$$

Solving for r, one obtains

$$r = i e^{ik\Delta x} \tan \frac{k\Delta x}{2},$$

in which case

$$|r| = \left| \tan \frac{k\Delta x}{2} \right|.$$

If the incident wave is well-resolved, then $k\Delta x \ll 1$, and there is very little reflection. The magnitude of the reflection coefficient rises to unity for the $4\Delta x$ wave, and it approaches infinity for a $2\Delta x$ wave. It may appear that the large reflection coefficients associated with very short waves make the zero gradient condition unusable, but if (8.15) is applied at the outflow boundary, no waves shorter than $4\Delta x$ will ever reach that boundary because the group velocities of these short waves are directed upstream. The zero-gradient condition is therefore a possibly useful outflow boundary condition for the advection equation. On the other hand, the large reflection coefficients associated with the shortest waves do render (8.15) unsuitable as an inflow boundary condition because waves with wavelengths between $2\Delta x$ and $4\Delta x$ propagate upstream and rapidly amplify when they encounter the inflow boundary.

Nitta (1962) and Matsuno (1966a) used a similar procedure to analyze the reflections generated by a wide variety of boundary conditions. The magnitude of the reflection coefficient generated by a fixed boundary value, such as $\phi_0 = 0$, is unity. Although fixed boundary values are not appropriate at outflow, they are useful at the inflow boundary, where the only waves that encounter the boundary will be low-amplitude computational modes. Second-order extrapolation,

$$\phi_0 = 2\phi_1 - \phi_2, \tag{8.18}$$

has reflection $|r| = |\tan k\Delta x/2|^2$. The general nth-order extrapolation

$$\sum_{m=0}^{n} (-1)^m \left(\frac{n!}{m!(n-m)!} \right) \phi_m = 0$$

reflects with amplitude $|r| = |\tan k\Delta x/2|^n$. Higher-order extrapolation reduces the reflection of well-resolved physical waves, but it increases the reflection of computational modes.

As an alternative to extrapolation, one might employ a one-sided finite difference at an outflow boundary, replacing (8.14) by

$$\frac{d\phi_0}{dt} + U \left(\frac{\phi_1 - \phi_0}{\Delta x} \right) = 0. \tag{8.19}$$

The reflections generated by this scheme may once again be analyzed by substituting the general solution (8.17) into (8.19). The result is

$$-i\omega(1+r) + \frac{U}{\Delta x}(e^{ik\Delta x} - re^{-ik\Delta x}) - \frac{U}{\Delta x}(1+r) = 0.$$

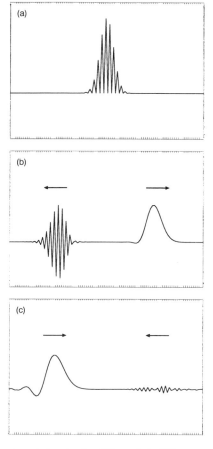

FIGURE 8.2. Approximate solution to the differential–difference equation (8.14) at time: (a) 0, (b) 0.25, and (c) 0.75.

After substituting for ω from the dispersion relation (8.16), the preceding simplifies to

$$|r| = \left| \frac{1 - \cos k\Delta x}{1 + \cos k\Delta x} \right| = \tan^2\left(\frac{k\Delta x}{2}\right).$$

Thus, the magnitude of the reflection coefficient generated by one-sided differencing is identical to that produced by second-order extrapolation.

The interaction of physical and computational modes with the lateral boundaries is illustrated in Fig. 8.2, which is patterned after an example in Trefethen (1985). The differential–difference advection equation was solved on the domain $0 \leq x \leq 1$, with $U = 1$, $\Delta x = 0.01$, and the initial distribution of ϕ_j was given by the rectified Gaussian

$$\phi_j = 0.5\left[1 + (-1)^j\right]e^{-400(x-0.5)^2}.$$

The inflow condition was $\phi_0 = 0$, and the outflow condition $\phi_N = \phi_{N-1}$.

The time derivative was approximated using trapezoidal time-differencing with a small Courant number. The initial condition is shown in Fig. 8.2a. Figure 8.2b shows the solution shortly before the disturbances encounter the lateral boundaries; the direction in which each disturbance is propagating is indicated by an arrow. The long-wave component of the solution travels toward the outflow boundary in a physically correct manner, but the $2\Delta x$ wave packet propagates upwind at speed $-U$. After the disturbances have reflected off the lateral boundaries, the solution appears as shown in Fig. 8.2c. Physical modes are reflected into computational modes by the outflow boundary condition, but since the initial disturbance is well-resolved, the reflection is relatively weak. The left-moving computational mode is reflected, without loss of amplitude, into a right-moving physical mode at the inflow boundary. The strong reflection at the right boundary is not necessarily indicative of a deficiency in the inflow boundary condition. Only computational modes lead to difficulties at the inflow boundary, and even those modes are not amplified. One can eliminate the problem at the upstream boundary by filtering the computational modes out of the interior solution.

In addition to the evaluation of the reflection coefficient, one can also examine the formal accuracy of the numerical boundary condition by substituting Taylor series expansions into the discretized formula in the usual manner (see Section 2.1). Gustafsson (1975) has shown that there is generally no degradation in the overall order of convergence if the boundary conditions are formulated with one order less accuracy than that used for the interior finite differences. In many practical problems the order of accuracy of the discretized approximation to the radiation condition is actually a relatively minor concern. The numerical errors generated at the boundary often involve reflections into poorly resolved short waves, and the accuracy achieved in the representation of such poorly resolved features is not reliably determined by the leading-order term in the truncation error. Moreover, in complex multidimensional problems the most serious errors are often introduced in the mathematical formulation of approximate radiation boundary conditions, and it is not necessary to implement these approximate conditions with highly accurate numerical formulae.

In fact, the optimal nonreflecting boundary condition for a discretized problem is not generally a high-order approximation to the exact one-way wave equation for the continuous problem, but rather the one-way wave equation whose discrete dispersion relation best replicates the propagation characteristics of the interior finite-difference scheme (Engquist and Majda 1979). As an example, suppose that solutions are sought to the one-dimensional advection equation using the semidiscrete approximation (8.14). The optimal boundary condition would be one that masks the apparent change in the propagation medium generated by the difference between the finite-difference formulae at the boundary and the interior of the domain. The dispersion relation for the one-way wave equation that will pass solutions to this semidiscrete system through the boundary with zero reflection would be $\omega = U \sin(k\Delta x)/\Delta x$ rather than the exact condition $\omega = Uk$. Unfortunately, techniques are not currently available to create a practical boundary condition with either of these dispersion relations.

8.1.6   Stability in the Presence of Boundaries

Pure initial-value, or *Cauchy*, problems are posed on infinite or periodic domains. Methods for analyzing the stability of finite-difference approximations to Cauchy problems were presented in Section 2.2. Unfortunately, the imposition of a boundary condition at the edge of a limited domain can destabilize numerical methods that are stable approximations to the Cauchy problem. In this section we will examine the additional properties that must be satisfied to ensure the stability of numerical approximations to initial-boundary-value problems (IBVP).

As in Section 2.2, the present analysis will be restricted to linear problems with constant coefficients, for which the theory is simplest and most complete. The easiest way to determine the stability of a linear constant-coefficient initial value problem on the unbounded domain $-\infty < x < \infty$ is to use Von Neumann's method, which examines the behavior of individual waves of the form

$$e^{i(kj\Delta x - \omega n \Delta t)}. \tag{8.20}$$

According to the Von Neumann criterion, a numerical method is stable if it does not amplify any of the modes resolved on the numerical mesh. Defining $\kappa = e^{ik\Delta x}$ and $z = e^{-i\omega\Delta t}$, the grid-point values associated with each wave may be written $\kappa^j z^n$. The Von Neumann stability of the Cauchy problem requires $|z| \leq 1$ for all $|\kappa| = 1$. Spatial distributions other than pure sinusoidal waves may also be represented in the form $\kappa^j z^n$. If $|\kappa| < 1$, the expression (8.20) represents an oscillation whose amplitude grows exponentially as $x \to -\infty$. There is no need to worry about the stability of this type of mode in the Cauchy problem, because it is not an admissible solution on a periodic or unbounded domain. If, however, the same equations are to be solved as an IBVP on the half-infinite domain $x \geq 0$, then the mode is admissible (since $|\kappa|^j \to 0$ as $j \to \infty$). The stability requirement for IBVP on the half-bounded domain $x \geq 0$ must therefore be extended to the *Godunov–Ryabenki* condition that $|z| \leq 1$ for all $|\kappa| \leq 1$. (If the domain is $-\infty < x \leq 0$, the modes to be examined are those for which $|\kappa| \geq 1$.)

Although the Godunov–Ryabenki condition is necessary for stability, it is not sufficient to guarantee that numerical solutions to IBVP are stable. General conditions for stability were given by Gustafsson, Kreiss, and Sundström (1972), who noted that in practice, the most important stability question involves the behavior of modes for which $|z| = 1$ and $|\kappa| = 1$. These modes are undamped waves; they can lead to instability if the interior finite-difference scheme together with the numerical boundary condition permits unforced waves to propagate inward through the boundary. Trefethen (1983) has shown that the Gustafsson–Kreiss–Sundström (GKS) stability condition is essentially equivalent to the requirement that (1) the interior difference formula be stable for the Cauchy problem; (2) the model (including the boundary conditions) admit no eigensolutions that amplify with each time step by a constant factor z with $|z| > 1$ (i.e., the Godunov–Ryabenki stability condition is satisfied), and (3) the model (including the boundary conditions) admit no unforced waves with group velocities directed inward through the boundaries of the domain.

In order to see how a numerical method might allow waves to spontaneously propagate inward through a lateral boundary, suppose that the advection equation (8.1) is to be solved on the semi-infinite domain $0 \le x \le \infty$, and that $U < 0$, so $x = 0$ is an outflow boundary. The interior solution will be approximated using leapfrog-time centered-space differencing

$$\phi_j^{n+1} - \phi_j^{n-1} + \mu(\phi_{j+1}^n - \phi_{j-1}^n) = 0, \qquad (8.21)$$

where $\mu = U \Delta t / \Delta x$. The outflow boundary condition will be determined by extrapolation, $\phi_0^n = \phi_1^n$. The numerical dispersion relation for the interior finite-difference scheme (8.21) is

$$\sin \omega \Delta t = \mu \sin k \Delta x. \qquad (8.22)$$

Substituting a wave of the form (8.20) into the extrapolation boundary condition yields

$$1 = e^{ik\Delta x}. \qquad (8.23)$$

The free modes of this finite-difference IBVP consist of those waves for which ω and k simultaneously satisfy (8.22) and (8.23), and they include the mode $(\omega, k) = (\pi/\Delta t, 0)$, whose group velocity,

$$\frac{\partial \omega}{\partial k} = U \frac{\cos k\Delta x}{\cos \omega \Delta t},$$

is $-U$. The mode $(\pi/\Delta t, 0)$ is therefore a free mode of the discretized problem that can propagate inward through the downstream boundary, and according to Trefethen's interpretation of the GKS stability criteria, the scheme should be unstable. This instability is demonstrated in Fig. 8.3, which shows a numerical integration of (8.21) on the interval $0 \le x \le 1$. The initial data, plotted in Fig. 6.3a, are random numbers with amplitudes between 0 and 0.2. The wind speed is $U = -1$, so $x = 1$ is an inflow boundary at which ϕ_{100}^n is fixed at zero. The extrapolation condition $\phi_0^n = \phi_1^n$ is applied at outflow; $\Delta x = 0.01$, and the time step is fixed such that the Courant number is 0.9. Figures 6.3b and 6.3c show the solution at two adjacent time steps after the instability has developed at the downstream boundary. As suggested by the preceding analysis, the growing mode has a period of $2\Delta t$ and a long horizontal wavelength.

In order to stabilize the numerical solution, it is necessary to change either the interior finite-difference equation or the boundary condition. Suppose the boundary condition is replaced by $\phi_0^n = \phi_1^{n-1}$, which is a backwards extrapolation in both space and time. Substitution of the wave solution (8.20) into this boundary condition yields

$$1 = e^{ik\Delta x} e^{i\omega\Delta t},$$

or

$$k\Delta x + \omega\Delta t = 2n\pi, \qquad n = 0, \pm1, \pm2, \ldots. \qquad (8.24)$$

FIGURE 8.3. Solution to (8.1) at (a) the initial time, and after (b) 530 and (c) 531 time steps.

This condition is not satisfied by the previously troublesome mode $(\omega, k) = (\pi/\Delta t, 0)$. The only resolvable modes that simultaneously satisfy (8.24) and the dispersion relation (8.22) are $(\omega, k) = (0, 0)$ and $(\pi/\Delta t, \pi/\Delta x)$, both of which have outward-directed group velocities. The numerical IBVP is therefore stable.

The numerical solution may be alternatively stabilized without modifying the original boundary condition $\phi_0^n = \phi_1^n$ if the leapfrog time difference is replaced by the trapezoidal scheme:

$$\phi_j^{n+1} - \phi_j^n + \frac{\mu}{4}\left(\phi_{j+1}^{n+1} - \phi_{j-1}^{n+1} + \phi_{j+1}^n - \phi_{j-1}^n\right) = 0.$$

The dispersion relation and group velocity for the preceding trapezoidal scheme are

$$\tan\left(\frac{\omega\Delta t}{2}\right) = \frac{\mu}{2}\sin k\Delta x \quad \text{and} \quad \frac{\partial\omega}{\partial k} = U\cos(k\Delta x)\cos^2\left(\frac{\omega\Delta t}{2}\right).$$

The only resolvable modes satisfying the boundary condition (8.23) are those for which $k\Delta x = 0$. The group velocities of all such modes have the same sign as U and are directed outward through the downstream boundary. Since the numerical scheme does not support free modes with inward-directed group velocities, it is stable.

Observe that the extrapolation condition $\phi_0^n = \phi_1^n$ would never be stable for an inflow boundary, because the group velocity of the zero-wave-number physical mode $(\omega, k) = (0, 0)$ is directed inward through the boundary, and this mode satisfies both the boundary condition and the interior finite-difference scheme. Such instability might have been anticipated from the reflection-coefficient analysis in the preceding section, which indicated that the reflection coefficient associated with zero-order extrapolation at an inflow boundary becomes infinite as $k\Delta x \to \pi$. The connection between the reflection coefficient and stability is, however, somewhat complex. Boundary conditions associated with infinite coefficients are always unstable, but GKS stability does not require $|r| \leq 1$. Further discussion of the relation between reflection coefficients and instability appears in Trefethen (1985). The stability, or lack thereof, of several basic methods for the numerical solution of the advection equation is presented in a series of examples by Goldberg and Tadmor (1985).

In most situations, two boundaries are present, and one needs stable methods for closed spatial domains such as $0 \leq x \leq L$. Gustafsson et al. (1972) showed that the stability of the two-boundary problem can be determined by analyzing each boundary separately. If the boundary condition at $x = 0$, and the interior finite-difference formulae, are GKS-stable approximations to an IBVP on the domain $0 \leq x < \infty$, and if the boundary condition at $x = L$ is also GKS stable for problems on the domain $-\infty < x \leq 0$, then the two-boundary problem is GKS stable. The basic reason that each boundary can be considered separately is that the simultaneous presence of two boundaries does not admit new types of eigensolutions beyond those whose stability was already tested in the pair of single-boundary problems. Nevertheless, GKS stability does not guarantee that the solution to the two-boundary problem will be completely satisfactory. In particular, it is possible for a method to be GKS stable but to generate reflections with $|r| > 1$. If the reflection coefficient at both boundaries exceeds unity, the solution may grow with time. Such growth is not a true instability, in the sense that it need not prevent the convergence of integrations performed over some *fixed time interval*, since in theory, the error can be made arbitrarily small in the limit $\Delta x \to 0$, $\Delta t \to 0$. Nevertheless, in most practical situations it is advisable to choose a method with $|r| \leq 1$ at both boundaries.

8.2   Two-Dimensional Shallow-Water Flow

Exact and practically useful mathematical formulae for the specification of nonreflecting boundary conditions are seldom available in multidimensional problems. The difficulties are readily apparent in two relatively simple examples: two-dimensional shallow-water flow and two-dimensional vertically stratified flow. Boundary conditions for the two-dimensional shallow-water system will be discussed in this section. Stratified flow will be considered in Section 8.3.

The two-dimensional shallow-water equations, linearized about a basic state with constant mean flow (U, V) and depth H, may be written

$$
\begin{pmatrix} u \\ v \\ \eta \end{pmatrix}_t + \begin{pmatrix} U & 0 & 1 \\ 0 & U & 0 \\ c^2 & 0 & U \end{pmatrix} \begin{pmatrix} u \\ v \\ \eta \end{pmatrix}_x + \begin{pmatrix} V & 0 & 0 \\ 0 & V & 1 \\ 0 & c^2 & V \end{pmatrix} \begin{pmatrix} u \\ v \\ \eta \end{pmatrix}_y = 0,
$$

where the notation follows that used in Section 8.1.1. Suppose that a solution is sought in the limited domain $0 \le x \le L$, $0 \le y \le L$. In contrast to the one-dimensional case, the two-dimensional system cannot be reduced to a set of three scalar equations because no transformation of variables will simultaneously diagonalize both coefficient matrices. Nevertheless, the coefficient matrix multiplying the x-derivative can be diagonalized through the same change of variables used in the one-dimensional problem. Defining $d \equiv u - \eta/c$ and $e \equiv u + \eta/c$ as before, the two-dimensional system becomes

$$
\begin{pmatrix} d \\ v \\ e \end{pmatrix}_t + \begin{pmatrix} U-c & 0 & 0 \\ 0 & U & 0 \\ 0 & 0 & U+c \end{pmatrix} \begin{pmatrix} d \\ v \\ e \end{pmatrix}_x
$$
$$
+ \begin{pmatrix} V & -c & 0 \\ -c/2 & V & c/2 \\ 0 & c & V \end{pmatrix} \begin{pmatrix} d \\ v \\ e \end{pmatrix}_y = 0.
$$

This equation is useful for determining the number of boundary conditions that should be specified at the x-boundaries. In the case $c > U > 0$, the signal in v and e is propagating inward through the boundary at $x = 0$, and the signal in d is propagating inward through the boundary at $x = L$. In order to obtain a well-posed problem, one might therefore attempt to specify two conditions at $x = 0$ of the form

$$
e(0, t) = \alpha_1 d(0, t) + f_1(t), \qquad v(0, t) = \alpha_2 d(0, t) + f_2(t),
$$

and one condition at $x = L$ of the form

$$
d(L, t) = \alpha_3 e(L, t) + \alpha_4 v(L, t) + f_3(t). \tag{8.25}
$$

This approach follows the guideline that the number of conditions specified at each boundary should be equal to the number of eigenvalues associated with outward propagation through each boundary. Even so, not all choices of $\alpha_1, \ldots, \alpha_4$ yield a well-posed problem. Oliger and Sundström (1978) and Sundström and Elvius (1979) provide details about various allowable values for $\alpha_1, \ldots, \alpha_4$. One way to obtain a well-posed problem is by choosing $\alpha_1 = \alpha_2 = \alpha_3 = \alpha_4 = 0$, which specifies e and v at inflow and d at outflow.

Unlike the one-dimensional case, it is not clear what values of e, v, and d should be prescribed to prevent waves impinging on the boundary from partially reflecting into an inward-propagating mode. Suppose that the incoming variable d is set to zero at $x = L$. Then the transformed equation for d reduces to

$$(U - c)\frac{\partial d}{\partial x} - c\frac{\partial v}{\partial y} = 0$$

along $x = L$. Since $\partial d/\partial y = 0$ along $x = L$, the preceding implies that $\partial^2 v/\partial y^2$ is also zero along $x = L$ and thus either $\ell = 0$, or v must be zero throughout the entire domain. The only nontrivial solution that is consistent with these conditions ($\ell = 0$ or $v = 0$) is a wave propagating exactly parallel to the x-axis. The condition $d(L, t) = 0$ must therefore produce spurious reflection whenever a wave strikes the boundary at some angle other than $90°$.

8.2.1 One-Way Wave Equations

Engquist and Majda (1977) suggested that an effective approximation to the true radiation boundary condition in the two-dimensional shallow-water system could be obtained using a one-way wave equation whose solutions are waves with group velocities directed outward through the boundary. In order to minimize spurious reflection at the boundary, this one-way wave equation should be designed such that its solutions approximate the outward-directed waves in the shallow-water system as closely as possible.

Consider the linearized shallow-water equations for a basic state with no mean flow, which reduce to

$$\frac{\partial^2 \eta}{\partial t^2} - c^2 \left(\frac{\partial^2 \eta}{\partial x^2} + \frac{\partial^2 \eta}{\partial y^2} \right) = 0. \tag{8.26}$$

Substituting solutions of the form

$$\eta(x, y, t) = \eta_0 e^{i(kx + \ell y - \omega t)}$$

(where k and ℓ may be positive or negative, but $\omega \geq 0$ to avoid redundancy) into (8.26) gives the shallow-water dispersion relation

$$\omega^2 = c^2 \left(k^2 + \ell^2 \right), \tag{8.27}$$

or equivalently,

$$k = \pm \frac{\omega}{c} \left(1 - \frac{c^2 \ell^2}{\omega^2} \right)^{1/2}. \tag{8.28}$$

The group velocity parallel to the x-axis is

$$c_{g_x} = \frac{\partial \omega}{\partial k} = kc\left(k^2 + \ell^2\right)^{-1/2}.$$

No plus or minus sign appears in the preceding because ω is nonnegative—the sign of the group velocity is determined by the sign of k. Spurious reflection can be eliminated at the right boundary by requiring that all waves present at $x = L$ propagate energy in the positive x direction, or equivalently, that their dispersion relation be given by the positive root of (8.28), so that

$$k = \frac{\omega}{c}\left(1 - \frac{c^2\ell^2}{\omega^2}\right)^{1/2}. \tag{8.29}$$

If ℓ were zero, (8.29) would reduce to $\omega = kc$, which is the dispersion relation associated with a one-way wave equation of the form

$$\frac{\partial \eta}{\partial t} + c\frac{\partial \eta}{\partial x} = 0. \tag{8.30}$$

This is the radiation boundary condition obtained in Section 8.1.2 for the one-dimensional shallow-water system, and it is also an exact radiation condition for two-dimensional waves propagating directly parallel to the x-axis, but it is only an approximation to the correct boundary condition for those waves that strike the boundary at nonnormal angles of incidence.

When ℓ is not zero, (8.29) ceases to be the dispersion relation for any differential equation, because it contains a square root.[2] Engquist and Majda proposed approximating (8.29) with an algebraic expression that *is* the dispersion relation for some differential equation and using that differential equation as an approximate nonreflecting boundary condition. Let

$$\gamma^2 = \frac{c^2\ell^2}{\omega^2} = \frac{\ell^2}{k^2 + \ell^2}.$$

Under the assumption that γ is small, the lowest-order approximation to the square root in (8.29) is simply

$$(1 - \gamma^2)^{1/2} \approx 1,$$

which reduces (8.29) to the dispersion relation for one-dimensional shallow-water flow and yields the boundary condition (8.30).

Engquist and Majda's second-order approximation is

$$(1 - \gamma^2)^{1/2} \approx 1 - \frac{\gamma^2}{2},$$

---

[2]The relation (8.29) is sometimes described as a dispersion relation for a *pseudodifferential* operator.

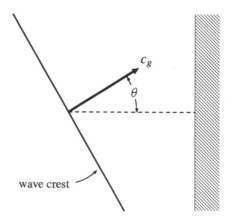

FIGURE 8.4. Wave crest approaching the "east" boundary at $x = L$.

which yields the dispersion relation

$$\omega^2 - ck\omega - \frac{c^2 \ell^2}{2} = 0.$$

The partial differential equation associated with this dispersion relation,

$$\frac{\partial^2 \eta}{\partial t^2} + c \frac{\partial^2 \eta}{\partial t \partial x} - \frac{c^2}{2} \frac{\partial^2 \eta}{\partial y^2} = 0, \tag{8.31}$$

can be imposed as an approximate radiation boundary condition at $x = L$. The well-posedness of this boundary condition was demonstrated by Engquist and Majda (1977); see also Trefethen and Halpern (1986).

The benefits associated with the use of (8.31), instead of the first-order scheme (8.30), can be assessed by computing the magnitude of the spurious reflection generated by each scheme as a function of the angle at which outward-propagating waves strike the lateral boundary. Consider the situation shown in Fig. 8.4, in which a wave is approaching the boundary at $x = L$. The lines of constant phase are parallel to the wave troughs and crests and satisfy $F(x, y) = 0$, where

$$F(x, y) = kx + \ell y - C,$$

C is a constant, and in the case shown in Fig. 8.4, k and ℓ are positive. The group velocity of the wave is both perpendicular to the lines of constant phase and parallel to the wave-number vector (k, ℓ), since

$$\left(\frac{\partial \omega}{\partial k}, \frac{\partial \omega}{\partial \ell} \right) = \frac{c^2}{\omega}(k, \ell) = \frac{c^2}{\omega} \nabla F.$$

Let θ be the angle by which the propagation of the wave deviates from the direction normal to the boundary. Then $\tan \theta = \ell/k$, and from the dispersion relation (8.27),

$$k = \frac{\omega}{c} \cos \theta, \quad \ell = \frac{\omega}{c} \sin \theta. \tag{8.32}$$

Suppose that a wave of unit amplitude in η strikes the boundary and is reflected as a wave of amplitude r. Since both the first- and second-order one-way wave equations are linear functions of η, they cannot be satisfied at $x = L$ unless the incident and reflected waves are linearly dependent functions of y and t. The frequency and the wave number parallel to the y-axis must therefore be the same in the incident and reflected waves. Then it follows from the dispersion relation (8.27) that the wave numbers parallel to the x-axis have the same magnitude and opposite sign and that the sum of the incident and reflected waves may be expressed in the form

$$\eta(x, y, t) = e^{i(kx+\ell y-\omega t)} + re^{i(-kx+\ell y-\omega t)}. \tag{8.33}$$

Here the first term represents the wave approaching the boundary, and the second term represents the reflected wave.

In order to evaluate the magnitude of the reflected wave, (8.33) may be substituted into the first-order boundary condition (8.30) to obtain

$$r = -\left(\frac{\omega - kc}{\omega + kc}\right)e^{2ikL}.$$

Using (8.32),

$$|r| = \left|\frac{w - kc}{w + kc}\right| = \left|\frac{1 - \cos\theta}{1 + \cos\theta}\right|,$$

showing that the first-order condition is perfectly nonreflecting when the waves approach the boundary along a line normal to the boundary.

The second-order Engquist and Majda boundary condition can be rewritten in the form

$$\left(\frac{\partial}{\partial t} + c\frac{\partial}{\partial x}\right)^2 \eta = 0 \tag{8.34}$$

by eliminating η_{yy} from (8.31) using the shallow-water wave equation (8.26). Substituting (8.33) into the preceding, one obtains the reflection coefficient for the second-order scheme,

$$|r| = \left|\frac{1 - \cos\theta}{1 + \cos\theta}\right|^2,$$

which is once again perfectly nonreflecting when the waves are propagating perpendicular to the boundary. Higdon (1986) observed that if the first-order Engquist–Majda boundary condition (8.30) is modified to require

$$\cos\alpha\frac{\partial\eta}{\partial t} + c\frac{\partial\eta}{\partial x} = 0,$$

the reflection coefficient becomes

$$|r| = \left|\frac{\cos\alpha - \cos\theta}{\cos\alpha + \cos\theta}\right|,$$

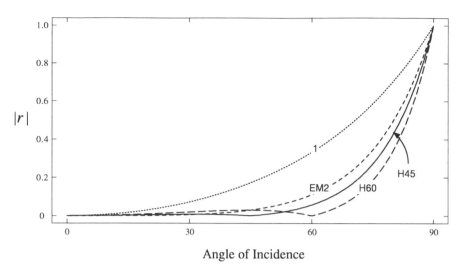

FIGURE 8.5. Dependence of the reflection coefficient on the angle of incidence (θ) for shallow-water waves in which the lateral boundary condition is given by the first-order condition (1), the second-order Engquist and Majda condition (EM2), or by Higdon's second-order formulation with perfect transmission at $\theta = 0°$ and $45°$ (H45) or $0°$ and $60°$ (H60).

which is perfectly nonreflecting for waves striking the boundary at an angle α. Higdon also noted that higher-order nonreflecting boundary conditions may be written in the form

$$\left[\prod_{j=1}^{p}\left(\cos\alpha_j\frac{\partial}{\partial t} + c\frac{\partial}{\partial x}\right)\right]\eta = 0. \tag{8.35}$$

The preceding family of schemes includes the second-order Engquist and Majda formulation (for which $p = 2$ and $\alpha_1 = \alpha_2 = 0$). The advantage of (8.35) arises from the fact that

$$|r| = \prod_{j=1}^{p}\left|\frac{\cos\alpha_j - \cos\theta}{\cos\alpha_j + \cos\theta}\right|,$$

and thus the generalized scheme is perfectly nonreflecting for waves arriving at each of the angles α_j, whereas the original Engquist and Majda formulations are only perfectly nonreflecting for waves propagating perpendicular to the boundary.

The reflectivity of several schemes is illustrated in Fig. 8.5, in which the magnitude of the reflection coefficient is plotted as a function of the angle at which the waves propagate into the boundary. The advantages of the various second-order methods over the first-order method (8.30) are clearly evident; however, the difference between the various second-order schemes is more subtle. The second-order method with perfect transmission at $\theta = 0°$ and $45°$ (H45) appears to offer the best overall performance for waves striking the boundary at angles between $0°$ and $50°$.

8.2.2   Numerical Implementation

The second-order approximation of Engquist and Majda (8.31) requires the evaluation of derivatives parallel to the boundary, which can be a problem near the corners of a rectangular domain. Engquist and Majda (1979) suggested using a first-order approximation at the corner and at the two mesh points closest to the corner. They assumed that the waves propagated into the corner along a diagonal, in which case the boundary condition at the "northeast" corner of a rectangular domain would be

$$\frac{\partial}{\partial t} + \frac{c}{\sqrt{2}}\left(\frac{\partial}{\partial x} + \frac{\partial}{\partial y}\right)\eta = 0.$$

This equation can be approximated using upstream differencing.

Higdon's higher-order boundary condition (8.35) does not involve derivatives parallel to the boundary and thereby avoids problems in the corners. Care must nevertheless be taken to ensure the stability of the numerical approximation to the higher-order derivatives that appear in (8.35) when $p \geq 2$. Define $\phi_{r,s}^n$ as the numerical approximation to $\eta(r\Delta x, s\Delta y, n\Delta t)$ and S_j^- as a shift operator with respect to the jth coordinate such that $S_x^-(\phi_{r,s}^n) = \phi_{r-1,s}^n$ and $S_t^-(\phi_{r,s}^n) = \phi_{r,s}^{n-1}$. The pth-order radiation boundary condition at the "east" boundary may be approximated as

$$\left[\prod_{j=1}^{p}\left\{\left(\frac{I - S_t^-}{\Delta t}\right) + c_j\left(\frac{I - S_x^-}{\Delta x}\right)\right\}\right]\phi_{N,s}^n = 0, \tag{8.36}$$

where N is the x index of the eastern boundary point and $c_j = c/\cos\alpha_j$. The preceding may be solved to yield a formula for $\phi_{N,s}^n$. For $p \geq 2$ there is an implicit coupling between the interior and boundary values at time $n\Delta t$. This coupling does not, however, lead to a loss of efficiency, provided that the solution on the interior points can be updated before the points on the boundary, as would be the case if an explicit finite-difference scheme were used to approximate the governing equation in the interior. Higdon (1987) has shown that (8.36) is stable when used in conjunction with a centered second-order approximation to the interior finite-difference equation of the form

$$\delta_t^2\phi - c^2\left(\delta_x^2\phi + \delta_y^2\phi\right) = 0.$$

8.3   Two-Dimensional Stratified Flow

The incompressible Boussinesq equations linearized about a reference state with a uniform horizontal wind U can be written in the form

$$\left(\frac{\partial}{\partial t} + U\frac{\partial}{\partial x}\right)u + \frac{\partial P}{\partial x} = 0, \tag{8.37}$$

$$\left(\frac{\partial}{\partial t} + U\frac{\partial}{\partial x}\right)w + \frac{\partial P}{\partial z} = b, \tag{8.38}$$

$$\left(\frac{\partial}{\partial t} + U\frac{\partial}{\partial x}\right)b + N^2 w = 0, \tag{8.39}$$

$$\frac{\partial u}{\partial x} + \frac{\partial w}{\partial z} = 0, \tag{8.40}$$

where b, P, and N^2 are defined according to (7.42). Suppose that solutions to these equations are sought in the limited domain $-L \le x \le L$ and $-H \le z \le H$ and that open boundary conditions are to be imposed at the edges of the domain. As in the two-dimensional shallow-water system, difficulties are immediately encountered in trying to formulate exact open boundary conditions for the continuous equations. Our first goal will be, therefore, to derive approximate open boundary conditions for the nondiscretized problem, after which we will consider numerical methods for using these boundary conditions in conjunction with the fully discretized equations.

8.3.1 Lateral Boundary Conditions

The chief difficulty in formulating open lateral boundary conditions for two-dimensional stratified flow is usually attributed to the dispersive nature of internal gravity waves. The incompressible Boussinesq system (8.37)–(8.40) supports solutions of the form

$$\begin{pmatrix} u \\ w \\ b \\ P \end{pmatrix} = \begin{pmatrix} \hat{u} \\ \hat{w} \\ \hat{b} \\ \hat{P} \end{pmatrix} e^{i(kx+mz-\omega t)}, \tag{8.41}$$

provided that the wave numbers and frequencies satisfy the dispersion relation

$$\omega = Uk \pm \frac{Nk}{(k^2 + m^2)^{1/2}}. \tag{8.42}$$

These waves are dispersive, since their phase speed $\omega(k^2 + m^2)^{-1/2}$ is a function of the spatial wave numbers k and m. (See Whitham 1974 for further details on dispersive waves.) The difficulties that arise in formulating open lateral boundary conditions for these waves are, however, very similar to those encountered in formulating boundary conditions for nondispersive shallow-water waves in two dimensions. Those difficulties do not arise from wave dispersion per se, but rather from the fact that the x trace speed[3] for the outward-propagating wave is a function of k and m that cannot be manipulated to form the exact dispersion relation for a partial differential equation.

---

[3]The x trace speed, ω/k, is the apparent phase speed of the wave parallel to the x-axis. Unless the wave is propagating parallel to the x-axis, the x trace speed exceeds the true phase speed $\omega(k^2 + m^2)^{-1/2}$.

In order to determine the dispersion relation governing the outward-propagating wave, note that the horizontal group velocity for internal gravity waves is

$$\frac{\partial \omega}{\partial k} = U \pm \frac{Nm^2}{(m^2 + k^2)^{1/2}}. \tag{8.43}$$

Temporarily assume that $U = 0$; then the mode associated with the positive root in (8.42) will have positive group velocity, and a perfect open boundary condition at $x = L$ will require that all waves present at the boundary satisfy the dispersion relation

$$\omega = \frac{Nk}{(k^2 + m^2)^{1/2}}. \tag{8.44}$$

Since the preceding is not the dispersion relation for a partial differential equation, it cannot be used to express the open boundary condition at $x = L$ in terms of the physical variables on the computational mesh. In order to obtain an approximate open boundary condition, define $\tilde{c}$ as the x trace speed, ω/k. If the dependence of $\tilde{c}$ on m and k could be neglected, the dispersion relation for the correct open boundary condition would become $\omega = \tilde{c}k$, which is the dispersion relation for the familiar one-way wave equation

$$\left(\frac{\partial}{\partial t} + \tilde{c} \frac{\partial}{\partial x} \right) \psi = 0. \tag{8.45}$$

Of course, the dependence of $\tilde{c}$ on m and k cannot be properly ignored, and as a consequence, (8.45) is only an approximate open boundary condition that will induce spurious reflection in all waves for which

$$\tilde{c} \neq \frac{N}{(k^2 + m^2)^{1/2}}.$$

In most practical applications several different waves are simultaneously present at the boundary, and no single value of $\tilde{c}$ will correctly radiate all of the waves.

Considerable effort has, nevertheless, been devoted to devising estimates for $\tilde{c}$ that minimize the reflections generated by (8.45). In the hydrostatic limit, which often applies in geophysical problems, $k^2 \ll m^2$ and $\tilde{c} \to N/m$, so the task of estimating $\tilde{c}$ reduces to that of estimating the vertical wave number of those modes striking the boundary. Pearson (1974) suggested using a fixed value of $\tilde{c}$ equal to the x trace speed of the dominant vertical mode. As an alternative, Orlanski (1976) suggested calculating $\tilde{c}$ at a point just inside the boundary using the relation

$$\tilde{c} = -\frac{\partial \psi/\partial t}{\partial \psi/\partial x}. \tag{8.46}$$

The results of this calculation must be limited to values in the interval $0 \leq \tilde{c} \leq \Delta x/\Delta t$ in order to preserve the stability of the commonly used upstream approximation to (8.45). Orlanski's approach has the virtue of avoiding the specification of arbitrary parameters, but it amounts to little more than an extrapolation procedure, since the same equation is applied at slightly different locations on the space–time grid to determine both $\tilde{c}$ and $\partial \psi / \partial t$. Moreover, (8.45) has no a priori validity in the continuously stratified problem, since there is generally no correct value of $\tilde{c}$ to diagnose via (8.46).

Durran et al. (1993) investigated the effectiveness with which (8.46) diagnosed the phase velocity of numerically simulated shallow-water waves, for which $\tilde{c}$ has the well-defined value of $\sqrt{gH}$. Except during the passage of a trough or crest, a simple finite-difference approximation to (8.46) proved capable of diagnosing a reasonable approximation to $\sqrt{gH}$ in a control simulation on a very large periodic domain in which all waves propagating past the point of the calculation were really traveling in the positive x direction. Attempts to perform the same diagnostic calculation for $\tilde{c}$ in conjunction with the imposition of a radiation boundary condition in a second simulation were, however, a complete failure. Small errors in the initial diagnosis of $\tilde{c}$ generated weak reflected waves. These reflected waves generated increasing errors in the calculation of $\tilde{c}$ because (8.45) does not apply at locations where both rightward- and leftward-propagating waves are present. The additional error in $\tilde{c}$ increased the amplitude of the spurious reflected waves and induced a positive feedback that rapidly destroyed the reliability of the $\tilde{c}$ calculation. The majority of the values computed from (8.46) were outside the stability limits for the upstream method and had to be reset to either zero or $\Delta t / \Delta x$. Durran et al. (1993) also considered tests in which the physical system supported several different modes moving at different trace speeds and concluded that it is best to use a fixed $\tilde{c}$.

When the mean horizontal wind is not zero, $\tilde{c}$ is replaced by the Doppler-shifted trace speeds $U + \tilde{c}$ and $U - \tilde{c}$ at the upstream and downstream boundaries, respectively. One should also ensure that these Doppler-shifted trace speeds are actually directed out of the domain by choosing $|\tilde{c}| > |U|$ at the inflow boundary. In some applications the dominant upstream-propagating mode may have a different intrinsic trace speed from the dominant downstream-propagating mode, and in such circumstances it can be useful to specify the Doppler-shifted trace speeds as $U + \tilde{c}_1$ and $U - \tilde{c}_2$ without requiring $\tilde{c}_1 = \tilde{c}_2$.

Higdon (1994) has suggested that an improved approximation to the radiation boundary condition for dispersive waves can be obtained by replacing the basic one-way wave equation (8.45) with the product of a series of one-way operators of the form

$$\left[\prod_{j=1}^{p} \left(\frac{\partial}{\partial t} + \tilde{c}_j \frac{\partial}{\partial x} \right) \right] \psi = 0, \tag{8.47}$$

where the set of $\tilde{c}_j$ is chosen to span the range of x trace speeds associated with the waves appearing at $x = L$. The preceding is a generalization of Higdon's radiation boundary condition for shallow-water waves (8.35). Although this scheme does not seem to have been used as a boundary condition for limited-area models of stably stratified flow, Higdon has successfully used to it simulate dispersive shallow-water waves on an f-plane.

The spurious reflection generated by the Higdon boundary condition can be determined by considering a unit-amplitude wave striking the boundary at $x = L$.

In order to satisfy (8.47) exactly, a reflected wave of amplitude r must also be present, in which case the total solution may be expressed as

$$\psi(x, z, t) = e^{i(kx+mz-\omega t)} + r e^{i(\tilde{k}x+\tilde{m}z-\tilde{\omega}t)}.$$

The frequencies and vertical wave numbers of the incident and reflected waves must be identical, or the preceding expression will not satisfy (8.47) for arbitrary values of z and t. Since $\tilde{m} = m$ and $\tilde{\omega} = \omega$, the dispersion relation implies that $\tilde{k} = -k$. Using these results to substitute

$$\psi(x, z, t) = e^{i(kx+mz-\omega t)} + r e^{i(-kx+mz-\omega t)}$$

into (8.47) and defining $c = \omega/k$, one obtains

$$|r| = \prod_{j=1}^{p} \left| \frac{c - c_j}{c + c_j} \right|.$$

If the waves are hydrostatic, $c = N/m$, and the reflection coefficient may be alternatively expressed as

$$|r| = \prod_{j=1}^{p} \left| \frac{m - m_j}{m + m_j} \right|, \tag{8.48}$$

where m_j is the vertical wave number of a hydrostatic wave moving parallel to the x-axis at speed c_j. The reflections generated by the Higdon boundary condition are plotted as a function of the vertical wave number of the incident wave in Fig. 8.6. The waves are assumed hydrostatic, so that the reflection coefficient is determined by (8.48). Perfect transmission is achieved whenever the vertical wave number of the incident wave matches one of the m_j. When the perfectly transmitted waves in the two-operator scheme are chosen such that $m_2 = 3m_1$, the range of vertical wave numbers that are transmitted with minimal reflection is much larger than that obtained using the standard one-way wave equation. The range of wave numbers that undergo minimal spurious reflection can be further increased by using the three-operator scheme, which was configured such that $m_2 = 3m_1$ and $m_3 = 9m_1$ in the example plotted in Fig. 8.6.

The Higdon boundary condition can be implemented in the numerical approximations to the momentum and buoyancy equations (8.37)–(8.39) using the finite-difference formula (8.36). The numerical formulation of the boundary condition for pressure is less obvious, because (8.36) is implicit whenever $p \geq 2$. This implicitness need not reduce the efficiency of the numerical integration when the solution in the interior can be updated prior to the evaluation of the boundary condition, as will be the case if (8.37)–(8.39) are approximated using explicit time-differencing. The pressure, however, is evaluated by solving a Poisson equation, and the derivation of a stable pressure boundary condition for the Poisson equation from (8.36) may be nontrivial. An alternative approach for formulating the pressure boundary condition is investigated in Problems 8 and 9.

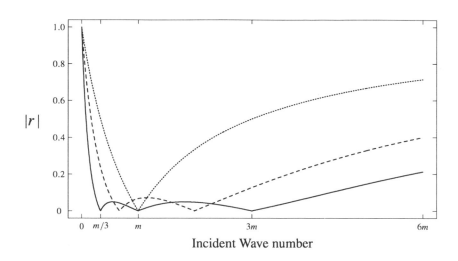

FIGURE 8.6. Reflection coefficients generated by the Higdon lateral boundary condition (8.47) as a function of the incident vertical wave number for p equal to one (dotted curve), two (dashed), and three (solid). The waves are assumed to be hydrostatic.

8.3.2   Upper Boundary Conditions

Open upper or lower boundary conditions for the incompressible Boussinesq equations cannot be formulated in the same manner as the lateral boundary conditions because the vertical trace speed of an internal gravity wave is frequently directed opposite to its vertical group velocity. The difference in the direction of the vertical trace speed and group velocity is particularly evident when the waves are hydrostatic and $U = 0$. In such circumstances the dispersion relation simplifies to $\omega = \pm Nk/m$; the vertical trace speed is

$$\frac{\omega}{m} = \pm \frac{Nk}{m^2},\tag{8.49}$$

and the vertical group velocity is

$$\frac{\partial \omega}{\partial m} = \mp \frac{Nk}{m^2},\tag{8.50}$$

which is equal in magnitude to the trace speed but opposite in sign.

Let us attempt to follow the same procedure used in the preceding section to obtain an approximate open boundary condition at the upper boundary. As a first step, the correct dispersion relation is approximated as $\omega = \tilde{c}m$, where $\tilde{c}$ is a constant equal to the estimated vertical trace speed of the dominant mode in the physical system. The relation $\omega = \tilde{c}m$ is the exact dispersion relation for a one-way wave equation of the form

$$\left(\frac{\partial}{\partial t} + \tilde{c}\frac{\partial}{\partial z}\right)\psi = 0. \tag{8.51}$$

In order to avoid reflections from the upper boundary, the value of $\tilde{c}$ used in this one-way wave equation must be the trace speed for a mode whose group velocity is directed upward. According to (8.49) and (8.50), any mode with a positive group velocity will have a negative trace speed, so all the acceptable values for $\tilde{c}$ are negative. Yet only one-sided numerical approximations to the spatial derivative can be evaluated at the upper boundary, and these one-sided differences yield unstable finite-difference approximations to (8.51) whenever $\tilde{c} < 0$, because the numerical domain of dependence does not include the domain of dependence of the true solution.

Boundary Conditions That Are Nonlocal in Space, but Local in Time

The problems with the one-way wave equation (8.51) can be avoided by following the strategy of Klemp and Durran (1983) and Bougeault (1983), in which a radiation upper boundary condition is imposed through a diagnostic relationship between the pressure and the vertical velocity at the top boundary of the domain. This relationship is derived by substituting a wave of the form (8.41) into the horizontal momentum equation (8.37) to yield

$$-i(\omega - Uk)\hat{u} + ik\hat{P} = 0. \tag{8.52}$$

Substituting the same wave into the continuity equation (8.40), one obtains

$$ik\hat{u} + im\hat{w} = 0.$$

Eliminating $\hat{u}$ between the two preceding equations gives

$$\hat{P} = -\left(\frac{\omega - Uk}{k}\right)\left(\frac{m}{k}\right)\hat{w}. \tag{8.53}$$

After making the hydrostatic approximation, the dispersion relation (8.42) can be written

$$\frac{\omega - Uk}{k} = \pm\frac{N}{m}.$$

The vertical group velocity for hydrostatic waves, which is given by (8.50), will depend on both the sign of k and the choice of the positive or negative root. The choice of sign required to limit the dispersion relation to those waves with upward group velocity is

$$\frac{\omega - Uk}{k} = -\text{sgn}(k)\frac{N}{m}. \tag{8.54}$$

Substituting the preceding into (8.53) yields

$$\hat{P} = \frac{N}{|k|}\hat{w}. \tag{8.55}$$

This relationship between $\hat{P}$ and $\hat{w}$ is nonlocal, in the sense that it cannot be imposed as an algebraic or differential relation involving P and w on the physical mesh.

The one-way dispersion relations (8.44) and (8.54) are also nonlocal formulae that might be useful as radiation boundary conditions if it were possible to transform the values of $w(x, z, t)$ on the computational mesh to and from the space of dual variables $\hat{w}(k, m, \omega)$. There is, however, no way to determine the frequency dependence of the dual variables, because the grid-point data are never simultaneously available at more than one or two time levels, and as a consequence, (8.44) and (8.54) have no direct practical utility. In contrast, the nonlocal condition (8.55) can be easily used in practical computations, because neither ω nor m appear in that formula. All that is required to use this boundary condition in a numerical model with periodic lateral boundaries is to compute the Fourier transform of the w values along the top row of the computational mesh, evaluate $\hat{P}$ using (8.55), and then obtain the values of P along the top row of the computational mesh from an inverse Fourier transform. Further details about the numerical implementation of this boundary condition are provided in Section 8.3.3.

The boundary condition (8.55) perfectly transmits linear hydrostatic waves through the upper boundary, but nonhydrostatic waves will be partially reflected back into the domain. The strength of the partial reflection can be determined as follows. Since the vertical group velocity obtained without making the hydrostatic approximation is

$$\frac{\partial \omega}{\partial m} = \mp \frac{Nkm}{(k^2 + m^2)^{3/2}},$$

the gravity-wave dispersion relation can be limited to those waves with upward group velocities by taking the negative root in (8.42) when $\text{sgn}(k) = \text{sgn}(m)$ and the positive root when $\text{sgn}(k) = -\text{sgn}(m)$. In both cases, using the dispersion relation for the wave with upward group velocity to substitute for $(\omega - Uk)/k$ in (8.53) yields the correct radiation condition for nonhydrostatic waves,

$$\hat{P} = \frac{N}{|k|} s \hat{w}, \tag{8.56}$$

where

$$s = \frac{|m|}{(k^2 + m^2)^{1/2}}.$$

A similar derivation shows that the pressure and vertical velocity in downward-propagating waves are correlated such that

$$\hat{P} = -\frac{N}{|k|} s \hat{w}. \tag{8.57}$$

Now suppose that a nonhydrostatic wave of the form (8.41) encounters the top boundary. In order to satisfy the hydrostatic open boundary condition, a second downward-propagating wave with the same horizontal wave number must also be present at the upper boundary. Letting the subscripts 1 and 2 denote the incident and reflected waves, respectively, (8.55) becomes

$$(\hat{P}_1 + \hat{P}_2) = \frac{N}{|k|}(\hat{w}_1 + \hat{w}_2). \tag{8.58}$$

The horizontal momentum equation (8.52), which is also satisfied at the upper boundary, implies that both of these waves have the same frequency because they have the same horizontal wave number. Since both waves have the same ω and k, the dispersion relation requires their vertical wave numbers to differ by a factor of -1, and as a consequence, s is identical for both waves. Substituting for $\hat{P}_1$ and $\hat{P}_2$ from (8.56) and (8.57), respectively, (8.58) becomes

$$s(\hat{w}_1 - \hat{w}_2) = \hat{w}_1 + \hat{w}_2.$$

Let r be the ratio of the vertical-velocity amplitude in the reflected wave to that in the incident wave; then the preceding implies that

$$|r| = \left| \frac{s-1}{s+1} \right|.$$

In the hydrostatic limit, $s \to 1$ and there is no reflection. The reflection increases as the waves become less hydrostatic, but even when $k = m$, the reflection coefficient remains a relatively modest 0.17.

In fully three-dimensional problems (8.55) generalizes to

$$\hat{P} = \frac{N}{\sqrt{k^2 + \ell^2}}\hat{w}, \tag{8.59}$$

where ℓ is the wave number parallel to the y-axis. This boundary condition is evaluated in the same manner as that for the two-dimensional problem (8.55), except that two-dimensional Fourier transforms are computed with respect to the x and y coordinates along the top boundary of the domain. Extensions of the preceding approach to include the effects of a constant Coriolis force were suggested by Garner (1986). Rasch (1986) provides a further generalization suitable for both gravity and Rossby waves.

Boundary Conditions That Are Local in Both Space and Time

The relations (8.55) and (8.59) are best suited to problems in laterally periodic domains, for which the horizontal wave numbers present on the numerical mesh can be computed exactly using fast Fourier transforms. These boundary conditions can also be used in conjunction with open lateral boundaries, but some type of periodic completion must be assumed to allow the computation of the Fourier transforms. A simple assumption of false periodicity in the w and p fields at the topmost level usually gives adequate results when the main disturbance is located in the central portion of the domain. Nevertheless, this assumption introduces a modest erroneous coupling between the upstream and downstream boundaries.

The use of Fourier transforms and the assumption of false periodicity can be eliminated by approximating the factor of $|k|$ in (8.55) with an algebraic expression that converts (8.55) into a dispersion relation for a partial differential equation that can be solved on the physical mesh. Let $|k|$ be replaced by the rational function

$$\frac{a_1 + a_2 k^2}{1 + a_3 k^2},\tag{8.60}$$

where a_1, a_2, and a_3 are constants chosen to ensure that (8.60) is a good approximation to $|k|$ over the entire range of wave numbers that need to be transmitted through the upper boundary. It is convenient to express the arbitrary constants a_1, a_2, and a_3 in terms of the three wave numbers, k_1, k_2, and k_3, for which (8.60) would be exactly equal to $|k|$. Then

$$a_1 = \frac{k_1 k_2 k_3}{D}, \qquad a_2 = \frac{k_1 + k_2 + k_3}{D}, \qquad a_3 = \frac{1}{D},$$

where

$$D = k_1 k_2 + k_1 k_3 + k_2 k_3.$$

Numerical tests have suggested that an effective strategy for choosing appropriate values for k_1, k_2, and k_3 (and thereby specifying a_1, a_2, and a_3) is to let k_3 be the largest horizontal wave number likely to appear with significant amplitude in the perturbations at the upper boundary and then choose $k_2 = k_3/3$ and $k_1 = k_3/9$. Replacing $|k|$ by (8.60) in (8.55) and taking an inverse Fourier transform yields the local differential equation

$$\left(a_1 - a_2 \frac{\partial^2}{\partial x^2} \right) P = N \left(1 - a_3 \frac{\partial^2}{\partial x^2} \right) w.\tag{8.61}$$

The reflection generated by (8.61) in a vertically propagating hydrostatic gravity wave of horizontal wave number k is

$$|r| = \left| \frac{k - k_1}{k + k_1} \right| \left| \frac{k - k_2}{k + k_2} \right| \left| \frac{k - k_3}{k + k_3} \right|.$$

This reflection coefficient has the same form as that for the three-operator Higdon scheme, except that the wave number parallel to the boundary is k in the preceding and is m in (8.48). The dependence of r on the horizontal wave number of an incident hydrostatic wave is given by the solid curve in Fig. 8.6, except that the m's appearing in the labels for the horizontal axis should be replaced with k's. Recall, however, that Higdon's one-way wave equation is not suitable for use at the upper boundary because the phase speed and group velocity of internal gravity waves are generally opposite in sign. Equation (8.61) provides an alternative that avoids instability while achieving the same degree of wave transmission through the upper boundary.

The preceding local boundary condition is generalized to three-dimensional problems by approximating the factor $\sqrt{k^2 + \ell^2}$ in (8.59) with

$$\frac{a_1 + a_2(k^2 + \ell^2)}{1 + a_3(k^2 + \ell^2)} \, .$$

The approximate radiation upper boundary condition that results is

$$\left[a_1 - a_2 \left(\frac{\partial^2}{\partial x^2} + \frac{\partial^2}{\partial y^2} \right) \right] P = N \left[1 - a_3 \left(\frac{\partial^2}{\partial x^2} + \frac{\partial^2}{\partial y^2} \right) \right] w.$$

8.3.3 Numerical Implementation of the Radiation Upper Boundary Condition

First consider the nonlocal formulation (8.55) and suppose that the numerical solution of the Boussinesq system is to be obtained using the projection method on the staggered mesh shown in Fig. 3.6. An upper boundary condition can be obtained for the Poisson equation (7.9) as follows. Fourier transform w^n along the top computational level of the domain; compute the Fourier coefficients for pressure from the relation $\hat{p}^n = N\rho_0 \hat{w}^n / |k|$; and inverse transform to obtain p^n at a level $\Delta z/2$ above the uppermost row of p-points.[4] These values of p^n are used as approximations to $\tilde{p}^{n+1}$ along the boundary and provide a Dirichlet boundary condition for (7.9). Ideally, one would formulate the Dirichlet boundary condition using an exact expression for the boundary values of $\tilde{p}^{n+1}$, but recall that $\tilde{p}^{n+1}$ does not represent the actual pressure at any given time. Approximating $\tilde{p}^{n+1}$ with p^n has yielded satisfactory results in tests conducted by this author. After solving the Poisson equation, the interior velocities are updated from (7.8). As a final step, the w^{n+1} along the upper boundary can be obtained from the non-divergence condition, or by linear extrapolation from below. In most applications, the mean vertical velocity is zero, and no other upper boundary conditions are required to obtain numerical solutions to the linearized equations.

As noted by Bougeault (1983), stability considerations require that (8.55) be implemented in a compressible model using implicit time-differencing. As an example, suppose that solutions are to be obtained using the partial time-split approximation to the "compressible" Boussinesq equations (7.82)–(7.85) on the staggered grid shown in Fig. 3.6 and that N is the vertical index of the topmost row of pressure and buoyancy points. The boundary condition can be cast into an implicit expression for $P_{N-\frac{1}{2}}$, which represents the pressure at the same vertical level as $w_{N-\frac{1}{2}}$. A vertically discretized approximation to (7.83) at level $N - \frac{1}{2}$

---

[4]Here P is replaced by p/ρ_0 to match the terminology in Section 7.3.2.

may be written as

$$\frac{w_{N-\frac{1}{2}}^{m+1} - w_{N-\frac{1}{2}}^{m}}{\Delta \tau} + \frac{1}{2}\left(\frac{P_{N-\frac{1}{2}}^{m+1} - P_{N-1}^{m+1}}{\Delta z/2} + \frac{P_{N}^{m} - P_{N-1}^{m}}{\Delta z}\right) = b_{N-\frac{1}{2}}^{m} - U\frac{\partial w_{N-\frac{1}{2}}^{n}}{\partial x},$$

(8.62)

where m and n are the indices for the small and large time steps, respectively. The vertically discretized pressure equation for grid level $N - 1$ is

$$\frac{P_{N-1}^{m+1} - P_{N-1}^{m}}{\Delta \tau} + \frac{c_s^2}{2}\left(\frac{w_{N-\frac{1}{2}}^{m+1} - w_{N-\frac{3}{2}}^{m+1}}{\Delta z} + \frac{w_{N-\frac{1}{2}}^{m} - w_{N-\frac{3}{2}}^{m}}{\Delta z}\right) =$$

$$- c_s^2 \frac{\partial u_{N-1}^{m+1}}{\partial x} - U\frac{\partial P_{N-1}^{n}}{\partial x},$$

which can be used to substitute for P_{N-1}^{m+1} in (8.62) to obtain

$$\left(1 + \frac{\tilde{c}^2}{2}\right)w_{N-\frac{1}{2}}^{m} - \frac{\tilde{c}^2}{2}w_{N-\frac{3}{2}}^{m+1} = F_{N-\frac{1}{2}} - \frac{\Delta \tau}{\Delta z}P_{N-\frac{1}{2}}^{m+1},$$

(8.63)

where $\tilde{c} = c_s \Delta \tau / \Delta z$ and $F_{N-\frac{1}{2}}$ is the sum of all those terms that can be explicitly evaluated at this stage of the integration cycle. Equation (8.63) closes the tridiagonal system for w^{m+1} generated by the implicit coupling between the discretized pressure and vertical momentum equations throughout each vertical column in the interior of the domain. After the forward elimination sweep of the tridiagonal solver described in Appendix A.2.1,

$$w_{N-\frac{1}{2}}^{m+1} = \gamma - \beta P_{N-\frac{1}{2}}^{m+1},$$

(8.64)

where

$$\beta = p\Delta \tau / \Delta z, \qquad \gamma = p\left(F_{N-\frac{1}{2}} + \tilde{c}^2 f/2\right),$$

p is as defined at the last iteration in the loop ($j = jmx$) and f is the array element $f(jmx - 1)$. Taking the Fourier transform of (8.64) and using (8.55), the radiation condition becomes

$$\hat{P}_{k,N-\frac{1}{2}}^{m+1} = \left(\frac{|k|}{N} + \beta\right)^{-1}\hat{\gamma}_k,$$

which allows the computation of $\hat{P}_{N-\frac{1}{2}}^{m+1}$ from the Fourier transform of γ. After inverse transforming to obtain $P_{N-\frac{1}{2}}^{m+1}$, the computation of $w_{N-\frac{1}{2}}^{m+1}$ is completed using (8.64), and the remaining w^{m+1} are updated during the backward pass of the tridiagonal solver. The P^{m+1} in the interior of the domain are updated using these w^{m+1}, and finally, P_N^{m+1} is computed from $P_{N-\frac{1}{2}}^{m+1}$ and P_{N-1}^{m+1} by linear extrapolation.

The approximate local radiation condition (8.61) is implemented in essentially the same manner as the nonlocal condition (8.55), except that instead of Fourier transforming w, applying (8.55), and then inverse transforming to obtain the boundary values for the pressure, (8.61) is solved to compute the grid-point values of P directly from the grid-point values of w. A tridiagonal system for the grid-point values of P along the top boundary is obtained when the second derivatives in (8.61) are approximated by a standard three-point finite difference. Special conditions are required at those points adjacent to the lateral boundaries, where it can be advantageous to use the less accurate approximation

$$\left(b_1 - b_2 \frac{\partial^2}{\partial x^2}\right) P = Nw.$$

This relation, which is derived from (8.55) using the approximation $|k| \approx b_1 + b_2 k^2$, does not require any assumption about the horizontal variation of w at the lateral boundaries. Some assumption is nevertheless required about the variation of P near the boundary, and satisfactory results have been obtained by setting $\partial P / \partial x$ to zero in the upper corners of the domain.

8.4   Wave-Absorbing Layers

One way to prevent outward-propagating disturbances from reflecting back into the domain when they encounter the boundary is to place a wave-absorbing layer at the edge of the domain. Wave-absorbing layers are conceptually simple and are particularly attractive in applications for which appropriate radiation boundary conditions have not been determined. Wave-absorbing layers also allow "large-scale" time tendencies to be easily imposed at the lateral boundaries of the domain. These large-scale tendencies might be generated by a previous or concurrent coarse-resolution simulation on a larger spatial domain. The chief disadvantage of the absorbing-layer approach is that the absorber often needs to be rather thick in order to be effective, and significant computational effort may be required to compute the solution on the mesh points within a thick absorbing layer. Considerable engineering may also be required to ensure that a wave-absorbing layer performs adequately in a given application.

Suppose that a radiation upper boundary condition for the two-dimensional linearized Boussinesq equations (8.37)–(8.40) is to be approximated using a wave-absorbing layer of thickness D. This absorbing layer can be created by defining a vertically varying viscosity $\alpha(z)$ and adding viscous terms of the form $\alpha(z)\partial^2 u/\partial x^2$, $\alpha(z)\partial^2 w/\partial x^2$, and $\alpha(z)\partial^2 b/\partial x^2$ to the right sides of (8.37), (8.38), and (8.39), respectively. The viscosity is zero in the region $z \leq H$ within which the solution is to be accurately approximated and increases gradually with height throughout the layer $H < z \leq H + D$. A simple rigid-lid condition, $w = 0$, can be imposed at the top of the absorbing layer. The performance of this absorbing layer is largely determined by the vertical profile of the artificial viscosity $\alpha(z)$ and the total absorbing-layer depth D. The total viscosity in the absorbing layer must be sufficient to dissipate a wave before it has time to propagate upward through the absorbing layer, reflect off the rigid upper boundary, and travel back down through the depth of the layer. It might, therefore, appear advantageous simply to set α to the maximum value permitted by the stability constraints of the finite-difference scheme. Reflections will also occur, however, when a wave encounters a rapid change in the propagation characteristics of its medium, and as a consequence, reflection will be produced if the artificial viscosity increases too rapidly with height. The only way to make the total damping within the wave absorber large while keeping the gradient of $\alpha(z)$ small is to use a relatively thick wave-absorbing layer.

The reflectivity of a wave-absorbing layer also depends on the characteristics of the incident wave. Klemp and Lilly (1978) examined the reflections produced by a wave-absorbing layer at the upper boundary in a problem where hydrostatic vertically propagating gravity waves were generated by continuously stratified flow over topography. They found that although a very thin absorbing layer could be tuned to efficiently remove a single horizontal wave number, considerably deeper layers were required to uniformly minimize the reflection over a broad range of wave numbers. In order to ensure that the absorbing layer was sufficiently deep, and to guarantee that the numerical solution was adequately resolved within the wave-absorbing layer, Klemp and Lilly devoted the entire upper half of their computational domain to the absorber. The efficiency of their numerical model could have been increased by a factor of two if the wave-absorbing layer had been replaced with the radiation upper boundary condition described in the preceding section.

The finding that effective wave-absorbing layers must often be rather thick is also supported by Israeli and Orszag (1981), who examined both viscous and Rayleigh-damping absorbing layers for the linearized shallow-water system of the form

$$\frac{\partial u}{\partial t} + \frac{\partial \eta}{\partial x} = \nu(x)\frac{\partial^2 u}{\partial x^2} - R(x)u, \tag{8.65}$$

$$\frac{\partial \eta}{\partial t} + c^2\frac{\partial u}{\partial x} = 0 \tag{8.66}$$

on the domain $-L \leq x \leq L$. Boundary conditions were specified for $u(-L, t)$ and $u(L, t)$; no boundary conditions were specified for η. Since there is one inward-directed characteristic at each boundary, the specification of u at each boundary yields a well-posed problem for all nonnegative ν. Numerical solutions to the preceding system can be conveniently obtained without requiring numerical boundary conditions for η by using a staggered mesh where the outermost u points are located on the boundaries and the outermost η points are $\Delta x/2$ inside those boundaries (see Section 3.1.2). Israeli and Orszag demonstrated that better results could be obtained using Rayleigh damping ($R > 0$, $\nu = 0$) than by using viscous damping ($v > 0$, $R = 0$) because the erroneous backward reflection induced by the Rayleigh damping is less scale dependent than that generated by viscous damping. Israeli and Orszag also suggested that superior results could be obtained by using a wave-absorbing layer in combination with a one-way wave equation at the actual boundary. When using both techniques in combination, the one-way wave equation must be modified to account for the dissipation near the boundary. For example, an approximate one-way wave equation for the right-moving wave supported by (8.65) and (8.66) with $v = 0$ and R constant is

$$\frac{\partial u}{\partial t} + c\frac{\partial u}{\partial x} = -\frac{R}{2}u. \tag{8.67}$$

The problem considered by Israeli and Orszag is somewhat special in that it can be numerically integrated without specifying any boundary conditions for η. If a mean current were present, so that the unapproximated linear system is described by (8.2), then numerical boundary conditions would also be required for η in order to evaluate $U\partial\eta/\partial x$. The specification of η at the boundary where the mean wind is directed inward, together with specification of u at each boundary, will lead to an overdetermined problem. Nevertheless, Davies (1976, 1983) has suggested that wave-absorbing layers can have considerable practical utility even when they require overspecification of the boundary conditions.

As a simple example of overspecification, consider the one-dimensional scalar advection equation with $U > 0$. Davies's absorbing boundary condition for the outflow boundary corresponds to the mathematical problem of solving

$$\frac{\partial \psi}{\partial t} + U\frac{\partial \psi}{\partial x} = -R(x)(\psi - \psi_b) \tag{8.68}$$

on the domain $-L \leq x \leq L$ subject to the boundary conditions

$$\psi(-L, t) = s(t), \qquad \psi(L, t) = \psi_b(t).$$

The Rayleigh damper is constructed such that $R(x)$ is zero except in a narrow region in the vicinity of $x = L$. The boundary condition at $x = -L$ is required to uniquely determine the solution, since the characteristic curves intersecting that boundary are directed into the domain. The boundary condition at $x = L$ is, however, redundant, since no characteristics are directed inward through that boundary, and as noted by Oliger and Sundström (1978), the imposition of a boundary condition at $x = L$ renders the problem ill-posed.

The practical ramifications of this ill-posedness are, however, somewhat subtle. One certainly cannot expect to obtain a numerical solution that converges to the unique solution to an ill-posed problem. This behavior is illustrated by the test problem shown in Fig. 8.7, in which (8.68) was approximated as

$$\frac{\phi_j^{n+1} - \phi_j^{n-1}}{2\Delta t} + U\frac{\phi_{j+1}^n - \phi_{j-1}^n}{2\Delta x} = R_j(\phi_j^{n+1} - \phi_N)$$

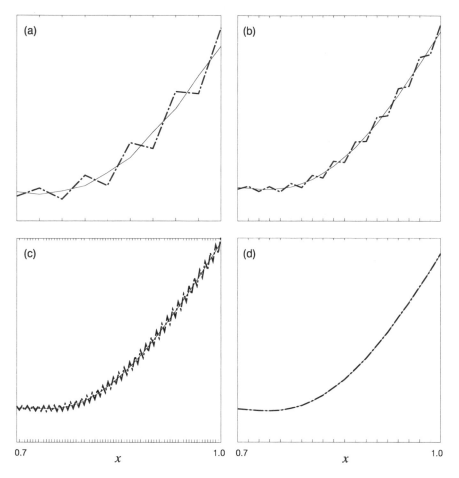

FIGURE 8.7. Comparison of numerical solutions to the advection equation obtained using a Rayleigh damping wave absorber (dot-dashed curve) or linear extrapolation (thin solid curve) and Δx equal to (a) 1/32, (b) 1/64, or (c) 1/256. In (d) the magnitude of the Rayleigh damping coefficient is doubled and $\Delta x = 1/64$.

on the interval $-1 \le x \le 1.125$. Here N is the spatial index of the grid point on the right-boundary, and the Rayleigh damping coefficient is defined as

$$R(x) = \begin{cases} 0, & \text{if } x \le 1; \\ \alpha\big(1 + \cos[8\pi(x-1)]\big), & \text{otherwise,} \end{cases}$$

so that the region $1 \le x \le 1.125$ contains the wave absorber. The Courant number was one-half, U was 1, and ϕ was fixed at a constant value of zero at the right and left boundaries. The initial condition was

$$\psi(x,0) = \begin{cases} 0, & \text{if } |x - \tfrac{1}{2}| \ge 1; \\ \cos^2[\pi(x-\tfrac{1}{2})], & \text{otherwise.} \end{cases}$$

A second solution, indicated by the thin solid line in Fig. 8.7, was obtained using the linear extrapolation boundary condition (8.18) at $x = L$ instead of using a wave-absorbing layer.

Figure 8.7 focuses on the trailing edge of the disturbance in the subdomain $0.7 \leq x \leq 1$. The time shown is $t = \frac{3}{4}$, at which time three-quarters of the initial pulse has passed into the wave-absorbing layer. The interface between the absorbing layer and the interior domain coincides with the right edge of each plot. The horizontal grid spacing for the simulation shown in Fig. 8.7a is $1/32$ and α is chosen such that $R_N \Delta t$ is unity. Considerable reflection is produced by the absorbing layer, which is only four grid points wide. Weaker reflection is also produced by the extrapolation boundary condition. Figure 8.7b shows the solutions to the same physical problem obtained after halving Δx, which increases the width of the absorbing layer to eight grid points. Since the Courant number is fixed, Δt is also halved, and the maximum value of $R_n \Delta t$ is reduced to one-half. Both solutions are improved by this increase in resolution, but the sponge layer continues to produce significantly more reflection than that generated by the extrapolation boundary condition. The grid spacing is reduced by an additional factor of four in Fig. 8.7c, so that $\Delta x = 1/256$. This increase in numerical resolution continues to improve the solution obtained with the extrapolation boundary condition, but does not improve the solution obtained with the wave-absorbing layer. Further increases in the resolution do not make the solution computed with the wave-absorbing layer converge toward the correct solution. Such convergence is, however, exhibited by the solution obtained using the extrapolation boundary condition.

If α is doubled, the wave-absorbing layer does perform much better in the two preceding higher-resolution simulations. Figure 8.7d shows a simulation with $\Delta x = 1/64$ and $R_n \Delta t = 1$ in which the performance of the wave-absorbing layer is very similar to that obtained using the extrapolation boundary condition. Additional high-resolution simulations suggest that although the ill-posedness of the underlying mathematical problem prevents the solution from uniformly converging to the correct solution within the absorbing layer as $\Delta x \to 0$ and $\Delta t \to 0$, the error *outside the absorbing layer* remains small and may be acceptable in some applications. Of course, the use of a Rayleigh-damping absorber is not actually recommended in situations, such as this, where an exact open boundary condition can be formulated for the original partial differential equation. It is harder to make a clear-cut recommendation in the vast majority of cases, for which exact open boundary conditions are not available. It is certainly possible that over some range of numerical parameters, the errors generated by the ill-posed Rayleigh damping absorber can be less than those obtained using well-posed numerical approximations to overly reflective boundary conditions. Indeed, the Rayleigh-damping absorber appears to have been used successfully in a variety of studies. Caution is, nevertheless, advised.

8.5 Summary

When simulating the evolution of localized disturbances within a large body of fluid, it is often necessary to limit the computational domain to an arbitrary subset of the total fluid. The conditions that are imposed at artificial boundaries within the fluid are often only approximate descriptions of the true physical behavior of the system and can be a significant source of error. Unless special information is available describing the solution outside the limited domain, the only practical boundary condition that can be specified is one that attempts to radiate disturbances through the boundary without spurious reflection. Nevertheless, the radiation condition is not always the correct physical boundary condition, because nonlinear interactions among the waves that have passed through an artificial boundary can generate a new wave that propagates backwards and should reenter the domain.

Even when the radiation boundary condition correctly describes the physical processes occurring at an open boundary, it is frequently impossible to exactly express that boundary condition in a useful mathematical form. In many practical problems, the radiation boundary condition can be expressed only in a form that involves the frequencies and wave numbers of the incident disturbance, or equivalently, temporal and spatial integrals of the solution along the boundary. Such boundary conditions are nonlocal because the condition that must be imposed on the fields at a point (x_0, t_0) cannot be exactly evaluated from the data available in a limited neighborhood of (x_0, t_0). Boundary conditions that are nonlocal in time are particularly unsuitable because only data from a few time levels are routinely stored during the numerical solution of time-dependent problems.

Exact local radiation boundary conditions can be obtained for some one-dimensional problems, such as the linearized one-dimensional shallow-water system. In more complicated situations, approximate radiation boundary conditions can be obtained using approximate one-way wave equations. Exact radiation boundary conditions that are local in time but nonlocal in space are also available for a limited class of multidimensional problems. Examples include the upper boundary condition for the linearized hydrostatic waves discussed in connection with (8.55) and (8.59), and the radiation condition proposed by Grote and Keller (1996) for the solution of the three-dimensional wave equation

$$\frac{\partial^2 \psi}{\partial t^2} - \nabla \cdot (\nabla \psi) = 0$$

outside a spherical domain. Spherical boundaries are attractive in applications where the wave propagation is isotropic in the region external to the boundary. Spherical boundaries are less well suited to many geophysical problems in which gravity introduces a fundamental anisotropy in the waves that makes it advantageous to use different mathematical formulae at the lateral and the upper or lower boundaries. A comprehensive review the wave-permeable boundary conditions employed in many different disciplines is given by Givoli (1991, 1992).

Once exact or approximate boundary conditions have been formulated for the continuous problem, they must be approximated for use in the numerical integration. Extra boundary conditions beyond those required to yield a well-posed problem may also be needed to evaluate finite differences near the boundary. Although these numerical boundary conditions can have a significant impact on the stability and accuracy of the solution, it can be easier to reduce these finite-differencing errors than the errors that originate from inadequate approximations to the correct open boundary condition.

Problems

1. Explain how to choose α_1 and f_1 in (8.4) to enforce a rigid-side-wall condition at the lateral boundary of the shallow-water system. (This is appropriate only when $U = 0$.)

2. Downstream differencing is obviously an unstable way to approximate an advection term at an inflow boundary. Suppose that one attempts to use backwards time-differencing to stabilize the approximation. As an indicator of the probable stability of the result, use a Von Neumann analysis to determine the stability of the backwards approximation to the advection equation

$$\frac{\phi_j^{n+1} - \phi_j^n}{\Delta t} + U \frac{\phi_j^{n+1} - \phi_{j-1}^{n+1}}{2\Delta x} = 0$$

on the unbounded domain $-\infty < x < \infty$. Consider the case $U < 0$.

3. Suppose that the one-dimensional constant-wind-speed advection equation is approximated as

$$(\delta_{2t} + U\delta_{2x})\,\phi_j^n = 0$$

in the interior of the domain and as

$$\left(\langle\delta_t\rangle^x + \tilde{U}\,\langle\delta_x\rangle^t\right)\phi_{N-\frac{1}{2}}^{n+\frac{1}{2}} = 0,$$

where N is the x index of the point on the downstream boundary. Show that setting

$$\tilde{U} = U\,\frac{\cos^2(k\Delta x/2)}{\cos^2(\omega\Delta t/2)}$$

will allow the discretized mode with wave number k and frequency ω to pass through the boundary without reflection. Explain why this technique cannot be used in practice to create a perfectly nonreflecting boundary.

4. Show that the second-order Higdon boundary condition (8.35) can be expressed in a form similar to the second-order Engquist and Majda condition (8.31) as

$$(1 + \cos\alpha_1 \cos\alpha_2)\frac{\partial^2\eta}{\partial t^2} + c(\cos\alpha_1 + \cos\alpha_2)\frac{\partial^2\eta}{\partial t\partial x} - c^2\frac{\partial^2\eta}{\partial y^2} = 0.$$

5. Show that if c is the shallow-water gravity-wave phase speed and $|U| < c$, then

$$\frac{\partial^2\psi}{\partial t^2} + (U + c)\left(\frac{\partial^2\psi}{\partial t\partial x} - \frac{c}{2}\frac{\partial^2\psi}{\partial y^2}\right) = 0$$

is a one-way wave equation that admits only waves with x-component group velocities greater than zero. How do wave solutions to this equation compare with the solutions to the two-dimensional shallow-water equations linearized about a constant basic-state flow U parallel to the x-axis,

$$\left(\frac{\partial}{\partial t} + U\frac{\partial}{\partial x}\right)^2\psi - c^2\left(\frac{\partial^2}{\partial x^2} + \frac{\partial^2}{\partial y^2}\right)\psi = 0.$$

Consider the dispersion relations for each system. Also discuss the dependence of the sign of the x-component group velocity on U and the horizontal wave numbers.

6. Derive the approximate one-way wave equation (8.67) for the right-moving wave satisfying the Rayleigh damped shallow-water equations (8.65) and (8.66) with $\nu = 0$ and R constant. Describe the conditions under which this is an accurate approximation.

7. Derive the conditions under which the vertical trace speed (ω/m) of internal gravity waves satisfying the dispersion relation (8.42) is in the same direction as the vertical group velocity.

8. Consider linear Boussinesq flow in the x-z plane. Show that a hydrostatic internal gravity wave propagating relative to the mean flow in the positive x-direction satisfies the relation $\hat{P} = N\hat{w}/|m|$, where the hat denotes a Fourier transform along the z coordinate and m is the vertical wave number. Use these results to derive a partial differential equation relating P to u that could, in principle, be evaluated using the grid-point values of pressure and velocity.

9. Suppose that the linearized Boussinesq equations are to be solved in the domain $-L \le x \le L$. The relation derived in Problem 9 might serve as a radiation boundary condition along the boundary at $x = L$ when the basic-state horizontal wind speed is zero. Explain how this boundary condition might prove unsatisfactory if $U < 0$.

Appendix
Numerical Miscellany

A.1   Finite-Difference Operator Notation

Complex finite-difference formulae are written in compact form using the following operator notation, which is similar to that used in Shuman and Hovermale (1968):

$$\delta_{nx} f(x) = \frac{f(x + n\Delta x/2) - f(x - n\Delta x/2)}{n\Delta x} \tag{A.1}$$

and

$$\langle f(x)\rangle^{nx} = \frac{f(x + n\Delta x/2) + f(x - n\Delta x/2)}{2}. \tag{A.2}$$

The grid-point approximation to the value of a continuous function $\psi(x, t)$ at the point $(j\Delta x, n\Delta t)$ is denoted by ϕ_j^n. Thus,

$$\delta_{2x}\phi_j^n = \frac{\phi_{j+1}^n - \phi_{j-1}^n}{2\Delta x}, \qquad \delta_x^2\phi_j^n = \frac{\phi_{j+1}^n - 2\phi_j^n - \phi_{j-1}^n}{(\Delta x)^2},$$

and

$$\langle (c_j)^x \delta_x \phi_j^n \rangle^x = \frac{1}{4}\left[(c_{j+1} + c_j)\left(\frac{\phi_{j+1}^n - \phi_j^n}{\Delta x}\right) + (c_j + c_{j-1})\left(\frac{\phi_j^n - \phi_{j-1}^n}{\Delta x}\right) \right].$$

When it does not lead to ambiguity, the grid-point indices are omitted from finite-difference formulae expressed in operator notation. As an example,

$$\delta_{2t}\phi + \delta_{2x}\phi = 0$$

expands to

$$\frac{\phi_j^{n+1} - \phi_j^{n-1}}{2\Delta t} + \frac{\phi_{j+1}^n - \phi_{j-1}^n}{2\Delta x} = 0.$$

Simple finite-difference formulae are generally written out in expanded form.

A.2   Tridiagonal Solvers

Implicit finite-difference schemes for one-dimensional problems often lead to tridiagonal systems of linear algebraic equations. Tridiagonal systems can be solved very efficiently using the Thomas tridiagonal algorithm (Isaacson and Keller 1966). The algorithm can be extended to the periodic case as discussed in Strikwerda (1989). FORTRAN programs for both the standard and periodic cases are given below. A well-behaved solution will be obtained using the following algorithms whenever the tridiagonal systems are diagonally dominant, which will be the case if for all j, $|b_j| > |a_j| + |c_j|$, where a_j, b_j, and c_j are, respectively, the subdiagonal, diagonal, and superdiagonal entries in row j.

A.2.1   Code for a Tridiagonal Solver

```
      subroutine tridiag(jmx,a,b,c,f,q)

c     Solves a standard tridiagonal system
c
c     Definition of the variables:
c        jmx  = dimension of all the following arrays
c        a    = sub (lower) diagonal
c        b    = center diagonal
c        c    = super (upper) diagonal
c        f    = right hand side
c        q    = work array provided by calling program
c
c        a(1) and c(jmx) need not be initialized
c     The output is in f; a, b, and c are unchanged

      real a(*),b(*),c(*),f(*),q(*),p
      integer j,jmx

      c(jmx)=0.
```

```
c Forward elimination sweep

      q(1)=-c(1)/b(1)
      f(1)= f(1)/b(1)

      do j=2,jmx
        p= 1.0/( b(j)+a(j)*q(j-1) )
        q(j)= -c(j)*p
        f(j)=( f(j)-a(j)*f(j-1) )*p
      end do

c Backward pass

      do j=jmx-1,1,-1
        f(j)=f(j)+q(j)*f(j+1)
      end do

      return
      end
```

A.2.2   Code for a Periodic Tridiagonal Solver

```
      subroutine tridiag_per(jmx,a,b,c,f,q,s)
c
c     Solves a periodic tridiagonal system
c
c     Definition of the variables:
c        jmx  = dimension of all arrays
c        a    = sub (lower) diagonal
c        b    = center diagonal
c        c    = super (upper) diagonal
c        f    = right hand side
c        q    = work array provided by calling program
c        s    = work array provided by calling program
c
c     Output is in f; a, b, and c are unchanged

      real a(*),b(*),c(*),f(*),q(*),s(*),p,fmx
      integer j,jmx

      fmx=f(jmx)
```

c Forward elimination sweep

```
      q(1)=-c(1)/b(1)
      f(1)= f(1)/b(1)
      s(1)=-a(1)/b(1)

      do j=2,jmx
        p=1.0/(b(j)+a(j)*q(j-1))
        q(j)=-c(j)*p
        f(j)=(f(j)-a(j)*f(j-1))*p
        s(j)=-a(j)*s(j-1)*p
      end do
```

c Backward pass

```
      q(jmx)=0.0
      s(jmx)=1.0

      do j=jmx-1,1,-1
        s(j)=s(j)+q(j)*s(j+1)
        q(j)=f(j)+q(j)*q(j+1)
      end do
```

c Final pass

```
      f(jmx)=( fmx-c(jmx)*q(1)-a(jmx)*q(jmx-1) )/
     &        ( c(jmx)*s(1)+a(jmx)*s(jmx-1)+b(jmx) )

      do j=1,jmx-1
       f(j)=f(jmx)*s(j)+q(j)
      end do

      return
      end
```

Bibliography

Abarbanel, S. and D. Gottlieb (1976). A note on the leap-frog scheme in two and three dimensions. *J. Comp. Phys.* **21**, 351–355.

Adams, J., R. Garcia, B. Gross, J. Hack, D. Haidvogel, and V. Pizzo (1992). Applications of multigrid software in the Atmospheric sciences. *Mon. Wea. Rev.* **120**, 1447–1458.

Arakawa, A. (1966). Computational design for long-term numerical integration of the equations of fluid motion: Two-dimensional incompressible flow. Part I. *J. Comp. Phys.* **1**, 119–143.

Arakawa, A. and C. S. Konor (1996). Vertical differencing of the primitive equations based on the Charney-Phillips grid in hybrid σ-p vertical coordinates. *Mon. Wea. Rev.* **124**, 511–528.

Arakawa, A. and V. R. Lamb (1977). Computational design of the basic dynamical processes of the UCLA general circulation model. *Methods in Computational Physics* **17**, 173–265.

Asselin, R. (1972). Frequency filter for time integrations. *Mon. Wea. Rev.* **100**, 487–490.

Batchelor, G. K. (1967). *An Introduction to Fluid Dynamics*. Cambridge: Cambridge University Press.

Bates, J. R., Y. Li, A. Brandt, S. McCormick, and J. Ruge (1995). A global shallow-water numerical model based on the semi-Lagrangian advection of potential vorticity. *Quart. J. Roy. Meteor. Soc.* **121**, 1981–2005.

Bates, J. R. and A. McDonald (1982). Multiply-upstream, semi-Lagrangian advective schemes: analysis and application to a multi-level primitive equation model. *Mon. Wea. Rev.* **110**, 1831–1842.

Bird, R. B., W. Stewart, and E. Lightfoot (1960). *Transport Phenomena.* New York: John Wiley and Sons.

Bjerknes, V. (1904). Das Problem der Wettervorhersage, betrachtet vom Stanpunkt der Mechanik und der Physik. *Meteor. Zeits.* **21**, 1–7.

Blum, E. (1962). A modification of the Runge–Kutta fourth-order method. *Math. Comp.* **16**, 176–187.

Boris, J. and D. Book (1973). Flux-corrected transport I: SHASTA, a fluid transport algorithm that works. *J. Comp. Phys.* **11**, 38–69.

Bougeault, P. (1983). A nonreflective upper boundary condition for limited-height hydrostatic models. *Mon. Wea. Rev.* **111**, 420–429.

Bourke, W. (1972). An efficient, one-level, primitive-equation spectral model. *Mon. Wea. Rev.* **100**, 683–689.

Bourke, W. (1974). A multi-level spectral model. I. Formulation and hemispheric integrations. *Mon. Wea. Rev.* **102**, 687–701.

Boussinesq, J. (1903). *Théorie analytique de la chaleur*, Volume II. Paris: Gauthier-Villars.

Boyd, J. P. (1989). *Chebyshev and Fourier Spectral Methods.* Berlin: Springer-Verlag.

Briggs, W. L. (1987). *A Multigrid Tutorial.* Philadelphia: SIAM.

Burks, A. and A. Burks (1981). The ENIAC: first general purpose electronic computer. *Annals of the History of Computing* **3**, 310–389.

Carrier, G. F. and C. E. Pearson (1988). *Partial Differential Equations: Theory and Technique* (Second ed.). Boston: Academic Press.

Charney, J. G., R. Fjörtoft, and J. von Neumann (1950). Numerical integration of the barotropic vorticity equation. *Tellus* **2**, 237–254.

Chorin, A. J. (1968). Numerical solution of the Navier–Stokes equations. *Math. Comp.* **22**, 745–762.

Clark, T. (1979). Numerical simulations with a three-dimensional cloud model: lateral boundary conditions and multicellular severe storm simulations. *J. Atmos. Sci.* **36**, 2191–2215.

Colella, P. (1990). Multidimensional upwind methods for hyperbolic conservation laws. *J. Comp. Phys.* **87**, 171–200.

Colella, P. and P. Woodward (1984). The piecewise-parabolic method (PPM) for gas-dynamical simulations. *J. Comp. Phys.* **54**, 174–201.

Corby, G. A., A. Gilchrist, and R. L. Newson (1972). A general circulation model of the atmosphere suitable for long period integration. *Quart. J. Roy. Meteor. Soc.* **98**, 809–832.

Courant, R. and D. Hilbert (1953). *Methods of Mathematical Physics*, Volume I. New York: Interscience Publishers.

Courant, R., E. Isaacson, and M. Rees (1952). On the solution of nonlinear hyperbolic differential equations by finite differences. *Comm. Pure Appl. Math.* **5**, 243–255.

Crandall, M. G. and A. Majda (1980a). The method of fractional steps for conservation laws. *Math. Comp.* **34**, 285–314.

Crandall, M. G. and A. Majda (1980b). Monotone difference approximations for scalar conservation laws. *Math. Comp.* **34**, 1–21.

Cullen, M. J. P. (1974). Integrations of the primitive equations on a sphere using the finite element method. *Quart. J. Roy. Meteor. Soc.* **100**, 555–562.

Cullen, M. J. P. (1979). The finite element method. In *Numerical Methods Used in Atmospheric Models*, Volume II of *GARP Publication Series No. 17*, pp. 300–337. Geneva: World Meteorological Organization.

Cullen, M. J. P. (1982). The use of quadratic finite element methods and irregular grids in the solution of hyperbolic problems. *J. Comp. Phys.* **45**, 221–245.

Cullen, M. J. P. (1990). A test of a semi-implicit integration technique for a fully compressible non-hydrostatic model. *Quart. J. Roy. Meteor. Soc.* **116**, 1253–1258.

Cullen, M. J. P. and C. Hall (1979). Forecasting and general circulation results from finite element models. *Quart. J. Roy. Meteor. Soc.* **105**, 571–592.

Cullen, M. J. P. and K. Morton (1980). Analysis of evolutionary error in finite element and other methods. *J. Comp. Phys.* **34**, 245–267.

Dahlquist, G. and A. Björck (1974). *Numerical Methods*. Englewood Cliffs, New Jersey: Prentice-Hall.

Davies, H. (1976). A lateral boundary formulation for multi-level prediction models. *Quart. J. Roy. Meteor. Soc.* **102**, 405–418.

Davies, H. (1983). Limitations of some common lateral boundary conditions used in regional NWP models. *Mon. Wea. Rev.* **111**, 1002–1012.

Donea, J. (1984). A Taylor-Galerkin method for convective transport problems. *Int. J. Numer. Methods Eng.* **20**, 101–119.

Donea, J., L. Quartapelle, and V. Selmin (1987). An analysis of time discretization in the finite element solution of hyperbolic problems. *J. Comp. Phys.* **70**, 463–499.

Dupont, T. (1973). Galerkin methods for first order hyperbolics: an example. *SIAM J. Numer. Anal.* **10**, 890–899.

Durran, D. R. (1989). Improving the anelastic approximation. *J. Atmos. Sci.* **46**, 1453–1461.

Durran, D. R. (1991). The third-order Adams-Bashforth method: an attractive alternative to leapfrog time differencing. *Mon. Wea. Rev.* **119**, 702–720.

Durran, D. R. and J. B. Klemp (1983). A compressible model for the simulation of moist mountain waves. *Mon. Wea. Rev.* **111**, 2341–2361.

Durran, D. R., M.-J. Yang, D. N. Slinn, and R. G. Brown (1993). Toward more accurate wave-permeable boundary conditions. *Mon. Wea. Rev.* **121**, 604–620.

Eliasen, E., B. Machenhauer, and E. Rasmussen (1970). On a numerical method for integration of the hydrodynamical equations with a spectral representation of the horizontal fields. Report no. 2, Institut for Teoretisk Meteorologi, University of Copenhagen.

Engquist, B. and A. Majda (1977). Absorbing boundary conditions for the numerical simulation of waves. *Math. Comp.* **31**, 629–651.

Engquist, B. and A. Majda (1979). Radiation boundary conditions for acoustic and elastic wave calculations. *Comm. Pure Appl. Math.* **32**, 313–357.

Ferziger, J. H. and M. Perić (1997). *Computational Methods for Fluid Dynamics*. Berlin: Springer-Verlag.

Fjørtoft, R. (1953). On the changes in the spectral distribution of kinetic energy for two-dimensional nondivergent flow. *Tellus* **5**, 225–230.

Foreman, M. and R. Thomson (1997). Three-dimensional model simulations of tides and buoyancy currents along the west coast of Vancouver Island. *J. Phys. Oceanogr.* **27**, 1300–1325.

Fornberg, B. (1973). On the instability of leap-frog and Crank-Nicolson approximations to a nonlinear partial differential equation. *Math. Comp.* **27**, 45–57.

Fornberg, B. (1996). *A Practical Guide to Pseudospectral Methods*. Cambridge: Cambridge University Press.

Gal-Chen, T. and R. C. J. Somerville (1975). On the use of a coordinate transformation for the solution of the Navier-Stokes equations. *J. Comp. Phys.* **17**, 209–228.

Garner, S. (1986). A radiative upper boundary condition for f-plane models. *Mon. Wea. Rev.* **114**, 1570–1577.

Gary, J. (1979). Nonlinear instability. In *Numerical Methods Used in Atmospheric Models*, Volume II of *GARP Publication Series No. 17*, pp. 476–499. Geneva: World Meteorological Organization.

Gary, J. M. (1973). Estimate of truncation error in transformed coordinate, primitive equation atmospheric models. *J. Atmos. Sci.* **30**, 223–233.

Gear, C. W. (1971). *Numerical Initial Value Problems in Ordinary Differential Equations*. Englewood Cliffs: Prentice-Hall.

Gill, A. E. (1982). *Atmosphere—Ocean Dynamics*. Orlando: Academic Press.

Givoli, D. (1991). Non-reflecting boundary conditions. *J. Comp. Phys.* **94**, 1–29.

Givoli, D. (1992). *Numerical Methods for Problems in Infinite Domains*. Amsterdam: Elsevier.

Godlewski, E. and P.-A. Raviart (1996). *Numerical Approximation of Hyperbolic Systems of Conservation Laws*. New York: Springer-Verlag.

Godunov, S. K. (1959). A difference scheme for numerical computation of discontinuous solutions of equations in fluid dynamics. *Math. Sb.* **47**, 271. Also: Cornell Aero. Lab. translation.

Goldberg, M. and E. Tadmor (1985). New stability criteria for difference approximations of hyperbolic initial-boundary value problems. *Lectures in Appl. Math.* **22**, 177–192.

Golub, G. H. and C. F. van Loan (1996). *Matrix Computations* (third ed.). Baltimore: John-Hopkins University Press.

Goodman, J. B. and R. J. LeVeque (1985). On the accuracy of stable schemes for 2D scalar conservation laws. *Math. Comp.* **45**, 15–21.

Gottlieb, D. and S. A. Orszag (1977). *Numerical Analysis of Spectral Methods: Theory and Applications*. Philadelphia: SIAM.

Gresho, P. M., R. L. Lee, and R. L. Sani (1978). Advection-dominated flows with emphasis on the consequences of mass lumping. In *Finite Elements in Fluids*, Volume 3, pp. 335–350. New York: John Wiley and Sons.

Grote, M. and J. Keller (1996). Nonreflecting boundary conditions for time-dependent scattering. *J. Comp. Phys.* **127**, 52–65.

Gustafsson, B. (1975). The convergence rate for difference approximations to mixed initial boundary value problems. *Math. Comp.* **29**, 396–406.

Gustafsson, B., H.-O. Kreiss, and J. Oliger (1995). *Time Dependent Problems and Difference Methods*. New York: John Wiley and Sons.

Gustafsson, B., H.-O. Kreiss, and A. Sundström (1972). Stability theory of difference approximations for initial boundary value problems. II. *Math. Comp.* **26**, 649–686.

Hackbusch, W. (1985). *Multi-grid Methods and Applications*. Berlin: Springer-Verlag.

Harten, A. (1983). High resolution schemes for hyperbolic conservation laws. *J. Comp. Phys.* **49**, 357–393.

Harten, A., J. Hyman, and P. Lax (1976). On finite-difference approximations and entropy conditions for shocks. *Comm. Pure Appl. Math* **29**, 297–322.

Hedley, M. and M. Yau (1988). Radiation boundary conditions in numerical modelling. *Mon. Wea. Rev.* **116**, 1721–1736.

Hedstrom, G. (1979a). The Galerkin method based on Hermite cubics. *SIAM J. Numer. Anal.* **16**, 385–393.

Hedstrom, G. (1979b). Nonreflecting boundary conditions for nonlinear hyperbolic systems. *J. Comp. Phys.* **30**, 222–237.

Héreil, P. and R. Laprise (1996). Sensitivity of internal gravity wave solutions to the time step of a semi-implicit semi-Lagrangian nonhydrostatic model. *Mon. Wea. Rev.* **124**, 972–999.

Higdon, R. (1986). Absorbing boundary conditions for difference approximations to the multi-dimensional wave equation. *Math. Comp.* **47**, 437–459.

Higdon, R. (1987). Numerical absorbing boundary conditions for the wave equation. *Math. Comp.* **49**, 65–90.

Higdon, R. (1994). Radiation boundary conditions for dispersive waves. *SIAM J. Numer. Anal.* **31**, 64–100.

Holton, J. R. (1992). *An Introduction to Dynamic Meteorology* (Third ed.). San Diego: Academic Press.

Hong, H., R. Liska, and S. Steinberg (1997). Testing stability by quantifier elimination. *J. Symbolic Computation* **24**, 161–187.

Hoskins, B. J. and F. P. Bretherton (1972). Atmospheric frontogenesis models: Mathematical formulation and solution. *J. Atmos. Sci.* **29**, 11–37.

Hoskins, B. J. and A. J. Simmons (1975). A multi-layer spectral model and the semi-implicit method. *Quart. J. Roy. Meteor. Soc.* **101**, 637–655.

Isaacson, E. and H. B. Keller (1966). *Analysis of Numerical Methods*. New York: John Wiley and Sons.

Iserles, A. (1996). *A First Course in the Numerical Analysis of Differential Equations*. Cambridge: Cambridge University Press.

Israeli, M. and S. Orszag (1981). Approximation of radiation boundary conditions. *J. Comp. Phys.* **41**, 115–135.

Karniadakis, G. E., M. Israeli, and S. Orszag (1991). High-order splitting methods for the incompressible Navier-Stokes equations. *J. Comp. Phys.* **97**, 414–443.

Kasahara, A. (1974). Various vertical coordinate systems used for numerical weather prediction. *Mon. Wea. Rev.* **102**, 509–522.

Kevorkian, J. (1990). *Partial Differential Equations: Analytical Solution Techniques*. Pacific Grove, California: Wadsworth and Brooks Cole.

Kiehl, J., J. Hack, G. Bonan, B. Boville, B. Briegleb, D. Williamson, and P. Rasch (1996). Description of the NCAR community climate model (CCM3). NCAR Technical Note NCAR/TN-420+STR, National Center for Atmospheric Research.

Klemp, J. B. and D. Durran (1983). An upper boundary condition permitting internal gravity wave radiation in numerical mesoscale models. *Mon. Wea. Rev.* **111**, 430–444.

Klemp, J. B. and D. Lilly (1978). Numerical simulation of hydrostatic mountain waves. *J. Atmos. Sci.* **35**, 78–107.

Klemp, J. B. and R. Wilhelmson (1978). The simulation of three-dimensional convective storm dynamics. *J. Atmos. Sci.* **35**, 1070–1096.

Kreiss, H. (1962). Über die Stabilitätsdefinition für Differenzengleichungen die partielle Differentialgleichungen approximieren. *Nordisk Tidskrift Informationsbehandling (BIT)* **2**, 153–181.

Krishnamurti, T., A. Kumar, K. Yap, A. Dastoor, N. Davidson, and J. Shieng (1990). Performance of a high-resolution mesoscale tropical prediction model. *Advances in Geophysics* **32**, 133–286.

Kurihara, Y. (1965). On the use of implicit and iterative methods for the time integration of the wave equation. *Mon. Wea. Rev.* **93**, 33–46.

Kuznecov, N. N. and S. A. Vološin (1976). On monotone difference approximations for a first-order quasi-linear equation. *Soviet Math. Dokl.* **17**, 1203–1206.

Kwizak, M. and A. J. Robert (1971). A semi-implicit scheme for grid point atmospheric models of the primitive equations. *Mon. Wea. Rev.* **99**, 32–36.

Lander, J. and B. Hoskins (1997). Believable scales and parameterizations in a spectral transform model. *Mon. Wea. Rev.* **125**, 292–303.

Lax, P. (1971). Shock waves and entropy. In E. H. Zarantonello (Ed.), *Contributions to Nonlinear Analysis*, New York, pp. 603–634. Academic Press.

Lax, P. and R. D. Richtmyer (1956). Survey of the stability of linear finite difference equations. *Comm. Pure Appl. Math.* **9**, 267–293.

Lax, P. and B. Wendroff (1960). Systems of conservation laws. *Comm. Pure Appl. Math.* **13**, 217–237.

Lele, S. (1992). Compact finite difference schemes with spectral-like resolution. *J. Comp. Phys.* **103**, 16–42.

Leonard, B. P., M. K. MacVean, and A. P. Lock (1993). Positivity-preserving numerical schemes for multi-dimensional advection. Technical Memorandum 106055, ICOMP-93-05, NASA.

LeVeque, R. J. (1992). *Numerical Methods for Conservation Laws*. Basel: Birkhäuser Verlag.

LeVeque, R. J. (1996). High-resolution conservative algorithms for advection in incompressible flow. *SIAM J. Numer. Anal.* **33**, 627–665.

Lipps, F. (1990). On the anelastic approximation for deep convection. *J. Atmos. Sci.* **47**, 1794–1798.

Lipps, F. and R. Hemler (1982). A scale analysis of deep moist convection and some related numerical calculations. *J. Atmos. Sci.* **29**, 2192–2210.

Lorenz, E. N. (1967). *The Nature and Theory of the General Circulation of the Atmosphere*. Geneva: World Meteorological Organization.

Lynch, D. and W. Gray (1979). A wave equation model for finite element tidal computations. *Computers and Fluids* **7**, 207–228.

MacCormack, R. (1969). The effect of viscosity in hypervelocity impact cratering. *AIAA Paper 69-352*.

Machenhauer, B. (1979). The spectral method. In *Numerical Methods Used in Atmospheric Models*, Volume II of *GARP Publication Series No. 17*, pp. 121–275. Geneva: World Meteorological Organization.

Magazenkov, L. (1980). *Trudy Glavnoi Geofizicheskoi Observatorii* **410**, 120–129 (Transactions of the Main Geophysical Observatory).

Mahlman, J. D. and R. W. Sinclair (1977). Tests of various numerical algorithms applied to a simple trace constituent air transport problem. In I. H. Suffet (ed.), *Fate of Pollutants in Air and Water Environments*, Volume 8. New York: John Wiley and Sons. Part 1.

Matsuno, T. (1966a). False reflection of waves at the boundary due to the use of finite differences. *J. Meteor. Soc. Japan, Ser. 2* **44**, 145–157.

Matsuno, T. (1966b). Numerical integrations of the primative equations by a simulated backward difference method. *J. Meteor. Soc. Japan, Ser. 2* **44**, 76–84.

McDonald, A. (1984). Accuracy of multiply-upstream, semi-Lagrangian advective schemes. *Mon. Wea. Rev.* **112**, 1267–1275.

McDonald, A. and J. Bates (1989). Semi-Lagrangian integration of a gridpoint shallow-water model on the sphere. *Mon. Wea. Rev.* **117**, 130–137.

Merilees, P. E. and S. A. Orszag (1979). The pseudospectral method. In *Numerical Methods Used in Atmospheric Models*, Volume II of *GARP Publication Series No. 17*, pp. 276–299. Geneva: World Meteorological Organization.

Mesinger, F. (1984). A blocking technique for representation of mountains in atmospheric models. *Riv. Meteor. Aeronautica* **44**, 195–202.

Mesinger, F. and Z. I. Janić (1985). Problems and numerical methods of the incorporation of mountains in atmospheric models. *Lectures in Appl. Math.* **22**, 81–120.

Miyakoda, K. (1962). Contribution to the numerical weather prediction. Computation with finite difference. *Japanese J. Geophys.* **3**, 75–190.

Morton, K. and A. Parrott (1980). Generalised Galerkin methods for first-order hyperbolic equations. *J. Comp. Phys.* **36**, 249–270.

Nakamura, H. (1978). Dynamical effects of mountains on the general circulation of the atmosphere. I. Development of finite-difference schemes suitable for incorporating mountains. *J. Meteor. Soc. Japan* **56**, 317–339.

Nance, L. B. (1997). On the inclusion of compressibility effects in the scorer parameter. *J. Atmos. Sci.* **54**, 362–367.

Nance, L. B. and D. Durran (1994). A comparison of three anelastic systems and the pseudo-incompressible system. *J. Atmos. Sci.* **51**, 3549–3565.

Nitta, T. (1962). The outflow boundary condition in numerical time integration of advective equations. *J. Meteor. Soc. Japan* **40**, 13–24.

Ogura, Y. and N. Phillips (1962). Scale analysis for deep and shallow convection in the atmosphere. *J. Atmos. Sci.* **19**, 173–179.

Oleinik, O. (1957). Discontinuous solutions of nonlinear differential equations. *Amer. Math. Soc. Tranl., Ser. 2* **26**, 95–172.

Oliger, J. and A. Sundström (1978). Theoretical and practical aspects of some initial boundary value problems in fluid dynamics. *SIAM J. Appl. Math* **35**, 419–446.

Oliveira, A. and A. M. Baptista (1995). A comparison of integration and interpolation Eulerian-Lagrangian methods. *Int. J. Numer. Methods Fluids* **21**, 183–204.

Orlanski, I. (1976). A simple boundary condition for unbounded hyperbolic flows. *J. Comp. Phys.* **21**, 251–269.

Orszag, S. A. (1970). Transform method for calculation of vector-coupled sums: Application to the spectral form of the vorticity equation. *J. Atmos. Sci.* **27**, 890–895.

Orszag, S. A., M. Israeli, and M. O. Deville (1986). Boundary conditions for incompressible flows. *J. Scientific Comp.* **1**, 75–111.

Pearson, R. (1974). Consistent boundary conditions for numerical models of systems that admit dispersive waves. *J. Atmos. Sci.* **31**, 1481–1489.

Pedlosky, J. (1987). *Geophysical Fluid Dynamics* (second ed.). New York: Springer-Verlag.

Phillips, N. A. (1957). A coordinate system having some special advantages for numerical forecasting. *J. Meteor.* **14**, 184–185.

Phillips, N. A. (1959). An example of non-linear computational instability. In *The Atmosphere and Sea in Motion, Rossby Memorial Volume*, pp. 501–504. New York: Rockefeller Institute Press.

Phillips, N. A. (1973). Principles of large scale numerical weather prediction. In P. Morel (Ed.), *Dynamic Meteorology*, pp. 1–96. Dordrecht: Riedel.

Pinty, J., R. Benoit, E. Richard, and R. Laprise (1995). Simple tests of a semi-implicit semi-Lagrangian model on 2D mountain wave problems. *Mon. Wea. Rev.* **123**, 3042–3058.

Priestley, A. (1992). The Taylor-Galerkin method for the shallow-water equations on the sphere. *Mon. Wea. Rev.* **120**, 3003–3015.

Rasch, P. (1986). Towards atmospheres without tops: absorbing upper boundary conditions for numerical models. *Quart. J. Roy. Meteor. Soc.* **112**, 1195–1218.

Richardson, L. (1922). *Weather Prediction by Numerical Process*. Cambridge: Cambridge University Press. (Reprinted by Dover Publications, New York, 1965.)

Ritchie, H. (1986). Eliminating the interpolation associated with the semi-Lagrangian scheme. *Mon. Wea. Rev.* **114**, 135–146.

Ritchie, H. (1987). Semi-Lagrangian advection on a Gaussian grid. *Mon. Wea. Rev.* **115**, 608–619.

Ritchie, H., C. Temperton, A. Simmons, M. Hortal, T. Davies, D. Dent, and M. Hamrud (1995). Implementation of the semi-Lagrangian method in a high-resolution version of the ECMWF forecast model. *Mon. Wea. Rev.* **123**, 489–514.

Robert, A. (1966). The integration of a low order spectral form of the primitive meteorological equations. *J. Meteor. Soc. Japan* **44**, 237–244.

Robert, A. (1981). A stable numerical integration scheme for the primitive meteorological equations. *Atmos. Ocean* **19**, 35–46.

Robert, A. (1982). A semi-Lagrangian and semi-implicit numerical integration scheme for the primitive meteorological equations. *J. Meteor. Soc. Japan* **60**, 319–324.

Roe, P. L. (1985). Some contributions to the modeling of discontinuous flows. In *Lecture Notes in Applied Mathematics*, Volume 22, pp. 163–193. New York: Springer-Verlag.

Rood, R. B. (1987). Numerical advection algorithms and their role in atmospheric transport and chemistry models. *Reviews of Geophysics* **25**, 71–100.

Sadourny, R., A. Arakawa, and Y. Mintz (1968). Integration of the nondivergent barotropic vorticity equation with an iscosahedral-hexagonal grid for the sphere. *Mon. Wea. Rev.* **96**, 351–356.

Saltzman, J. (1994). An unsplit 3D upwind method for hyperbolic conservation laws. *J. Comp. Phys.* **115**, 153–168.

Sangster, W. E. (1960). A method of representing the horizontal pressure force without reduction of station pressures to sea level. *J. Meteor.* **17**, 166–176.

Schär, C. and P. K. Smolarkiewicz (1996). A synchronous and iterative flux-correction formalism for coupled transport equations. *J. Comp. Phys.* **128**, 101–120.

Schlesinger, R., L. Uccellini, and D. Johnson (1983). The effects of the asselin time filter on numerical solutions to the linearized shallow-water wave equations. *Mon. Wea. Rev.* **111**, 455–467.

Shuman, F. G. and J. B. Hovermale (1968). An operational six-layer primitive equation model. *J. Appl. Meteor.* **7**, 525–547.

Simmons, A. J. and D. M. Burridge (1981). An energy and angular-momentum conserving vertical finite-difference scheme and hybrid vertical coordinates. *Mon. Wea. Rev.* **109**, 758–766.

Simmons, A. J., B. Hoskins, and D. Burridge (1978). Stability of the semi-implicit method of time integration. *Mon. Wea. Rev.* **106**, 405–412.

Simmons, A. J. and C. Temperton (1997). Stability of a two-time-level semi-implicit integration scheme for gravity wave motion. *Mon. Wea. Rev.* **125**, 600–615.

Skamarock, W. C. and J. B. Klemp (1992). The stability of time-split numerical methods for the hydrostatic and nonhydrostatic elastic equations. *Mon. Wea. Rev.* **120**, 2109–2127.

Skamarock, W. C., P. K. Smolarkiewicz, and J. B. Klemp (1997). Precondition conjugate-residual solvers for Helmholtz equations in nonhydrostatic models. *Mon. Wea. Rev.* **125**, 587–599.

Smolarkiewicz, P. K. (1983). A simple positive definite advection scheme with small implicit diffusion. *Mon. Wea. Rev.* **111**, 479–486.

Smolarkiewicz, P. K. (1984). A fully multidimensional positive definite advection transport algorithm with small implicit diffusion. *J. Comp. Phys.* **54**, 325–362.

Smolarkiewicz, P. K. (1989). Comment on "A positive definite advection scheme obtained by nonlinear renormalization of the advective fluxes". *Mon. Wea. Rev.* **117**, 2626–2632.

Smolarkiewicz, P. K. and W. Grabowski (1990). The multidimensional positive definite advection transport algorithm: nonoscillatory option. *J. Comp. Phys.* **86**, 355–375.

Smolarkiewicz, P. K. and L. G. Margolin (1993). On forward in time differencing for fluids: Extension to a curvilinear framework. *Mon. Wea. Rev.* **121**, 1847–1859.

Smolarkiewicz, P. K. and L. G. Margolin (1994). Variational solver for elliptic problems in atmospheric flows. *Appl. Math. and Comp. Sci.* **4**, 527–551.

Smolarkiewicz, P. K. and J. Pudykiewicz (1992). A class of semi-Lagrangian algorithms for fluids. *J. Atmos. Sci.* **49**, 2082–2096.

Smolarkiewicz, P. K. and R. Rasch (1991). Monotone advection on the sphere: an Eulerian versus semi-Lagrangian approach. *J. Atmos. Sci.* **48**, 793–810.

Sommerfeld, A. (1949). *Partial Differential Equations in Physics*. New York: Academic Press.

Staniforth, A. N. (1987). Review: formulating efficient finite-element codes for flows in regular domains. *Int. J. Numer. Methods Fluids* **7**, 1–16.

Staniforth, A. N. and J. Côté (1991). Semi-Lagrangian integration schemes for atmospheric models—A review. *Mon. Wea. Rev.* **119**, 2206–2223.

Staniforth, A. N. and R. W. Daley (1977). A finite-element formulation for vertical discretization of sigma-coordinate primitive equation models. *Mon. Wea. Rev.* **105**, 1108–1118.

Stevens, D. E. and C. J. Bretherton (1996). A forward-in-time advection scheme and adaptive multilevel flow solver for nearly incompressible atmospheric flow. *J. Comp. Phys.* **129**, 284–295.

Stoker, J. J. (1957). *Water Waves; the Mathematical Theory with Applications.* New York: Interscience.

Strang, G. (1964). Accurate partial difference methods. II. Nonlinear problems. *Numer. Mathematik* **6**, 37–46.

Strang, G. (1968). On the construction and comparison of difference schemes. *SIAM J. Numer. Anal.* **5**, 506–517.

Strang, G. and G. J. Fix (1973). *An Analysis of the Finite Element Method.* New York: Prentice-Hall.

Strikwerda, J. C. (1989). *Finite Difference Schemes and Partial Differential Equations.* Pacific Grove, California: Wadsworth and Brooks/Cole.

Sundström, A. and T. Elvius (1979). Computational problems related to limited area modelling. In *Numerical Methods Used in Atmospheric Models*, Volume II of *GARP Publication Series No. 17*, pp. 379–416. Geneva: World Meteorological Organization.

Sweby, P. K. (1984). High resolution schemes using flux limiters for hyperbolic conservation laws. *SIAM J. Numer. Anal.* **21**, 995–1011.

Tanguay, M., A. Robert, and R. Laprise (1990). A semi-implicit semi-Lagrangian fully compressible regional forecast model. *Mon. Wea. Rev.* **118**, 1970–1980.

Tapp, M. C. and P. W. White (1976). A non-hydrostatic mesoscale model. *Quart. J. Roy. Meteor. Soc.* **102**, 277–296.

Tatsumi, Y. (1983). An economical explicit time integration scheme for a primitive model. *J. Meteor. Soc. Japan* **61**, 269–287.

Temperton, C. and A. Staniforth (1987). An efficient two-time level semi-Lagrangian semi-implicit integration scheme. *Quart. J. Roy. Meteor. Soc.* **113**, 1025–1039.

Thomée, V. and B. Wendroff (1974). Convergence estimates for Galerkin methods for variable coefficient initial value problems. *SIAM J. Numer. Anal.* **11**, 1059–1068.

Thompson, P. (1983). A history of numerical weather prediction in the United States. *Bull. Amer. Meteor. Soc.* **64**, 755–769.

Thuburn, J. (1996). Multidimensional flux-limited advection schemes. *J. Comp. Phys.* **123**, 74–83.

Trefethen, L. (1983). Group velocity interpretation of the stability theory of Gustafsson, Kreiss and Sundström. *J. Comp. Phys.* **49**, 199–217.

Trefethen, L. (1985). Stability of hyperbolic finite-difference models with one or two boundaries. *Lectures in Appl. Math.* **22**, 311–326.

Trefethen, L. and L. Halpern (1986). Well-posedness of one-way wave equations and absorbing boundary conditions. *Math. Comp.* **47**, 421–435.

Turkel, E. (1977). Symmetric hyperbolic difference schemes and matrix problems. *Linear Algebra Appl.* **16**, 109–129.

van Leer, B. (1974). Towards the ultimate conservative difference scheme II. Monotonicity and conservation combined in a second order scheme. *J. Comp. Phys.* **14**, 361–370.

van Leer, B. (1977). Towards the ultimate conservative difference scheme IV. A new approach to numerical convection. *J. Comp. Phys.* **23**, 276–299.

Warming, R. and R. Beam (1976). Upwind second-order difference schemes and applications in aerodynamic flows. *AIAA Journal* **14**, 1241–1249.

Warming, R. and B. Hyett (1974). The modified equation approach to the stability and accuracy analysis of finite-difference methods. *J. Comp. Phys.* **14**, 159–179.

Whitham, G. B. (1974). *Linear and Nonlinear Waves.* New York: John Wiley and Sons.

Wiin-Nielsen, A. (1959). On the application of trajectory methods in numerical forecasting. *Tellus* **11**, 180–196.

Williamson, D. L. (1968). Integration of the barotropic vorticity equation on a spherical geodesic grid. *Tellus* **20**, 642–653.

Williamson, D. L. (1979). Difference approximations for fluid flow on a sphere. In *Numerical Methods Used in Atmospheric Models*, Volume II of *GARP Publication Series No. 17*, pp. 53–120. Geneva: World Meteorological Organization.

Williamson, D. L. and J. G. Olson (1994). Climate simulations with a semi-Lagrangian version of the NCAR community climate model. *Mon. Wea. Rev.* **122**, 1594–1610.

Williamson, D. L. and P. Rasch (1989). Two-dimensional semi-Lagrangian transport with shape-preserving interpolation. *Mon. Wea. Rev.* **117**, 102–129.

Williamson, J. (1980). Low-storage Runge–Kutta schemes. *J. Comp. Phys.* **35**, 48–56.

Yih, C.-S. (1977). *Fluid Mechanics: A Concise Introduction to the Theory.* Ann Arbor: West River Press.

Zalesak, S. T. (1979). Fully multidimensional flux-corrected transport algorithms for fluids. *J. Comp. Phys.* **31**, 335–362.

Index

$\langle\ \rangle^{nx}$ (averaging operator), 119
δ_{nx} (finite-difference operator), 37, 439

A-stable finite-difference schemes, 141, 142
absolute stability, 139–143
accelerating time difference, 49
Adams–Bashforth
    time-differencing
  higher-order, 65–71, 338
  second-order, 57, 59–60
  for semi-Lagrangian schemes, 314–318
Adams–Moulton methods, 67
advection equation
  defined, 12
  for nonuniform flow, 147–154
  semi-discrete finite-difference approximation to, 72–80
  two-dimensional
    for nondivergent flow, 154–158

phase-speed error in finite-difference approximations to, 120–122
  stability of finite-difference approximations to, 117–120, 122
advection–diffusion equation, 138–144
advection–source–sink equation, 144–147
advective form, 157, 241–243, 297
aliasing error
  defined, 148, 149
  and numerical instability, 149–152, 154, 166
  in the pseudospectral method, 192–195
  and the spectral transform method, 186–188, 205–207
  in the variable wind-speed advection equation, 149

alternating direction method, 133
amplification factor, 43, 49
amplitude error, 49, 71, 72, 81
analytic function, 10
anelastic approximation, 21, 25
anelastic equations, 24–25
antidiffusion via upstream
         differencing, 294–296
antidiffusive flux, 258
Arakawa Jacobian, 166
artificial dissipation, 82–86
Asselin filter, 62–64, 368

backward time-differencing, 50–52,
         55, 56
barotropic vorticity equation,
         163–167
block cyclic reduction, 342
boundary conditions
      accuracy of, 408
      influence on stability, 409–412
      nonlocal, 426, 436
      nonreflecting, *see* radiation
         boundary condition, *see*
         wave-absorbing layers,
      reflections generated by,
         402–408, 416–418,
         422–423, 426–428
      well-posed, 397–400
Boussinesq approximation, 20, 23
Boussinesq equations, 23–24, 337
      compressible, 351–352, 356,
         371
      solution via the projection
         method, 337–342
Brunt–Väisälä frequency, 18, 24
Burgers's equation
      conservative differential-
         difference
         approximations to,
         160–163
      discontinuous solutions to,
         159–160, 246–249
      finite-element approximations
         to, 230–231

generalized, 241
spectral and pseudospectral
         approximations to,
         193–194
and transonic rarefaction
         waves, 274

C grid, 127
Cauchy problem, 409
CFL condition, *see*
         Courant-Fredrichs-Lewy
         condition
chapeau function, 213
characteristic curve, 4, 5, 9
      and boundary conditions for
         1-D hyperbolic equations,
         398–402
      for Burgers's equation,
         159–160, 247–249, 274
      and tracer transport, 12
chemically reacting flow,
         291–292
collocation, 175, 191
compact differencing, 86–89
      for the 2-D advection
         equation, 287
      phase-speed error of, 87–88
computational mode
      of finite-element
         approximations to the
         1-D advection equation,
         222–223, 228
      of the leapfrog scheme, 57–59,
         61–64
      of the second-order
         Adams–Bashforth
         scheme, 59–60
      of the third-order
         Adams–Bashforth
         method, 70
conjugate gradient method, 342
conservation and stability,
         154–157
conservation form
      defined, 250

conservation law, 157, 243–244
consistency
    of a finite-difference method,
      39
    of numerically approximated
      fluxes to a conservation
      law, 250
convergence, 40–41
    to the entropy consistent weak
      solution, 251
    to a false solution, 241–243
    to a weak solution, 250
convolution sum, 186
Courant–Fredrichs–Lewy
    condition, 45–47, 306
Crank–Nicolson method, 137
CTU method, 123–125, 280–282
    and semi-Lagrangian schemes,
      309
curvilinear coordinates,
    296–297

decelerating time difference, 49
diffusion equation, 136–138
dispersion relation
    for differential-difference
      approximations to the
      1-D advection equation,
      73–77, 79
    exact for
      the 1-D advection equation,
      72
      the 2-D advection equation,
      120
      the middle-latitude β-plane
      equations, 390
    for finite-difference
      approximations to
      the 1-D advection equation,
      91–94
      the 1-D shallow-water
      equations, 112–113, 117
      the 2-D advection equation,
      117–120

    the Boussinesq equations,
      126–128
    for hydrostatic gravity waves,
      374
    for leapfrog approximations to
      the compressible
      Boussinesq equations,
      352–353
    for linear finite-element
      approximations to the
      1-D advection equation,
      215
    for numerical approximations
      to the 2-D shallow-water
      equations of the sphere,
      201
    for semi-implicit
      approximations to
      the 1-D shallow-water
      equations, 348–349
      the compressible
      Boussinesq equations,
      354–356
divergence damper, 369–370
domain of dependence, 45
domain of influence, 45
donor cell method, *see* upstream
    differencing
Dufort–Frankel method, 102

eddy diffusivity, 13, 139, 167
energy method, 41–42, 154–158
entropy condition, 246–249
Euler equations, 337
    approximation via leapfrog
      time differencing,
      352–353
    approximation via the
      semi-implicit method,
      353–359
    hyperbolic nature of,
      16–19
    simplified, 350–352
Euler's method, *see* forward
    time-differencing

Eulerian form, 303
Eulerian–Lagrangian methods, *see*
      semi-Lagrangian methods
Exner function, 17
explicit time-differencing, 53

fast Fourier transform (FFT), 180,
      235
   efficiency of, 185–186
filtered equations, 11, 20–26
finite difference, 36
   on an irregular mesh, 101
finite Fourier transform, 179–180
finite-difference method, 26–27
finite-difference operator notation,
      439–440
finite-element method, 29
   comparison with compact
         differencing, 215, 232
   in two dimensions, 231–234
   using the Galerkin
         approximation
      with cubic expansion
            functions, 226–231
      with linear expansion
            functions, 214–216
      with quadratic expansion
            functions, 219–225
   using the Petrov–Galerkin
         approximation, 216–217
   using the Taylor–Galerkin
         approximation, 217–219
   using triangular elements,
         232–234
finite-volume method, 27–28
   defined, 250
Fjørtoft's theorem, 164–165
flux form, 157, 241–243, 297
flux-corrected transport (FCT),
      257–263, 269–271
   in two dimensions, 277, 278
   iterative flux correction, 263
   and positive definite advection,
         293–294
   the Zalesak corrector, 260–262

flux-limiter methods, 263–272
   for 2-D flow, 279–290
   in curvilinear coordinates,
         296–297
forward time-differencing, 50–52,
      55, 56
   stability of, 51, 90
forward–backward
      time-differencing, 54
forward-backward approximation
      to the shallow-water
      equations, 115, 117, 200,
      360
Fourier series, 177
   and poor resolution, 178–179
fractional-step methods, 129–136
   for scalar conservation laws,
         277
   stability of, 134–136
   time splitting, 359–372
   truncation error of, 130–133
frequency, 72
frozen-coefficient stability analysis,
      147, 169

Galerkin approximation
   conservation properties of,
         189–190
   defined, 175
   equivalence to minimizing the
         ℓ_2-norm of the residual,
         175–176
   and the finite-element-method,
         213
Gaussian quadrature, 203–207
Godunov's method, 272–274
Godunov–Ryabenki condition,
      409
gravity waves, 11, 335–336, 347,
      352–356, 390–391
   dispersion relation, 19
group velocity, 72
   for the semidiscrete 1-D
         advection equation,
         73–77

and the stability of
    finite-difference
    approximations to IBVP,
    410–412, 424
Gustafsson–Kreiss–Sundström
    (GKS) condition,
    409–410

Hermite-cubic polynomials, 226
Heun method, 53
hydrostatic approximation,
    372–374
hydrostatic reference state, 23

implicit time-differencing, 53

Jacobian operator, 164

Kreiss matrix theorem, 110

Lagrange interpolating polynomial,
    204, 220, 308
Lagrangian form, 303
Lamb wave, 374
Laplacian operator on the sphere,
    196–197
Lax equivalence theorem, 40
Lax–Fredrichs method, 102, 254,
    299–300
Lax–Wendroff method, 95–99
    for the advection-diffusion
        equation, 168
    for a general conservation law,
        98
    for the 1-D advection
        equation, 95–98, 275–276
    flux form, 261, 265–272
    modified equation
        associated with, 104
    and semi-Lagrangian schemes,
        308
    and the Tayler–Galerkin
        approximation, 217–219
    two-step form of, 96, 98
    for the 2-D advection
        equation, 98, 125–126

for 2-D conservation laws,
    281–282, 284
leapfrog time-differencing, 57–59,
    110–111, 343
    for the Euler equations,
        352–353
    and the projection method,
        339–340
    for semi-Lagrangian schemes,
        313, 315, 320, 321
leapfrog-time centered-space
    schemes, 92–93
leapfrog-trapezoidal method, 64
Legendre functions, 195, 198
Legendre polynomials, 195,
    204–205

Mach number, 21
Magazenkov method, 64–66
mass lumping, 232
mass matrix, 213, 219, 225
matrix norm, 109
Matsuno's method, 54–56
MC limiter, 268–271, 300
midpoint method, 54, 61
minmod limiter, 267–271, 276
modified equation
    for the discrete advection
        equation, 94–95
    for the semi-discrete advection
        equation, 81
    for the semidiscrete
        advection–diffusion
        equation, 138
    for the Taylor–Galerkin
        approximation to the
        advection equation,
        217–219
monotone method, 251, 255–256,
    292
monotonicity-preserving method,
    252–253, 292
multigrid method, 342
multistage time-differencing,
    53–56

multistep time-differencing, 56–60
    initialization of, 60

Newton polynomial, 308, 331–332
nondispersive wave, 72
nonlinear instability, 159–167
norm
    of a grid-point function, 39–40
    of a matrix, 109
    of a real-valued continuous
        function, 174
numerical dispersion, 82
numerical dissipation, 81, 138–139
numerical resolution, 51
numerical smoothing, 82–86
numerical weather prediction, 1

one-way wave equation, 5, 401,
    414–418, 433
open boundary condition, see
    radiation boundary
    condition
order of accuracy, 36
    of a monotone finite-difference
        scheme, 253
orthogonality, 175, 177–178
oscillation equation, 49

pagoda function, 231
partial differential equation
    elliptic
        approximation by marching
            schemes, 31–32
        defined, 7–11
    hyperbolic
        approximation by marching
            schemes, 31–32
        defined, 4–9, 11
        and tracer transport, 12
    linear, 3
    order of, 2
    parabolic
        defined, 7–11
        and tracer transport, 12
    quasi-linear, 3

Peaceman–Rachford method, 133
Péclet number, 138–139
Petrov–Galerkin method, 216–217
phase speed, 15, 72
phase-speed error, 49, 71–72, 91
    in finite-difference
        approximations to the
        2-D advection equation,
        120–122
    in Hermite-cubic
        finite-element
        approximations to the
        1-D advection equation,
        227–229
    in quadratic finite-element
        approximations to the
        1-D advection equation,
        222–225
    produced by compact
        finite-differencing, 87–88
    produced by second- and
        fourth-order spatial
        differencing, 73–75
    for the semidiscrete 1-D
        advection equation,
        73–77
    on a staggered mesh, 114, 115
physical mode, 57
pole problem, 200–202
positive definite advection schemes,
    292–296
    antidiffusion via upstream
        differencing, 294–296
    and flux-corrected transport,
        293, 294
potential temperature, 16
Prandtl number, 138
pressure equation
    for the anelastic system, 342
    for the Boussinesq system,
        337, 338, 340
        boundary conditions for,
            338–339
    for the pseudo-incompressible
        system, 342

for the semi-implicit Euler
equations, 357, 358
solution of, 340–342
pressure gradient computation over
topography, 378–379
primitive equations
in generalized vertical
coordinates, 375–377
in height coordinates,
374–375
in σ coordinates, 377–378
energy conservation,
383–387
semi-implicit
time-differencing,
387–389
vertical differencing,
381–383
projection method, 337, 371–372
pseudo-incompressible
approximation, 21, 24
pseudo-incompressible system, 22
pseudospectral methods, 191–195

radiation boundary condition
definition and applicability of,
395–397
for the 1-D shallow-water
equations, 400–401
for the 2-D shallow-water
equations, 414–419
and time-dependent boundary
data, 401–402
for 2-D stratified flow
lateral boundaries, 419–423
upper boundary, 424–431
Rankine–Hugoniot condition, 245
rarefaction wave, 246–248,
273–274
residual, 174
Reynolds number, 138, 167
Riemann invariant, 399
Riemann problem, 244, 272–274
Robert-Asselin filter, *see* Asselin
filter

Rossby waves, 11, 15–16, 20,
335–336, 347,
390–391
Runge–Kutta methods
for trajectory calculations in
semi-Lagrangian
schemes, 311–313
higher-order, 65–71,360
second order, 53, 56

Schwarz inequality, 42
semi-implicit method, 371–372
comparison with the projection
method, 357
for the Euler equations,
353–359
for the oscillation equation,
345–347
for the primitive equations in
σ coordinates, 387–389
for the shallow-water
equations, 347–350
semi-implicit semi-Lagrangian
methods, 320–324
semi-Lagrangian methods
for the advection equation,
305–313
accuracy of, 306–313
stability of, 306
comparison with Eulerian
schemes, 330–331
comparison with the method of
characteristics, 318–320
computation of the backward
trajectory, 310–313
defined, 304
for the forced advection
equation, 313–318
using generalized trajectories,
324–329
noninterpolating, 325–327
and positive definite advection,
327–329
for the shallow-water
equations, 320–324

shallow-water equations
 hyperbolic nature of, 13–16
 for the middle-latitude β
  plane, 389–390
 1-D linearized, 107–108
 on the sphere
  linear, 200–202
  nonlinear, 207–212
 radiation boundary conditions
  for the 2-D system,
  413–419
 semi-implicit approximations
  to, 347–350
 semi-implicit semi-Lagrangian
  approximations to, 321,
  324
shock, speed of, 245
σ coordinate, 377
slant derivative method, 104
slope limiters, 275–276
sound waves, 11, 335–337,
  352–356
 dispersion relation, 19
 filtering of, 20–26
 speed of, 18
spatial smoothing, 82–86
spectral method, 29
 accuracy of, 182–184
 comparison with
  finite-difference methods,
  177–184
 conservation properties of,
  189–190
 and the Galerkin
  approximation, 176–177
 improving efficiency of,
  184–189
 remote influence in, 181–182
 for spherical domains,
  195–212, 379–381
spectral radius, 109
spherical harmonic functions
 comparison of triangular and
  rhomboidal truncations,
  197–200

defined, 195–196
 in global primitive equation
  models, 379–381
 in global shallow-water
  models, 207–209
 recurrence relations for, 196
 transforming between physical
  and wave-number space,
  202–203
sponge layers, 431–435
stability
 and convergence, 40–41
 energy-method analysis of,
  41–42, 154–158
 problems at the poles, 200–202
 of systems of finite-difference
  equations, 108–113
 Von Neumann analysis of, 43,
  45, 91, 108–111
staggered meshes, 113–117
 C grid, 127
Strang perturbation theorem,
  146–147
Strang splitting, 131–132
stream function, 163, 208
superbee limiter, 267–271,
  285–290

Taylor–Galerkin method, 217–219
total variation, 252
 in two dimensions, 256–257
total variation diminishing method,
  252–254, 256–257,
  274–276, 292
 sufficient conditions for,
  264–268
transform method
 for the Fourier spectral
  method, 184–189
 for the spherical harmonic
  spectral method,
  202–207, 209–212
transonic rarefaction wave, 274
trapezoidal time-differencing, 50,
  53, 55, 56, 202

inaccuracy at large Δt,
343–345
for semi-Lagrangian schemes,
304, 314–315, 321–324
tridiagonal solvers
periodic, 441–442
standard, 440–441
truncation error
of a finite difference, 36
of a finite-difference method,
39
global, 50
and the modified equation,
80–82
one-step (local), 50
TVD, *see* total variation
diminishing
two-time-step method
with complete operator
splitting, 359–364
with partial operator splitting,
365–371

upstream differencing, 279,
294–296
and the semi-Lagrangian
method, 306–308

discrete dispersion relation for,
93–94
flux-form, 261
modified equation for, 94
order of accuracy, 38–39
stability of, 41–42, 44–45
in two-dimensions, 122–125

velocity potential, 208
von Neumann's method, 43, 45
for systems of equations,
108–110

Warming and Beam method, 99,
104, 267
wave number, 72
wave-absorbing layers, 431–435
wave-like flow, 2
wave-permeable boundary
condition, *see* radiation
boundary condition
weak solution, 244–245
entropy consistent, 246–249
well-posed problem, 31, 169,
397–400, 432–435